VITAMINS AND HORMONES

VOLUME 75

Editorial Board

TADHG P. BEGLEY

ANTHONY R. MEANS

BERT W. O'MALLEY

LYNN RIDDIFORD

ARMEN H. TASHJIAN, JR.

VITAMIN A

VITAMINS AND HORMONES
ADVANCES IN RESEARCH AND APPLICATIONS

Editor-in-Chief

GERALD LITWACK

Former Professor and Chair
Department of Biochemistry and Molecular Pharmacology
Thomas Jefferson University Medical College
Philadelphia, Pennsylvania

Former Visiting Scholar
Department of Biological Chemistry
David Geffen School of Medicine at UCLA
Los Angeles, California

VOLUME 75

AMSTERDAM • BOSTON • HEIDELBERG • LONDON
NEW YORK • OXFORD • PARIS • SAN DIEGO
SAN FRANCISCO • SINGAPORE • SYDNEY • TOKYO
Academic Press is an imprint of Elsevier

Academic Press is an imprint of Elsevier
525 B Street, Suite 1900, San Diego, California 92101-4495, USA
84 Theobald's Road, London WC1X 8RR, UK

This book is printed on acid-free paper.

Copyright © 2007, Elsevier Inc. All Rights Reserved.

No part of this publication may be reproduced or transmitted in any form or by any means, electronic or mechanical, including photocopy, recording, or any information storage and retrieval system, without permission in writing from the Publisher.

The appearance of the code at the bottom of the first page of a chapter in this book indicates the Publisher's consent that copies of the chapter may be made for personal or internal use of specific clients. This consent is given on the condition, however, that the copier pay the stated per copy fee through the Copyright Clearance Center, Inc. (www.copyright.com), for copying beyond that permitted by Sections 107 or 108 of the U.S. Copyright Law. This consent does not extend to other kinds of copying, such as copying for general distribution, for advertising or promotional purposes, for creating new collective works, or for resale. Copy fees for pre-2007 chapters are as shown on the title pages. If no fee code appears on the title page, the copy fee is the same as for current chapters.
0083-6729/2007 $35.00

Permissions may be sought directly from Elsevier's Science & Technology Rights Department in Oxford, UK: phone: (+44) 1865 843830, fax: (+44) 1865 853333, E-mail: permissions@elsevier.com. You may also complete your request on-line via the Elsevier homepage (http://elsevier.com), by selecting "Support & Contact" then "Copyright and Permission" and then "Obtaining Permissions."

For information on all Elsevier Academic Press publications
visit our Web site at www.books.elsevier.com

ISBN-13: 978-0-12-709875-3
ISBN-10: 0-12-709875-5

PRINTED IN THE UNITED STATES OF AMERICA
07 08 09 10 9 8 7 6 5 4 3 2 1

**Working together to grow
libraries in developing countries**

www.elsevier.com | www.bookaid.org | www.sabre.org

ELSEVIER   BOOK AID International   Sabre Foundation

Former Editors

ROBERT S. HARRIS
Newton, Massachusetts

JOHN A. LORRAINE
*University of Edinburgh
Edinburgh, Scotland*

PAUL L. MUNSON
*University of North Carolina
Chapel Hill, North Carolina*

JOHN GLOVER
*University of Liverpool
Liverpool, England*

GERALD D. AURBACH
*Metabolic Diseases Branch
National Institute of Diabetes
and Digestive and Kidney Diseases
National Institutes of Health
Bethesda, Maryland*

KENNETH V. THIMANN
*University of California
Santa Cruz, California*

IRA G. WOOL
*University of Chicago
Chicago, Illinois*

EGON DICZFALUSY
*Karolinska Sjukhuset
Stockholm, Sweden*

ROBERT OLSEN
*School of Medicine
State University of New York
at Stony Brook
Stony Brook, New York*

DONALD B. MCCORMICK
*Department of Biochemistry
Emory University School of Medicine
Atlanta, Georgia*

Contents

Contributors xv
Preface xix

1

RXR: From Partnership to Leadership in Metabolic Regulations

Béatrice Desvergne

I. Introduction 2
II. RXRs and Their Many Partners Belong to the Nuclear Receptor Superfamily 3
III. RXR in Partnership: The Permissive Heterodimers as Metabolic Sensors 7
IV. The Rexinoid-Signaling Pathways: From Partnership to Leadership 15
V. Conclusions 21
References 22

2

THE INTERSECTION BETWEEN THE ARYL HYDROCARBON RECEPTOR (AhR)- AND RETINOIC ACID-SIGNALING PATHWAYS

KYLE A. MURPHY, LOREDANA QUADRO, AND LORI A. WHITE

I. Introduction 34
II. Retinoid Signaling 38
III. The AhR/Arnt Pathway 40
IV. AhR and RA Availability 45
V. Molecular Interactions Between the RA and AhR Pathways 52
References 56

3

ROLE OF RETINOIC ACID IN THE DIFFERENTIATION OF EMBRYONAL CARCINOMA AND EMBRYONIC STEM CELLS

DIANNE ROBERT SOPRANO, BRYAN W. TEETS, AND KENNETH J. SOPRANO

I. Introduction 70
II. Molecular Mechanism of Action of RA 71
III. Model Systems to Study Differentiation 73
IV. Role of RARs 77
V. RA-Regulated Genes 78
VI. Role of Specific RA-Regulated Genes 80
VII. Conclusions 86
References 87

4

METABOLISM OF RETINOL DURING MAMMALIAN PLACENTAL AND EMBRYONIC DEVELOPMENT

GEOFFROY MARCEAU, DENIS GALLOT, DIDIER LEMERY, AND VINCENT SAPIN

I. General Aspects of Retinol Transport and Metabolism in Mammalian Species 98
II. Placental Transport and Metabolism of Retinol During Mammalian Development 101

III. Embryonic Metabolism of Retinol During
Mammalian Development   105
References   109

5

CONVERSION OF β-CAROTENE TO RETINAL PIGMENT

HANS K. BIESALSKI, GURUNADH R. CHICHILI, JÜRGEN FRANK,
JOHANNES VON LINTIG, AND DONATUS NOHR

I. General Aspects of Vitamin A Metabolism   118
II. Conversion of β-Carotene to Vitamin A   120
III. β-Carotene as Provitamin A in Retinal Pigment Epithelial Cells   124
IV. Alternative Routes of Vitamin A Supply   126
References   128

6

VITAMIN A-STORING CELLS (STELLATE CELLS)

HARUKI SENOO, NAOSUKE KOJIMA, AND MITSURU SATO

I. Introduction   132
II. Morphology of HSCs   133
III. Regulation of Vitamin A Homeostasis by HSCs   134
IV. HSCs in Arctic Animals   137
V. Roles of HSCs During Liver Regeneration   141
VI. Production and Degradation of ECM Components by HSCs   143
VII. Reversible Regulation of Morphology, Proliferation, and Function of the HSCs by 3D Structure of ECM   146
VIII. Stimulation of Proliferation of HSCs and Tissue Formation of the Liver by a Long-Acting Vitamin C Derivative   149
IX. Extrahepatic Stellate Cells   151
X. Conclusions   152
References   153

7

USE OF MODEL-BASED COMPARTMENTAL ANALYSIS TO STUDY VITAMIN A KINETICS AND METABOLISM

CHRISTOPHER J. CIFELLI, JOANNE BALMER GREEN, AND MICHAEL H. GREEN

I. Introduction 162
II. Highlights of Whole-Body Vitamin A Metabolism 163
III. Early Kinetic Studies of Vitamin A Metabolism 164
IV. Overview of Compartmental Analysis 166
V. Use of Model-Based Compartmental Analysis to Study Vitamin A Kinetics 169
VI. Conclusions 191
References 192

8

VITAMIN A SUPPLEMENTATION AND RETINOIC ACID TREATMENT IN THE REGULATION OF ANTIBODY RESPONSES *IN VIVO*

A. CATHARINE ROSS

I. Introduction 199
II. Rationale for Interest in VA Supplementation and Antibody Production 199
III. VA and the Response to Immunization in Children 202
IV. Experimental Studies of VA or RA Supplementation and Antibody Production *In Vivo* 204
V. Innate Immune Cells and Factors Regulated by VA and RA That May Affect Immunization Outcome 211
VI. Discussion and Perspectives 215
References 218

9

PHYSIOLOGICAL ROLE OF RETINYL PALMITATE IN THE SKIN

PETER P. FU, QINGSU XIA, MARY D. BOUDREAU, PAUL C. HOWARD, WILLIAM H. TOLLESON, AND WAYNE G. WAMER

I. Introduction   224
II. Structure and Physiological Functions of the Skin   226
III. Cutaneous Absorption and Deposition of Dietary and Topically Applied Retinol and Retinyl Esters   228
IV. Mobilization and Metabolism of Retinol and Retinyl Esters in the Skin   241
V. Effects on Selected Biological Responses of the Skin   243
VI. Summary   249
References   249

10

RETINOIC ACID AND THE HEART

JING PAN AND KENNETH M. BAKER

I. Introduction   258
II. Role of RA in Heart Development and Congenital Heart Defects   259
III. Postnatal Development Effects of RA in the Heart   268
IV. Conclusions   273
References   274

11

TOCOTRIENOLS IN CARDIOPROTECTION

SAMARJIT DAS, KALANITHI NESARETNAM, AND DIPAK K. DAS

I. Introduction   285
II. A Brief History of Vitamin   286
III. Tocotrienols and Cardioprotection   292
IV. Atherosclerosis   293
V. Tocotrienols in Free Radical Scavenging and Antioxidant Activity   294

VI. Tocotrienols in Ischemic Heart Disease 295
VII. Conclusions 296
   References 296

12

CYTODIFFERENTIATION BY RETINOIDS, A NOVEL THERAPEUTIC OPTION IN ONCOLOGY: RATIONAL COMBINATIONS WITH OTHER THERAPEUTIC AGENTS

ENRICO GARATTINI, MAURIZIO GIANNI', AND MINEKO TERAO

I. Premise and Scope: Differentiation Therapy with Retinoids Is a Significant Goal in the Management of the Neoplastic Diseases 303
II. The Classical Nuclear RAR Pathway Is Complex and Has Led to the Development of Different Types of Synthetic Retinoids 305
III. Retinoids Promote Differentiation in Numerous Types of Neoplastic Cells 307
IV. Retinoids Exert Pleiotropic Effects Interacting with Multiple Intracellular Pathways: An Opportunity for Combination Therapy 313
V. Retinoid-Based Differentiation Therapy, General Observations, and Conclusion 334
   References 336

13

EFFECTS OF VITAMINS, INCLUDING VITAMIN A, ON HIV/AIDS PATIENTS

SAURABH MEHTA AND WAFAIE FAWZI

I. Introduction 357
II. Vitamins and Immune Function 358
III. Vitamins, HIV Transmission, and Pregnancy Outcomes 360
IV. Vitamins and HIV Disease Progression in Adults 366
V. Vitamins, Growth, and Disease Progression in HIV-Infected Children and HIV-Negative Children Born to HIV-Infected Mothers 372
VI. Comment 374
VII. Future Research 377
   References 377

14

Vitamin A and Emphysema
Richard C. Baybutt and Agostino Molteni

  I. Does Vitamin A Protect Against Pulmonary Emphysema?   386
 II. Conclusions   397
     References   398

Index   403

Contributors

Numbers in parentheses indicate the pages on which the authors' contributions begin.

Kenneth M. Baker (257) Division of Molecular Cardiology, The Texas A&M University System Health Science Center, Cardiovascular Research Institute, College of Medicine, Central Texas Veterans Health Care System, Temple, Texas 76504.

Richard C. Baybutt (385) Department of Human Nutrition, Kansas State University, Manhattan, Kansas 66506.

Hans K. Biesalski (117) Department of Biological Chemistry and Nutrition, University of Hohenheim, Stuttgart, Germany.

Mary D. Boudreau (223) National Center for Toxicological Research, Food and Drug Administration, Jefferson, Arkansas 72079.

Gurunadh R. Chichili (117) Institute of Biology I, Animal Physiology and Neurobiology, University of Freiburg, Freiburg, Germany.

Christopher J. Cifelli (161) Department of Nutritional Sciences, The Pennsylvania State University, University Park, Pennsylvania 16801.

Dipak K. Das (285) Cardiovascular Research Center, University of Connecticut School of Medicine, Farmington, Connecticut 06030.

Samarjit Das (285) Cardiovascular Research Center, University of Connecticut School of Medicine, Farmington, Connecticut 06030.

Béatrice Desvergne (1) Center for Integrative Genomics, Building Génopode, University of Lausanne, CH-1015 Lausanne, Switzerland.

Wafaie Fawzi (355) Department of Epidemiology, Harvard School of Public Health, 677 Huntington Avenue, Boston, Massachusetts 02115; Department of Nutrition, Harvard School of Public Health, 677 Huntington Avenue, Boston, Massachusetts 02115.

Jürgen Frank (117) Department of Biological Chemistry and Nutrition, University of Hohenheim, Stuttgart, Germany.

Peter P. Fu (223) National Center for Toxicological Research, Food and Drug Administration, Jefferson, Arkansas 72079.

Denis Gallot (97) Université d'Auvergne, JE 2447, ARDEMO, F-63000, Clermont-Ferrand, France; CHU Clermont-Ferrand, Maternité, Hôtel-Dieu, F-63000, Clermont-Ferrand, France.

Enrico Garattini (301) Laboratorio di Biologia Molecolare, Centro Catullo e Daniela Borgomainerio, Istituto di Ricerche Farmacologiche "Mario Negri," via Eritrea 62, 20157 Milano, Italy.

Maurizio Gianni' (301) Laboratorio di Biologia Molecolare, Centro Catullo e Daniela Borgomainerio, Istituto di Ricerche Farmacologiche "Mario Negri," via Eritrea 62, 20157 Milano, Italy.

Joanne Balmer Green (161) Department of Nutritional Sciences, The Pennsylvania State University, University Park, Pennsylvania 16801.

Michael H. Green (161) Department of Nutritional Sciences, The Pennsylvania State University, University Park, Pennsylvania 16801.

Paul C. Howard (223) National Center for Toxicological Research, Food and Drug Administration, Jefferson, Arkansas 72079.

Naosuke Kojima (131) Department of Cell Biology and Histology, Akita University School of Medicine, 1-1-1 Hondo, Akita 010-8543, Japan.

Didier Lemery (97) Université d'Auvergne, JE 2447, ARDEMO, F-63000, Clermont-Ferrand, France; CHU Clermont-Ferrand, Maternité, Hôtel-Dieu, F-63000, Clermont-Ferrand, France.

Johannes von Lintig (117) Molecular Immunogenetics Program, Oklahoma Medical Research Foundation, Oklahoma City, Oklahoma 73104.

Geoffroy Marceau (97) Université d'Auvergne, JE 2447, ARDEMO, F-63000, Clermont-Ferrand, France; INSERM, U.384, Laboratoire de Biochimie, Faculté de Médecine, F-63000 Clermont-Ferrand, France.

Saurabh Mehta (355) Department of Epidemiology, Harvard School of Public Health, 677 Huntington Avenue, Boston, Massachusetts 02115; Department of Nutrition, Harvard School of Public Health, 677 Huntington Avenue, Boston, Massachusetts 02115.

Agostino Molteni (385) Department of Pathology and Pharmacology, University of Missouri, Kansas City Medical School, Kansas City, Missouri 64108.

Kyle A. Murphy (33) Department of Biochemistry and Microbiology, Rutgers, The State University of New Jersey, New Brunswick, New Jersey 08901.

Kalanithi Nesaretnam (285) Malaysian Palm Oil Board, Kuala Lumpur, Malaysia.

Donatus Nohr (117) Department of Biological Chemistry and Nutrition, University of Hohenheim, Stuttgart, Germany.

Jing Pan (257) Division of Molecular Cardiology, The Texas A&M University System Health Science Center, Cardiovascular Research Institute, College of Medicine, Central Texas Veterans Health Care System, Temple, Texas 76504.

Loredana Quadro (33) Department of Food Science, Rutgers, The State University of New Jersey, New Brunswick, New Jersey 08901.

A. Catharine Ross (197) Department of Nutritional Sciences and Huck Institute for Life Sciences, Pennsylvania State University, University Park, Pennsylvania 16802.

Vincent Sapin (97) Université d'Auvergne, JE 2447, ARDEMO, F-63000, Clermont-Ferrand, France; INSERM, U.384, Laboratoire de Biochimie, Faculté de Médecine, F-63000, Clermont-Ferrand, France.

Mitsuru Sato (131) Department of Cell Biology and Histology, Akita University School of Medicine, 1-1-1 Hondo, Akita 010-8543, Japan.

Haruki Senoo (131) Department of Cell Biology and Histology, Akita University School of Medicine, 1-1-1 Hondo, Akita 010-8543, Japan.

Dianne Robert Soprano (69) Department of Biochemistry, Temple University School of Medicine, Philadelphia, Pennsylvania 19140; Fels Institute for Cancer Research and Molecular Biology, Temple University School of Medicine, Philadelphia, Pennsylvania 19140.

Kenneth J. Soprano (69) Fels Institute for Cancer Research and Molecular Biology, Temple University School of Medicine, Philadelphia, Pennsylvania 19140; Department of Microbiology and Immunology, Temple University School of Medicine, Philadelphia, Pennsylvania 19140.

Bryan W. Teets (69) Department of Biochemistry, Temple University School of Medicine, Philadelphia, Pennsylvania 19140.

Mineko Terao (301) Laboratorio di Biologia Molecolare, Centro Catullo e Daniela Borgomainerio, Istituto di Ricerche Farmacologiche "Mario Negri," via Eritrea 62, 20157 Milano, Italy.

William H. Tolleson (223) National Center for Toxicological Research, Food and Drug Administration, Jefferson, Arkansas 72079.

Wayne G. Wamer (223) Center for Food Safety and Applied Nutrition, Food and Drug Administration, College Park, Maryland 20740.

Lori A. White (33) Department of Biochemistry and Microbiology, Rutgers, The State University of New Jersey, New Brunswick, New Jersey 08901.

Qingsu Xia (223) National Center for Toxicological Research, Food and Drug Administration, Jefferson, Arkansas 72079.

Preface

Retinoic acid and its relatives, natural products of vitamin A metabolism, have been discovered to be critical ligands for key receptors of the steroid receptor superfamily, such as the retinoic acid receptor and the RXR receptor. *Vitamins and Hormones* has not had a review of this important fat-soluble vitamin for some time, and it seemed important not only to touch on the modern aspects of the retinoic acid receptor and its relatives but also of the roles of vitamin A in development and differentiation as well as its activity in certain tissues and in disease states. Chapters in this volume focus on many of these topics and should be of interest to biochemists, pharmacologists, nutritionists, and medical scientists as well as students in a wide variety of fields.

The volume begins with chapters on receptors and signaling pathways followed by chapters on development and differentiation. Vitamin A-storing cells are considered and then studies on compartmental analysis and metabolism. Continuing chapters discuss the regulation of antibody responses, the skin, the heart, therapeutics, and the effects of vitamin A in various disease states such as HIV/AIDS and emphysema.

The first chapter is entitled: "RXR: From Partnership to Leadership in Metabolic Regulations" by B. Desvergne. K. A. Murphy, L. Quadro, and L. A. White discuss: "The Intersection Between the Aryl Hydrocarbon Receptor (AhR) and Retinoic Acid Signaling Pathways." "Role of Retinoic Acid in the Differentiation of Embryonal Carcinoma and Embryonic Stem Cells" is contributed by D. R. Soprano, B. W. Teets, and K. J. Soprano. G. Marceau, D. Gallot, D. Lemery, and V. Sapin produce a review on: "Metabolism of Retinol During Mammalian Placental and Embryonic Development." "Conversion of β-Carotene to Retinal Pigment" is a chapter prepared by

H. K. Biesalski, G. R. Chichili, J. Frank, J. von Lintig, and D. Nohr, H. Senoo, N. Kojima, and M. Sato contribute a study entitled: "Vitamin A-Storing Cells (Stellate Cells)." The next article shifts to kinetics and metabolism in a chapter entitled: "Use of Model-Based Compartmental Analysis to Study Vitamin A Kinetics and Metabolism" by C. J. Cifelli, J. B. Green, and M. H. Green. A. C. Ross discusses: "Vitamin A Supplementation and Retinoic Acid Treatment in the Regulation of Antibody Responses *In Vivo*." A detailed chapter on vitamin A and the skin bears the title: "Physiological Role of Retinyl Palmitate in the Skin" by P. P. Fu, Q. Xia, M. D. Boudreau, P. C. Howard, W. H. Tolleson, and W. G. Wamer.

Shifting to organs and disease states, the remainder of the volume is filled by the chapters whose titles follow. "Retinoic Acid and the Heart" is the chapter reviewed by J. Pan and K. M. Baker. "Tocotrienols in Cardioprotection" is written by S. Das, K. Nesaretam, and D. K. Das, E. Garattini, M. Gianni, and M. Terao discuss: "Cytodifferentiation by Retinoids, a Novel Therapeutic Option in Oncology: Rational Combinations with Other Therapeutic Agents." "Effects of Vitamins, Including Vitamin A, on HIV/AIDS Patients" is contributed by S. Mehta. The final chapter written by R. C. Baybutt and A. Molteni focuses on "Vitamin A and Emphysema."

The Editor-in-Chief is grateful to Renske van Dijk and Tari Broderick of Academic Press/Elsevier whose cooperation in furthering this Serial is most appreciated.

Gerald Litwack
Toluca Lake, California
August, 2006

1

RXR: From Partnership to Leadership in Metabolic Regulations

Béatrice Desvergne

Center for Integrative Genomics, Building Génopode
University of Lausanne, CH-1015 Lausanne, Switzerland

I. Introduction
II. RXRs and Their Many Partners Belong to the Nuclear Receptor Superfamily
    A. Overview of the Nuclear Receptor Superfamily
    B. Classifying Nuclear Receptors: An Informative Mean for Positioning RXR-Related Signaling Pathways
III. RXR in Partnership: The Permissive Heterodimers as Metabolic Sensors
    A. An Overview of LXR:RXR Physiological Activities
    B. An Overview of FXR:RXR Physiological Activities
    C. An Overview of PPAR:RXR Physiological Activities
IV. The Rexinoid-Signaling Pathways: From Partnership to Leadership
    A. The Receptors
    B. The Nature of the Endogenous Ligand(s)
    C. The Nature of the Functional Complexes
    D. RXR Functional Activities
V. Conclusions
    References

Vitamin A signaling occurs through nuclear receptors recognizing diverse forms of retinoic acid (RA). The retinoic acid receptors (RARs) bind all-*trans* RA and its 9-*cis* isomer (9-*cis* RA). They convey most of the activity of RA, particularly during embryogenesis. The second subset of receptors, the rexinoid receptors (RXRs), binds 9-*cis* RA only. However, RXRs are obligatory DNA-binding partners for a number of nuclear receptors, broadening the spectrum of their biological activity to the corresponding nuclear receptor-signaling pathways.

The present chapter more particularly focuses on RXR-containing transcriptional complexes for which RXR is not only a structural component necessary for DNA binding but also acts as a ligand-activated partner. After positioning RXR among the nuclear receptor superfamily in the first part, we will give an overview of three major signaling pathways involved in metabolism, which are sensitive to RXR activation: LXR:RXR, FXR:RXR, and PPAR:RXR. The third and last part is focused on RXR signaling and its potential role in metabolic regulation. Indeed, while the nature of the endogenous ligand for RXR is still in question, as we will discuss herein, a better understanding of RXR activities is necessary to envisage the potential therapeutic applications of synthetic RXR ligands. © 2007 Elsevier Inc.

I. INTRODUCTION

Vitamin A or retinol is required at many stages during vertebrate development and remains crucial for the body homeostasis in the adult organism. Most of its numerous activities are due to the action of all-*trans* retinoic acid (atRA) and its 9-*cis* isomer (9-*cis* RA). Two classes of receptors convey the activity of retinoic acid (RA). The retinoic acid receptors (RARs) bind atRA and 9-*cis* RA. This class is well-characterized for its predominant but not exclusive role in embryogenesis and organogenesis (reviewed in Mark *et al.*, 2006). The second class corresponds to the rexinoid receptors (RXRs), which bind 9-*cis* RA only. Both RARs and RXRs belong to the nuclear receptor superfamily, the largest class of transcription factors. Within this family, RXRs occupy a particular place as they are obligatory DNA-binding partners for a number of other nuclear receptors. RXRs work in three different configurations: (1) as a structural component of the heterodimer complex, required for DNA binding but not acting as a receptor per se, these are the so-called "nonpermissive" heterodimers; (2) as both a structural and a functional component of the heterodimer, allowing 9-*cis* RA to signal through the corresponding heterodimer, forming the so-called permissive heterodimers; and (3) as conveyers of a 9-*cis* RA signal, independently of other nuclear receptors, by forming functional homodimers. However, one lingering question about RXR signaling concerns the true nature of the endogenous ligand, as will be discussed herein.

The present chapter is focused on the properties of RXR-containing transcriptional complexes, more particularly those playing a role in metabolic regulation. After positioning RXR among the nuclear receptor superfamily, we will discuss the three major signaling pathways involved in metabolism that are sensitive to RXR activation, that is, LXR:RXR, FXR:RXR, and PPAR:RXR pathways. The third and last part is focused on RXR signaling and its potential-specific role in metabolic regulations. Indeed, while the nature of the endogenous ligand for RXR is still in question, RXR synthetic ligands are already in use in clinical trials for their antitumorigenic activity. Experimental data also indicated a potential use in the treatment of metabolic diseases such as the type 2 diabetes. However, an extended understanding of RXR activities is still necessary to envisage the potential and safe therapeutic applications of synthetic RXR ligands.

II. RXRs AND THEIR MANY PARTNERS BELONG TO THE NUCLEAR RECEPTOR SUPERFAMILY

A. OVERVIEW OF THE NUCLEAR RECEPTOR SUPERFAMILY

Nuclear receptors are transcription factors characterized by two important properties: first, they are activated on the binding of specific ligands and second, they bind to specific response elements mainly located within the promoters of their target genes. Thus, in a simplified view, the effector function of nuclear receptors in a cell is to adapt gene expression according to signals received in the form of specific ligands. An official nomenclature for these receptors across species is now used, organized according to their phylogeny (Nuclear Receptors Nomenclature Committee, 1999; reviewed in Aranda and Pascual, 2001).

Nuclear receptors share a common structural organization and functional behavior. The poorly structured N-terminal domain encompasses, depending on the receptor, a very weak to strong ligand-independent transactivation domain. The DNA-binding domain (DBD) is folded in two zinc fingers and is the hallmark of the nuclear receptor family. The hinge region links the DBD to the ligand-binding domain (LBD). The general fold of the LBD is structured by 12 α-helices and 3 β-sheets defining the ligand-binding pocket.

Most nuclear receptors function as dimers, either homodimers, such as the glucocorticoid receptor (GR) or the estrogen receptor (ER), but more often as heterodimers with RXR. The DNA response element of nuclear receptors comprises two motifs corresponding to or closely related to the hexamer AGGTCA. The organization of these two motifs in direct, inverted, or palindromic repeats and the spacing between the two hexamers determine the specificity of these response elements toward each receptor dimer.

The general scheme for transactivation via nuclear receptors is thought to occur in at least two steps. In absence of ligand, nuclear receptor dimers may bind a corepressor protein that inhibits their transactivation properties. In the presence of ligand, or due to an alternative pathway of activation such as phosphorylation, the corepressor is released and a coactivator is recruited, allowing interactions with the transcription initiation complex as well as local histone modifications, eventually triggering or enhancing transcription. These cofactors are shared by numerous transcription factors, among which are the nuclear receptors, and might play key roles in the integration and coordination of the response of multiple genes to a variety of signals (Nettles and Greene, 2005; Tsai and Fondell, 2004).

B. CLASSIFYING NUCLEAR RECEPTORS: AN INFORMATIVE MEAN FOR POSITIONING RXR-RELATED SIGNALING PATHWAYS

Analyses of the human genome identified 48 nuclear receptor genes, some of them generating more than 1 receptor isoform. Different classifications have been proposed according to different criteria.

1. Classification According to the Phylogenetic Tree

A classification according to the position along the phylogenetic tree provided a practical and significant tool for unifying the nomenclature of all nuclear receptors across species. This system is based on the evolution of the two well-conserved domains of nuclear receptors (the DBD and the LBD) and distinguishes six subfamilies (Nuclear Receptors Nomenclature Committee, 1999). Interestingly, besides their phylogenic relationship, some common functional properties may be found within each group. All receptors contained in subfamily I are forming heterodimers with RXR. The three RARs (RARα, RARβ, and RARγ), the thyroid hormone receptors (TRα and TRβ), the vitamin D receptor (VDR), and the peroxisome proliferator-activated receptors (PPARα, PPARβ, and PPARγ) belong to this subfamily. Steroid receptors, which comprise the estrogen receptor, androgen receptor, progesterone receptor, mineralocorticoid receptor, and glucocorticoid receptor (ER, AR, PR, MR, and GR, respectively), mostly function as homodimers and are all found in subfamily III. Intriguingly, RXR that can function in both configurations (homodimer and heterodimer) is linked to yet another subfamily (subfamily II) together with HNF4.

This phylogeny-based classification can be in part superimposed to a structurally based classification, where nuclear receptors are ordered regarding the conservation of an amino acid located in the main dimer interface of the LBD and determinant for homodimer versus RXR-containing heterodimer formation (Brelivet et al., 2004). This ability to structurally distinguish the two classes, homodimers versus heterodimers, might become a very useful tool to further understand ligand interference, as will be discussed below, or some properties of orphan nuclear receptors. Intriguingly, RXR along this structural partitioning falls into the class of homodimers, consistent with its ability to function as a homodimer, while being the required partner for nuclear receptor heterodimers.

2. Classification According to the Ligand-Binding Properties

Ordering nuclear receptors according to their ligand-binding properties offers a functional classification in which they fall into three groups.

The orphan receptors possess the structural characteristics of nuclear receptors, including a sequence consistent with the presence of an LBD. However, no ligand has been identified so far for these receptors. Interestingly, the tight structure of the 12 helices in the Nurr1 LBD has been shown to preclude the formation of a ligand-binding pocket (Wang et al., 2003b), suggesting that at least some of the orphan receptors function in a ligand-independent manner. Interestingly, orphan receptors are mainly grouped in the phylogenetic subfamilies IV, V, and VI (the latter containing only one nuclear receptor, GCNF1). As for now, the functions of many orphan receptors remain elusive.

On the opposite, the "classic" hormone receptors bind a narrow range of molecules with very high affinity. They comprise the steroid receptors (ER, AR, PR, MR, and GR), which mediate the corresponding endocrine functions. It also includes TRα and TRβ, which bind triiodothyronine and VDR. Finally, the three RARs (RARα, RARβ, and RARγ) also belong to this class. While steroid receptors form homodimers, TRs, VDR, and RARs are functioning as RXR heterodimer.

The receptors of the "intermediary" class are metabolic sensors. This group comprises receptors binding to a broad range of molecules with, as a corollary, a relatively poor affinity. Rather than responding to hormones secreted by endocrine glands with tight feedback controls, these receptors can bind to molecules that are components of metabolic pathways as substrates, intermediates, or end-products. In this class are the PPARs, which are involved in many aspects of lipid metabolism, and more generally in energy metabolism. They can bind a wide variety of fatty acids, from dietary lipids to lipids derivatives such as eicosanoids. The liver X receptors (LXRα and LXRβ) recognize cholesterol metabolites such as oxysterols. Together with the farnesoid X receptor (FXR), which binds bile acid derivatives, they are closely involved in cholesterol metabolism. The pregnane X receptor (PXR) is activated by many endobiotic and xenobiotic compounds, a property shared with its close relative, the constitutive androstane receptor (CAR). Their activities induce the expression of multiple genes from the

CYP family, forming a redundant network for the detoxification and excretion of potentially harmful molecules, including therapeutic drugs (Jacobs et al., 2005; Xie et al., 2000). In addition, they also contribute to the enterohepatic circulation of bile acids as well as bile acid detoxification (Cao et al., 2004; Kullak-Ublick et al., 2004). In summary, the receptors of this class are sensors of the metabolic status, respond both to incoming dietary signals and to metabolites, and orchestrate the metabolic adaptation at the cell, organ, and whole organism levels. Interestingly, all these receptors act as heterodimers with RXR.

RXR itself is difficult to assign to a distinct class following this criterion. While it behaves as a classic receptor with respect to 9-*cis* RA, the nature of its endogenous ligand(s) is still unclear (see Section IV.B).

3. Classification According to the Active Regulatory Role of RXR

Among the RXR heterodimer-forming receptors, a further functional distinction may be done between nonpermissive and permissive receptors, very important with respect to rexinoid signaling. In permissive heterodimers, the DNA-binding complex is active in the presence of an agonist for either RXR or the partner receptor (Mangelsdorf and Evans, 1995; Minucci and Ozato, 1996). PPAR:RXR, LXR:RXR, and FXR:RXR are the best characterized permissive heterodimers. The mechanism underlying such a property is not well known, especially regarding cofactor recruitment. In the context of PPAR:RXR, the AF2 domain of RXR is not required for permissiveness, suggesting that transcriptional activation by RXR agonists results from an indirect conformational change of the PPAR LBD, transmitted through the heterodimerization interface (Schulman et al., 1998). However, it might not be a general rule as the AF2 domain of RXR is required for the permissiveness of PPARβ:RXR bound to the Hmgcs2 promoter (Calleja et al., 2006). TR and VDR form nonpermissive complexes, in which the RXR ligand cannot trigger transcription, except in the context of the prolactin gene promoter where RXR may also be an active partner of TR (Castillo et al., 2004; Li et al., 2004). The fact that permissive receptors belong to the "sensor" receptors whereas TR and VDR are classic high-affinity hormonal receptors led to a detailed molecular analysis of the structural determinant for receptor permissiveness. One attractive hypothesis is that loss of this property in TR and VDR arose during evolution in parallel with the ability to recognize endocrine ligands with high affinity (Shulman et al., 2004). CAR:RXR heterodimers seem to be neither strictly permissive nor nonpermissive for RXR signaling, as rexinoids have distinct effects depending on the context, particularly that of the response element (Tzameli et al., 2003). Finally, in this mode of action, RAR is not truly permissive since an RXR agonist is only active when the RAR agonist is previously bound to RAR (Germain et al., 2002), creating a so-called subordination mechanism.

Besides direct transcriptional activation, RXR may play an active modulating role by regulating the subcellular localization of its partners. This is convincingly described for VDR:RXR (Prufer and Barsony, 2002; Yasmin et al., 2005) and for Nurr/TR3/NGF1-B:RXR (Cao et al., 2004; Jacobs and Paulsen, 2005), but has not been reported for other RXR-dependent nuclear receptors.

We have emphasized in these classifications the particular place occupied by RXR heterodimers. We will now focus our attention on the RXR permissive complexes which are key metabolic regulators (Desvergne et al., 2006). These heterodimers are ideal targets in drug research on metabolic diseases, and understanding their main activities underlines the possible positive or negative therapeutic interference that rexinoids may provoke.

III. RXR IN PARTNERSHIP: THE PERMISSIVE HETERODIMERS AS METABOLIC SENSORS

A. AN OVERVIEW OF LXR:RXR PHYSIOLOGICAL ACTIVITIES

The endogenous activators of LXRs (NR1H3) are oxysterols and other derivatives of cholesterol metabolism. As such, they participate in cholesterol sensing and regulate important aspects of cholesterol and fatty acid metabolism (Tontonoz and Mangelsdorf, 2003).

Two isotypes, LXRα and LXRβ, share 77% amino acid identity in their DBD and LBD, and are highly conserved between rodents and human. LXRα is highly expressed in the liver but is also found in kidney, intestine, adipose tissue, and macrophages, whereas LXRβ is expressed ubiquitously. LXRs heterodimerize with RXR to bind to their DNA response element, formed of a direct repeat of two hexamers related to the sequence AGTTCA, separated by four nucleotides.

Mono-oxidized derivatives of cholesterol are potent LXR ligands. The most potent of these are 22(R)-hydroxycholesterol, 24(S)-hydroxycholesterol, and 24(S),25-epoxycholesterol, which activate both LXRα and LXRβ (Janowski et al., 1996; Lehmann et al., 1997). Little is known about the sterol hydroxylases that produce these metabolites, but it is assumed that oxysterol concentrations parallel those of cholesterol. Importantly, oxysterols are found at micromolar concentrations in tissues that express high levels of LXRα or LXRβ.

LXR was initially characterized by its role in the positive regulation of the gene encoding cholesterol 7α-hydroxylase (CYP7A), the rate-limiting enzyme in the neutral bile acid biosynthetic pathway that diverts cholesterol into bile acids. The nature of LXR endogenous ligands further emphasized the importance of this receptor in cholesterol metabolism. This function has been confirmed by the phenotype of *LXR*α null mice, which display a severely impaired cholesterol and bile acid metabolism when fed with a cholesterol-enriched diet. Indeed, these mutant mice fail to induce *CYP7A* and consequently suffer from a dramatic accumulation of cholesteryl esters in the liver with no increase in bile acid production (Peet *et al.*, 1998). However, it is now clear that the human *CYP7A* is not responsive to LXR and might even be repressed by LXRα activation (Chen *et al.*, 2002; Goodwin *et al.*, 2003). This difference between mouse and human is of interest as it might explain at least in part both the higher capability of mouse to face high-cholesterol diet and its increased resistance to the development of atherosclerosis.

As a consequence, the research focuses now more on other important LXR-mediated regulations that converge on the reverse cholesterol transport pathway. This pathway limits the exposure of peripheral cells to cholesterol excess and its modulation by LXR has been reported in both human and mouse, at least at three levels. First, LXR upregulates the expression of several genes encoding members of the ABC transporter family. ABCG5 and ABCG8, which are expressed almost exclusively in the liver and small intestine, favor the secretion of sterols from the liver epithelial cells to the bile duct and from the gut epithelial cells to the intestinal lumen (Berge *et al.*, 2000). Activation of these two genes by LXR is considered as the main mechanism by which an LXR agonist in mice causes a total blockade of cholesterol absorption (Plosch *et al.*, 2002; Repa *et al.*, 2002). Another important target is the widespread ABCA1 transporter, which promotes efflux of intracellular and plasma membrane cholesterol to the nascent high-density lipoprotein (HDL) particles via interaction with ApoA1, thereby increasing HDL levels (Repa *et al.*, 2000b). Second, LXR increases ApoE expression. Effluxed cholesterol from cell membrane can also be charged on HDL particles by ApoE, increasing their total capacity of accepting cholesterol. In addition, ApoE containing particles can interact with the scavenger receptor that increases the uptake of these particles in the liver. Third, LXR increases the expression of the cholesteryl ester transfer protein (CETP), which promotes cholesteryl ester transfer from VLDL to HDL and from HDL to low-density lipoprotein (LDL), a lipoprotein which is also efficiently taken up by the liver. By these means, LXR increases cholesterol clearance from the blood (reviewed in Tall *et al.*, 2002). Some of these regulations are shared by LXRα and LXRβ. If *LXRβ* null mice do not have the dramatic phenotype described for *LXRα*, the double mutant mice are more strongly affected than the *LXRα* null mice (Repa and Mangelsdorf, 1999). However, no functional compensation by LXRβ is seen in *LXRα* null mutant mice and a specific role for LXRβ has not been clearly defined yet.

Activated LXR also acts on fatty acid synthesis mainly by mediating the insulin-induced expression of SREBP-1c (Chen *et al.*, 2004), a transcription factor playing a major lipogenic role in hepatocytes (Repa *et al.*, 2000a; Schultz *et al.*, 2000). This lipogenic activity is also reinforced by the

LXR-dependent increase of fatty acid synthase (FAS) expression (Joseph et al., 2002). Together, these effects might explain the steatosis and the massive increase in VLDL and triglyceride blood levels observed in mice treated with pharmacological doses of an LXR ligand (Grefhorst et al., 2002).

In summary, LXR is a major transcription factor, which acts as a sensor of cholesterol levels via its interaction with oxysterols and, in turn, drives the disposal of the excess of cholesterol. It also acts at the level of individual cells, by increasing the ABC transporter molecules responsible for cholesterol efflux, and at the level of the organism by decreasing the cholesterol uptake from the diet. In mouse liver, it also increases the conversion of cholesterol into bile acids. These positive activities on cholesterol metabolism are, however, minored by the strong lipogenic and hypertriglyceridemia effects of LXR agonists.

B. AN OVERVIEW OF FXR:RXR PHYSIOLOGICAL ACTIVITIES

FXR acts as a bile acid sensor and is involved in a negative feedback regulation controlling bile acid production (reviewed in Edwards et al., 2002). FXR was initially cloned as an RXR-interacting protein called RIP14 (Seol et al., 1995). Its expression is restricted to adrenal cortex, intestine, colon, kidney, and liver. FXR forms heterodimers with RXR to bind to DNA response elements (FXREs), most often consisting of two hexamers (GGGTCA or close derivatives), organized in an inverted palindromic configuration, and spaced by one nucleotide.

Initially proposed as a receptor of farnesol metabolites, FXR was shown to bind and be activated by physiological concentrations of free and conjugated bile acids that are the end-products of the neutral and acidic bile acid biosynthetic pathway: chenodeoxycholic acid, lithocholic acid, and deoxycholic acid (Makishima et al., 1999; Parks et al., 1999; Wang et al., 1999). FXR senses bile acids and responds by inhibiting further bile acid synthesis, as illustrated in *FXR* null mice. These mice, which have no overt phenotype except increased bile acid levels in the blood, cannot sustain a cholic acid-enriched diet. They suffer from a severe wasting syndrome with hypothermia, and around 30% of them die by day 7 on such a diet (Sinal et al., 2000).

FXR acts at multiple levels of bile acid metabolism. First, it negatively regulates *CYP7A* whose expression positively controls the neutral pathway of bile acid synthesis (Goodwin et al., 2000; Lu et al., 2000). Second, it decreases the expression of the sodium taurocholate cotransporting polypeptide (NTCP), which mediates the uptake of bile acids in hepatocytes along the enterohepatic cycle (Denson et al., 2001; Sinal et al., 2000). These two inhibitions of gene expression occur via an indirect mechanism, involving the upregulation of the expression of the short heterodimerization partner (SHP-1), which inhibits in turn the activity of several transcription factors. Third, FXR:RXR positively regulates the gene encoding the bile salt export pump (BSEP), which belongs to the ABC transporter superfamily and allows the extrusion of bile acids from hepatocytes into the biliary canaliculus (Ananthanarayanan *et al.*, 2001; Schuetz *et al.*, 2001). These coordinated actions of FXR on *CYP7A*, *NTCP*, and *BSEP* result in lowering the potentially deleterious high levels of bile acids to which hepatocytes are exposed (reviewed in Francis *et al.*, 2003). In contrast, the FXR-mediated induction of the ileal bile acid-binding protein (IBABP), an intracellular carrier of bile acids expressed in the ileal epithelial cells, favors the reuptake of bile acids from the gut lumen (Grober *et al.*, 1999). Finally, FXR might also act directly on circulating lipoproteins by inducing the expression and secretion of hepatic ApoCII (Kast *et al.*, 2001), and by increasing the levels of the secreted enzyme phospholipid transfer protein (PLTP) that facilitates the transfer of cholesterol and phospholipids from triglyceride-rich lipoproteins to HDL (Urizar *et al.*, 2000). However, because of the more prominent effects of FXR on bile acid metabolism, it remains uneasy to identify the specific outcome of these increased gene expressions.

In summary, FXR is the transcription factor that senses the intracellular levels of bile acids and is required for limiting bile acid accumulation in the liver. It inhibits bile acid synthesis via the downregulation of *CYP7A* and increases bile acid efflux in the bile via increased *BSEP* expression. In the ileal enterocytes, the reabsorbed bile acids are taken in charge by the cytosolic IBABP whose expression is also increased by FXR.

C. AN OVERVIEW OF PPAR:RXR PHYSIOLOGICAL ACTIVITIES

1. General Properties of PPARs

PPARs were the first nuclear receptors identified as "sensors" rather than classic hormone receptors. They are nuclear, lipid-activable molecules that control a variety of genes in several pathways of lipid metabolism (reviewed in Desvergne and Wahli, 1999; Feige *et al.*, 2006). Three isotypes of PPAR, PPARα, PPARβ (also called PPARδ, NUCI, and FAAR), and PPARγ, have been cloned in *Xenopus*, rodents, and human. Two PPARγ isoforms, PPARγ1 and PPARγ2, are splice variants in their N-terminal domain. PPARα is highly expressed in tissues with high-lipid catabolism, for example liver, brown adipose tissue, skeletal, and heart muscle. PPARβ is ubiquitously expressed. PPARγ1 is mainly expressed in adipose tissues but is also present in the colon, spleen, retina, hematopoietic cells, and skeletal muscle. PPARγ2 has been found mainly in the brown and white adipose tissue (Braissant *et al.*, 1996; Escher *et al.*, 2001). Their modular structure is that of all nuclear receptors. The less conserved N-terminal region bears a ligand-independent activation domain at least in PPARα and PPARγ. The DBD is extremely well conserved. The ligand-binding pocket of PPARs is much larger than that of the other nuclear receptors and relatively easily accessible (Xu *et al.*, 2001 and reference therein).

PPARs bind to DNA as heterodimers with RXR on PPAR response elements (PPRE) comprising a direct repeat of two hexamers, closely related to the sequence AGGTCA and separated by one nucleotide (DR-1 sequence). The five nucleotides flanking the 5′ end of this core sequence are also important for the efficiency of PPARα:RXR binding.

The first molecules to be recognized as PPARα activators, and later on characterized as ligands, belong to a group of molecules that induce peroxisome proliferation in rodents, thus explaining the name of PPAR given to this receptor. This diverse group of substances includes, for example, some plasticizers and herbicides. More interestingly, various fatty acids, more particularly unsaturated fatty acids and some eicosanoids mainly derived from arachidonic acid and linoleic acid, bind to PPARα, PPARβ, and PPARγ with varying affinities. In addition to being activated by fatty acids, PPARα responds to fibrates which are hypolipidemic drugs, and PPARγ responds to thiazolidinediones which are insulin sensitizers, demonstrating their potential as drug targets.

In the process of transcriptional regulation, ligand-bound PPARs recruit coactivators, most likely organized in large complexes (Surapureddi *et al.*, 2002). Cofactor recruitment may be PPAR isotype specific and may ensure the specificity of target gene activation. In addition to PPAR ligand binding, PPARs can also be activated by phosphorylation of serines located in the A/B domain (Gelman *et al.*, 2005).

As can be expected from sensors, PPARs, which recognize and bind a variety of fatty acids, regulate in turn most of the pathways linked to lipid metabolism. Most fascinating is their balanced regulatory actions between fatty acid oxidation in the liver and other organs via PPARα, and fatty acid storage in the adipose tissue via PPARγ. In contrast, the role of PPARβ in metabolism remains elusive, albeit evidence is emerging for its function in lipid and cholesterol metabolism and transport (reviewed in Michalik *et al.*, 2003).

2. PPARγ: A Major Regulator of Fatty Acid Storage and Adipogenesis

PPARγ is a late marker of adipocyte differentiation, and its artificial expression is sufficient to force fibroblasts to undergo adipogenesis. Whereas *PPARγ* null mice are not viable, due to defects in placenta formation (Barak *et al.*, 1999), the lack of adipocytes carrying the genotype $PPAR\gamma^{-/-}$ in chimeric $PPAR\gamma^{+/+}:PPAR\gamma^{-/-}$ mice has demonstrated the importance, *in vivo*, of PPARγ for adipogenesis (Rosen *et al.*, 1999). In addition, mice with an adipose tissue-specific deletion of *PPARγ* exhibit a severe reduction of the number of mature adipocytes both in white and brown adipose tissues, while small and likely nascent adipocytes are appearing. In this model, the expression of the CRE enzyme responsible for the gene deletion is under the activity of the *aP2* promoter, thus triggering the deletion only after adipogenesis has taken place. This suggests that PPARγ is also essential for the survival of mature adipocytes (He *et al.*, 2003; Imai *et al.*, 2004b). Among PPARγ target genes are those encoding the adipocyte fatty acid-binding protein (aP2), the lipoprotein lipase (LPL), the acyl-CoA synthase (ACS), the fatty acid transport protein (FAT/CD36), the glycerol kinase, and the adipose differentiation-related protein (reviewed in Rosen and Spiegelman, 2001). Hence, PPARγ target genes are involved in each step of lipid entry and storage in the cells. Finally, PPARγ increases the expression of the uncoupling protein, thereby promoting increased energy expenditure via a futile cycle (Guan *et al.*, 2002).

A puzzling observation made a decade ago was that glitazones, which were developed for the treatment of insulin resistance, are PPARγ-selective ligands. The link between the promotion of adipocyte differentiation and lipid storage by PPARγ and the antidiabetic effects of these compounds is not fully understood. One hypothesis is fat redistribution from muscle to adipose tissue more particularly to subcutaneous fat, which is itself more sensitive to insulin than visceral fat (Gurnell *et al.*, 2003; Wajchenberg, 2000). Alternately, some data support the hypothesis that adiponectin, an adipokine with insulin-sensitizing property and a PPARγ target gene, might be a crucial component connecting PPARγ activation in the adipose tissue and the metabolic response of the peripheral organs (Gurnell *et al.*, 2003). Other possibilities are the inhibition of hepatic neoglucogenesis or induction of a futile cycle, as mentioned above. Unexpectedly, $PPAR\gamma^{+/-}$ heterozygous mice, rather than being prone to insulin resistance, are partially protected from high-fat diet-induced or monosodiumglutamate-induced weight gain and insulin resistance (Kubota *et al.*, 1999; Miles *et al.*, 2000). A similar protection is obtained via the use of PPARγ partial antagonists (Rieusset *et al.*, 2002). Besides the fact that glitazones have nonnegligible side effects, these observations led to the current approaches searching for PPARγ modulators rather than for full agonists.

3. PPARβ: From Adipogenesis to Fatty Acid Oxidation

The identity of PPARβ natural ligands remains the most elusive. Carbaprostacyclin (cPGI), a stable analogue of prostacyclin (PGI$_2$), acts as an agonist of PPARβ, supporting the notion that the cyclooxygenase-2 arachidonate metabolite PGI$_2$ might itself act as a *bona fide* natural ligand for PPARβ (Shao *et al.*, 2002). In addition, like PPARα and PPARγ, PPARβ binds fatty acids and, therefore, is also most likely a sensor of dietary lipids and lipid derivatives.

PPARβ may play a role in the early steps of adipogenesis. Together with two additional transcription factors, C/EBPβ and C/EBPδ, PPARβ appears to be implicated in the induction of PPARγ expression (Bastie *et al.*, 1999;

Holst *et al.*, 2003). In turn, high expression of PPARγ and C/EBPα in adipocytes establishes and maintains the terminal differentiation program. However, an adipose tissue-specific deletion of *PPARβ* does not alter fat mass (Barak *et al.*, 2002), whereas overexpression of a constitutively active form of PPARβ in brown and white adipose tissues generates lean mice and increases the mobilization and oxidation of fatty acids (Wang *et al.*, 2003a). Such activities are also found in the muscle (reviewed in Bedu *et al.*, 2005). *In vivo*, transgenic mice that overexpress PPARβ or a constitutively active PPARβ-VP16 fusion protein in muscle exhibit an enrichment of the muscle in red oxidative fibers, with an increased oxidative capacity assessed both at the gene expression and functional levels (Luquet *et al.*, 2003; Wang *et al.*, 2004). Similar results were obtained in wild-type mice treated with the PPARβ agonist (GW501516). Such a treatment results in a dose-dependent activation of fatty acid β-oxidation in the muscles, sustained by the higher expression of genes encoding enzymes involved in mitochondrial fatty acid catabolism, such as fatty acid transport proteins (FAT and LCAD) as well as uncoupling protein UCP2 and UCP3, associated to an increased energy expenditure (Tanaka *et al.*, 2003). Interestingly, the energy source of heart muscle cells depends on fatty acid oxidation, and a tissue-specific gene deletion of *PPARβ* in heart led to cardiomyopathy (Cheng *et al.*, 2004). The activity of PPARβ also directly affects glucose metabolism, increasing glucose transport in muscle cells (Kramer *et al.*, 2005) and favoring glucose utilization via the pentose phosphate pathway in the liver (Lee *et al.*, 2006). Whereas these results are very encouraging and have prompted the search for PPARβ agonists for the treatment of obesity and/or the metabolic syndrome, how and when PPARβ is activated in muscle cells, in the liver, or in adipocytes in the physiological context remain to be elucidated. In addition, there is accumulating evidence for an important role of PPARβ both in development and in wound healing. The latter involves an active PPARβ-dependent prosurvival activity (Di-Poi *et al.*, 2002; Michalik *et al.*, 2001) and raises concerns over the potential but still debated protumorigenic activity of PPARβ (reviewed in Michalik *et al.*, 2004).

4. PPARα: A Major Regulator of Fatty Acid Oxidation

PPARα target genes constitute a comprehensive set of genes that participate in many if not all aspects of lipid catabolism. This includes fatty acid transport across the cell membrane (fatty acid transporter protein genes), intracellular binding (liver fatty acid binding protein gene), activation via the formation of acyl-CoA (long-chain fatty acid acyl-CoA synthase gene), catabolism by β-oxidation in peroxisomes and mitochondria, and catabolism by ω-oxidation in microsomes (acyl-CoA oxidase gene, CYP4A1 and CYP4A6 genes, medium-chain acyl-CoA dehydrogenase gene, and 3-hydroxy-3-methylglutaryl-CoA synthase gene) (reviewed in Desvergne and Wahli, 1999). The role of PPARα in fatty acid oxidation is particularly highlighted during fasting that results in an enhanced load of fatty acids in the liver, then used as the source of energy. Food deprivation provokes an increased expression and activity of PPARα, which stimulates β-oxidation. *PPARα* null mice, which are viable and exhibit only subtle abnormalities in lipid metabolism when kept under normal laboratory confinement and diet (Lee *et al.*, 1995; Patel *et al.*, 2001), cannot sustain fasting. Their inability to enhance fatty acid oxidation results in hypoketonemia, associated with severe hypothermia and hypoglycemia (Kersten *et al.*, 1999; Leone *et al.*, 1999). Thus, PPARα is crucial for the organism to adapt to an increased demand in fatty acid oxidation, while it seems to play a marginal role in the basal situation with normal diets.

The contribution of PPARα in fatty acid oxidation in muscle tissues, where it is also well expressed, is possibly hidden by other factors, notably PPARβ (Muoio *et al.*, 2002). However, high levels of PPARα activity in muscles have been observed in diabetic mice both in heart and skeletal muscle (Finck *et al.*, 2002, 2003; Yechoor *et al.*, 2002). In parallel, overexpression of PPARα in skeletal muscle increases fatty acid oxidation and decreases insulin-stimulated glucose uptake via inhibition of Glut4 expression (Finck *et al.*, 2005), suggesting that the increased activity of PPARα in the muscles of diabetic patients may contribute to insulin resistance. A contrasting pattern is described in the pathological cardiac hypertrophy. In this context, the expression of PPARα is downregulated, the utilization of fatty acids as energy substrate is decreased, and the genes implicated in the utilization of glucose as the main energy source are reinduced (Barger *et al.*, 2000). It is presently unclear whether the decline in PPARα activity and fatty acid oxidation is a cause or a consequence of cardiac hypertrophy. Interestingly, *PPARα* null mice exhibit a decreased contractile and metabolic reserve in heart, rescued by favoring glucose transport and utilization (Luptak *et al.*, 2005). At present, it is still unclear whether PPARα ligands in human would be beneficial or detrimental to the heart in the context of cardiac hypertrophy.

5. The Role of PPARs in Lipoprotein Metabolism

As a consequence of their activities in lipid metabolism, all three PPARs act on blood lipid levels. PPARα increases reverse cholesterol transport by upregulating the expression of the genes encoding the cholesterol acceptor apolipoprotein ApoAI and ApoAII, and that of the hepatic expression of the scavenger receptor BI (SR-BI)/CLA-1, thereby increasing the selective uptake of HDL cholesteryl esters from the blood. Indeed, fibrates have been used for the treatment of dyslipidemia, much before the discovery of their mechanism of action via PPARα. In addition, both PPARα and PPARγ upregulate the expression of the lipoprotein lipase gene (reviewed in Bocher *et al.*, 2002). Together with a decreased expression of ApoCIII, these effects increase free fatty acid delivery to peripheral tissues. This explains in part why thiazolidinediones, which potently activate PPARγ, not only act on insulin sensitivity but also decrease plasma free fatty acid concentrations and improve the overall lipoprotein profile (Mayerson et al., 2002; Oakes et al., 1997).

Treatment with a PPARβ agonist of obese rhesus monkeys, used as a relevant animal model for human obesity and the associated metabolic disorders, caused an increase in the level of serum HDL cholesterol, while lowering the level of small-dense LDL, fasting triglycerides, and fasting insulin (Oliver et al., 2001). Similar results were obtained in *db/db* mice (Leibowitz et al., 2000). Further studies in non-obese mice demonstrate that PPARβ activation results in both increased HDL concentration in the blood and accelerated fecal cholesterol removal from the body via downregulation of the intestinal gene expression of cholesterol absorption protein Niemann-Pick C1-like 1 (NPC1L1) (van der Veen et al., 2005).

In summary, each PPAR acts as a lipid sensor with distinct activities that adjust at the cellular and organism levels the metabolic status, with balancing actions on fatty acid oxidation and fatty acid storage processes. These metabolic regulations also involve a PPAR-dependent coordination of glucose metabolism.

IV. THE REXINOID-SIGNALING PATHWAYS: FROM PARTNERSHIP TO LEADERSHIP

As mentioned above, RXR is itself a nuclear receptor that can be activated by 9-*cis* RA, an isomer of atRA. The present chapter discusses the observations more specifically linked to RXR activities.

A. THE RECEPTORS

There are three isotypes of RXR, α, β, and γ, and several isoforms for each of them (Mangelsdorf et al., 1992; reviewed in Chambon, 1996). Each isotype and isoform has its specific expression pattern. RXRα is widely expressed, with a high expression in the liver, kidney, spleen, placenta, and epidermis. *RXRα* null mutation is embryonic lethal, consistent with its crucial implication in development. RXRβ is ubiquitously present. However, the defective spermatogenesis in *RXRβ* null male mice underscores its specific role in the testis, more particularly in Sertoli cells. RXRγ expression is more restricted to the muscle and the brain, and the phenotype of *RXRγ* null mice is mainly characterized by metabolic and behavioral defects (reviewed in Mark et al., 2006).

Thus, while at least one functional *RXRα* allele is strictly required for a successful development, analyses of the simple, double, and triple RXR knockout mice demonstrated a partial redundancy in RXRα, RXRβ, and RXRγ functions.

B. THE NATURE OF THE ENDOGENOUS LIGAND(S)

RXR can be activated by 9-*cis* RA, an isomer of atRA (Heyman *et al.*, 1992). While the occurrence of this molecule *in vivo* has been questioned, the identification of two enzymes that participate in the isomerization of atRA to form 9-*cis* RA lends support for its relevance in the whole organism (Mertz *et al.*, 1997; Romert *et al.*, 1998). However, some doubts linger about the nature of the major natural RXR ligand, if any. More particularly, a contradiction is raised by the subordination mechanism. As mentioned above, this mechanism allows ligand-activated RXR to enhance RAR:RXR transcriptional activity only if RAR is first liganded. In this mechanism, it is indeed quite essential to ensure that the RAR:RXR activity is initially driven by RAR signals and then amplified by an RXR agonist, thereby avoiding signaling pathway promiscuity. This control, however, must consider that the ligand for RXR is different from the ligand for RAR. This is clearly not the case for 9-*cis* RA, which binds RAR more efficiently than RXR. This question was more precisely raised in the context of the formation of lamellar granules in the epidermal keratinocytes. In this context, it is proposed that RAR:RXR must remain unliganded to maintain repression of yet uncharacterized target genes. In the very same context, RXR agonist is required for an efficient activity of PPARβ:RXR on the 3-hydroxy-3-methylglutaryl coenzyme A (*Hmgcs2*) gene expression, thus suggesting that 9-*cis* RA cannot be the RXR-activating ligand in suprabasal keratinocytes (Calleja *et al.*, 2006). It must be noted that independently of the nature of the rexinoid present *in vivo*, the promiscuity on the signaling via permissive but not subordinate partners cannot be avoided.

Phytanic acid or dietary phytol metabolites have been proposed as possible natural RXR ligands (Kitareewan *et al.*, 1996; Lemotte *et al.*, 1996). An oleic acid molecule was found in the RXR ligand-binding pocket in a crystal of RAR:RXR LBDs (Bourguet *et al.*, 2000), while a cell-based reporter assay identified in lipid cellular extracts some unsaturated fatty acids as active components for RXR activation (Goldstein *et al.*, 2003; Lengqvist *et al.*, 2004). In support of this notion, RXR is activated in the adult mouse brain by the long-chain polyunsaturated fatty acid docohexaenoic acid (de Urquiza *et al.*, 2000). A comparison of high-resolution structures of the RXRα LBD bound to 9-*cis* RA or to docohexaenoic acid favors the latter as having the highest number of ligand-protein contacts (Egea *et al.*, 2002). This binding activity seems to be specific to *n-3* PUFA, as no functional activity via RXR is found with *n-6* PUFA (Fan *et al.*, 2003). These results complement those obtained with transgenic mice and *Xenopus* engineered to detect RXR ligands, which identified the presence of an RXR ligand in the developing spinal cord (Luria and Furlow, 2004; Solomin *et al.*, 1998).

However, even though a specific deletion of the AF2 domain in RXRα clearly emphasizes the importance of a fully functional RXR LBD for the proper function of RXR *in vivo* (Calleja *et al.*, 2006; Mascrez *et al.*, 1998), one can still question whether any RXR endogenous ligand is biologically active *in vivo*.

C. THE NATURE OF THE FUNCTIONAL COMPLEXES

RXR behavior in cells is unusual in that homotetramer, homodimers, and heterodimers are proposed to mediate its activity. Indeed, RXR can self-associate to form in solution inactive homotetramers that are transcriptionally inactive (Gampe *et al.*, 2000; Kersten *et al.*, 1995, 1998). Their dissociation on ligand binding allows the formation of transcriptionally active homodimers and heterodimers (Chen *et al.*, 1998). A possible role of the RXR homotetramer is to act as an architectural complex, allowing the looping of DNA between two RXR response elements positioned in tandem along a promoter sequence (Yasmin *et al.*, 2004). To which extent RXR tetramers form in the living cells remains unsettled. That an RXR ligand is required to direct the dissociation would indicate that this ligand is ubiquitously present, since permissive and nonpermissive RXR heterodimers are formed in absence of known ligand for RXR. In addition, fluorescence studies in living cells demonstrate that the formation of RXR heterodimers do not depend on the presence of an exogenous RXR agonist (Feige *et al.*, 2005). Alternately, the dissociation from homotetramers might operate according to an equilibrium shifted in the presence of a ligand for a partner, increasing the affinity and formation of the corresponding heterodimers. This seems to apply for the PPARα:RXR heterodimer (Feige *et al.*, 2005; Tudor *et al.*, 2006), although the effect is weak and such a mechanism remains to be explored for other RXR partners.

Regardless the functional relevance of RXR tetramers, RXR can form homodimers. The formation and DNA binding of RXR homodimers were demonstrated *in vitro* (Zhang *et al.*, 1992) as well as *in vivo* (IJpenberg *et al.*, 2004). Accordingly, fluorescence correlation spectroscopy further demonstrated the formation of homodimers of RXR LBD in living cells exposed to an RXR agonist (Chen *et al.*, 2005). Indeed, binding of 9-*cis* RA triggers the formation of RXR homodimers, followed by the recruitment of specific coactivators, such as SRC1 and TIF2. These key steps stabilize the RXR:RXR complex onto DR1 elements, that is, similar elements as for PPAR:RXR complexes, increasing the promiscuity between these two signaling pathways (IJpenberg *et al.*, 2004). However, as long as the nature of the endogenous ligand for RXR is not solved, the extent of the physiological relevance of the homodimer activity cannot be properly evaluated.

Finally, and as previously discussed, the various permissive heterodimers are the main characterized conveyors of RXR agonist activity, as these complexes may be turned on by an RXR ligand, independently of the partner activation. Interestingly, a structural analysis of ligand binding and protein dynamics revealed a dynamic range of possible conformations, persisting after ligand binding, in line with the multiple partnership of RXR (Lu et al., 2006).

We have discussed in Chapter 3 the physiological activities of the permissive RXR heterodimers, as viewed from the agonist-activated partner. In the next section, we will review the activities of RXR more specifically from the RXR point of view.

D. RXR FUNCTIONAL ACTIVITIES

Insights in specific RXR activity come from loss of function experiments in mouse lacking *RXR* alleles, and from gain of function experiments in mice treated with RXR specific agonists. The first approach demonstrates the confounding actions of RXR on the permissive partner signaling pathways; the second may more specifically addresses the rexinoid pathway *in vivo*, assuming that a natural ligand for RXR do exist *in vivo*. Independently of the physiological questions, this latter approach is essential for determining the range of possible therapeutic activities of RXR agonists.

1. Lessons from Knockout Mice

RXR is an obligatory partner for many nuclear receptors important during development (Mangelsdorf and Evans, 1995). Thus, its constitutive deletion likely affects a very diverse array of developmental and physiological pathways. This has been clearly underlined by the thorough phenotypic analyses of the developmental defects occurring in mice carrying various *RXR* gene deletions or mutations. Whereas *RXRβ* and *RXRγ* null mutations give rise to rather minor developmental defects, *RXRα* null mice die at early embryonic stages (Kastner *et al.*, 1994; Sucov *et al.*, 1994). This lethality might be due to defects in the PPARβ- and PPARγ-signaling pathways, since the invalidation of any of these genes leads to embryonic lethality with placental defects that are timely overlapping with those observed in *RXRα* null placenta (Barak *et al.*, 1999; Nadra *et al.*, 2006; Wendling *et al.*, 1999). Other defects analyzed in the various *RXR* mutants are seen in various tissues, such as the skin, the eye, the heart, and the testis, and reflect alterations in the pathways of other receptors. While RARs, which transduce the vitamin A activities through the RA signal, play a major role, TR- and VDR-dependent pathways are also affected (Mark *et al.*, 1999; Wendling *et al.*, 1999; reviewed in Mark *et al.*, 2006).

From the metabolic point of view, analyzing the importance of RXRα in "metabolic" adult tissues required the generation of tissue-specific knockouts. Specific invalidation of *RXRα* has been generated in liver, and metabolic studies were performed to identify which pathways were most affected. As could be expected, many PPARα-mediated functions in fatty acid oxidation were altered due to the lack of RXRα. However, other pathways that include LXR and FXR pathways were also compromised, at least partially, by the absence of RXRα. These effects could not be compensated for by RXRβ and RXRγ (Imai et al., 2001b; Wan et al., 2000a,b). Invalidation of *RXRα* in the adipose tissue of adult animals resulted in an alteration of preadipocyte differentiation as well as in resistance to induced obesity (Imai et al., 2001a). These results are reminiscent of the observations obtained with *PPARγ* heterozygous mice (Kubota et al., 1999) and with an adipose-specific deletion of *PPARγ* (Imai et al., 2004a), suggesting that most of the effects due to the lack of RXRα expression reflect altered PPARγ functions (Imai et al., 2001a). However, the impaired lipolysis observed in these mice might be related to an alteration of LXR:RXR heterodimer signaling.

2. The Pharmacological Activity of RXR Agonists

As mentioned above, RXR is a *bona fide* receptor but the nature of its endogenous ligand, if any, remains elusive. Thus, all observations made so far were obtained in a pharmacological context, treating mice with 9-*cis* RA or synthetic RXR agonists. On such treatments, RXR may regulate transcription as a homodimer (RXR:RXR), binding to DR-1 like response elements. However, most of RXR agonist activities likely rely on the activation of permissive heterodimers, raising a major interest in the exploitation of this ability for therapeutic purposes.

The first application of rexinoids in clinical studies took advantage of their efficacy in triggering apoptosis, in contrast to cell differentiation seen with retinoids (Mehta et al., 1996; Nagy et al., 1995). This led to their successful use since the 1980s in the treatment of refractory or persistent early-stage cutaneous T-cell lymphoma. Presently, a number of cancer types and cell types are being tested for their possible responsiveness to rexinoids, such as acute myeloid leukemia (Altucci et al., 2005), aerodigestive tract cancer (Dragnev et al., 2005), human breast cancer cells (Toma et al., 1998), or pancreatic cancer cells (Balasubramanian et al., 2004). Along the same line, chemopreventive *n-3* fatty acids in colon were shown to activate RXR in colonocytes (Fan et al., 2003).

However, initial studies as well as phase 2 and 3 clinical trials with bexarotene (Targretin® capsules corresponding to the well-characterized rexinoid LG1069) reported high triglyceridemia, hypothyroidism, and hypercholesterolemia (Duvic et al., 2001; Rizvi et al., 1999). In parallel, global gene expression profiles of various tissues from rats treated with bexarotene further underscore its action on metabolic pathways (Wang et al., 2006).

Actually, numerous studies have reported the broad impact of RXR synthetic ligands on metabolic regulations in the adult organism. The first seminal observation was that *in vivo* administration of the synthetic specific RXR ligands mimics—and increases when given in combination with TZD—the metabolic effects of PPARγ ligands, by decreasing hyperglycemia, and improving insulin sensitivity (Mukherjee et al., 1997, 1998).

The crucial role of muscle physiology in insulin sensitivity oriented the search for the mechanism on this tissue, where rexinoids were shown to activate a number of genes related to fatty acid uptake (CD36) and desaturation (SCD1), while increasing glycogen synthase activity in muscle cells in culture and improving glucose disposal in skeletal muscles (Cha et al., 2001; Shen et al., 2004; Singh Ahuja et al., 2001; reviewed in Szanto et al., 2004b). Because PPARγ agonists are also insulin sensitizer, the first hypothesis was that PPARγ is the partner of RXR in these rexinoid activities. However, careful and extensive gene expression analyses comparing TZD and rexinoids activities revealed some common but also some clearly distinct target genes and tissue-specific activities. More particularly, rexinoids act primarily in the liver and the skeletal muscle, in contrast to TZD which exert their effects mainly on the adipose tissue and to a lesser extent in muscle (Shen et al., 2004; Singh Ahuja et al., 2001). Thus, PPARα, PPARβ, and LXR must be considered as likely partners accounting for rexinoid action, since all were found to have some antidiabetic activity (Cao et al., 2003; Lee et al., 2006; Park et al., 2006). In addition, the regulation of many genes by rexinoids in the liver was also PPARα dependent (Ouamrane et al., 2003).

With respect to lipid metabolism, rexinoids provoke a very efficient inhibition of cholesterol absorption. A dual mechanism for this is proposed: repression of bile acid production via inhibition of Cyp7A in an FXR-dependent manner and increased efflux of cholesterol from enterocytes into the lumen via an LXR:RXR-dependent induced expression of the transport protein ABC1 (Repa et al., 2000b). Positive effects are also observed in the $apoE^{-/-}$ mouse model where rexinoids reduced the development of atherosclerosis (Claudel et al., 2001), likely through the concomitant activation of PPAR- and LXR-signaling pathways in macrophages (Szanto et al., 2004a).

Paradoxically, rexinoids may also antagonize FXR activity, an effect which may result from the disruption of the ability of the FXR:RXR heterodimer to interact with coactivators, as seen on the *BSEP* promoter (Kassam et al., 2003). Another paradoxical effect concerns a severe hypertriglyceridemia frequently observed in human and in some animal models (Miller et al., 1997). It has been linked to reduced LPL activity in skeletal and cardiac tissue (Davies et al., 2001), but the nature of corresponding heterodimer involved is elusive, as neither PPARs, nor LXR or RAR may explain this effect. Also unexpectedly, rexinoid treatment provokes a central hypothyroidism, due to the specific inhibition of TSH expression and secretion (Liu et al., 2002; Sharma et al., 2006), and which may explain some cases of rexinoid-associated hypercholesterolemia.

These observations underline the problem that faces the experimentalist with such a promiscuous agent. On the one hand, it is difficult to predict the scope of changes that a specific RXR ligand may provoke in the whole organism with respect to metabolic homeostasis. On the other hand, the pleiotropic action of RXR might also be an advantage in the context of complex and multifactorial diseases, still raising interest in the identification of RXR ligands specific for a given or a limited number of heterodimers, allowing more targeted therapeutic approaches of metabolic diseases. In that respect, new agonists but also antagonists are proposed (Cavasotto et al., 2004; Cesario et al., 2001; Deng et al., 2005). In parallel, intensive screening for RXR modulators are performed (Gernert et al., 2003, 2004; Haffner et al., 2004), more particularly searching for molecules active on type 2 diabetes (Leibowitz et al., 2006; Michellys et al., 2003a,b).

V. CONCLUSIONS

New insights in the regulation of expression, mechanisms of action, and physiological role of RXR-dependent signaling pathways have led to a better understanding of energy homeostasis regulation. While the perspectives are promising, it clearly appears that further exploration is needed to better define the potential of RXR as a therapeutic target. The difficulty comes first from the high levels of complexity generated by the multiplicity of the dimers in which RXR is engaged, including RXR homodimers. More particularly, each dimer may fulfill regulatory functions that antagonize the action of other dimers. This is, for example, seen with the opposite action of LXR:RXR and FXR:RXR on the neutral bile acid biosynthesis, or of PPARα:RXR and PPARγ:RXR on fatty acid oxidation and storage. A second level of difficulties comes from the fact that all intermediary metabolisms (glucose, lipid, and amino acid metabolisms) are highly connected, particularly through the cross talk between the numerous transcription factors involved (reviewed in Desvergne et al., 2006). One example is the positive regulation of *LXR* expression by PPARγ:RXR. A third difficulty is that most RXR-containing heterodimers are not only affecting metabolic regulation but also act on systemic regulations such as inflammation or immune processes. In addition, they act on various key cellular functions, such as cell migration, proliferation, differentiation, and survival, as particularly emphasized for PPARβ:RXR functions. Although these accumulating levels of complexity might deter from envisioning RXR as a therapeutic target, it also might be considered as a possibility to smoothly shift an overall equilibrium from a pathological to a more physiological status. This is indeed the road already taken when looking for modulators, partial agonists, or dual agonists, rather than highly specific activators of nuclear receptors.

ACKNOWLEDGMENTS

We gratefully thank Laurent Gelman for the critical reading of the manuscript. The author's works are benefiting from grants from the Swiss National Foundation for Research, the Etat de Vaud, and The European program "Eumorphia."

REFERENCES

Altucci, L., Rossin, A., Hirsch, O., Nebbioso, A., Vitoux, D., Wilhelm, E., Guidez, F., De Simone, M., Schiavone, E. M., Grimwade, D., Zelent, A., de The, H., et al. (2005). Rexinoid-triggered differentiation and tumor-selective apoptosis of acute myeloid leukemia by protein kinase A-mediated desubordination of retinoid X receptor. *Cancer Res.* **65**(19), 8754–8765.

Ananthanarayanan, M., Balasubramanian, N., Makishima, M., Mangelsdorf, D. J., and Suchy, F. J. (2001). Human bile salt export pump promoter is transactivated by the farnesoid X receptor/bile acid receptor. *J. Biol. Chem.* **276**(31), 28857–28865.

Aranda, A., and Pascual, A. (2001). Nuclear hormone receptors and gene expression. *Physiol. Rev.* **81**(3), 1269–1304.

Balasubramanian, S., Chandraratna, R. A., and Eckert, R. L. (2004). Suppression of human pancreatic cancer cell proliferation by AGN194204, an RXR-selective retinoid. *Carcinogenesis* **25**(8), 1377–1385.

Barak, Y., Nelson, M. C., Ong, E. S., Jones, Y. Z., Ruiz-Lozano, P., Chien, K. R., Koder, A., and Evans, R. M. (1999). PPAR gamma is required for placental, cardiac, and adipose tissue development. *Mol. Cell* **4**(4), 585–595.

Barak, Y., Liao, D., He, W., Ong, E. S., Nelson, M. C., Olefsky, J. M., Boland, R., and Evans, R. M. (2002). Effects of peroxisome proliferator-activated receptor delta on placentation, adiposity, and colorectal cancer. *Proc. Natl. Acad. Sci. USA* **99**(1), 303–308.

Barger, P. M., Brandt, J. M., Leone, T. C., Weinheimer, C. J., and Kelly, D. P. (2000). Deactivation of peroxisome proliferator-activated receptor-alpha during cardiac hypertrophic growth. *J. Clin. Invest.* **105**(12), 1723–1730.

Bastie, C., Holst, D., Gaillard, D., Jehl-Pietri, C., and Grimaldi, P. A. (1999). Expression of peroxisome proliferator-activated receptor PPARdelta promotes induction of PPARgamma and adipocyte differentiation in 3T3C2 fibroblasts. *J. Biol. Chem.* **274**(31), 21920–21925.

Bedu, E., Wahli, W., and Desvergne, B. (2005). Peroxisome proliferator-activated receptor beta/delta as a therapeutic target for metabolic diseases. *Expert Opin. Ther. Targets* **9**(4), 861–873.

Berge, K. E., Tian, H., Graf, G. A., Yu, L., Grishin, N. V., Schultz, J., Kwiterovich, P., Shan, B., Barnes, R., and Hobbs, H. H. (2000). Accumulation of dietary cholesterol in sitosterolemia caused by mutations in adjacent ABC transporters. *Science* **290**(5497), 1771–1775.

Bocher, V., Pineda-Torra, I., Fruchart, J. C., and Staels, B. (2002). PPARs: Transcription factors controlling lipid and lipoprotein metabolism. *Ann. NY Acad. Sci.* **967**, 7–18.

Bourguet, W., Vivat, V., Wurtz, J. M., Chambon, P., Gronemeyer, H., and Moras, D. (2000). Crystal structure of a heterodimeric complex of RAR and RXR ligand-binding domains. *Mol. Cell* **5**(2), 289–298.

Braissant, O., Foufelle, F., Scotto, C., Dauça, M., and Wahli, W. (1996). Differential expression of peroxisome proliferator-activated receptors: Tissue distribution of PPAR-α, -β and -γ in the adult rat. *Endocrinology* **137**, 354–366.

Brelivet, Y., Kammerer, S., Rochel, N., Poch, O., and Moras, D. (2004). Signature of the oligomeric behaviour of nuclear receptors at the sequence and structural level. *EMBO Rep.* **5**(4), 423–429.

Calleja, C., Messaddeq, N., Chapellier, B., Yang, H., Krezel, W., Li, M., Metzger, D., Mascrez, B., Ohta, K., Kagechika, H., Endo, Y., Mark, M., et al. (2006). Genetic and pharmacological evidence that a acid cannot be the RXR-activating ligand in mouse epidermis keratinocytes. *Genes Dev.* **20**(11), 1525–1538.

Cao, G., Liang, Y., Broderick, C. L., Oldham, B. A., Beyer, T. P., Schmidt, R. J., Zhang, Y., Stayrook, K. R., Suen, C., Otto, K. A., Miller, A. R., Dai, J., et al. (2003). Antidiabetic action of a liver x receptor agonist mediated by inhibition of hepatic gluconeogenesis. *J. Biol. Chem.* **278**(2), 1131–1136.

Cao, X., Liu, W., Lin, F., Li, H., Kolluri, S. K., Lin, B., Han, Y. H., Dawson, M. I., and Zhang, X. K. (2004). Retinoid X receptor regulates Nur77/TR3-dependent apoptosis by modulating its nuclear export and mitochondrial targeting. *Mol. Cell. Biol.* **24**(22), 9705–9725.

Castillo, A. I., Sanchez-Martinez, R., Moreno, J. L., Martinez-Iglesias, O. A., Palacios, D., and Aranda, A. (2004). A permissive retinoid X receptor/thyroid hormone receptor heterodimer allows stimulation of prolactin gene transcription by thyroid hormone and 9-*cis*-retinoic acid. *Mol. Cell. Biol.* **24**(2), 502–513.

Cavasotto, C. N., Liu, G., James, S. Y., Hobbs, P. D., Peterson, V. J., Bhattacharya, A. A., Kolluri, S. K., Zhang, X. K., Leid, M., Abagyan, R., Liddington, R. C., and Dawson, M. I. (2004). Determinants of retinoid X receptor transcriptional antagonism. *J. Med. Chem.* **47**(18), 4360–4372.

Cesario, R. M., Klausing, K., Razzaghi, H., Crombie, D., Rungta, D., Heyman, R. A., and Lala, D. S. (2001). The rexinoid LG100754 is a novel RXR:PPARgamma agonist and decreases glucose levels *in vivo*. *Mol. Endocrinol.* **15**(8), 1360–1369.

Cha, B. S., Ciaraldi, T. P., Carter, L., Nikoulina, S. E., Mudaliar, S., Mukherjee, R., Paterniti, J. R., Jr., and Henry, R. R. (2001). Peroxisome proliferator-activated receptor (PPAR) gamma and retinoid X receptor (RXR) agonists have complementary effects on glucose and lipid metabolism in human skeletal muscle. *Diabetologia* **44**(4), 444–452.

Chambon, P. (1996). A decade of molecular biology of retinoic acid receptors. *FASEB J.* **10**, 940–954.

Chen, G., Liang, G., Ou, J., Goldstein, J. L., and Brown, M. S. (2004). Central role for liver X receptor in insulin-mediated activation of Srebp-1c transcription and stimulation of fatty acid synthesis in liver. *Proc. Natl. Acad. Sci. USA* **101**(31), 11245–11250.

Chen, J. Y., Levy-Wilson, B., Goodart, S., and Cooper, A. D. (2002). Mice expressing the human CYP7A1 gene in the mouse CYP7A1 knock-out background lack induction of CYP7A1 expression by cholesterol feeding and have increased hypercholesterolemia when fed a high fat diet. *J. Biol. Chem.* **277**(45), 42588–42595.

Chen, Y., Wei, L. N., and Muller, J. D. (2005). Unraveling protein-protein interactions in living cells with fluorescence fluctuation brightness analysis. *Biophys. J.* **88**(6), 4366–4377.

Chen, Z. P., Iyer, J., Bourguet, W., Held, P., Mioskowski, C., Lebeau, L., Noy, N., Chambon, P., and Gronemeyer, H. (1998). Ligand- and DNA-induced dissociation of RXR tetramers. *J. Mol. Biol.* **275**(1), 55–65.

Cheng, L., Ding, G., Qin, Q., Huang, Y., Lewis, W., He, N., Evans, R. M., Schneider, M. D., Brako, F. A., Xiao, Y., Chen, Y. E., and Yang, Q. (2004). Cardiomyocyte-restricted peroxisome proliferator-activated receptor-delta deletion perturbs myocardial fatty acid oxidation and leads to cardiomyopathy. *Nat. Med.* **10**(11), 1245–1250.

Claudel, T., Leibowitz, M. D., Fievet, C., Tailleux, A., Wagner, B., Repa, J. J., Torpier, G., Lobaccaro, J. M., Paterniti, J. R., Mangelsdorf, D. J., Heyman, R. A., and Auwerx, J. (2001). Reduction of atherosclerosis in apolipoprotein E knockout mice by activation of the retinoid X receptor. *Proc. Natl. Acad. Sci. USA* **98**(5), 2610–2615.

Davies, P. J., Berry, S. A., Shipley, G. L., Eckel, R. H., Hennuyer, N., Crombie, D. L., Ogilvie, K. M., Peinado-Onsurbe, J., Fievet, C., Leibowitz, M. D., Heyman, R. A., and Auwerx, J. (2001). Metabolic effects of rexinoids: Tissue-specific regulation of lipoprotein lipase activity. *Mol. Pharmacol.* **59**(2), 170–176.

de Urquiza, A. M., Liu, S., Sjoberg, M., Zetterstrom, R. H., Griffiths, W., Sjovall, J., and Perlmann, T. (2000). Docosahexaenoic acid, a ligand for the retinoid X receptor in mouse brain. *Science* **290**(5499), 2140–2144.

Deng, T., Shan, S., Li, Z. B., Wu, Z. W., Liao, C. Z., Ko, B., Lu, X. P., Cheng, J., and Ning, Z. Q. (2005). A new retinoid-like compound that activates peroxisome proliferator-activated receptors and lowers blood glucose in diabetic mice. *Biol. Pharm. Bull.* **28**(7), 1192–1196.

Denson, L. A., Sturm, E., Echevarria, W., Zimmerman, T. L., Makishima, M., Mangelsdorf, D. J., and Karpen, S. J. (2001). The orphan nuclear receptor, shp, mediates bile acid-induced inhibition of the rat bile acid transporter, ntcp. *Gastroenterology* **121**(1), 140–147.

Desvergne, B., and Wahli, W. (1999). Peroxisome proliferator-activated receptors: Nuclear control of metabolism. *Endocr. Rev.* **20**, 649–688.

Desvergne, B., Michalik, L., and Wahli, W. (2006). Transcriptional regulation of metabolism. *Physiol. Rev.* **86**(2), 465–514.

Di-Poi, N., Tan, N. S., Michalik, L., Wahli, W., and Desvergne, B. (2002). Antiapoptotic role of PPARbeta in keratinocytes via transcriptional control of the Akt1 signaling pathway. *Mol. Cell* **10**(4), 721–733.

Dragnev, K. H., Petty, W. J., Shah, S., Biddle, A., Desai, N. B., Memoli, V., Rigas, J. R., and Dmitrovsky, E. (2005). Bexarotene and erlotinib for aerodigestive tract cancer. *J. Clin. Oncol.* **23**(34), 8757–8764.

Duvic, M., Martin, A. G., Kim, Y., Olsen, E., Wood, G. S., Crowley, C. A., and Yocum, R. C. (2001). Phase 2 and 3 clinical trial of oral bexarotene (Targretin capsules) for the treatment of refractory or persistent early-stage cutaneous T-cell lymphoma. *Arch. Dermatol.* **137**(5), 581–593.

Edwards, P. A., Kast, H. R., and Anisfeld, A. M. (2002). BAREing it all: The adoption of LXR and FXR and their roles in lipid homeostasis. *J. Lipid. Res.* **43**(1), 2–12.

Egea, P. F., Mitschler, A., and Moras, D. (2002). Molecular recognition of agonist ligands by RXRs. *Mol. Endocrinol.* **16**(5), 987–997.

Escher, P., Braissant, O., Basu-Modak, S., Michalik, L., Wahli, W., and Desvergne, B. (2001). Rat PPARs: Quantitative analysis in adult rat tissues and regulation in fasting and refeeding. *Endocrinology* **142**(10), 4195–4202.

Fan, Y. Y., Spencer, T. E., Wang, N., Moyer, M. P., and Chapkin, R. S. (2003). Chemopreventive n-3 fatty acids activate RXRalpha in colonocytes. *Carcinogenesis* **24**(9), 1541–1548.

Feige, J. N., Gelman, L., Tudor, C., Engelborghs, Y., Wahli, W., and Desvergne, B. (2005). Fluorescence imaging reveals the nuclear behavior of peroxisome proliferator-activated receptor/retinoid X receptor heterodimers in the absence and presence of ligand. *J. Biol. Chem.* **280**(18), 17880–17890.

Feige, J. N., Gelman, L., Michalik, L., Desvergne, B., and Wahli, W. (2006). From molecular action to physiological outputs: Peroxisome proliferator-activated receptors are nuclear receptors at the crossroads of key cellular functions. *Prog. Lipid Res.* **45**(2), 120–159.

Finck, B. N., Lehman, J. J., Leone, T. C., Welch, M. J., Bennett, M. J., Kovacs, A., Han, X., Gross, R. W., Kozak, R., Lopaschuk, G. D., and Kelly, D. P. (2002). The cardiac phenotype induced by PPARalpha overexpression mimics that caused by diabetes mellitus. *J. Clin. Invest.* **109**(1), 121–130.

Finck, B. N., Han, X., Courtois, M., Aimond, F., Nerbonne, J. M., Kovacs, A., Gross, R. W., and Kelly, D. P. (2003). A critical role for PPARalpha-mediated lipotoxicity in the pathogenesis of diabetic cardiomyopathy: Modulation by dietary fat content. *Proc. Natl. Acad. Sci. USA* **100**(3), 1226–1231.

Finck, B. N., Bernal-Mizrachi, C., Han, D. H., Coleman, T., Sambandam, N., LaRiviere, L. L., Holloszy, J. O., Semenkovich, C. F., and Kelly, D. P. (2005). A potential link between muscle peroxisome proliferator-activated receptor-alpha signaling and obesity-related diabetes. *Cell Metab.* **1**(2), 133–144.

Francis, G. A., Fayard, E., Picard, F., and Auwerx, J. (2003). Nuclear receptors and the control of metabolism. *Annu. Rev. Physiol.* **65**, 261–311.

Gampe, R. T., Jr., Montana, V. G., Lambert, M. H., Wisely, G. B., Milburn, M. V., and Xu, H. E. (2000). Structural basis for autorepression of retinoid X receptor by tetramer formation and the AF-2 helix. *Genes Dev.* **14**(17), 2229–2241.

Gelman, L., Michalik, L., Desvergne, B., and Wahli, W. (2005). Kinase signaling cascades that modulate peroxisome proliferator-activated receptors. *Curr. Opin. Cell Biol.* **17**(2), 216–222.

Germain, P., Iyer, J., Zechel, C., and Gronemeyer, H. (2002). Co-regulator recruitment and the mechanism of retinoic acid receptor synergy. *Nature* **415**(6868), 187–192.

Gernert, D. L., Ajamie, R., Ardecky, R. A., Bell, M. G., Leibowitz, M. D., Mais, D. A., Mapes, C. M., Michellys, P. Y., Rungta, D., Reifel-Miller, A., Tyhonas, J. S., Yumibe, N., *et al.* (2003). Design and synthesis of fluorinated RXR modulators. *Bioorg. Med. Chem. Lett.* **13**(19), 3191–3195.

Gernert, D. L., Neel, D. A., Boehm, M. F., Leibowitz, M. D., Mais, D. A., Michellys, P. Y., Rungta, D., Reifel-Miller, A., and Grese, T. A. (2004). Design and synthesis of benzofused heterocyclic RXR modulators. *Bioorg. Med. Chem. Lett.* **14**(11), 2759–2763.

Goldstein, J. T., Dobrzyn, A., Clagett-Dame, M., Pike, J. W., and DeLuca, H. F. (2003). Isolation and characterization of unsaturated fatty acids as natural ligands for the retinoid-X receptor. *Arch. Biochem. Biophys.* **420**(1), 185–193.

Goodwin, B., Jones, S. A., Price, R. R., Watson, M. A., McKee, D. D., Moore, L. B., Galardi, C., Wilson, J. G., Lewis, M. C., Roth, M. E., Maloney, P. R., Willson, T. M., *et al.* (2000). A regulatory cascade of the nuclear receptors FXR, SHP-1, and LRH-1 represses bile acid biosynthesis. *Mol. Cell* **6**(3), 517–526.

Goodwin, B., Watson, M. A., Kim, H., Miao, J., Kemper, J. K., and Kliewer, S. A. (2003). Differential regulation of rat and human CYP7A1 by the nuclear oxysterol receptor liver X receptor-alpha. *Mol. Endocrinol.* **17**(3), 386–394.

Grefhorst, A., Elzinga, B. M., Voshol, P. J., Plosch, T., Kok, T., Bloks, V. W., van der Sluijs, F. H., Havekes, L. M., Romijn, J. A., Verkade, H. J., and Kuipers, F. (2002). Stimulation of lipogenesis by pharmacological activation of the liver X receptor leads to production of large, triglyceride-rich very low density lipoprotein particles. *J. Biol. Chem.* **277**(37), 34182–34190.

Grober, J., Zaghini, I., Fujii, H., Jones, S. A., Kliewer, S. A., Willson, T. M., Ono, T., and Besnard, P. (1999). Identification of a bile acid-responsive element in the human ileal bile acid-binding protein gene. Involvement of the farnesoid X receptor/9-*cis*-retinoic acid receptor heterodimer. *J. Biol. Chem.* **274**(42), 29749–29754.

Guan, H. P., Li, Y., Jensen, M. V., Newgard, C. B., Steppan, C. M., and Lazar, M. A. (2002). A futile metabolic cycle activated in adipocytes by antidiabetic agents. *Nat. Med.* **8**(10), 1122–1128.

Gurnell, M., Savage, D. B., Chatterjee, V. K., and O'Rahilly, S. (2003). The metabolic syndrome: Peroxisome proliferator-activated receptor gamma and its therapeutic modulation. *J. Clin. Endocrinol. Metab.* **88**(6), 2412–2421.

Haffner, C. D., Lenhard, J. M., Miller, A. B., McDougald, D. L., Dwornik, K., Ittoop, O. R., Gampe, R. T., Jr., Xu, H. E., Blanchard, S., Montana, V. G., Consler, T. G., Bledsoe, R., *et al.* (2004). Structure-based design of potent retinoid X receptor alpha agonists. *J. Med. Chem.* **47**(8), 2010–2029.

He, W., Barak, Y., Hevener, A., Olson, P., Liao, D., Le, J., Nelson, M., Ong, E., Olefsky, J. M., and Evans, R. M. (2003). Adipose-specific peroxisome proliferator-activated receptor γ knockout causes insulin resistance in fat and liver but not in muscle. *Proc. Natl. Acad. Sci. USA* **100**, 15712–15717.

Heyman, R. A., Mangelsdorf, D. J., Dyck, J. A., Stein, R. B., Eichele, G., Evans, R. M., and Thaller, C. (1992). 9-*cis* retinoic acid is a high affinity ligand for the retinoid X receptor. *Cell* **68**(2), 397–406.

Holst, D., Luquet, S., Kristiansen, K., and Grimaldi, P. A. (2003). Roles of peroxisome proliferator-activated receptors delta and gamma in myoblast transdifferentiation. *Exp. Cell Res.* **288**(1), 168–176.

IJpenberg, A., Tan, N. S., Gelman, L., Kersten, S., Seydoux, J., Xu, J., Metzger, D., Canaple, L., Chambon, P., Wahli, W., and Desvergne, B. (2004). In vivo activation of PPAR target genes by RXR homodimers. *EMBO J.* **23**(10), 2083–2091.

Imai, T., Jiang, M., Chambon, P., and Metzger, D. (2001a). Impaired adipogenesis and lipolysis in the mouse upon selective ablation of the retinoid X receptor alpha mediated by a tamoxifen-inducible chimeric Cre recombinase (Cre-ERT2) in adipocytes. *Proc. Natl. Acad. Sci. USA* **98**(1), 224–228.

Imai, T., Jiang, M., Kastner, P., Chambon, P., and Metzger, D. (2001b). Selective ablation of retinoid X receptor alpha in hepatocytes impairs their lifespan and regenerative capacity. *Proc. Natl. Acad. Sci. USA* **98**(8), 4581–4586.

Imai, T., Takakuwa, R., Marchand, S., Dentz, E., Bornert, J. M., Messaddeq, N., Wendling, O., Mark, M., Desvergne, B., Wahli, W., Chambon, P., and Metzger, D. (2004a). Peroxisome proliferator-activated receptor gamma is required in mature white and brown adipocytes for their survival in the mouse. *Proc. Natl. Acad. Sci. USA* **101**(13), 4543–4547.

Imai, T., Takawuka, R., Marchand, S., Dentz, E., Bornert, J.-M., Messadedeq, N., Wendling, O., Mark, M., Desvergne, B., Wahli, W., Chambon, P., and Metzger, D. (2004b). PPAR gamma is required in mature white and brown adipocytes for their survival in the mouse. *Proc. Natl. Acad. Sci. USA* **101**, 4543–4547.

Jacobs, C. M., and Paulsen, R. E. (2005). Crosstalk between ERK2 and RXR regulates nuclear import of transcription factor NGFI-B. *Biochem. Biophys. Res. Commun.* **336**(2), 646–652.

Jacobs, M. N., Nolan, G. T., and Hood, S. R. (2005). Lignans, bacteriocides and organochlorine compounds activate the human pregnane X receptor (PXR). *Toxicol. Appl. Pharmacol.* **209**(2), 123–133.

Janowski, B. A., Willy, P. J., Devi, T. R., and Mangelsdorf, D. J. (1996). An oxysterol signalling pathway mediated by the nuclear receptor LXRa. *Nature* **383**, 728–731.

Joseph, S. B., Laffitte, B. A., Patel, P. H., Watson, M. A., Matsukuma, K. E., Walczak, R., Collins, J. L., Osborne, T. F., and Tontonoz, P. (2002). Direct and indirect mechanisms for regulation of fatty acid synthase gene expression by liver X receptors. *J. Biol. Chem.* **277**(13), 11019–11025.

Kassam, A., Miao, B., Young, P. R., and Mukherjee, R. (2003). Retinoid X receptor (RXR) agonist-induced antagonism of farnesoid X receptor (FXR) activity due to absence of coactivator recruitment and decreased DNA binding. *J. Biol. Chem.* **278**(12), 10028–10032.

Kast, H. R., Nguyen, C. M., Sinal, C. J., Jones, S. A., Laffitte, B. A., Reue, K., Gonzalez, F. J., Willson, T. M., and Edwards, P. A. (2001). Farnesoid X-activated receptor induces apolipoprotein C-II transcription: A molecular mechanism linking plasma triglyceride levels to bile acids. *Mol. Endocrinol.* **15**(10), 1720–1728.

Kastner, P., Grondona, J. M., Mark, M., Gansmuller, A., LeMeur, M., Decimo, D., Vonesch, J.-L., Dollé, P., and Chambon, P. (1994). Genetic analysis of RXRa developmental function: Convergence of RXR and RAR signalling pathways in heart and eye morphogenesis. *Cell* **78**, 987–1003.

Kersten, S., Kelleher, D., Chambon, P., Gronemeyer, H., and Noy, N. (1995). Retinoid X receptor α forms tetramers in solution. *Proc. Natl. Acad. Sci. USA* **92**, 8645–8649.

Kersten, S., Dong, D., Lee, W., Reczek, P. R., and Noy, N. (1998). Auto-silencing by the retinoid X receptor. *J. Mol. Biol.* **284**(1), 21–32.

Kersten, S., Seydoux, J., Peters, J. M., Gonzalez, F. J., Desvergne, B., and Wahli, W. (1999). Peroxisome proliferator-activated receptor α mediates the adaptive response to fasting. *J. Clin. Invest.* **103**, 1489–1498.

Kitareewan, S., Burka, L. T., Tomer, K. B., Parker, C. E., Deterding, L. J., Stevens, R. D., Forman, B. M., Mais, D. E., Heyman, R. A., McMorris, T., and Weinberger, C. (1996). Phytol metabolites are circulating dietary factors that activate the nuclear receptor RXR. *Mol. Biol. Cell* **7**(8), 1153–1166.

Kramer, D. K., Al-Khalili, L., Perrini, S., Skogsberg, J., Wretenberg, P., Kannisto, K., Wallberg-Henriksson, H., Ehrenborg, E., Zierath, J. R., and Krook, A. (2005). Direct activation of glucose transport in primary human myotubes after activation of peroxisome proliferator-activated receptor delta. *Diabetes* **54**(4), 1157–1163.

Kubota, N., Terauchi, Y., Miki, H., Tamemoto, H., Yamauchi, T., Komeda, K., Satoh, S., Nakano, R., Ishii, C., Sugiyama, T., Eto, K., Tsubamoto, Y., *et al.* (1999). PPAR gamma

mediates high-fat diet-induced adipocyte hypertrophy and insulin resistance. *Mol. Cell* **4**(4), 597–609.
Kullak-Ublick, G. A., Stieger, B., and Meier, P. J. (2004). Enterohepatic bile salt transporters in normal physiology and liver disease. *Gastroenterology* **126**(1), 322–342.
Lee, C. H., Olson, P., Hevener, A., Mehl, I., Chong, L. W., Olefsky, J. M., Gonzalez, F. J., Ham, J., Kang, H., Peters, J. M., and Evans, R. M. (2006). PPARdelta regulates glucose metabolism and insulin sensitivity. *Proc. Natl. Acad. Sci. USA* **103**(9), 3444–3449.
Lee, S. S., Pineau, T., Drago, J., Lee, E. J., Owens, J. W., Kroetz, D. L., Fernandez-Salguero, P. M., Westphal, H., and Gonzalez, F. J. (1995). Targeted disruption of the α isoform of the peroxisome proliferator-activated receptor gene in mice results in abolishment of the pleiotropic effects of peroxisome proliferators. *Mol. Cell. Biol.* **15**, 3012–3022.
Lehmann, J. M., Kliewer, S. A., Moore, L. B., Smith-Oliver, T. A., Oliver, B. B., Su, J. L., Sundseth, S. S., Winegar, D. A., Blanchard, D. E., Spencer, T. A., and Willson, T. M. (1997). Activation of the nuclear receptor LXR by oxysterols defines a new hormone response pathway. *J. Biol. Chem.* **272**(6), 3137–3140.
Leibowitz, M. D., Fievet, C., Hennuyer, N., Peinado-Onsurbe, J., Duez, H., Bergera, J., Cullinan, C. A., Sparrow, C. P., Baffic, J., Berger, G. D., Santini, C., Marquis, R. W., *et al.* (2000). Activation of PPARdelta alters lipid metabolism in db/db mice. *FEBS Lett.* **473**(3), 333–336.
Leibowitz, M. D., Ardecky, R. J., Boehm, M. F., Broderick, C. L., Carfagna, M. A., Crombie, D. L., D'Arrigo, J., Etgen, G. J., Faul, M. M., Grese, T. A., Havel, H., Hein, N., *et al.* (2006). Biological characterization of a heterodimer-selective retinoid X receptor modulator: Potential benefits for the treatment of type 2 diabetes. *Endocrinology* **147**(2), 1044–1053.
Lemotte, P. K., Keidel, S., and Apfel, C. M. (1996). Phytanic acid is a retinoid X receptor ligand. *Eur. J. Biochem.* **236**, 328–333.
Lengqvist, J., Mata De Urquiza, A., Bergman, A. C., Willson, T. M., Sjovall, J., Perlmann, T., and Griffiths, W. J. (2004). Polyunsaturated fatty acids including docosahexaenoic and arachidonic acid bind to the retinoid X receptor alpha ligand-binding domain. *Mol. Cell. Proteomics* **3**(7), 692–703.
Leone, T., Weinheimer, C., and Kelly, D. (1999). A critical role for the peroxisome proliferator-activated receptor α (PPARα) in the cellular fasting response: The PPARα-null mouse as a model of fatty acid oxidation disorders. *Proc. Natl. Acad. Sci. USA* **96**, 7473–7478.
Li, D., Yamada, T., Wang, F., Vulin, A. I., and Samuels, H. H. (2004). Novel roles of retinoid X receptor (RXR) and RXR ligand in dynamically modulating the activity of the thyroid hormone receptor/RXR heterodimer. *J. Biol. Chem.* **279**(9), 7427–7437.
Liu, S., Ogilvie, K. M., Klausing, K., Lawson, M. A., Jolley, D., Li, D., Bilakovics, J., Pascual, B., Hein, N., Urcan, M., and Leibowitz, M. D. (2002). Mechanism of selective retinoid X receptor agonist-induced hypothyroidism in the rat. *Endocrinology* **143**(8), 2880–2885.
Lu, J., Cistola, D. P., and Li, E. (2006). Analysis of ligand binding and protein dynamics of human retinoid X receptor alpha ligand-binding domain by nuclear magnetic resonance. *Biochemistry* **45**(6), 1629–1639.
Lu, T. T., Makishima, M., Repa, J. J., Schoonjans, K., Kerr, T. A., Auwerx, J., and Mangelsdorf, D. J. (2000). Molecular basis for feedback regulation of bile acid synthesis by nuclear receptors. *Mol. Cell* **6**(3), 507–515.
Luptak, I., Balschi, J. A., Xing, Y., Leone, T. C., Kelly, D. P., and Tian, R. (2005). Decreased contractile and metabolic reserve in peroxisome proliferator-activated receptor-alpha-null hearts can be rescued by increasing glucose transport and utilization. *Circulation* **112**(15), 2339–2346.
Luquet, S., Lopez-Soriano, J., Holst, D., Fredenrich, A., Melki, J., Rassoulzadegan, M., and Grimaldi, P. A. (2003). Peroxisome proliferator-activated receptor delta controls muscle development and oxidative capability. *FASEB J.* **17**(15), 2299–2301.
Luria, A., and Furlow, J. D. (2004). Spatiotemporal retinoid-X receptor activation detected in live vertebrate embryos. *Proc. Natl. Acad. Sci. USA* **101**(24), 8987–8992.

Makishima, M., Okamoto, A. Y., Repa, J. J., Tu, H., Learned, R. M., Luk, A., Hull, M. V., Lustig, K. D., Mangelsdorf, D. J., and Shan, B. (1999). Identification of a nuclear receptor for bile acids. *Science* **284**(5418), 1362–1365.

Mangelsdorf, D. J., and Evans, R. M. (1995). The RXR heterodimers and orphan receptors. *Cell* **83**(6), 841–850.

Mangelsdorf, D. J., Borgmeyer, U., Heyman, R. A., Zhou, J. Y., Ong, E. S., Oro, A. E., Kakizuka, A., and Evans, R. M. (1992). Characterization of three RXR genes that mediate the action of 9-*cis* retinoic acid. *Genes Dev.* **6**(3), 329–344.

Mark, M., Ghyselinck, N. B., Wendling, O., Dupe, V., Mascrez, B., Kastner, P., and Chambon, P. (1999). A genetic dissection of the retinoid signalling pathway in the mouse. *Proc. Nutr. Soc.* **58**(3), 609–613.

Mark, M., Ghyselinck, N. B., and Chambon, P. (2006). Function of retinoid nuclear receptors: Lessons from genetic and pharmacological dissections of the retinoic acid signaling pathway during mouse embryogenesis. *Annu. Rev. Pharmacol. Toxicol.* **46**, 451–480.

Mascrez, B., Mark, M., Dierich, A., Ghyselinck, N., Kastner, P., and Chambon, P. (1998). The RXRα ligand-dependent activation function (AF-2) is important for mouse development. *Development* **125**, 4691–4707.

Mayerson, A. B., Hundal, R. S., Dufour, S., Lebon, V., Befroy, D., Cline, G. W., Enocksson, S., Inzucchi, S. E., Shulman, G. I., and Petersen, K. F. (2002). The effects of rosiglitazone on insulin sensitivity, lipolysis, and hepatic and skeletal muscle triglyceride content in patients with type 2 diabetes. *Diabetes* **51**(3), 797–802.

Mehta, K., McQueen, T., Neamati, N., Collins, S., and Andreeff, M. (1996). Activation of retinoid receptors RAR alpha and RXR alpha induces differentiation and apoptosis, respectively, in HL-60 cells. *Cell Growth Differ.* **7**(2), 179–186.

Mertz, J. R., Shang, E., Piantedosi, R., Wei, S., Wolgemuth, D. J., and Blaner, W. S. (1997). Identification and characterization of a stereospecific human enzyme that catalyzes 9-*cis*-retinol oxidation. A possible role in 9-*cis*-retinoic acid formation. *J. Biol. Chem.* **272**(18), 11744–11749.

Michalik, L., Desvergne, B., Tan, N. S., Basu-Modak, S., Escher, P., Rieusset, J., Peters, J. M., Kaya, G., Gonzalez, F. J., Zakany, J., Metzger, D., Chambon, P., *et al.* (2001). Impaired skin wound healing in peroxisome proliferator-activated receptor (PPAR)alpha and PPARbeta mutant mice. *J. Cell Biol.* **154**(4), 799–814.

Michalik, L., Desvergne, B., and Wahli, W. (2003). Peroxisome proliferator-activated receptors beta/delta: Emerging roles for a previously neglected third family member. *Curr. Opin. Lipidol.* **14**(2), 129–135.

Michalik, L., Desvergne, B., and Wahli, W. (2004). Peroxisome-proliferator-activated receptors and cancers: Complex stories. *Nat. Rev. Cancer* **4**(1), 61–70.

Michellys, P. Y., Ardecky, R. J., Chen, J. H., Crombie, D. L., Etgen, G. J., Faul, M. M., Faulkner, A. L., Grese, T. A., Heyman, R. A., Karanewsky, D. S., Klausing, K., Leibowitz, M. D., *et al.* (2003a). Novel (2E,4E,6Z)-7-(2-alkoxy-3,5-dialkylbenzene)-3-methylocta-2,4,6-trienoic acid retinoid X receptor modulators are active in models of type 2 diabetes. *J. Med. Chem.* **46**(13), 2683–2696.

Michellys, P. Y., Boehm, M. F., Chen, J. H., Grese, T. A., Karanewsky, D. S., Leibowitz, M. D., Liu, S., Mais, D. A., Mapes, C. M., Reifel-Miller, A., Ogilvie, K. M., Rungta, D., *et al.* (2003b). Design and synthesis of novel RXR-selective modulators with improved pharmacological profile. *Bioorg. Med. Chem. Lett.* **13**(22), 4071–4075.

Miles, P. D., Barak, Y., He, W., Evans, R. M., and Olefsky, J. M. (2000). Improved insulin-sensitivity in mice heterozygous for PPAR-gamma deficiency. *J. Clin. Invest.* **105**(3), 287–292.

Miller, V. A., Benedetti, F. M., Rigas, J. R., Verret, A. L., Pfister, D. G., Straus, D., Kris, M. G., Crisp, M., Heyman, R., Loewen, G. R., Truglia, J. A., and Warrell, R. P., Jr. (1997). Initial clinical trial of a selective retinoid X receptor ligand, LGD1069. *J. Clin. Oncol.* **15**(2), 790–795.

Minucci, S., and Ozato, K. (1996). Retinoid receptors in transcriptional regulation. *Curr. Opin. Genet. Dev.* **6,** 567–574.

Mukherjee, R., Davies, P. J. A., Cromble, D. L., Bischoff, E. D., Cesario, R. M., Jow, L., Hamann, L. G., Boehm, M. F., Mondon, C. E., Nadzan, A. M., Paterniti, J. R., Jr., and Heyman, R. A. (1997). Sensitization of diabetic and obese mice to insulin by retinoid X receptor agonists. *Nature* **386,** 407–410.

Mukherjee, R., Strasser, J., Jow, L., Hoener, P., Paterniti, J. R., Jr., and Heyman, R. A. (1998). RXR agonists activate PPARα-inducible genes, lower triglycerides, and raise HDL levels in vivo. *Arterioscler. Thromb. Vasc. Biol.* **18**(2), 272–276.

Muoio, D. M., MacLean, P. S., Lang, D. B., Li, S., Houmard, J. A., Way, J. M., Winegar, D. A., Corton, J. C., Dohm, G. L., and Kraus, W. E. (2002). Fatty acid homeostasis and induction of lipid regulatory genes in skeletal muscles of peroxisome proliferator-activated receptor (PPAR) alpha knock-out mice. Evidence for compensatory regulation by PPAR delta. *J. Biol. Chem.* **277**(29), 26089–26097.

Nadra, K., Anghel, S. I., Joye, E., Tan, N. S., Basu-Modak, S., Trono, D., Wahli, W., and Desvergne, B. (2006). Differentiation of trophoblast giant cells and their metabolic functions are dependent on peroxisome proliferator-activated receptor beta/delta. *Mol. Cell. Biol.* **26**(8), 3266–3281.

Nagy, L., Thomazy, V. A., Shipley, G. L., Fesus, L., Lamph, W., Heyman, R. A., Chandraratna, R. A. S., and Davies, P. J. A. (1995). Activation of retinoid X receptors induces apoptosis in HL-60 cell lines. *Mol. Cell. Biol.* **15,** 3540–3551.

Nettles, K. W., and Greene, G. L. (2005). Ligand control of coregulator recruitment to nuclear receptors. *Annu. Rev. Physiol.* **67,** 309–333.

Nuclear Receptors Nomenclature Committee (1999). A unified nomenclature system for the nuclear receptor superfamily. *Cell* **97,** 161–163.

Oakes, N. D., Camilleri, S., Furler, S. M., Chisholm, D. J., and Kraegen, E. W. (1997). The insulin sensitizer, BRL 49653, reduces systemic fatty acid supply and utilization and tissue lipid availability in the rat. *Metabolism* **46**(8), 935–942.

Oliver, W. R., Jr., Shenk, J. L., Snaith, M. R., Russell, C. S., Plunket, K. D., Bodkin, N. L., Lewis, M. C., Winegar, D. A., Sznaidman, M. L., Lambert, M. H., Xu, H. E., Sternbach, D. D., et al. (2001). A selective peroxisome proliferator-activated receptor delta agonist promotes reverse cholesterol transport. *Proc. Natl. Acad. Sci. USA* **98**(9), 5306–5311.

Ouamrane, L., Larrieu, G., Gauthier, B., and Pineau, T. (2003). RXR activators molecular signalling: Involvement of a PPARalpha-dependent pathway in the liver and kidney, evidence for an alternative pathway in the heart. *Br. J. Pharmacol.* **138**(5), 845–854.

Park, C. W., Zhang, Y., Zhang, X., Wu, J., Chen, L., Cha, D. R., Su, D., Hwang, M. T., Fan, X., Davis, L., Striker, G., Zheng, F., et al. (2006). PPARalpha agonist fenofibrate improves diabetic nephropathy in db/db mice. *Kidney Int.* **69**(9), 1511–1517.

Parks, D. J., Blanchard, S. G., Bledsoe, R. K., Chandra, G., Consler, T. G., Kliewer, S. A., Stimmel, J. B., Willson, T. M., Zavacki, A. M., Moore, D. D., and Lehmann, J. M. (1999). Bile acids: Natural ligands for an orphan nuclear receptor. *Science* **284**(5418), 1365–1368.

Patel, D. D., Knight, B. L., Wiggins, D., Humphreys, S. M., and Gibbons, G. F. (2001). Disturbances in the normal regulation of SREBP-sensitive genes in PPAR alpha-deficient mice. *J. Lipid Res.* **42**(3), 328–337.

Peet, D. J., Turley, S. D., Ma, W., Janowski, B. A., Lobaccaro, J. M., Hammer, R. E., and Mangelsdorf, D. J. (1998). Cholesterol and bile acid metabolism are impaired in mice lacking the nuclear oxysterol receptor LXR alpha. *Cell* **93**(5), 693–704.

Plosch, T., Kok, T., Bloks, V. W., Smit, M. J., Havinga, R., Chimini, G., Groen, A. K., and Kuipers, F. (2002). Increased hepatobiliary and fecal cholesterol excretion upon activation of the liver X receptor is independent of ABCA1. *J. Biol. Chem.* **277**(37), 33870–33877.

Prufer, K., and Barsony, J. (2002). Retinoid X receptor dominates the nuclear import and export of the unliganded vitamin D receptor. *Mol. Endocrinol.* **16**(8), 1738–1751.

Repa, J. J., and Mangelsdorf, D. J. (1999). Nuclear receptor regulation of cholesterol and bile acid metabolism. *Curr. Opin. Biotechnol.* **10**(6), 557–563.
Repa, J. J., Liang, G., Ou, J., Bashmakov, Y., Lobaccaro, J. M., Shimomura, I., Shan, B., Brown, M. S., Goldstein, J. L., and Mangelsdorf, D. J. (2000a). Regulation of mouse sterol regulatory element-binding protein-1c gene (SREBP-1c) by oxysterol receptors, LXRalpha and LXRbeta. *Genes Dev.* **14**(22), 2819–2830.
Repa, J. J., Turley, S. D., Lobaccaro, J. A., Medina, J., Li, L., Lustig, K., Shan, B., Heyman, R. A., Dietschy, J. M., and Mangelsdorf, D. J. (2000b). Regulation of absorption and ABC1-mediated efflux of cholesterol by RXR heterodimers. *Science* **289**(5484), 1524–1529.
Repa, J. J., Berge, K. E., Pomajzl, C., Richardson, J. A., Hobbs, H., and Mangelsdorf, D. J. (2002). Regulation of ATP-binding cassette sterol transporters ABCG5 and ABCG8 by the liver X receptors alpha and beta. *J. Biol. Chem.* **277**(21), 18793–18800.
Rieusset, J., Touri, F., Michalik, L., Escher, P., Desvergne, B., Niesor, E., and Wahli, W. (2002). A new selective peroxisome proliferator-activated receptor gamma antagonist with antiobesity and antidiabetic activity. *Mol. Endocrinol.* **16**(11), 2628–2644.
Rizvi, N. A., Marshall, J. L., Dahut, W., Ness, E., Truglia, J. A., Loewen, G., Gill, G. M., Ulm, E. H., Geiser, R., Jaunakais, D., and Hawkins, M. J. (1999). A Phase I study of LGD1069 in adults with advanced cancer. *Clin. Cancer Res.* **5**(7), 1658–1664.
Romert, A., Tuvendal, P., Simon, A., Dencker, L., and Eriksson, U. (1998). The identification of a *9-cis* retinol dehydrogenase in the mouse embryo reveals a pathway for synthesis of *9-cis* retinoic acid. *Proc. Natl. Acad. Sci. USA* **95**(8), 4404–4409.
Rosen, E. D., and Spiegelman, B. M. (2001). PPARgamma: A nuclear regulator of metabolism, differentiation, and cell growth. *J. Biol. Chem.* **276**(41), 37731–37734.
Rosen, E. D., Sarraf, P., Troy, A. E., Bradwin, G., Moore, K., Milstone, D. S., Spiegelman, B. M., and Mortensen, R. M. (1999). PPAR gamma is required for the differentiation of adipose tissue *in vivo* and *in vitro*. *Mol. Cell* **4**(4), 611–617.
Schuetz, E. G., Strom, S., Yasuda, K., Lecureur, V., Assem, M., Brimer, C., Lamba, J., Kim, R. B., Ramachandran, V., Komoroski, B. J., Venkataramanan, R., Cai, H., *et al.* (2001). Disrupted bile acid homeostasis reveals an unexpected interaction among nuclear hormone receptors, transporters, and cytochrome P450. *J. Biol. Chem.* **276**(42), 39411–39418.
Schulman, I. G., Shao, G., and Heyman, R. A. (1998). Transactivation by retinoid X receptor-peroxisome proliferator-activated receptor gamma (PPARgamma) heterodimers: Intermolecular synergy requires only the PPARgamma hormone-dependent activation function. *Mol. Cell. Biol.* **18**(6), 3483–3494.
Schultz, J. R., Tu, H., Luk, A., Repa, J. J., Medina, J. C., Li, L., Schwendner, S., Wang, S., Thoolen, M., Mangelsdorf, D. J., Lustig, K. D., and Shan, B. (2000). Role of LXRs in control of lipogenesis. *Genes Dev.* **14**(22), 2831–2838.
Seol, W., Choi, H. S., and Moore, D. D. (1995). Isolation of proteins that interact specifically with the retinoid X receptor: Two novel orphan receptors. *Mol. Endocrinol.* **9**(1), 72–85.
Shao, J., Sheng, H., and DuBois, R. N. (2002). Peroxisome proliferator-activated receptors modulate K-Ras-mediated transformation of intestinal epithelial cells. *Cancer Res.* **62**(11), 3282–3288.
Sharma, V., Hays, W. R., Wood, W. M., Pugazhenthi, U., St Germain, D. L., Bianco, A. C., Krezel, W., Chambon, P., and Haugen, B. R. (2006). Effects of rexinoids on thyrotrope function and the hypothalamic-pituitary-thyroid axis. *Endocrinology* **147**(3), 1438–1451.
Shen, Q., Cline, G. W., Shulman, G. I., Leibowitz, M. D., and Davies, P. J. (2004). Effects of rexinoids on glucose transport and insulin-mediated signaling in skeletal muscles of diabetic (db/db) mice. *J. Biol. Chem.* **279**(19), 19721–19731.
Shulman, A. I., Larson, C., Mangelsdorf, D. J., and Ranganathan, R. (2004). Structural determinants of allosteric ligand activation in RXR heterodimers. *Cell* **116**(3), 417–429.

Sinal, C. J., Tohkin, M., Miyata, M., Ward, J. M., Lambert, G., and Gonzalez, F. J. (2000). Targeted disruption of the nuclear receptor FXR/BAR impairs bile acid and lipid homeostasis. *Cell* **102**(6), 731–744.

Singh Ahuja, H., Liu, S., Crombie, D. L., Boehm, M., Leibowitz, M. D., Heyman, R. A., Depre, C., Nagy, L., Tontonoz, P., and Davies, P. J. (2001). Differential effects of rexinoids and thiazolidinediones on metabolic gene expression in diabetic rodents. *Mol. Pharmacol.* **59**(4), 765–773.

Solomin, L., Johansson, C. B., Zetterstrom, R. H., Bissonnette, R. P., Heyman, R. A., Olson, L., Lendahl, U., Frisen, J., and Perlmann, T. (1998). Retinoid-X receptor signalling in the developing spinal cord. *Nature* **395**(6700), 398–402.

Sucov, H. M., Dyson, E., Gumeringer, C. L., Price, J., Chien, K. R., and Evans, R. M. (1994). RXR alpha mutant mice establish a genetic basis for vitamin A signaling in heart morphogenesis. *Genes Dev.* **8**(9), 1007–1018.

Surapureddi, S., Yu, S., Bu, H., Hashimoto, T., Yeldandi, A. V., Kashireddy, P., Cherkaoui-Malki, M., Qi, C., Zhu, Y. J., Rao, M. S., and Reddy, J. K. (2002). Identification of a transcriptionally active peroxisome proliferator-activated receptor alpha-interacting cofactor complex in rat liver and characterization of PRIC285 as a coactivator. *Proc. Natl. Acad. Sci. USA* **99**(18), 11836–11841.

Szanto, A., Benko, S., Szatmari, I., Balint, B. L., Furtos, I., Ruhl, R., Molnar, S., Csiba, L., Garuti, R., Calandra, S., Larsson, H., Diczfalusy, U., et al. (2004a). Transcriptional regulation of human CYP27 integrates retinoid, peroxisome proliferator-activated receptor, and liver X receptor signaling in macrophages. *Mol. Cell. Biol.* **24**(18), 8154–8166.

Szanto, A., Narkar, V., Shen, Q., Uray, I. P., Davies, P. J., and Nagy, L. (2004b). Retinoid X receptors: X-ploring their (patho)physiological functions. *Cell Death Differ.* **11**(Suppl. 2), S126–S143.

Tall, A. R., Costet, P., and Wang, N. (2002). Regulation and mechanisms of macrophage cholesterol efflux. *J. Clin. Invest.* **110**(7), 899–904.

Tanaka, T., Yamamoto, J., Iwasaki, S., Asaba, H., Hamura, H., Ikeda, Y., Watanabe, M., Magoori, K., Ioka, R. X., Tachibana, K., Watanabe, Y., Uchiyama, Y., et al. (2003). Activation of peroxisome proliferator-activated receptor delta induces fatty acid beta-oxidation in skeletal muscle and attenuates metabolic syndrome. *Proc. Natl. Acad. Sci. USA* **100**(26), 15924–15929.

Toma, S., Isnardi, L., Riccardi, L., and Bollag, W. (1998). Induction of apoptosis in MCF-7 breast carcinoma cell line by RAR and RXR selective retinoids. *Anticancer. Res.* **18**(2A), 935–942.

Tontonoz, P., and Mangelsdorf, D. J. (2003). Liver X receptor signaling pathways in cardiovascular disease. *Mol. Endocrinol.* **17**(6), 985–993.

Tsai, C. C., and Fondell, J. D. (2004). Nuclear receptor recruitment of histone-modifying enzymes to target gene promoters. *Vitam. Horm.* **68**, 93–122.

Tudor, C., Feige, J. N., Pingali, H., Bhushan Lohray, V., Wahli, W., Desvergne, B., Engelborghs, Y., and Gelman, L. (2006). Association with coregulators is the major determinant governing PPAR mobility in living cells. *J. Biol. Chem.* (in press).

Tzameli, I., Chua, S. S., Cheskis, B., and Moore, D. D. (2003). Complex effects of rexinoids on ligand dependent activation or inhibition of the xenobiotic receptor, CAR. *Nucl. Recept.* **1**(1), 2.

Urizar, N. L., Dowhan, D. H., and Moore, D. D. (2000). The farnesoid X-activated receptor mediates bile acid activation of phospholipid transfer protein gene expression. *J. Biol. Chem.* **275**(50), 39313–39317.

van der Veen, J. N., Kruit, J. K., Havinga, R., Baller, J. F., Chimini, G., Lestavel, S., Staels, B., Groot, P. H., Groen, A. K., and Kuipers, F. (2005). Reduced cholesterol absorption upon PPARdelta activation coincides with decreased intestinal expression of NPC1L1. *J. Lipid Res.* **46**(3), 526–534.

Wajchenberg, B. L. (2000). Subcutaneous and visceral adipose tissue: Their relation to the metabolic syndrome. *Endocr. Rev.* **21**(6), 697–738.

Wan, Y. J., An, D., Cai, Y., Repa, J. J., Hung-Po Chen, T., Flores, M., Postic, C., Magnuson, M. A., Chen, J., Chien, K. R., French, S., Mangelsdorf, D. J., et al. (2000a). Hepatocyte-specific mutation establishes retinoid X receptor alpha as a heterodimeric integrator of multiple physiological processes in the liver. *Mol. Cell. Biol.* **20**(12), 4436–4444.

Wan, Y. J., Cai, Y., Lungo, W., Fu, P., Locker, J., French, S., and Sucov, H. M. (2000b). Peroxisome proliferator-activated receptor alpha-mediated pathways are altered in hepatocyte-specific retinoid X receptor alpha-deficient mice. *J. Biol. Chem.* **275**(36), 28285–28290.

Wang, H., Chen, J., Hollister, K., Sowers, L. C., and Forman, B. M. (1999). Endogenous bile acids are ligands for the nuclear receptor FXR/BAR. *Mol. Cell* **3**(5), 543–553.

Wang, Y., Yao, R., Maciag, A., Grubbs, C. J., Lubet, R. A., and You, M. (2006). Organ-specific expression profiles of rat mammary gland, liver, and lung tissues treated with targretin, *9-cis* retinoic acid, and 4-hydroxyphenylretinamide. *Mol. Cancer Ther.* **5**(4), 1060–1072.

Wang, Y. X., Lee, C. H., Tiep, S., Yu, R. T., Ham, J., Kang, H., and Evans, R. M. (2003a). Peroxisome-proliferator-activated receptor delta activates fat metabolism to prevent obesity. *Cell* **113**(2), 159–170.

Wang, Y. X., Zhang, C. L., Yu, R. T., Cho, H. K., Nelson, M. C., Bayuga-Ocampo, C. R., Ham, J., Kang, H., and Evans, R. M. (2004). Regulation of muscle fiber type and running endurance by PPARdelta. *PLoS Biol.* **2**(10), e294.

Wang, Z., Benoit, G., Liu, J., Prasad, S., Aarnisalo, P., Liu, X., Xu, H., Walker, N. P., and Perlmann, T. (2003b). Structure and function of Nurr1 identifies a class of ligand-independent nuclear receptors. *Nature* **423**(6939), 555–560.

Wendling, O., Chambon, P., and Mark, M. (1999). Retinoid X receptors are essential for early mouse development and placentogenesis. *Proc. Natl. Acad. Sci. USA* **96**(2), 547–551.

Xie, W., Barwick, J. L., Simon, C. M., Pierce, A. M., Safe, S., Blumberg, B., Guzelian, P. S., and Evans, R. M. (2000). Reciprocal activation of xenobiotic response genes by nuclear receptors SXR/PXR and CAR. *Genes Dev.* **14**(23), 3014–3023.

Xu, H. E., Lambert, M. H., Montana, V. G., Plunket, K. D., Moore, L. B., Collins, J. L., Oplinger, J. A., Kliewer, S. A., Gampe, R. T., Jr., McKee, D. D., Moore, J. T., and Willson, T. M. (2001). Structural determinants of ligand binding selectivity between the peroxisome proliferator-activated receptors. *Proc. Natl. Acad. Sci. USA* **98**(24), 13919–13924.

Yasmin, R., Yeung, K. T., Chung, R. H., Gaczynska, M. E., Osmulski, P. A., and Noy, N. (2004). DNA-looping by RXR tetramers permits transcriptional regulation "at a distance". *J. Mol. Biol.* **343**(2), 327–338.

Yasmin, R., Williams, R. M., Xu, M., and Noy, N. (2005). Nuclear import of the retinoid X receptor, the vitamin D receptor, and their mutual heterodimer. *J. Biol. Chem.* **280**(48), 40152–40160.

Yechoor, V. K., Patti, M. E., Saccone, R., and Kahn, C. R. (2002). Coordinated patterns of gene expression for substrate and energy metabolism in skeletal muscle of diabetic mice. *Proc. Natl. Acad. Sci. USA* **99**(16), 10587–10592.

Zhang, X. K., Lehmann, J., Hoffmann, B., Dawson, M. I., Cameron, J., Graupner, G., Hermann, T., Tran, P., and Pfahl, M. (1992). Homodimer formation of retinoid X-Receptor induced by *9-cis* retinoic acid. *Nature* **358**, 587–591.

2

The Intersection Between the Aryl Hydrocarbon Receptor (AhR)- and Retinoic Acid-Signaling Pathways

Kyle A. Murphy,* Loredana Quadro,[†] and Lori A. White*

*Department of Biochemistry and Microbiology, Rutgers, The State University of New Jersey, New Brunswick, New Jersey 08901
[†]Department of Food Science, Rutgers, The State University of New Jersey New Brunswick, New Jersey 08901

I. Introduction
II. Retinoid Signaling
III. The AhR/Arnt Pathway
IV. AhR and RA Availability
   A. RA Synthesis
   B. RA Catabolism
   C. Interconversion
   D. Storage and Transport
V. Molecular Interactions Between the RA and AhR Pathways
   References

Data from a variety of animal and cell culture model systems have demonstrated an interaction between the aryl hydrocarbon receptor (AhR)- and retinoic acid (RA)-signaling pathways. The AhR[1] was originally identified as the receptor for the polycyclic aromatic hydrocarbon family of environmental contaminants; however, recent data indicate that the AhR binds to a variety of endogenous and exogenous compounds, including some synthetic retinoids. In addition, activation of the AhR pathway alters the function of nuclear hormone-signaling pathways, including the estrogen, thyroid, and RA pathways. Activation of the AhR pathway through exposure to environmental compounds results in significant changes in RA synthesis, catabolism, transport, and excretion. Some effects on retinoid homeostasis mediated by the AhR pathway may result from the interactions of these two pathways at the level of activating or repressing the expression of specific genes. This chapter will review these two pathways, the evidence demonstrating a link between them, and the data indicating the molecular basis of the interactions between these two pathways. © 2007 Elsevier Inc.

I. INTRODUCTION

2,3,7,8-Tetrachlorodibenzo-*p*-dioxin (TCDD) and related polycyclic and halogenated aromatic hydrocarbons (PAH/HAH) are ubiquitous environmental contaminants that are the unintentional by-products of industrial combustion (Bertazzi *et al.*, 1989, 2001). Exposure to these compounds results in a variety of lesions in mammals, including alterations in liver function and lipid metabolism, weight loss, immune system suppression, endocrine and nervous system dysfunction, as well as severe skin lesions (Mukerjee, 1998). TCDD is of particular interest due to its persistence in biological tissues (DeVito *et al.*, 1995; Ott and Zober, 1996). TCDD exposure occurs mainly through oral ingestion and is concentrated through the food chain. As TCDD accumulates in the adipose tissue, an individual's

---

[1]Abbreviations: ADH, alcohol dehydrogenase; AhR, aryl hydrocarbon receptor; ALDH, aldehyde dehydrogenase; Arnt, AhR nuclear translocator; atRA, all-*trans* RA; CRABP, cellular retinoic acid-binding protein; CRBPI, cellular retinol-binding protein type I; CYP450, cytochrome P450; GST, glutathione *S*-transferases; HAT, histone acetyltransferase; HDACs, histone deacetylases; HMTs, histone methyltransferases; LRAT, lecithin:retinol acyltransferase; MMPs, matrix metalloproteinases; RAL, retinal; RALDH, retinaldehyde dehydrogenase; *RAR*, RAR gene; RAREs, retinoic acid response elements; RARs, retinoic acid receptors; RBP, retinol-binding protein; RDH, retinol dehydrogenase; RE, retinyl ester; REHs, retinyl ester hydrolases; ROH, retinol; *RXR*, RXR gene; RXRs, retinoid X receptors; SCADs, short-chain alcohol dehydrogenases; SMRT, silencing mediator of retinoid and thyroid receptors; TCDD, 2,3,7, 8-tetrachlorodibenzo-*p*-dioxin; UGTs, UDP-glucuronosyltransferases; XREs, xenobiotic response elements; N-CoR, nuclear receptors corepressor; RA, retinoic acid.

body burden increases with age (DeVito et al., 1995). Although the exact mechanism underlying TCDD-mediated pathologies has not been completely elucidated, it is accepted that TCDD mediates the majority of these effects through activation of the aryl hydrocarbon receptor (AhR)-signaling pathway. However some AhR-independent effects of TCDD have been reported (Ahmed et al., 2005; Kondraganti et al., 2003; Park et al., 2003, 2005a,b; Sanders et al., 2005).

Retinoic acid (RA) is a natural product (lipid soluble hormone) derived from the metabolism of vitamin A. Vitamin A is an essential nutrient obtained from food either as preformed vitamin A (retinyl ester, retinol, and small amounts of RA) from animal products (eggs, liver, and milk) or as provitamin A (carotenoids) from fruits and vegetables (Fisher and Voorhees, 1996; Sporn et al., 1994). Vitamin A and its natural and synthetic derivatives are also known as retinoids. Dietary-derived all-*trans* RA (atRA) is the main signaling retinoid in the body and is vital for biological functions such as embryogenesis, growth and differentiation, as well as for vision and reproduction (Dragnev et al., 2000). Levels of atRA in the tissue are tightly regulated through its biosynthesis, metabolism, and storage in the liver (Fig. 1).

The observation that TCDD exposure results in lesions that are reminiscent of those observed in vitamin A-deficient animals of several species, including reduced growth, abnormal immune function, and developmental abnormalities, was the first suggestion that TCDD and related compounds had an impact on retinoid homeostasis and the RA-signaling pathway (Table I). In addition, the low endogenous retinoid levels in the kidney are increased by both exposure to TCDD (Hakansson and Ahlborg, 1985) and vitamin A deficiency (Morita and Nakano, 1982). These observations led to the hypothesis that TCDD and the AhR pathway were altering retinoid metabolism to mimic a vitamin A-deficient state. Indeed, a reduction in hepatic retinoid storage following exposure to TCDD was observed in a variety of species (Fletcher et al., 2001; Hakansson et al., 1991). Evidence for a link between TCDD exposure and vitamin A deficiency is further strengthened by findings demonstrating that rats pretreated with TCDD store and metabolize an oral dose of vitamin A as if they were deficient, despite considerable retinoids in storage (Hakansson and Ahlborg, 1985). Further, vitamin A-administered post-TCDD exposure accumulates to a lesser extent than in control rats, and endogenously stored retinoids are released more rapidly following TCDD treatment (Hakansson and Ahlborg, 1985; Hakansson and Hanberg, 1989; Kelley et al., 1998, 2000).

However, not all data support the conclusion that exposure to TCDD and related compounds results in a vitamin A-deficient state. Some data indicate that exposure perpetuates a vitamin A-excess state, particularly in reference to bone lesions (Jamsa et al., 2001; Lind et al., 2000) and teratogenesis (Abbott and Birnbaum, 1990; Peters et al., 1999). Therefore, although it is clear that

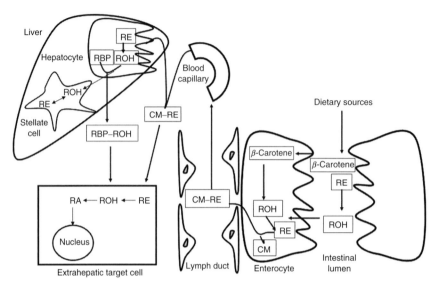

FIGURE 1. Vitamin A is obtained from food either as preformed vitamin A (retinyl ester), retinol, and small amount of RA from animal products (eggs, liver, milk) or as provitamin A (carotenoids) from fruits and vegetables. In the small intestine, retinyl esters (REs) are hydrolyzed to retinol (ROH) by retinyl ester hydrolases (REHs) on the cell surface of the enterocyte or in the intestinal lumen. While in the enterocyte, the ROH is bound to cellular retinol-binding protein II (CRBPII) and is reesterified back to RE by lecithin:retinol acyltransferase (LRAT). The REs are incorporated along with other dietary lipids into chylomicrons and transferred into the lymphatic system. These chylomicrons are specifically internalized into the hepatocytes of the liver, where the RE is converted to ROH through the action of REHs. Within the hepatocytes and stellate cells, the ROH is bound to CRBPI which is thought to transfer the ROH to the RBP for transport out of the liver, where the RBP–ROH complex is transferred to the circulation for use in extrahepatic tissues. In situations where vitamin A is in excess, it is stored in the stellate cells as RE.

TCDD has an impact on retinoid homeostasis, it is less clear as to whether it pushes the system toward a vitamin A-deficient state or simulates a vitamin A-excess state.

A situation where TCDD appears to mimic vitamin A excess is demonstrated by the synergistic effect of TCDD and atRA on palatal development in mice. Both excess atRA and TCDD cause developmental defects in mice and share a common target, the developing palate (Birnbaum et al., 1989), and data demonstrate a synergistic effect of TCDD and atRA on palate defects. These studies show that coadministration of atRA and TCDD result in 100% cleft palate formation at lower concentrations than required when atRA or TCDD is administered separately. The increase in cleft palate is attributed to increased expression of growth factors such as transforming growth factor (TGF)-β1 (Abbott and Birnbaum, 1990). This effect can be recapitulated in

TABLE I. TCDD Exposure Produces Lesions That Are Similar to Vitamin A Deficiency in a Variety of Animal Model Systems

	Vitamin A deficiency	TCDD
General effects		
Loss of appetite	r, mu, g, h, mo	r, mu, g, h, mo
Growth inhibition	r, mu, g, h, mo	r, mu, g, h, mo
Adipose reduction	r, mu, mo	r, mu, g, h
Inactivity/listlessness	r, mu	r, mu, g
Rough coat	r, mu, g, h, mo	r, mu, g, h
Death	r, mu, g, h, mo	r, mu, g, h, mo
Hyperplasia/metaplasia		
Gastrointestinal	r, g, mo	g, h, mo
Urinary tract	r, mu, g, mo	g, mo
Bile duct/gall bladder	r	r, mu, mo
Respiratory system	r, mu, g, h, mo	r
Uterus	r, g	r
Reproduction		
Testes degeneration	r, mo, h	r, mu, g, mo
Spermatogenesis		mo
Abnormal estrous cycle	r, mu	r, mu, mo
Fetus resorption	r, mu	r, mu, mo
Congenital abnormalities		
Cleft palate	r	mu, mo*
Abnormal kidneys	r	r*, mu, mo*
Immunosuppression		
Thymic atrophy	r, mu*, g	r, mu, g, h, mo
Impaired cellular immunity	r, mu	r, mu, g
Impaired humoral immunity	r*, mu	r#, mu, g*
Eye lesions		
Xerophthalmia	r, mu, g*, h, mo	Not tested
Closed eyes/exudate	r, mu	r, mo

Effects of TCDD exposure and vitamin A deficiency. This table is modified from Nilsson and Hakansson (2002). Many of the lesions observed following TCDD exposure resemble those seen in vitamin A deficiency. The animal model demonstrating the effect is indicated by r—rat, mu—mouse, g—guinea pig, h—hamster, and mo—monkey. All denote an increase in the described lesion, except when noted by an *: reduction or a #: moderate to small effect.

embryonic palatal cultures, where TCDD and atRA synergistic effects are mediated by different cellular processes: atRA-induced cleft palates were a result of abnormal epithelial proliferation and differentiation, whereas TCDD altered only the medial epithelial differentiation (Abbott and Buckalew, 1992). Together these data demonstrate that the interactions between these two pathways can have significant effects on developmental processes. Furthermore, the mechanisms of this synergy may involve parallel and intersecting pathways. Therefore, exposure to TCDD results in lesions reminiscent of both vitamin A deficiency and vitamin A excess. Given the complex regulation of retinoid homeostasis, these data suggest that TCDD and the AhR-signaling pathway have a variety of effects on the regulation of retinoid metabolism and directly on the signaling of atRA in the cell.

II. RETINOID SIGNALING

The majority of RA's biological activity is mediated through regulation of gene expression. RA binds to two types of nuclear receptors, the retinoic acid receptors (RARs), which bind both 9-*cis* and all-*trans* forms of RA, and retinoid X receptors (RXRs), which bind 9-*cis* RA (reviewed in Chambon, 1996; Mangelsdorf *et al.*, 1995). The RARs and RXRs each contain a well-conserved DNA-binding domain, a well-conserved ligand-binding domain, and three or four domains that are not as well conserved (Renaud and Moras, 2000). Both RAR and RXR have three subtypes (α, β, and γ), and gene knockout experiments suggest that the RAR subtypes may be functionally redundant (Mark *et al.*, 1999). Further, RAR and RXR subtypes form multiple isotypes through the use of alternate promoters, alternative splicing, and alternative initiation of translation (Chambon, 1996). Although these isoforms appear to have tissue-specific expression, it is unclear whether they differentially regulate gene transcription. RAR and RXR homo- and heterodimers stimulate transcription of target genes through binding to retinoic acid response elements (RAREs) consisting of direct repeats of 5'-AGGTCA-3' separated by one to five spaces (termed DR-1 or DR-5, respectively) (Glass *et al.*, 1997). Although RARs and RXRs can form homodimers, heterodimerization of RARs with RXRs increases the affinity of these receptors for the RARE. In addition to directly altering gene expression, atRA also can reduce the expression of certain genes through interference with other transcription factors, most notably AP-1 (Schule *et al.*, 1991). In an uninduced state, the RAR/RXR heterodimers are in complex with the RARE and are bound to the nuclear corepressors silencing mediator of retinoid and thyroid receptors (SMRT) or nuclear receptors corepressor (N-CoR) (Li *et al.*, 1997; Yoh and Privalsky, 2001). These corepressors function by recruiting complexes containing histone deacetylases to the

TABLE II. Coactivator and Corepressor Proteins Interacting with the AhR- and RA-Signaling Pathways

	Pathway	References
Coactivators		
SRC-1	AhR	Beischlag et al., 2002
	RA	Yao et al., 1996
RIP 140	AhR	Kumar et al., 1999
	RA	Vincenti et al., 1996
p300	AhR	Tohkin et al., 2000
	RA	Yao et al., 1996
p/CIP	AhR	Beischlag et al., 2002
NCoA-2	AhR	Beischlag et al., 2002
Brg-1	AhR	Wang and Hankinson, 2002
Corepressors		
SMRT	AhR	Nguyen et al., 1999; Rushing and Denison, 2002
	RA	Li et al., 1997
N-CoR	RA	Yoh and Privalsky, 2001

promoter. On binding of ligand, the RAR/RXR heterodimers are released from this complex and interact with coactivator complexes (SRC-1/TIF2/RAC3 and CBP/p300) that mediate transactivation (Chen and Li, 1998). Coactivators and corepressors that associate with RARs/RXRs are listed in Table II.

Although atRA is considered the primary signaling retinoid in the body, data indicate that other RA isoforms also contribute to signaling (Fig. 2). 9-cis RA has long been accepted as the other potent signaling retinoid, binding effectively to the RXRs. No binding activity is associated with the 13-cis RA isomer, and it is believed that the biological effects of 13-cis RA were mediated through isomerization to 9-cis RA. However, data now suggest that, although the 13-cis isomer is unable to activate transcription through the retinoid receptors, it may function through inhibition of enzymes involved in steroid metabolism or through other membrane receptors (Blaner, 2001). The 9,13-di-cis-RA binds to the RARα and increases cellular fibrogenesis by formation of TGF-β (Imai et al., 1997; Okuno et al., 1999). A novel RA metabolite, 9-cis-4-oxo-13,14-dihydroRA, was identified with particularly high hepatic levels (Schmidt et al., 2002); however, the function of this isoform is unclear. In addition, 4-oxo-RA, 4-oxo-ROH, and 4-oxo-retinaldehyde are able to activate transcription through the RAR and RXR receptors, and are thought to be important in *Xenopus* development (Achkar et al., 1996; Blumberg et al., 1996; Pijnappel et al., 1998) (Fig. 3).

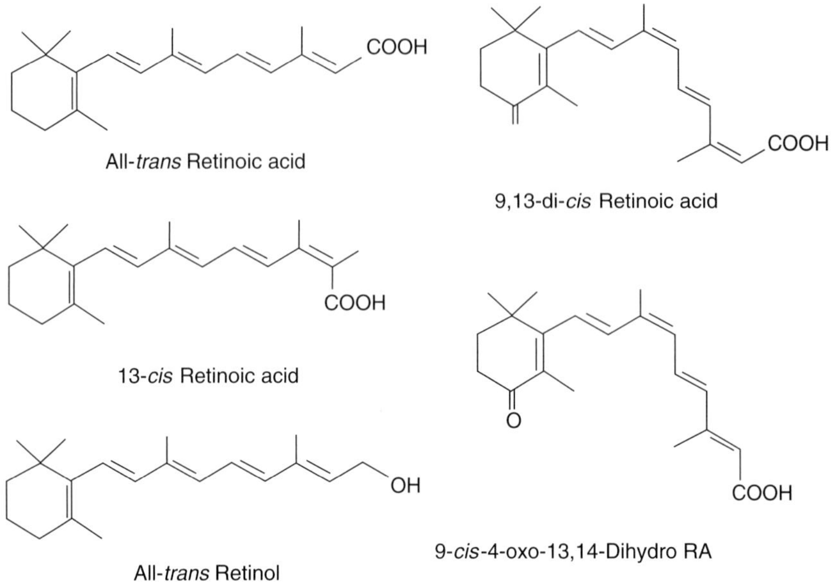

FIGURE 2. Structures of vitamin A metabolites.

III. THE AHR/ARNT PATHWAY

The AhR and its dimerization partner AhR nuclear translocator (Arnt) are members of the basic helix-loop-helix per-Arnt-sim (bHLH-PAS) domain family of transcription factors. Members of this family have diverse biological roles ranging from developmental regulation to environmental sensing (Crews and Fan, 1999). The DNA-binding and heterodimerization domains are located in the N-terminal portion of AhR and Arnt, which contains the bHLH motif and the PAS domain. The C-terminal regions of AhR and Arnt contain the transactivation domain sequences (TADs). In addition to ligand activation, AhR activity is also modulated by phosphorylation of serine residues in the C-terminal portion of the protein (Long et al., 1998; Mahon and Gasiewicz, 1995).

The latent form of AhR resides in the cytoplasm in complex with accessory proteins, including two heat shock protein 90 (HSP90) molecules, a cochaperon p-23, and an immunophilin-like protein, ARA9 (XAP2; AIP) (Carver et al., 1998; Kazlauskas et al., 2001). ARA9 appears to stabilize AhR protein levels in the cytoplasm by protecting the receptor from proteasome-mediated degradation and by altering AhR cytoplasmic distribution (LaPres et al., 2000). Binding of ligand to the AhR results in the release of the p23/ARA9 molecules and translocation of the ligand/AhR/HSP90 complex into the nucleus. Once in the nucleus, AhR releases the

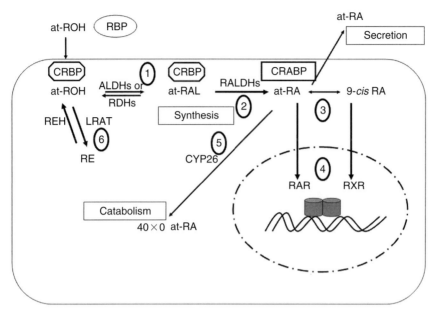

FIGURE 3. AhR- and TCDD-mediated changes in RA synthesis. Points where the TCDD and the AhR pathway alter RA cellular metabolism are indicated by the numbered ovals. Retinol (ROH) bound to retinol-binding protein (RBP) in the blood is taken up by the cell, and becomes associated with cellular retinol-binding protein (CRBP). (1) Conversion of ROH to retinal (RAL) is catalyzed by either alcohol dehydrogenases (ADH) or retinol dehydrogenases (RDHs). *In vitro*, CYP1A1, 1A2, and 1B1 are able to catalyze this reaction. (2) RAL is further oxidized to RA through the action of RALDHs. At this point, the retinoid is released from CRBP, and becomes bound to the cellular retinoic acid-binding proteins (CRABP). There are two XREs in the promoter for RALDH2, indicating that it may be a target for the AhR pathway (Wang *et al.*, 2001). Also, several CYP450s are known to catalyze this reaction *in vitro* and *in vivo* (Raner *et al.*, 1996; Roberts *et al.*, 1992; Tomita *et al.*, 1996; Zhang *et al.*, 2000). (3) Interconversion between the atRA and 9-*cis* RA can alter its RAR-binding specificity and alter RA signaling. TCDD reduces the levels of glutathione and nonprotein sulfhydryl contents in the liver, which are involved in catalyzing RA interconversion (Shertzer *et al.*, 1998; Stohs, 1990; Stohs *et al.*, 1990). Further, TCDD exposure results in increased glutathione *S*-transferase (GST) expression and activity, which is also implicated in RA isomerization (Aoki, 2001; Safe, 2001). (4) To mediate changes in gene expression, RA binds specific receptors in the nucleus, the RARs and RXRs. Data indicate that TCDD and the AhR pathway alter receptor expression (Murphy *et al.*, 2004; Weston *et al.*, 1995), and alter binding of the receptors to the RARE perhaps by altering availability of coactivators and corepressors (Rushing and Denison, 2002; Widerak *et al.*, 2005). (5) RA metabolism and excretion involve hydroxylation and glucuronidation. Several CYP450s are known to mediate the hydroxylation of RA, including CYP26 (Andreola *et al.*, 1997; Loudig *et al.*, 2000; Ray *et al.*, 1997). Although UDP-glucuronosyltransferase 1A1 (UGT1A1) is induced by TCDD in rodent livers and human cell lines (Munzel *et al.*, 1994; Yueh *et al.*, 2003, 2005), its involvement in metabolism of RA is unclear. (6) For storage, ROH bound to CRBP is converted to retinyl esters (REs) through the activity of lecithin:retinol acyltransferase (LRAT). RE in storage is hydrolyzed to ROH through the action of retinol ester hydrolases (REHs). TCDD exposure reduces LRAT activity in hepatic stellate cells (Nilsson *et al.*, 1997). Further, LRAT mRNA levels are reduced in whole liver homogenates from TCDD-exposed rats (Hoegberg *et al.*, 2003).

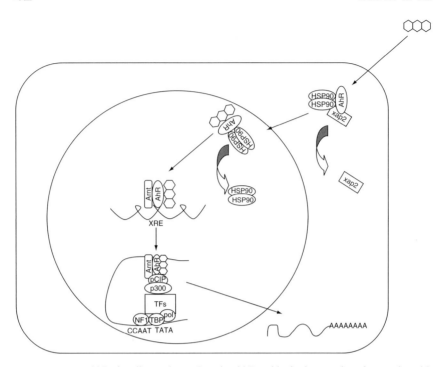

FIGURE 4. AhR-signaling pathway. Inactive AhR resides in the cytoplasm in complex with accessory proteins, including two HSP90 molecules, a cochaperon p-23, and an immunophilin-like protein, ARA9 (XAP2; AIP) (Carver et al., 1998; Kazlauskas et al., 2001). Binding of ligand to the AhR results in dissociation from the HSP90/p23/ARA9 complex and translocation of the ligand/AhR/HSP90 complex into the nucleus. Once in the nucleus, AhR dissociates from the HSP90 molecules and dimerizes with Arnt (Reyes et al., 1992; Sogawa et al., 1995). The AhR/Arnt heterodimer binds to specific DNA sequences in the 5' regions of AhR-responsive genes termed xenobiotic response elements (XRE: 5'-GCGTG-3') (Matsushita et al., 1993; Watson and Hankinson, 1992). Data now indicate that the AhR/Arnt complex recruits coactivator proteins to the transcriptional start site, and alters nucleosomal configuration to facilitate transcriptional activation (Hankinson, 2005).

HSP90 molecules and dimerizes with another bHLH-PAS family member, Arnt (Reyes et al., 1992; Sogawa et al., 1995). The AhR/Arnt heterodimer is a transcription factor, binding to specific DNA sequences in the 5' regions of AhR-responsive genes termed xenobiotic response elements (XRE: 5'-GCGTG-3') (Matsushita et al., 1993; Watson and Hankinson, 1992) (Fig. 4). Data indicate that binding to the XRE does not require the TAD of AhR or Arnt; however, interaction with the CCAAT and TATA box for transcriptional activation requires the AhR TAD (Ko et al., 1997; Sogawa et al., 1995). Heterodimer binding to the XRE results in nucleosomal disruption and recruitment of transcription activation factors to the promoter region. This is mediated through direct binding to the transcriptional activation machinery: mouse AhR binds directly to TFIIB, whereas human AhR has been demonstrated to bind to both TBP and TFIIF (Rowlands *et al.*, 1996; Swanson and Yang, 1998; Watt *et al.*, 2005).

AhR binding to the XRE and transcriptional regulatory proteins is associated with interactions with a variety of coactivators that mediate nucleosomal disruption. Coactivator CBP, a histone acetyltransferase (HAT), physically interacts with Arnt, as does SRC-1 and RIP140 (Beischlag *et al.*, 2002; Kobayashi *et al.*, 1997). The p160 HAT coactivators SRC-1, NCoA-1, and p/CIP associate with the cytochrome P450 1A1 (CYP1A1) promoter region following TCDD exposure (Hankinson, 2005). This, along with other data, indicates that the p160 coactivators are physiologically relevant coactivators for the AhR/Arnt-signaling pathway. Data also indicate a role for the Brahma/SWI-related gene protein (Brg-1) in AhR/Arnt-mediated changes in chromatin structure (Wang and Hankinson, 2002). A list of coactivator and corepressor proteins that interact with AhR/Arnt are shown in Table II.

Studies of the molecular mechanisms of the AhR/Arnt heterodimer have focused on the transcriptional activation of xenobiotic metabolizing genes, including Phase I drug-metabolizing enzymes, such as the cytochrome P450 (CYP450) family of monooxygenase enzymes (Fujii-Kuriyama *et al.*, 1992; Watson and Hankinson, 1992), as well as Phase II enzymes, including UGT1A1, GST-Ya subunit, and NADPH-quinone-oxido-reductase (reviewed in Mimura and Fujii-Kuriyama, 2003). A number of genes unrelated to xenobiotic metabolism are also activated by TCDD exposure. These include genes involved in growth control, such as TGF-α (Hankinson, 1995), TGF-β2 (Hankinson, 1995), ∂-aminolevulinic acid synthetase (Hankinson, 1995), Bax (Matikainen *et al.*, 2001), and p27kip1 (Kolluri *et al.*, 1999); cytokines such as interleukin-1β (Sutter *et al.*, 1991; Yin *et al.*, 1994); other nuclear transcription factors such as c-Fos, Jun-B, c-Jun, and Jun-D (Hoffer *et al.*, 1996; Puga *et al.*, 1992); and plasminogen activator inhibitor-2 (PAI-2) and several matrix metalloproteinases, regulator of ECM proteolysis (Murphy *et al.*, 2004; Sutter *et al.*, 1991; Villano *et al.*, 2006; Yin *et al.*, 1994).

Although originally identified as the receptor for the PAH family of environmental contaminants, data indicate that the AhR binds to a variety of endogenous and exogenous compounds, including flavonoids, UV photoproducts of tryptophan, as well as some synthetic retinoids (Carver and Bradfield, 1997; Denison *et al.*, 2002; Oberg *et al.*, 2005; Song *et al.*, 2002; Soprano and Soprano, 2003; Soprano *et al.*, 2001) (Fig. 5). The ability of synthetic retinoids to bind to and activate the AhR pathway has interesting implications for the cross talk between the AhR and RA pathways. Further, these retinoids were developed as therapies for skin disorders, inflammatory diseases, and for use as chemopreventatives (Nagpal and Chandraratna, 2000; Sporn and Suh, 2000; Thacher *et al.*, 2000); therefore, their ability to activate pathways other than the RA pathway is important to determining potential side effects. Experiments have shown that the pan-RAR antagonist

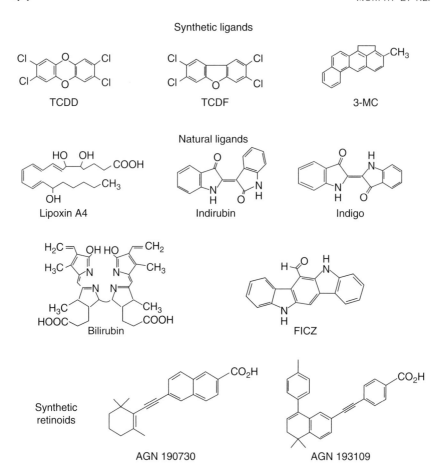

FIGURE 5. AhR ligands. The AhR was originally identified as the receptor for polycyclic aromatic hydrocarbons (PAHs). The most potent of these compounds is 2,3,7,8-TCDD. Other synthetic ligands include 2,3,7,8-tetrachlorodibenzofuran (TCDF) and 3-methylcholanthrene (3-MC). Data indicate that the AhR binds to and is activated by a variety of unrelated ligands (reviewed in Denison *et al.*, 2002). Some natural ligands for the AhR include lipoxin A4 (an arachidonic acid metabolite), indirubin and indigo (components of human urine and blood), bilirubin (a heme degradation product), and FICZ (6-formylindolo[3,2-*b*]carbazole, a tryptophan photoproduct). In addition, several synthetic retinoids also effectively bind to the AhR (Soprano and Soprano, 2003; Soprano *et al.*, 2001).

AGN 193109 induces CYP1A1 expression in mouse embryos (Soprano *et al.*, 2001). The activation of CYP1A1, a CYP450 and an AhR-responsive gene, by AGN 193109 was confirmed in Hepa-1c1c7 cells (Soprano *et al.*, 2001). No similar increase in CYP1A1 was observed following exposure of mice or Hepa1c7c cells with atRA, suggesting that AGN 193109 may not be activating CYP1A1 expression via the RAR/RXRs. Exposure of

Hepa1c7c cells that are lacking AhR or Arnt activity to AGN 193109 demonstrates a requirement for AhR/Arnt for AGN 193109-induced CYP1A1 expression (Soprano et al., 2001). To demonstrate whether other synthetic retinoids can also function through the AhR pathway, these compounds were tested for their ability to transactivate a luciferase reporter linked to the portion of the CYP1A1 promoter containing four XREs (Garrison et al., 1996). Of the 23 retinoids tested, 3 (AGN 190730, AGN 193109, and AGN 192837) were found to greatly stimulate activity of this promoter construct, in addition several others were shown to have moderate or minimal effects (Soprano and Soprano, 2003). Transcriptional activation by AGN 190730 was inhibited by the AhR agonist, α-naphthoflavone, suggesting that activation is AhR-dependent, and mobility shift assays demonstrate that this retinoid can stimulate binding of the AhR/Arnt complex to the XRE (Gambone et al., 2002; Soprano and Soprano, 2003). Further, competitive binding experiments demonstrated that AGN 190730 can inhibit TCDD interactions with the AhR and the XRE (Gambone et al., 2002).

IV. AHR AND RA AVAILABILITY

Endogenous levels of atRA in the body are maintained through a balance between atRA biosynthesis, metabolism, and storage. Data indicate that TCDD and the AhR pathway alter all three of these processes, resulting in substantial changes in the levels of atRA in the liver, as well as in extrahepatic tissues. TCDD-mediated decrease in hepatic retinoid levels is a well-characterized effect that is described in all rodent models (reviewed in Nilsson and Hakansson, 2002). Further, fish and other wild-life exposed to dioxin-like compounds in their environment also demonstrate changes in retinoid levels (Rolland, 2000). TCDD exposure results in an increased mobilization of retinoids from retinyl ester stores (Brouwer et al., 1989; Jurek et al., 1990; Kelley et al., 2000; Van Birgelen et al., 1995a,b), altered retinol esterification (Hoegberg et al., 2005; Nilsson et al., 1997), altered tissue levels of atRA (Hoegberg et al., 2005; Kelley et al., 1998; Nilsson et al., 2000), increased turnover of retinoids (Kelley et al., 1998), and increased metabolism and excretion (Brouwer et al., 1989; Hakansson and Ahlborg, 1985) (Fig. 3). The effects on hepatic retinoids are observed not only in directly exposed animals but also in animals exposed to TCDD *in utero* (Morse and Brouwer, 1995) and lactationally (Hakansson et al., 1987). Most of these data are from animals treated with a high acute dose of TCDD or related hydrocarbon (10 μg/kg). However, data indicate that reduction of hepatic retinoid stores is also a result of a single "no-effect" dose (0.1 μg/kg) of TCDD (Nilsson et al., 2000). These data indicate that TCDD-mediated alterations in hepatic retinoid stores may be the most sensitive measure of TCDD exposure.

Although TCDD and the AhR pathway alter vitamin A processing in a variety of species, there are some species-specific differences in the effects of TCDD on retinoid processing. One in particular concerns the tissue levels of atRA. In the rat, TCDD exposure causes a dose-dependant increase in serum, liver, and kidney levels of atRA levels (Hoegberg *et al.*, 2003; Kelley *et al.*, 2000; Nilsson *et al.*, 2000; Schmidt *et al.*, 2003). Concomitantly, a substantial decrease in the 9-*cis*-4-oxo-13,14-dihydroRA is observed in the liver (Schmidt *et al.*, 2003). There is no observable increase in hepatic atRA levels in mice following exposure to TCDD or 3-methylcholanthrene (3-MC) (Hoegberg *et al.*, 2005; Kalin *et al.*, 1984). However, the decrease in 9-*cis*-4-oxo-13,14-dihydroRA in mouse liver in response to TCDD is observed, indicating that this metabolite is a sensitive measure of TCDD/PAH exposure in both rats and mice (Fletcher *et al.*, 2005). The reason for the difference in response between the rat and mouse models is unclear. However, it is postulated to result from a difference in substrate specificities of, yet unidentified, dioxin-induced retinoid-metabolizing enzymes.

Given the observation that TCDD exposure alters RA synthesis and storage, it is not entirely surprising that the AhR null mouse demonstrated elevated levels of atRA and its derivatives in the liver (Andreola *et al.*, 1997). RA, retinol, and retinyl palmitate were found to be two to three times higher in AhR null animals in comparison to the wild-type controls. This is postulated to result from a reduced ability of the AhR null animals to catabolize RA (see below). In confirmation of the elevated levels of atRA, there was also an observed increase in tissue type transglutaminase-2, an atRA-responsive gene. Indeed, increased expression and activity of transglutaminase in the AhR null animals are believed to be related to the liver fibrosis observed in these animals. In support of this hypothesis, liver fibrosis in AhR null mice can be reversed by feeding mice a vitamin A-deficient diet (Andreola *et al.*, 2004).

A. RA SYNTHESIS

The involvement of TCDD and the AhR pathway in the synthesis of retinoids is indicated by increased retinoid metabolism in TCDD-treated animals, and in the reduced retinoid metabolism observed in AhR null mice (Andreola *et al.*, 1997). TCDD induction of atRA metabolism is seen both in hepatic and extrahepatic tissues, demonstrated by an increase in RA metabolism using microsomes isolated from animals exposed to TCDD (Fiorella *et al.*, 1995). Although increased atRA metabolism was observed in a variety of tissues, there was variation in the magnitude, with the highest induction in liver microsomes, followed by lung, kidney, and testes (Fiorella *et al.*, 1995). Changes in RA metabolism are observed following both acute and chronic TCDD exposure of rats. Rats exposed to long-term low dose of TCDD demonstrated dose-dependant decreases in retinyl esters and atRA in liver (Fletcher et al., 2005). Indeed, a 60% decrease in 9-cis-4-oxo-13, 14-dihydroRA levels was observed at a very low dose of TCDD (1 ng/kg per day), indicating that this metabolite is a sensitive measure of TCDD/PAH exposure (Fletcher et al., 2005).

A variety of metabolizing enzymes including CYPs, UGTs, and ALDHs are potentially involved in the synthesis and catabolism of RA (Fig. 3). Under normal physiological conditions, the cytosolic medium-chain alcohol dehydrogenases (ADHs) and the microsomal short-chain alcohol dehydrogenases (SCADs) are responsible for the oxidation of ROH to RAL. Alternatively, retinol dehydrogenase (RDH) can catalyze the oxidation of 9-cis ROH, but not atROH (Gamble et al., 1999). The irreversible oxidation of retinal to atRA is catalyzed by members of the aldehyde dehydrogenase (ALDH) family (Duester, 2000), including retinal dehydrogenase type 2 (RALDH2). Interestingly, two potential XRE-binding sites have been identified in the promoter of the mouse RALDH2 gene (Wang et al., 2001), suggesting that activation of RALDH2 by the AhR pathway may contribute to accumulation of atRA in TCDD-exposed animals. In addition, TCDD activates expression of ALDH3; however, there are no data to suggest the involvement of this enzyme in RA metabolism (Lindros et al., 1998). *In vitro* and *in vivo* evidence suggests that the CYP450s may play a critical role in the changes in atRA biosynthesis following TCDD exposure (Ahmad et al., 2000; Leo et al., 1984; Martini and Murray, 1994; McSorley and Daly, 2000; Roberts et al., 1992; Van Wauwe et al., 1990; Vanden Bossche et al., 1988).

The CYP450 superfamily of hemoproteins is composed of more than 3000 molecules distributed over species ranging from bacteria to vertebrates. These enzymes catalyze monooxygenation of various endogenous and exogenous substrates (Nebert and Russell, 2002; Nelson et al., 1996). Members of the families are classified according to structure similarity; members of families 1–4 are mainly involved in the metabolism of exogenous chemicals, including drugs, food additives, and environmental pollutants. CYP450s are typically inducible, with members of the CYP1 family being targets for the AhR-signaling pathway and members of the CYP2 family being regulated by the CAR/PXR/PPAR-signaling pathways. *In vitro*, TCDD-induced CYP1A1, 1A2, and 1B1 are able to catalyze the oxidative conversion of free atROH to atRAL (Chen et al., 2000) and atRAL to atRA (Raner et al., 1996; Roberts et al., 1992; Tomita et al., 1996; Zhang et al., 2000). It is thought that under normal conditions the contribution of CYPs may not be significant to RA synthesis. However, the activation of the CYP450s by TCDD or related congeners may result in changes in atRA availability (Hoegberg et al., 2003; Tomita et al., 1996). It is unclear what the effect of TCDD stimulation of CYP1A1 would be in humans, as the data on the activity of CYP1A1 toward retinoids is inconclusive at this time.

Some data indicate that the human CYP1A1 is involved in retinal oxidation and the synthesis of atRA (Chen et al., 2000), while other data suggest that CYP1A1 activity results in the oxidation of atRA and ultimately its excretion (Lampen *et al.*, 2000).

Although the CYP1 enzymes have demonstrated atRA biosynthetic activity, neither 1A1 nor 1A2 metabolize atRA in the mouse liver (Andreola *et al.*, 1997). However, some novel CYP450 enzymes with activity toward retinoids have been recently identified that may provide a definitive link between the AhR pathway and atRA biosynthesis. One potential enzyme for atRA synthesis is CYP2C39 which is expressed at lower levels in AhR null animals in comparison to their wild-type counterparts, suggesting that this enzyme may contribute to the altered retinoid levels in the AhR null animals (Andreola *et al.*, 2004). However, CYP2C39 expression is not altered following TCDD exposure, indicating that this enzyme is not likely involved in biosynthesis and metabolism changes in retinoid levels in TCDD-exposed animals (Andreola *et al.*, 2004).

Another CYP450 with activity toward atRA is CYP2S1, the sole member of an identified branch of the CYP2 family (Saarikoski *et al.*, 2005). CYP2S1 is inducible by AhR activation directly through XREs in its promoter region (Saarikoski *et al.*, 2005). Exposure to retinoids or UV light also results in increased expression and activity of CYP2S1. To date, the only endogenous substrates identified for CYP2S1 are retinoids; CYP2S1 can metabolize atRA to 4-hydroxy RA and 5,6-epoxy RA. Further, this enzyme is highly expressed in epithelial tissues, suggesting that CYP2S1 may play an important role in maintaining retinoid levels in differentiating skin (Du *et al.*, 2005). Furthermore, CYP2S1 is expressed in all fetal stages suggesting a role in development/teratogenesis (Choudhary *et al.*, 2003, 2005).

B. RA CATABOLISM

The primary pathway of atRA metabolism and excretion in vertebrates begins with a hydroxylation reaction followed by glucuronidation. The NADPH-dependant 4-hydroxylation of atRA is inhibited by carbon monoxide, suggesting the involvement of the CYP450s (Gonzalez and Fernandez-Salguero, 1998; Napoli, 1999). In humans, CYP2C8 appears to be the major hepatic RA 4-hydroxylase with CYP3A4 playing a minor role (McSorley and Daly, 2000). In mice, it is unclear which CYP enzyme is the primary hydroxylase for retinoid metabolism. However, CYP2C39 shows similar kinetic properties as CYP2C8 in humans, and it is suggested that it may be the prominent murine retinoid hydroxylase (Andreola *et al.*, 2004).

A variety of CYP450s are known to have metabolizing activity toward RAs, including members of the CYP1A, 2, 3A, and CYP26 families (Honkakoski and Negishi, 2000). Assays using microsomal extracts implicate CYP1A1 and 1A2 as the enzymes that mediate the hydroxylation of RA following TCDD exposure (Ahmad *et al.*, 2000; Schmidt *et al.*, 2003). However, data implicate other CYP450 families in atRA metabolism. An early study suggested that

CYP2C7 and 2B1 may have significant contributions in rat (Leo et al., 1984). However, these enzymes are not TCDD inducible (Chu et al., 2001), indicating that 2B1 does not contribute to changes following TCDD exposure. CYP26 is known to catabolize atRA; however, its levels are the same in wild-type and AhR null animals, indicating that this enzyme is not contributing to changes in RA levels in the AhR null animals (Andreola et al., 1997; Loudig et al., 2000; Ray et al., 1997). These data suggest that the effect of AhR on RA catabolism is perhaps mediated through a yet undescribed oxidative enzyme, or through changes in glucuronidation.

Glucuronidation is catalyzed by a superfamily of UDP-glucuronosyl-transferases (UGTs). TCDD exposure results in increased UGT activity in rodents (Bank et al., 1989; Kessler and Ritter, 1997; Malik and Owens, 1981) and cultured human cells (Munzel et al., 1999). In the human HEPG2 cells, UGT1A1 expression is mediated through an XRE in its promoter (Yueh et al., 2003, 2005). UGT1A1 has been shown to be able to catalyze glucuronidation of atRA in rat liver; however, the UGT1A1 Km for atRA is high (59 μM) (Radominska et al., 1997). These data suggest that TCDD activation of UGT expression and activity could result in increased metabolism of atRA in liver; however, given its high Km for atRA, the relative contribution of UGT1A1 is less clear.

C. INTERCONVERSION

Steric interconversions of atRA occur *in vitro* and *in vivo* (Kojima et al., 1994; Marchetti et al., 1997; Sundaresan and Bhat, 1982; Vane et al., 1982; Zile et al., 1967), and the conversion between atRA and 9- and 13-*cis* RAs alters the available binding forms. Further, glucuronidation is more effective on *cis*-isomers than on *trans*-isomers (Genchi et al., 1996; Marchetti et al., 1997), therefore changes in isomerization will alter metabolism and clearance. It is unclear how atRA interconversion is regulated; however, glutathione S-transferases (GSTs) and a variety of sulfhydryl compounds are able to catalyze these reactions (Chen and Juchau, 1997, 1998a,b; Urbach and Rando, 1994a,b). Additionally, retinol saturase converts atROH to at13,14-dihydroretinol (Moise et al., 2005) which can activate transcription through the RAR/RXR heterodimer (Moise et al., 2005). Exposure to TCDD results in decreased levels of glutathione and nonprotein sulfhydryl contents in the liver, through activation of oxidative stress (Shertzer et al., 1998; Stohs, 1990; Stohs et al., 1990). Further, TCDD exposure results in increased GST expression and activity (Aoki, 2001; Safe, 2001). These data suggest that TCDD and the AhR pathway may alter isomerization of atRA in the liver as well, which would influence atRA activity, metabolism, and clearance. However, the ultimate consequence of TCDD and the AhR pathway on RA isomerization is unclear. For example, 9,13-di-*cis* RA can transactivate through RARα. The production of this isomer could result from reduced *trans–cis* isomerization (from decreased glutathione and sulfhydryl contents) or from increased *cis–trans* isomerization (from increased GST expression).

D. STORAGE AND TRANSPORT

Vitamin A is obtained from the diet, either as preformed vitamin A (retinyl ester, retinol, and small amount of RA) or as provitamin A (carotenoids) (Fig. 1). In the lumen of the small intestine or in the intestinal mucosa, dietary retinyl esters are hydrolyzed to retinol, through the action of retinyl ester hydrolases (REHs). Provitamin A (mainly in the form of β-carotene) absorbed by the mucosal cells is converted to retinaldehyde through the actions of carotene-15,15′-dioxygenase, and this form is further reduced to retinol by retinaldehyde reductase. Within the enterocyte, retinol, independently of its dietary origin, is reesterified by the enzyme lecithin:retinol acyltransferase (LRAT), and retinyl esters are packaged into chylomicrons, together with other dietary lipids. Although LRAT is considered the primary enzyme for esterification of retinoids, LRAT null mice maintain some ability to convert retinol to retinyl esters, supporting the notion that another enzyme, namely acyl-CoA:retinol acyltransferase is involved in this process (O'Byrne *et al.*, 2005).

Once packaged into chylomicrons, the bulk of dietary retinoid is taken up by the liver, while the remaining 25% is taken up by extrahepatic tissues (Goodman, 1962; Goodman *et al.*, 1965). Upon uptake in the liver, chylomicron retinyl esters are once again hydrolyzed to retinol, which can either be secreted by the hepatocyte bound to retinol-binding protein (RBP) or it can be transferred to the hepatic stellate cells for storage (Vogel *et al.*, 1999). At this time the mechanism of transfer of retinol from the hepatocytes to the stellate cells for storage is still not completely elucidated. However, when dietary vitamin A is abundant, $\sim$80–90% of the stored retinyl esters are in the stellate cells. Both hepatocyte and stellate cells produce significant amounts of REH and LRAT, as well as the cellular retinol-binding protein type I (CRBPI). CRBPI is a chaperone protein necessary to solubilize retinol in the aqueous environment of the cell (Vogel *et al.*, 1999).

One of the earliest observations of dioxin toxicity was that exposure to these compounds altered hepatic retinyl ester storage in a variety of mammalian model systems (Hakansson *et al.*, 1991). In rats exposed to an acute dose of TCDD, storage of retinyl ester in the stellate cells was reduced, until the TCDD was eliminated from the liver (Hakansson and Hanberg, 1989; Hakansson *et al.*, 1994; Thunberg *et al.*, 1979, 1980). This is postulated to be somewhat influenced by a TCDD-induced mobilization of retinoids from hepatic and extrahepatic storage, as well as increased elimination in the urine (Brouwer *et al.*, 1989). The inhibition of retinyl ester storage following

TCDD exposure is most likely responsible for the reduction of retinoids in the rest of the animal (Hakansson et al., 1991).

Although it is clear that TCDD and related congeners alter hepatic retinyl ester storage, the mechanism of this reduction is less clear. The reduction in storage in the hepatic stellate cell population is not a result of a reduction in the number of stellate cells, or from any toxicity of TCDD on the stellate cell population (Hanberg et al., 1996). However, exposure to TCDD does appear to reduce LRAT activity in stellate cells (Nilsson et al., 1997). Further, LRAT mRNA levels are reduced in whole liver homogenates from TCDD-exposed rats (Hoegberg et al., 2003). These data indicate that TCDD may directly alter the expression of LRAT in hepatic stellate cells, and thereby reduce LRAT-induced retinyl ester formation for storage. Interestingly, TCDD treatment results in an increase in retinyl esters in the kidney, and this is preceded by an increase in the expression of LRAT (Hoegberg et al., 2003).

To maintain solubility in an aqueous environment, retinoids are bound to retinoid-specific binding proteins. The cellular retinol-binding proteins (CRBPs) and the cellular retinoic acid-binding proteins (CRABPs) are entirely intracellular; whereas the RBP and the intracellular retinol-binding proteins (IRBP) are extracellular. CRBPI is thought to have a critical role in regulation of retinoid storage by regulating ROH esterification by LRAT in the stellate cells (Ghyselinck et al., 1999; Nilsson et al., 1997). Further, several microsomal enzymes that are involved in retinoid metabolism prefer ROH bound to CRBPI, including RDHs I, II, and III. Also, SCADs prefer retinol that is bound to CRBPI (Napoli, 1999). There are also RALDH that are more effective toward retinal in complex with CRBPI. It has been suggested that retinol bound to CRBPI is protected from metabolism from liver enzymes such as ADH and the CYP450s (Napoli, 1999). CRBPI also acts as an atRA chaperone which may result in the metabolism of the low levels of free atROH, thus preventing excessive ROH oxidation but allowing a small amount to be converted to RAL for atRA synthesis (Duester, 2000). CRBPI is also critical to delivery of retinol to newly synthesized RBP for secretion from the liver into circulation, and it is implicated in facilitating uptake of retinol–RBP complexes by the extrahepatic cells. Therefore, CRBPI is an essential intracellular transporter of retinol and forms a link between retinol mobilization, metabolism, and uptake.

Although it is compelling to postulate that TCDD and the AhR pathway may alter the expression of the CRBPs, no change in CRBPI expression in the liver of TCDD-treated rats is observed (Schmidt et al., 2003). However, data from knockout animals suggest that the CRBP proteins somehow modulate the effect of TCDD on RA storage. TCDD exposure of mice that lack CRBPI, CRABPI, and CRABPII results in complete depletion of total retinoids in the liver. However, exposure of mice that are null for only CRABPI and CRABPII maintained 60–70% of the total hepatic retinoids

(Hoegberg et al., 2005). This suggests that loss of CRBPI may account for the increased susceptibility of the triple knockout mice to TCDD-induced retinoid depletion.

As discussed above, maintenance of whole-body retinoid metabolism is a complex process involving several organ systems, with the main storage of retinoids in the liver. This system of regulation provides the body with optimal amounts of retinoids despite changes in retinoid dietary intake. It has been proposed that there exists a feedback mechanism involving the hepatic and renal retinoid pools in conjunction with the circulating retinoids and that this feedback mechanism is regulated by a yet identified set point, which may be a critical level of a specific retinoid in the liver or kidney. This mechanism would allow for maintaining retinoid homeostasis even in times of vitamin A deficiency or excess. It has been proposed that TCDD, by the significant alterations in liver retinoid levels, may alter the set point for the feedback system, thereby resulting in a cascade of mis-regulation of retinoid acid synthesis, metabolism, and storage (Nilsson and Hakansson, 2002). Although it is unclear how TCDD and the AhR pathway may alter the set point, there are several candidates suggested for this mechanism. One is the apo:holo ratio of the CRBPI, which may be involved in maintaining a balance between retinoid hydrolysis, esterification, as well as conversion of retinol to RA (Boerman and Napoli, 1991, 1996; Herr and Ong, 1992). The importance of the binding proteins in the retinoid homeostasis is supported by the finding that CRBPI null animals do not store retinyl esters properly (Ghyselinck et al., 1999). Further, RBP knockouts fail to efficiently mobilize stored hepatic retinoids (Quadro et al., 1999).

V. MOLECULAR INTERACTIONS BETWEEN THE RA AND AhR PATHWAYS

Cross talk between the AhR and RA pathway extends beyond effects on retinoid metabolism, also affecting transcriptional regulation. Both the AhR and RA pathways regulate transcription of a variety of genes that are critical for the physiological effects mediated by these pathways. Like the numerous interactions observed for these pathways in retinoid metabolism, there are also several levels of molecular interactions between these pathways, including direct inhibition, alteration of receptor availability, and competition for transcriptional coactivators.

One of the first indications that the RA and AhR pathways interact at the level of gene expression was that TCDD exposure of SCC-4 keratinocytes inhibits atRA-induced activation of transglutaminase, an enzyme critical for proper differentiation of skin. The role of the AhR pathway in mediating this inhibition is indicated by two AhR-activating compounds, methylcholanthrene or benzo[a]pyrene, preventing transglutaminase activation

(Rubin and Rice, 1988). TCDD inhibition of transglutaminase in the SCC-4 cells is mediated primarily at the level of transcription, and does not result from a change in mRNA stability (Krig and Rice, 2000). Interestingly, the data also indicate that TCDD does not alter binding to and activation of the RARE, as there was no effect of TCDD on an RARE–luciferase construct transfected into these cells (Krig and Rice, 2000). Therefore, the mechanism of TCDD-induced interference of atRA-induced transglutaminase expression is still unknown. TCDD also demonstrates an inhibitory action toward other atRA target genes, including RDH9 (Tijet et al., 2006), and CRABPII (Weston et al., 1995). TCDD activates expression of retinal oxidase, the enzyme that catalyzes the conversion of retinal to RA. However, cotreatment with atRA and TCDD results in the downregulation of retinal oxidase expression and activity (Yang et al., 2005). Although the majority of data indicate that TCDD/AhR inhibit RA-mediated gene expression, there is growing evidence indicating that the interaction is more complex and may be tissue and cell-type specific.

Data from AhR knockout mice support the hypothesis that the AhR pathway interferes with expression of RA pathway target genes. For example, the expression of CRBPI is higher in the livers of AhR null animals than in their wild-type counterparts (Andreola et al., 1997). Interestingly, TCDD exposure does not appear to alter CRBPI expression in mice (Hoegberg et al., 2005). In addition, atRA levels are elevated in the livers of AhR null mice in comparison to wild-type mice (Andreola et al., 1997), which is coupled to a downregulation in CYP2C39 mRNA expression in the AhR null animals (Andreola et al., 2004). These data suggest that the AhR pathway, in the absence of exogenous ligand, is inhibitory toward the basal expression of genes that encode for proteins critical for retinoid homeostasis.

Conversely, the RA pathway also has an inhibitory effect on AhR-mediated transcription, and one of the most extensively studied is the effect of atRA on expression of CYP1A1. Because of the presence of an RARE in the human CYP1A1 promoter, it was originally postulated that atRA would enhance CYP1A1 expression. This was supported by findings demonstrating that the CYP1A1 RARE is able to bind nuclear proteins as well as mediate atRA-induced expression of a CYP1A1 reporter construct (Vecchini et al., 1994). However, studies in the expression of the endogenous CYP1A1 gene did not support this conclusion: neither mouse embryos nor Hepa-1c1c7 cells exposed to atRA show induction of endogenous CYP1A1 (Soprano et al., 2001). It is now accepted that atRA exposure is inhibitory to xenobiotic-induced CYP1A1 expression and activity, and that this inhibition is mediated through the RARE in the promoter (Wanner et al., 1996). In support of this conclusion, RAR$\alpha^{-/-}$ null animals display an increase in hepatic CYP1A1 activity after TCDD treatment compared to wild-type mice, suggesting that RARα may play an inhibitory role in TCDD-mediated CYP1A1 gene regulation (Hoegberg et al., 2005).

In addition to the effects of these pathways on target genes containing either the XRE or RARE elements, other target genes for both pathways have been identified that may be important for mediating some of the lesions observed following exposure to TCDD and atRA. Examples of potential target genes are the enzymes that mediate matrix metabolism, including the matrix metalloproteinases (MMPs). MMPs are a family of endopeptidases that mediate the cleavage of proteins involved in tissue structure, such as type I and type IV collagen. Further, these enzymes are also involved in the regulation of proliferation and angiogenesis through the cleavage and release of growth factors and receptors from the extracellular matrix and the cell surface (reviewed in Brinckerhoff and Matrisian, 2002). It is long established that atRA exposure alters the expression of MMPs in a variety of cell types, primarily downregulating expression through interference with the AP-1-signaling pathway (Vincenti et al., 1996). Data indicate that TCDD also modulates the expression and activity of the MMPs (Murphy et al., 2004; Villano et al., 2006). Interestingly, cotreatment with TCDD and atRA in normal human keratinocytes results in an enhancement of MMP-1 expression over exposure to TCDD alone. The coactivation of atRA and TCDD was also observed for PAI-2, a regulator of matrix remodeling, indicating that atRA/TCDD coactivation is not limited to MMPs. The induction of MMP-1 by cotreatment with atRA and TCDD does not rely on transcriptional interaction between the RARs and AhR, but instead is mediated through two distinct mechanisms: TCDD-induced transcription and atRA enhancement of MMP-1 mRNA stability (Murphy et al., 2004).

Although it is clear that there are interactions between these pathways at the level of transcriptional activation, it is unclear how these interactions are accomplished. Although atRA inhibition of CYP1A1 is most likely mediated by steric interference between proteins binding to the XRE and RARE, not all genes that are coregulated have both XREs and RAREs. A potential mechanism underlying changes in target gene expression by AhR and RA pathway interaction may be through changes in receptor availability. AhR availability in the cell is mediated by transcriptional and posttranslational mechanisms. Further, the targeted degradation of the AhR protein is also considered an important mechanism in regulating this pathway (Pollenz, 2002). Studies using the murine AhR promoter demonstrate that treatment of a murine epidermal cell line with atRA results in reduced AhR promoter activity (FitzGerald et al., 1996). Data from human keratinocytes did not demonstrate any change in either AhR or Arnt mRNA expression following atRA exposure (Murphy et al., 2004). This difference may be a consequence of species-specific differences between the human and murine AhR promoter activity or from a difference in atRA responsiveness of the human versus the murine cell line tested. In support of this idea is the fact that the murine cell line (JB6-C1 41-5a) used in the reported studies is highly responsive to RA (FitzGerald et al., 1996).

TCDD and the AhR pathway are known to alter the expression of the RARs and RXRs, although the effect of TCDD on *RAR* and *RXR* gene expression is receptor and cell-type dependent. RARβ expression is inhibited by TCDD exposure of embryonic palate mesenchymal cells (Weston et al., 1995). However, in SCC12Y cells TCDD treatment results in a decreased binding of atRA to RARα without any change in *RARα* gene expression (Lorick et al., 1998). In normal human keratinocytes, TCDD treatment results in an increase in RARγ and RXRα mRNA levels (Murphy et al., 2004). Therefore, one way in which TCDD may alter atRA target gene expression in some cell types may be through alterations of the receptor availability.

Activation of gene expression by the AhR/Arnt- or RA-signaling pathways requires the recruitment of coactivators and general transcription factors to the promoter region (Hankinson, 2005; Wei, 2003). The recruitment of coactivators to target genes can either enable chromatin remodeling or aid in recruiting basal transcription machinery to the promoter of the target gene (Chen, 2000). Coactivators are classified based on the mechanism used to induce transcription. HAT coactivators transfer an acetyl group onto specific lysine residues of histone tails destabilizing chromatin structure while histone methyltransferases (HMTs) modify arginines (to enhance) or lysines (to repress) of histone tails with methyl groups. Phosphorylation of histone tail serine residues as well as ubiquitination also serves as potential signals for altering chromatin structure. These signals serve to facilitate access to the DNA by destabilizing local chromatin structure through histone modification as well as being markers to recruit other coactivator proteins to the modified sites. Corepressors function by recruiting a complex of silencing proteins to the promoter region, including histone deacetylases (HDACs) (Baniahmad, 2005). A list of coactivators known to interact with the RA and AhR pathways are shown in Table II.

The importance of nucleosomal structure on TCDD-mediated CYP1A1 transcription was demonstrated in a study by Morgan and Whitlock (1992) identifying a nucleosome structure associated with CYP1A1 mouse promoter/enhancer region that is altered following TCDD treatment. There is also a concomitant increase in protection of this area located at -40 and -60 bp of the promoter which contains the TATAAA box and an NF-1-like recognition motif. This indicates that activated AhR alters nucleosome structure to facilitate transcriptional activation and suggests that coactivator or corepressors may be involved in TCDD-mediated CYP1A1 expression. Indeed, coactivator estrogen receptor associating protein 140 (ERAP 140) and the corepressor SMRT both physically interact with AhR/Arnt transcription factor complex and are able to increase AhR/Arnt binding to an XRE (Nguyen et al., 1999). However, the exact nature of the involvement of SMRT in AhR-mediated transcription is not yet elucidated. One study indicates that SMRT acts to inhibit AhR-mediated transcription (Nguyen et al., 1999);

while data from another study demonstrates that overexpression of SMRT activates TCDD-mediated transcription in some cell types and inhibits it in others (Rushing and Denison, 2002). These data indicate that SMRT–AhR interaction has a role in mediating AhR transcriptional activation, and suggest that the interaction of AhR and SMRT is dependent on cell-type-specific signals or factors.

Recent evidence indicates that corepressors may be a link between the AhR and RA expression pathways. The SMRT corepressor is known to interact with both the AhR and RARs and modulate their transactivating function (Nguyen et al., 1999; Rushing and Denison, 2002; Widerak et al., 2005). Further, SMRT may also be involved in TCDD-mediated effects on RAR binding and transactivation through the RARE. It has been known for some time that in some cell types, TCDD is able to activate expression of RARE-driven reporter constructs (Vecchini et al., 1994; Widerak et al., 2005). However, the mechanism of this activation was unknown. Data show that TCDD activation of the RARE–CAT construct is inhibited by cotransfection with an expression vector containing the SMRT corepressor (Widerak et al., 2005). Taken together these data suggest that the involvement of the corepressor as well as the coactivator proteins may provide a molecular pathway for the transcriptional cross talk between the AhR and RA pathways.

The data presented in this chapter demonstrate both direct and indirect interactions between the AhR- and RA-signaling pathways. These interactions include changes in the availability of atRA in the liver and extrahepatic tissues by AhR-mediated regulation of atRA synthesis and metabolism, as well as on storage and transport. Further, these two pathways directly impact each others signaling pathways through alterations in receptor availability and modulation of transcriptional regulation. Although it appears that the intersection of these two pathways may be mediated by specific coactivator and corepressor proteins, the exact mechanism is yet undefined. However, it is clear that a portion of toxicity related to TCDD and related congeners is mediated through their effect on RA homeostasis and on the atRA-signaling pathway.

REFERENCES

Abbott, B. D., and Birnbaum, L. S. (1990). Retinoic acid-induced alterations in the expression of growth factors in embryonic mouse palatal shelves. *Teratology* **42**, 597–610.

Abbott, B. D., and Buckalew, A. R. (1992). Embryonic palatal responses to teratogens in serum-free organ culture. *Teratology* **45**, 369–382.

Achkar, C. C., Derguini, F., Blumberg, B., Langston, A., Levin, A. A., Speck, J., Evans, R. M., Bolado, J., Jr., Nakanishi, K., Buck, J., and Gudas, L. J. (1996). 4-Oxoretinol, a new natural ligand and transactivator of the retinoic acid receptors. *Proc. Natl. Acad. Sci. USA* **93**, 4879–4884.

Ahmad, M., Nicholls, P. J., Smith, H. J., and Ahmadi, M. (2000). Effect of P450 isozyme-selective inhibitors on *in-vitro* metabolism of retinoic acid by rat hepatic microsomes. *J. Pharm. Pharmacol.* **52**, 311–314.

Ahmed, S., Shibazaki, M., Takeuchi, T., and Kikuchi, H. (2005). Protein kinase Ctheta activity is involved in the 2,3,7,8-tetrachlorodibenzo-p-dioxin-induced signal transduction pathway leading to apoptosis in L-MAT, a human lymphoblastic T-cell line. *FEBS J.* **272**, 903–915.

Andreola, F., Fernandez-Salguero, P. M., Chiantore, M. V., Petkovich, M. P., Gonzalez, F. J., and De Luca, L. M. (1997). Aryl hydrocarbon receptor knockout mice (AHR–/–) exhibit liver retinoid accumulation and reduced retinoic acid metabolism. *Cancer Res.* **57**, 2835–2838.

Andreola, F., Calvisi, D. F., Elizondo, G., Jakowlew, S. B., Mariano, J., Gonzalez, F. J., and De Luca, L. M. (2004). Reversal of liver fibrosis in aryl hydrocarbon receptor null mice by dietary vitamin A depletion. *Hepatology* **39**, 157–166.

Aoki, Y. (2001). Polychlorinated biphenyls, polychlorinated dibenzo-p-dioxins, and polychlorinated dibenzofurans as endocrine disrupters—what we have learned from Yusho disease. *Environ. Res.* **86**, 2–11.

Baniahmad, A. (2005). Nuclear hormone receptor co-repressors. *J. Steroid Biochem. Mol. Biol.* **93**, 89–97.

Bank, P. A., Salyers, K. L., and Zile, M. H. (1989). Effect of tetrachlorodibenzo-p-dioxin (TCDD) on the glucuronidation of retinoic acid in the rat. *Biochim. Biophys. Acta* **993**, 1–6.

Beischlag, T. V., Wang, S., Rose, D. W., Torchia, J., Reisz-Porszasz, S., Muhammad, K., Nelson, W. E., Probst, M. R., Rosenfeld, M. G., and Hankinson, O. (2002). Recruitment of the NCoA/SRC-1/p160 family of transcriptional coactivators by the aryl hydrocarbon receptor/aryl hydrocarbon receptor nuclear translocator complex. *Mol. Cell. Biol.* **22**, 4319–4333.

Bertazzi, P. A., Zocchetti, C., Pesatori, A. C., Guercilena, S., Sanarico, M., and Radice, L. (1989). Mortality in an area contaminated by TCDD following an industrial incident. *Med. Lav.* **80**, 316–329.

Bertazzi, P. A., Consonni, D., Bachetti, S., Rubagotti, M., Baccarelli, A., Zocchetti, C., and Pesatori, A. C. (2001). Health effects of dioxin exposure: A 20-year mortality study. *Am. J. Epidemiol.* **153**, 1031–1044.

Birnbaum, L. S., Harris, M. W., Stocking, L. M., Clark, A. M., and Morrissey, R. E. (1989). Retinoic acid and 2,3,7,8-tetrachlorodibenzo-p-dioxin selectively enhance teratogenesis in C57BL/6N mice. *Toxicol. Appl. Pharmacol.* **98**, 487–500.

Blaner, W. S. (2001). Cellular metabolism and actions of 13-cis-retinoic acid. *J. Am. Acad. Dermatol.* **45**, S129–S135.

Blumberg, B., Bolado, J., Jr., Derguini, F., Craig, A. G., Moreno, T. A., Chakravarti, D., Heyman, R. A., Buck, J., and Evans, R. M. (1996). Novel retinoic acid receptor ligands in *Xenopus* embryos. *Proc. Natl. Acad. Sci. USA* **93**, 4873–4878.

Boerman, M. H., and Napoli, J. L. (1991). Cholate-independent retinyl ester hydrolysis. Stimulation by Apo-cellular retinol-binding protein. *J. Biol. Chem.* **266**, 22273–22278.

Boerman, M. H., and Napoli, J. L. (1996). Cellular retinol-binding protein-supported retinoic acid synthesis. Relative roles of microsomes and cytosol. *J. Biol. Chem.* **271**, 5610–5616.

Brinckerhoff, C. E., and Matrisian, L. M. (2002). Matrix metalloproteinases: A tail of a frog that became a prince. *Nat. Rev. Mol. Cell. Biol.* **3**, 207–214.

Brouwer, A., Hakansson, H., Kukler, A., Van den Berg, K. J., and Ahlborg, U. G. (1989). Marked alterations in retinoid homeostasis of Sprague-Dawley rats induced by a single i.p. dose of 10 micrograms/kg of 2,3,7,8-tetrachlorodibenzo-p-dioxin. *Toxicology* **58**, 267–283.

Carver, L. A., and Bradfield, C. A. (1997). Ligand-dependent interaction of the aryl hydrocarbon receptor with a novel immunophilin homolog *in vivo*. *J. Biol. Chem.* **272**, 11452–11456.

Carver, L. A., LaPres, J. J., Jain, S., Dunham, E. E., and Bradfield, C. A. (1998). Characterization of the Ah receptor-associated protein, ARA9. *J. Biol. Chem.* **273**, 33580–33587.

Chambon, P. (1996). A decade of molecular biology of retinoic acid receptors. *FASEB J.* **10**, 940–954.

Chen, H., and Juchau, M. R. (1997). Glutathione S-transferases act as isomerases in isomerization of 13-cis-retinoic acid to all-trans-retinoic acid *in vitro*. *Biochem. J.* **327**(Pt. 3), 721–726.

Chen, H., and Juchau, M. R. (1998a). Biotransformation of 13-cis- and 9-cis-retinoic acid to all-trans-retinoic acid in rat conceptal homogenates. Evidence for catalysis by a conceptal isomerase. *Drug Metab. Dispos.* **26**, 222–228.

Chen, H., and Juchau, M. R. (1998b). Recombinant human glutathione S-transferases catalyse enzymic isomerization of 13-cis-retinoic acid to all-trans-retinoic acid *in vitro*. *Biochem. J.* **336**(Pt. 1), 223–226.

Chen, H., Howald, W. N., and Juchau, M. R. (2000). Biosynthesis of all-trans-retinoic acid from all-trans-retinol: Catalysis of all-trans-retinol oxidation by human P-450 cytochromes. *Drug Metab. Dispos.* **28**, 315–322.

Chen, J. D. (2000). Steroid/nuclear receptor coactivators. *Vitam. Horm.* **58**, 391–448.

Chen, J. D., and Li, H. (1998). Coactivation and corepression in transcriptional regulation by steroid/nuclear hormone receptors. *Crit. Rev. Eukaryot. Gene Expr.* **8**, 169–190.

Choudhary, D., Jansson, I., Schenkman, J. B., Sarfarazi, M., and Stoilov, I. (2003). Comparative expression profiling of 40 mouse cytochrome P450 genes in embryonic and adult tissues. *Arch. Biochem. Biophys.* **414**, 91–100.

Choudhary, D., Jansson, I., Stoilov, I., Sarfarazi, M., and Schenkman, J. B. (2005). Expression patterns of mouse and human CYP orthologs (families 1–4) during development and in different adult tissues. *Arch. Biochem. Biophys.* **436**, 50–61.

Chu, I., Lecavalier, P., Hakansson, H., Yagminas, A., Valli, V. E., Poon, P., and Feeley, M. (2001). Mixture effects of 2,3,7,8-tetrachlorodibenzo-p-dioxin and polychlorinated biphenyl congeners in rats. *Chemosphere* **43**, 807–814.

Crews, S. T., and Fan, C. M. (1999). Remembrance of things PAS: Regulation of development by bHLH-PAS proteins. *Curr. Opin. Genet. Dev.* **9**, 580–587.

Denison, M. S., Pandini, A., Nagy, S. R., Baldwin, E. P., and Bonati, L. (2002). Ligand binding and activation of the Ah receptor. *Chem. Biol. Interact.* **141**, 3–24.

DeVito, M. J., Birnbaum, L. S., Farland, W. H., and Gasiewicz, T. A. (1995). Comparisons of estimated human body burdens of dioxinlike chemicals and TCDD. *Environ. Health Perspect.* **103**, 820–831.

Dragnev, K. H., Rigas, J. R., and Dmitrovsky, E. (2000). The retinoids and cancer prevention mechanisms. *Oncologist* **5**, 361–368.

Du, L., Neis, M. M., Ladd, P. A., Lanza, D. L., Yost, G. S., and Keeney, D. S. (2005). Effects of the differentiated keratinocyte phenotype on expression levels of CYP1–4 family genes in human skin cells. *Toxicol. Appl. Pharmacol.* **213**(2), 135–144.

Duester, G. (2000). Families of retinoid dehydrogenases regulating vitamin A function: Production of visual pigment and retinoic acid. *Eur. J. Biochem.* **267**, 4315–4324.

Fiorella, P. D., Olson, J. R., and Napoli, J. L. (1995). 2,3,7,8-Tetrachlorodibenzo-p-dioxin induces diverse retinoic acid metabolites in multiple tissues of the Sprague-Dawley rat. *Toxicol. Appl. Pharmacol.* **134**, 222–228.

Fisher, G. J., and Voorhees, J. J. (1996). Molecular mechanisms of retinoid actions in skin. *FASEB J.* **10**, 1002–1013.

FitzGerald, C. T., Fernandez-Salguero, P., Gonzalez, F. J., Nebert, D. W., and Puga, A. (1996). Differential regulation of mouse Ah receptor gene expression in cell lines of different tissue origins. *Arch. Biochem. Biophys.* **333**, 170–178.

Fletcher, N., Hanberg, A., and Hakansson, H. (2001). Hepatic vitamin a depletion is a sensitive marker of 2,3,7,8-tetrachlorodibenzo-p-dioxin (TCDD) exposure in four rodent species. *Toxicol. Sci.* **62**, 166–175.

Fletcher, N., Giese, N., Schmidt, C., Stern, N., Lind, P. M., Viluksela, M., Tuomisto, J. T., Tuomisto, J., Nau, H., and Hakansson, H. (2005). Altered retinoid metabolism in female Long-Evans and Han/Wistar rats following long-term 2,3,7,8-tetrachlorodibenzo-p-dioxin (TCDD)-treatment. *Toxicol. Sci.* **86**, 264–272.

Fujii-Kuriyama, Y., Imataka, H., Sogawa, K., Yasumoto, K., and Kikuchi, Y. (1992). Regulation of CYP1A1 expression. *FASEB J.* **6**, 706–710.

Gamble, M. V., Shang, E., Zott, R. P., Mertz, J. R., Wolgemuth, D. J., and Blaner, W. S. (1999). Biochemical properties, tissue expression, and gene structure of a short chain dehydrogenase/reductase able to catalyze cis-retinol oxidation. *J. Lipid Res.* **40**, 2279–2292.

Gambone, C. J., Hutcheson, J. M., Gabriel, J. L., Beard, R. L., Chandraratna, R. A., Soprano, K. J., and Soprano, D. R. (2002). Unique property of some synthetic retinoids: Activation of the aryl hydrocarbon receptor pathway. *Mol. Pharmacol.* **61**, 334–342.

Garrison, P. M., Tullis, K., Aarts, J. M., Brouwer, A., Giesy, J. P., and Denison, M. S. (1996). Species-specific recombinant cell lines as bioassay systems for the detection of 2,3,7,8-tetrachlorodibenzo-p-dioxin-like chemicals. *Fundam. Appl. Toxicol.* **30**, 194–203.

Genchi, G., Wang, W., Barua, A., Bidlack, W. R., and Olson, J. A. (1996). Formation of beta-glucuronides and of beta-galacturonides of various retinoids catalyzed by induced and noninduced microsomal UDP-glucuronosyltransferases of rat liver. *Biochim. Biophys. Acta* **1289**, 284–290.

Ghyselinck, N. B., Bavik, C., Sapin, V., Mark, M., Bonnier, D., Hindelang, C., Dierich, A., Nilsson, C. B., Hakansson, H., Sauvant, P., Azais-Braesco, V., Frasson, M., *et al.* (1999). Cellular retinol-binding protein I is essential for vitamin A homeostasis. *EMBO J.* **18**, 4903–4914.

Glass, C. K., Rosenfeld, M. G., Rose, D. W., Kurokawa, R., Kamei, Y., Xu, L., Torchia, J., Ogliastro, M. H., and Westin, S. (1997). Mechanisms of transcriptional activation by retinoic acid receptors. *Biochem. Soc. Trans.* **25**, 602–605.

Gonzalez, F. J., and Fernandez-Salguero, P. (1998). The aryl hydrocarbon receptor: Studies using the AHR-null mice. *Drug Metab. Dispos.* **26**, 1194–1198.

Goodman, D. S. (1962). The metabolism of chylomicron cholesterol ester in the rat. *J. Clin. Invest.* **41**, 1886–1896.

Goodman, D. W., Huang, H. S., and Shiratori, T. (1965). Tissue distribution and metabolism of newly absorbed vitamin a in the rat. *J. Lipid Res.* **6**, 390–396.

Hakansson, H., and Ahlborg, U. G. (1985). The effect of 2,3,7,8-tetrachlorodibenzo-p-dioxin (TCDD) on the uptake, distribution and excretion of a single oral dose of [11,12-3H]retinyl acetate and on the vitamin A status in the rat. *J. Nutr.* **115**, 759–771.

Hakansson, H., and Hanberg, A. (1989). The distribution of [14C]-2,3,7,8-tetrachlorodibenzo-p-dioxin (TCDD) and its effect on the vitamin A content in parenchymal and stellate cells of rat liver. *J. Nutr.* **119**, 573–580.

Hakansson, H., Waern, F., and Ahlborg, U. G. (1987). Effects of 2,3,7,8-tetrachlorodibenzo-p-dioxin (TCDD) in the lactating rat on maternal and neonatal vitamin A status. *J. Nutr.* **117**, 580–586.

Hakansson, H., Johansson, L., Manzoor, E., and Ahlborg, U. G. (1991). Effects of 2,3,7,8-tetrachlorodibenzo-p-dioxin (TCDD) on the vitamin A status of Hartley guinea pigs, Sprague-Dawley rats, C57Bl/6 mice, DBA/2 mice, and Golden Syrian hamsters. *J. Nutr. Sci. Vitaminol. (Tokyo)* **37**, 117–138.

Hakansson, H., Manzoor, E., Trossvik, C., Ahlborg, U. G., Chu, I., and Villenueve, D. (1994). Effect on tissue vitamin A levels in the rat following subchronic exposure to four individual PCB congeners (IUPAC 77, 118, 126, and 153). *Chemosphere* **29**, 2309–2313.

Hanberg, A., Kling, L., and Hakansson, H. (1996). Effect of 2,3,7,8-tetrachlorodibenzo-P-dioxin (TCDD) on the hepatic stellate cell population in the rat. *Chemosphere* **32**, 1225–1233.

Hankinson, O. (1995). The aryl hydrocarbon receptor complex. *Ann. Rev. Pharmacol. Toxicol.* **35**, 307–340.

Hankinson, O. (2005). Role of coactivators in transcriptional activation by the aryl hydrocarbon receptor. *Arch. Biochem. Biophys.* **433**, 379–386.

Herr, F. M., and Ong, D. E. (1992). Differential interaction of lecithin-retinol acyltransferase with cellular retinol binding proteins. *Biochemistry* **31**, 6748–6755.

Hoegberg, P., Schmidt, C. K., Nau, H., Ross, A. C., Zolfaghari, R., Fletcher, N., Trossvik, C., Nilsson, C. B., and Hakansson, H. (2003). 2,3,7,8-Tetrachlorodibenzo-p-dioxin induces lecithin: Retinol acyltransferase transcription in the rat kidney. *Chem. Biol. Interact.* **145**, 1–16.

Hoegberg, P., Schmidt, C. K., Fletcher, N., Nilsson, C. B., Trossvik, C., Gerlienke Schuur, A., Brouwer, A., Nau, H., Ghyselinck, N. B., Chambon, P., and Hakansson, H. (2005). Retinoid status and responsiveness to 2,3,7,8-tetrachlorodibenzo-p-dioxin (TCDD) in mice lacking retinoid binding protein or retinoid receptor forms. *Chem. Biol. Interact.* **156**, 25–39.

Hoffer, A., Chang, C. Y., and Puga, A. (1996). Dioxin induces transcription of fos and jun genes by Ah receptor-dependent and -independent pathways. *Toxicol. Appl. Pharmacol.* **141**, 238–247.

Honkakoski, P., and Negishi, M. (2000). Regulation of cytochrome P450 (CYP) genes by nuclear receptors. *Biochem. J.* **347**, 321–337.

Imai, S., Okuno, M., Moriwaki, H., Muto, Y., Murakami, K., Shudo, K., Suzuki, Y., and Kojima, S. (1997). 9,13-di-cis-Retinoic acid induces the production of tPA and activation of latent TGF-beta via RAR alpha in a human liver stellate cell line, LI90. *FEBS Lett.* **411**, 102–106.

Jamsa, T., Viluksela, M., Tuomisto, J. T., Tuomisto, J., and Tuukkanen, J. (2001). Effects of 2,3,7,8-tetrachlorodibenzo-p-dioxin on bone in two rat strains with different aryl hydrocarbon receptor structures. *J. Bone Miner. Res.* **16**, 1812–1820.

Jurek, M. A., Powers, R. H., Gilbert, L. G., and Aust, S. D. (1990). The effect of TCDD on acyl CoA:retinol acyltransferase activity and vitamin A accumulation in the kidney of male Sprague-Dawley rats. *J. Biochem. Toxicol.* **5**, 155–160.

Kalin, J. R., Wells, M. J., and Hill, D. L. (1984). Effects of phenobarbital, 3-methylcholanthrene, and retinoid pretreatment on disposition of orally administered retinoids in mice. *Drug Metab. Dispos.* **12**, 63–67.

Kazlauskas, A., Sundstrom, S., Poellinger, L., and Pongratz, I. (2001). The hsp90 chaperone complex regulates intracellular localization of the dioxin receptor. *Mol. Cell. Biol.* **21**, 2594–2607.

Kelley, S. K., Nilsson, C. B., Green, M. H., Green, J. B., and Hakansson, H. (1998). Use of model-based compartmental analysis to study effects of 2,3,7,8-tetrachlorodibenzo-p-dioxin on vitamin A kinetics in rats. *Toxicol. Sci.* **44**, 1–13.

Kelley, S. K., Nilsson, C. B., Green, M. H., Green, J. B., and Hakansson, H. (2000). Mobilization of vitamin A stores in rats after administration of 2,3,7,8-tetrachlorodibenzo-p-dioxin: A kinetic analysis. *Toxicol. Sci.* **55**, 478–484.

Kessler, F. K., and Ritter, J. K. (1997). Induction of a rat liver benzo[a]pyrene-trans-7,8-dihydrodiol glucuronidating activity by oltipraz and beta-naphthoflavone. *Carcinogenesis* **18**, 107–114.

Ko, H. P., Okino, S. T., Ma, Q., and Whitlock, J. P., Jr. (1997). Transactivation domains facilitate promoter occupancy for the dioxin-inducible CYP1A1 gene *in vivo*. *Mol. Cell. Biol.* **17**, 3497–3507.

Kobayashi, A., Numayama-Tsuruta, K., Sogawa, K., and Fujii-Kuriyama, Y. (1997). CBP/p300 functions as a possible transcriptional coactivator of Ah receptor nuclear translocator (Arnt). *J. Biochem. (Tokyo)* **122**, 703–710.

Kojima, R., Fujimori, T., Kiyota, N., Toriya, Y., Fukuda, T., Ohashi, T., Sato, T., Yoshizawa, Y., Takeyama, K., Mano, H., Soiching, M., and Kato, S. (1994). In vivo isomerization of retinoic acids. Rapid isomer exchange and gene expression. *J. Biol. Chem.* **269,** 32700–32707.

Kolluri, S. K., Weiss, C., Koff, A., and Gottlicher, M. (1999). p27(Kip1) induction and inhibition of proliferation by the intracellular Ah receptor in developing thymus and hepatoma cells. *Genes Dev.* **13,** 1742–1753.

Kondraganti, S. R., Fernandez-Salguero, P., Gonzalez, F. J., Ramos, K. S., Jiang, W., and Moorthy, B. (2003). Polycyclic aromatic hydrocarbon-inducible DNA adducts: Evidence by 32P-postlabeling and use of knockout mice for Ah receptor-independent mechanisms of metabolic activation *in vivo*. *Int. J. Cancer* **103,** 5–11.

Krig, S. R., and Rice, R. H. (2000). TCDD suppression of tissue transglutaminase stimulation by retinoids in malignant human keratinocytes. *Toxicol. Sci.* **56,** 357–364.

Kumar, M. B., Tarpey, R. W., and Perdew, G. H. (1999). Differential recruitment of coactivator RIP140 by Ah and estrogen receptors. Absence of a role for LXXLL motifs. *J. Biol. Chem.* **274,** 22155–22164.

Lampen, A., Meyer, S., Arnhold, T., and Nau, H. (2000). Metabolism of vitamin A and its active metabolite all-trans-retinoic acid in small intestinal enterocytes. *J. Pharmacol. Exp. Ther.* **295,** 979–985.

LaPres, J. J., Glover, E., Dunham, E. E., Bunger, M. K., and Bradfield, C. A. (2000). ARA9 modifies agonist signaling through an increase in cytosolic aryl hydrocarbon receptor. *J. Biol. Chem.* **275,** 6153–6159.

Leo, M. A., Iida, S., and Lieber, C. S. (1984). Retinoic acid metabolism by a system reconstituted with cytochrome P-450. *Arch. Biochem. Biophys.* **234,** 305–312.

Li, H., Leo, C., Schroen, D. J., and Chen, J. D. (1997). Characterization of receptor interaction and transcriptional repression by the corepressor SMRT. *Mol. Endocrinol.* **11,** 2025–2037.

Lind, P. M., Larsson, S., Johansson, S., Melhus, H., Wikstrom, M., Lindhe, O., and Orberg, J. (2000). Bone tissue composition, dimensions and strength in female rats given an increased dietary level of vitamin A or exposed to 3,3′,4,4′,5-pentachlorobiphenyl (PCB126) alone or in combination with vitamin C. *Toxicology* **151,** 11–23.

Lindros, K. O., Oinonen, T., Kettunen, E., Sippel, H., Muro-Lupori, C., and Koivusalo, M. (1998). Aryl hydrocarbon receptor-associated genes in rat liver: Regional coinduction of aldehyde dehydrogenase 3 and glutathione transferase Ya. *Biochem. Pharmacol.* **55,** 413–421.

Long, W. P., Pray-Grant, M., Tsai, J. C., and Perdew, G. H. (1998). Protein kinase C activity is required for aryl hydrocarbon receptor pathway-mediated signal transduction. *Mol. Pharmacol.* **53,** 691–700.

Lorick, K. L., Toscano, D. L., and Toscano, W. A., Jr. (1998). 2,3,7,8-Tetrachlorodibenzo-p-dioxin alters retinoic acid receptor function in human keratinocytes. *Biochem. Biophys. Res. Commun.* **243,** 749–752.

Loudig, O., Babichuk, C., White, J., Abu-Abed, S., Mueller, C., and Petkovich, M. (2000). Cytochrome P450RAI(CYP26) promoter: A distinct composite retinoic acid response element underlies the complex regulation of retinoic acid metabolism. *Mol. Endocrinol.* **14,** 1483–1497.

Mahon, M. J., and Gasiewicz, T. A. (1995). Ah receptor phosphorylation: Localization of phosphorylation sites to the C-terminal half of the protein. *Arch. Biochem. Biophys.* **318,** 166–174.

Malik, N., and Owens, I. S. (1981). Genetic regulation of bilirubin-UDP-glucuronosyltransferase induction by polycyclic aromatic compounds and phenobarbital in mice. Association with aryl hydrocarbon (benzo[a]pyrene) hydroxylase induction. *J. Biol. Chem.* **256,** 9599–9604.

Mangelsdorf, D. J., Thummel, C., Beato, M., Herrlich, P., Schutz, G., Umesono, K., Blumberg, B., Kastner, P., Mark, M., Chambon, P., and Evans, R. M. (1995). The nuclear receptor superfamily: The second decade. *Cell* **83**, 835–839.

Marchetti, M. N., Sampol, E., Bun, H., Scoma, H., Lacarelle, B., and Durand, A. (1997). In vitro metabolism of three major isomers of retinoic acid in rats. Intersex and interstrain comparison. *Drug Metab. Dispos.* **25**, 637–646.

Mark, M., Ghyselinck, N. B., Wendling, O., Dupe, V., Mascrez, B., Kastner, P., and Chambon, P. (1999). A genetic dissection of the retinoid signalling pathway in the mouse. *Proc. Nutr. Soc.* **58**, 609–613.

Martini, R., and Murray, M. (1994). Retinal dehydrogenation and retinoic acid 4-hydroxylation in rat hepatic microsomes: Developmental studies and effect of foreign compounds on the activities. *Biochem. Pharmacol.* **47**, 905–909.

Matikainen, T., Perez, G. I., Jurisicova, A., Pru, J. K., Schlezinger, J. J., Ryu, H. Y., Laine, J., Sakai, T., Korsmeyer, S. J., Casper, R. F., Sherr, D. H., and Tilly, J. L. (2001). Aromatic hydrocarbon receptor-driven Bax gene expression is required for premature ovarian failure caused by biohazardous environmental chemicals. *Nat. Genet.* **28**, 355–360.

Matsushita, N., Sogawa, K., Ema, M., Yoshida, A., and Fujii-Kuriyama, Y. (1993). A factor binding to the xenobiotic responsive element (XRE) of P-4501A1 gene consists of at least two helix-loop-helix proteins, Ah receptor and Arnt. *J. Biol. Chem.* **268**, 21002–21006.

McSorley, L. C., and Daly, A. K. (2000). Identification of human cytochrome P450 isoforms that contribute to all-trans-retinoic acid 4-hydroxylation. *Biochem. Pharmacol.* **60**, 517–526.

Mimura, J., and Fujii-Kuriyama, Y. (2003). Functional role of AhR in the expression of toxic effects by TCDD. *Biochim. Biophys. Acta* **1619**, 263–268.

Moise, A. R., Kuksa, V., Blaner, W. S., Baehr, W., and Palczewski, K. (2005). Metabolism and transactivation activity of 13,14-dihydroretinoic acid. *J. Biol. Chem.* **280**, 27815–27825.

Morgan, J. E., and Whitlock, J. P., Jr. (1992). Transcription-dependent and transcription-independent nucleosome disruption induced by dioxin. *Proc. Natl. Acad. Sci. USA* **89**, 11622–11626.

Morita, A., and Nakano, K. (1982). Change in vitamin A content in tissues of rats fed on a vitamin A-free diet. *J. Nutr. Sci. Vitaminol. (Tokyo)* **28**, 343–350.

Morse, D. C., and Brouwer, A. (1995). Fetal, neonatal, and long-term alterations in hepatic retinoid levels following maternal polychlorinated biphenyl exposure in rats. *Toxicol. Appl. Pharmacol.* **131**, 175–182.

Mukerjee, D. (1998). Health impact of polychlorinated dibenzo-p-dioxins: A critical review. *J. Air Waste Manage. Assoc.* **48**, 157–165.

Munzel, P. A., Bruck, M., and Bock, K. W. (1994). Tissue-specific constitutive and inducible expression of rat phenol UDP-glucuronosyltransferase. *Biochem. Pharmacol.* **47**, 1445–1448.

Munzel, P. A., Schmohl, S., Heel, H., Kalberer, K., Bock-Hennig, B. S., and Bock, K. W. (1999). Induction of human UDP glucuronosyltransferases (UGT1A6, UGT1A9, and UGT2B7) by t-butylhydroquinone and 2,3,7,8-tetrachlorodibenzo-p-dioxin in Caco-2 cells. *Drug Metab. Dispos.* **27**, 569–573.

Murphy, K. A., Villano, C. M., Dorn, R., and White, L. A. (2004). Interaction between the aryl hydrocarbon receptor and retinoic acid pathways increases matrix metalloproteinase-1 expression in keratinocytes. *J. Biol. Chem.* **279**, 25284–25293.

Nagpal, S., and Chandraratna, R. A. (2000). Recent developments in receptor-selective retinoids. *Curr. Pharm. Des.* **6**, 919–931.

Napoli, J. L. (1999). Interactions of retinoid binding proteins and enzymes in retinoid metabolism. *Biochim. Biophys. Acta* **1440**, 139–162.

Nebert, D. W., and Russell, D. W. (2002). Clinical importance of the cytochromes P450. *Lancet* **360**, 1155–1162.

Nelson, D. R., Koymans, L., Kamataki, T., Stegeman, J. J., Feyereisen, R., Waxman, D. J., Waterman, M. R., Gotoh, O., Coon, M. J., Estabrook, R. W., Gunsalus, I. C., and Nebert, D. W.

(1996). P450 superfamily: Update on new sequences, gene mapping, accession numbers and nomenclature. *Pharmacogenetics* **6**, 1–42.
Nguyen, T. A., Hoivik, D., Lee, J. E., and Safe, S. (1999). Interactions of nuclear receptor coactivator/corepressor proteins with the aryl hydrocarbon receptor complex. *Arch. Biochem. Biophys.* **367**, 250–257.
Nilsson, A., Troen, G., Petersen, L. B., Reppe, S., Norum, K. R., and Blomhoff, R. (1997). Retinyl ester storage is altered in liver stellate cells and in HL60 cells transfected with cellular retinol-binding protein type I. *Int. J. Biochem. Cell Biol.* **29**, 381–389.
Nilsson, C. B., and Hakansson, H. (2002). The retinoid signaling system—a target in dioxin toxicity. *Crit. Rev. Toxicol.* **32**, 211–232.
Nilsson, C. B., Hoegberg, P., Trossvik, C., Azais-Braesco, V., Blaner, W. S., Fex, G., Harrison, E. H., Nau, H., Schmidt, C. K., van Bennekum, A. M., and Hakansson, H. (2000). 2,3,7,8-Tetrachlorodibenzo-p-dioxin increases serum and kidney retinoic acid levels and kidney retinol esterification in the rat. *Toxicol. Appl. Pharmacol.* **169**, 121–131.
Oberg, M., Bergander, L., Hakansson, H., Rannug, U., and Rannug, A. (2005). Identification of the tryptophan photoproduct 6-formylindolo[3,2-b]carbazole, in cell culture medium, as a factor that controls the background aryl hydrocarbon receptor activity. *Toxicol. Sci.* **85**, 935–943.
O'Byrne, S. M., Wongsiriroj, N., Libien, J., Vogel, S., Goldberg, I. J., Baehr, W., Palczewski, K., and Blaner, W. S. (2005). Retinoid absorption and storage is impaired in mice lacking lecithin:retinol acyltransferase (LRAT). *J. Biol. Chem.* **280**, 35647–35657.
Okuno, M., Sato, T., Kitamoto, T., Imai, S., Kawada, N., Suzuki, Y., Yoshimura, H., Moriwaki, H., Onuki, K., Masushige, S., Muto, Y., Friedman, S. L., *et al.* (1999). Increased 9,13-di-cis-retinoic acid in rat hepatic fibrosis: Implication for a potential link between retinoid loss and TGF-beta mediated fibrogenesis *in vivo. J. Hepatol.* **30**, 1073–1080.
Ott, M. G., and Zober, A. (1996). Cause specific mortality and cancer incidence among employees exposed to 2,3,7,8-TCDD after a 1953 reactor accident. *Occup. Environ. Med.* **53**, 606–612.
Park, J. H., Hahn, E. J., Kong, J. H., Cho, H. J., Yoon, C. S., Cheong, S. W., Oh, G. S., and Youn, H. J. (2003). TCDD-induced apoptosis in EL-4 cells deficient of the aryl hydrocarbon receptor and down-regulation of IGFBP-6 prevented the apoptotic cell death. *Toxicol. Lett.* **145**, 55–68.
Park, K. T., Mitchell, K. A., Huang, G., and Elferink, C. J. (2005a). The aryl hydrocarbon receptor predisposes hepatocytes to Fas-mediated apoptosis. *Mol. Pharmacol.* **67**, 612–622.
Park, S. J., Yoon, W. K., Kim, H. J., Son, H. Y., Cho, S. W., Jeong, K. S., Kim, T. H., Kim, S. H., Kim, S. R., and Ryu, S. Y. (2005b). 2,3,7,8-Tetrachlorodibenzo-p-dioxin activates ERK and p38 mitogen-activated protein kinases in RAW 264.7 cells. *Anticancer Res.* **25**, 2831–2836.
Peters, J. M., Narotsky, M. G., Elizondo, G., Fernandez-Salguero, P. M., Gonzalez, F. J., and Abbott, B. D. (1999). Amelioration of TCDD-induced teratogenesis in aryl hydrocarbon receptor (AhR)-null mice. *Toxicol. Sci.* **47**, 86–92.
Pijnappel, W. W., Folkers, G. E., de Jonge, W. J., Verdegem, P. J., de Laat, S. W., Lugtenburg, J., Hendriks, H. F., van der Saag, P. T., and Durston, A. J. (1998). Metabolism to a response pathway selective retinoid ligand during axial pattern formation. *Proc. Natl. Acad. Sci. USA* **95**, 15424–15429.
Pollenz, R. S. (2002). The mechanism of AH receptor protein down-regulation (degradation) and its impact on AH receptor-mediated gene regulation. *Chem. Biol. Interact.* **141**, 41–61.
Puga, A., Nebert, D. W., and Carrier, F. (1992). Dioxin induces expression of c-fos and c-jun proto-oncogenes and a large increase in transcription factor AP-1. *DNA Cell Biol.* **11**, 269–281.
Quadro, L., Blaner, W. S., Salchow, D. J., Vogel, S., Piantedosi, R., Gouras, P., Freeman, S., Cosma, M. P., Colantuoni, V., and Gottesman, M. E. (1999). Impaired retinal function and vitamin A availability in mice lacking retinol-binding protein. *EMBO J.* **18**, 4633–4644.

Radominska, A., Little, J. M., Lehman, P. A., Samokyszyn, V., Rios, G. R., King, C. D., Green, M. D., and Tephly, T. R. (1997). Glucuronidation of retinoids by rat recombinant UDP: Glucuronosyltransferase 1.1 (bilirubin UGT). *Drug Metab. Dispos.* **25**, 889–892.

Raner, G. M., Vaz, A. D., and Coon, M. J. (1996). Metabolism of all-trans, 9-cis, and 13-cis isomers of retinal by purified isozymes of microsomal cytochrome P450 and mechanism-based inhibition of retinoid oxidation by citral. *Mol. Pharmacol.* **49**, 515–522.

Ray, W. J., Bain, G., Yao, M., and Gottlieb, D. I. (1997). CYP26, a novel mammalian cytochrome P450, is induced by retinoic acid and defines a new family. *J. Biol. Chem.* **272**, 18702–18708.

Renaud, J. P., and Moras, D. (2000). Structural studies on nuclear receptors. *Cell. Mol. Life Sci.* **57**, 1748–1769.

Reyes, H., Reisz-Porszasz, S., and Hankinson, O. (1992). Identification of the Ah receptor nuclear translocator protein (Arnt) as a component of the DNA binding form of the Ah receptor. *Science* **256**, 1193–1195.

Roberts, E. S., Vaz, A. D., and Coon, M. J. (1992). Role of isozymes of rabbit microsomal cytochrome P-450 in the metabolism of retinoic acid, retinol, and retinal. *Mol. Pharmacol.* **41**, 427–433.

Rolland, R. M. (2000). A review of chemically-induced alterations in thyroid and vitamin A status from field studies of wildlife and fish. *J. Wildl. Dis.* **36**, 615–635.

Rowlands, J. C., McEwan, I. J., and Gustafsson, J. A. (1996). Trans-activation by the human aryl hydrocarbon receptor and aryl hydrocarbon receptor nuclear translocator proteins: Direct interactions with basal transcription factors. *Mol. Pharmacol.* **50**, 538–548.

Rubin, A. L., and Rice, R. H. (1988). 2,3,7,8-Tetrachlorodibenzo-p-dioxin and polycyclic aromatic hydrocarbons suppress retinoid-induced tissue transglutaminase in SCC-4 cultured human squamous carcinoma cells. *Carcinogenesis* **9**, 1067–1070.

Rushing, S. R., and Denison, M. S. (2002). The silencing mediator of retinoic acid and thyroid hormone receptors can interact with the aryl hydrocarbon (Ah) receptor but fails to repress Ah receptor-dependent gene expression. *Arch. Biochem. Biophys.* **403**, 189–201.

Saarikoski, S. T., Rivera, S. P., Hankinson, O., and Husgafvel-Pursiainen, K. (2005). CYP2S1: A short review. *Toxicol. Appl. Pharmacol.* **207**, 62–69.

Safe, S. (2001). Molecular biology of the Ah receptor and its role in carcinogenesis. *Toxicol. Lett.* **120**, 1–7.

Sanders, J. M., Burka, L. T., Smith, C. S., Black, W., James, R., and Cunningham, M. L. (2005). Differential expression of CYP1A, 2B, and 3A genes in the F344 rat following exposure to a polybrominated diphenyl ether mixture or individual components. *Toxicol. Sci.* **88**, 127–133.

Schmidt, C. K., Volland, J., Hamscher, G., and Nau, H. (2002). Characterization of a new endogenous vitamin A metabolite. *Biochim. Biophys. Acta* **1583**, 237–251.

Schmidt, C. K., Hoegberg, P., Fletcher, N., Nilsson, C. B., Trossvik, C., Hakansson, H., and Nau, H. (2003). 2,3,7,8-Tetrachlorodibenzo-p-dioxin (TCDD) alters the endogenous metabolism of all-trans-retinoic acid in the rat. *Arch. Toxicol.* **77**, 371–383.

Schule, R., Rangarajan, P., Yang, N., Kliewer, S., Ransone, L. J., Bolado, J., Verma, I. M., and Evans, R. M. (1991). Retinoic acid is a negative regulator of AP-1-responsive genes. *Proc. Natl. Acad. Sci. USA* **88**, 6092–6096.

Shertzer, H. G., Nebert, D. W., Puga, A., Ary, M., Sonntag, D., Dixon, K., Robinson, L. J., Cianciolo, E., and Dalton, T. P. (1998). Dioxin causes a sustained oxidative stress response in the mouse. *Biochem. Biophys. Res. Commun.* **253**, 44–48.

Sogawa, K., Iwabuchi, K., Abe, H., and Fujii-Kuriyama, Y. (1995). Transcriptional activation domains of the Ah receptor and Ah receptor nuclear translocator. *J. Cancer Res. Clin. Oncol.* **121**, 612–620.

Song, J., Clagett-Dame, M., Peterson, R. E., Hahn, M. E., Westler, W. M., Sicinski, R. R., and DeLuca, H. F. (2002). A ligand for the aryl hydrocarbon receptor isolated from lung. *Proc. Natl. Acad. Sci. USA* **99**, 14694–14699.

Soprano, D. R., and Soprano, K. J. (2003). Pharmacological doses of some synthetic retinoids can modulate both the aryl hydrocarbon receptor and retinoid receptor pathways. *J. Nutr.* **133,** 277S–281S.

Soprano, D. R., Gambone, C. J., Sheikh, S. N., Gabriel, J. L., Chandraratna, R. A., Soprano, K. J., and Kochhar, D. M. (2001). The synthetic retinoid AGN 193109 but not retinoic acid elevates CYP1A1 levels in mouse embryos and Hepa-1c1c7 cells. *Toxicol. Appl. Pharmacol.* **174,** 153–159.

Sporn, M. B., and Suh, N. (2000). Chemoprevention of cancer. *Carcinogenesis* **21,** 525–530.

Sporn, M. B., Roberts, A. B., and Goodman, D. S. (1994). "The Retinoids: Biology, Chemistry, and Medicine." Raven Press, New York.

Stohs, S. J. (1990). Oxidative stress induced by 2,3,7,8-tetrachlorodibenzo-p-dioxin (TCDD). *Free Radic. Biol. Med.* **9,** 79–90.

Stohs, S. J., Abbott, B. D., Lin, F. H., and Birnbaum, L. S. (1990). Induction of ethoxyresorufin-O-deethylase and inhibition of glucocorticoid receptor binding in skin and liver of haired and hairless HRS/J mice by topically applied 2,3,7,8-tetrachlorodibenzo-p-dioxin. *Toxicology* **65,** 123–136.

Sundaresan, P. R., and Bhat, P. V. (1982). Ion-pair high-pressure liquid chromatography of cis-trans isomers of retinoic acid in tissues of vitamin A-sufficient rats. *J. Lipid Res.* **23,** 448–455.

Sutter, T. R., Guzman, K., Dold, K. M., and Greenlee, W. F. (1991). Targets for dioxin: Genes for plasminogen activator inhibitor-2 and interleukin-1 beta. *Science* **254,** 415–418.

Swanson, H. I., and Yang, J. H. (1998). The aryl hydrocarbon receptor interacts with transcription factor IIB. *Mol. Pharmacol.* **54,** 671–677.

Thacher, S. M., Vasudevan, J., and Chandraratna, R. A. (2000). Therapeutic applications for ligands of retinoid receptors. *Curr. Pharm. Des.* **6,** 25–58.

Thunberg, T., Ahlborg, U. G., and Johnsson, H. (1979). Vitamin A (retinol) status in the rat after a single oral dose of 2,3,7,8-tetrachlorodibenzo-p-dioxin. *Arch. Toxicol.* **42,** 265–274.

Thunberg, T., Ahlborg, U. G., Hakansson, H., Krantz, C., and Monier, M. (1980). Effect of 2,3,7,8-tetrachlorodibenzo-p-dioxin on the hepatic storage of retinol in rats with different dietary supplies of vitamin A (retinol). *Arch. Toxicol.* **45,** 273–285.

Tijet, N., Boutros, P. C., Moffat, I. D., Okey, A. B., Tuomisto, J., and Pohjanvirta, R. (2006). Aryl hydrocarbon receptor regulates distinct dioxin-dependent and dioxin-independent gene batteries. *Mol. Pharmacol.* **69,** 140–153.

Tohkin, M., Fukuhara, M., Elizondo, G., Tomita, S., and Gonzalez, F. J. (2000). Aryl hydrocarbon receptor is required for p300-mediated induction of DNA synthesis by adenovirus E1A. *Mol. Pharmacol.* **58,** 845–851.

Tomita, S., Okuyama, E., Ohnishi, T., and Ichikawa, Y. (1996). Characteristic properties of a retinoic acid synthetic cytochrome P-450 purified from liver microsomes of 3-methylcholanthrene-induced rats. *Biochim. Biophys. Acta* **1290,** 273–281.

Urbach, J., and Rando, R. R. (1994a). Isomerization of all-trans-retinoic acid to 9-cis-retinoic acid. *Biochem. J.* **299**(Pt. 2), 459–465.

Urbach, J., and Rando, R. R. (1994b). Thiol dependent isomerization of all-trans-retinoic acid to 9-cis-retinoic acid. *FEBS Lett.* **351,** 429–432.

Van Birgelen, A. P., Smit, E. A., Kampen, I. M., Groeneveld, C. N., Fase, K. M., Van der Kolk, J., Poiger, H., Van den Berg, M., Koeman, J. H., and Brouwer, A. (1995a). Subchronic effects of 2,3,7,8-TCDD or PCBs on thyroid hormone metabolism: Use in risk assessment. *Eur. J. Pharmacol.* **293,** 77–85.

Van Birgelen, A. P., Van der Kolk, J., Fase, K. M., Bol, I., Poiger, H., Brouwer, A., and Van den Berg, M. (1995b). Subchronic dose-response study of 2,3,7,8-tetrachlorodibenzo-p-dioxin in female Sprague-Dawley rats. *Toxicol. Appl. Pharmacol.* **132,** 1–13.

Van Wauwe, J. P., Coene, M. C., Goossens, J., Cools, W., and Monbaliu, J. (1990). Effects of cytochrome P-450 inhibitors on the *in vivo* metabolism of all-trans-retinoic acid in rats. *J. Pharmacol. Exp. Ther.* **252,** 365–369.

Vanden Bossche, H., Willemsens, G., and Janssen, P. A. (1988). Cytochrome-P-450-dependent metabolism of retinoic acid in rat skin microsomes: Inhibition by ketoconazole. *Skin Pharmacol.* **1,** 176–185.

Vane, F. M., Stoltenborg, J. K., and Bugge, C. J. (1982). Determination of 13-cis-retinoic acid and its major metabolite, 4-oxo-13-cis-retinoic acid, in human blood by reversed-phase high-performance liquid chromatography. *J. Chromatogr.* **227,** 471–484.

Vecchini, F., Lenoir-Viale, M. C., Cathelineau, C., Magdalou, J., Bernard, B. A., and Shroot, B. (1994). Presence of a retinoid responsive element in the promoter region of the human cytochrome P4501A1 gene. *Biochem. Biophys. Res. Commun.* **201,** 1205–1212.

Villano, C. M., Murphy, K. A., Akintobi, A., and White, L. A. (2006). 2,3,7,8-Tetrachlorodibenzo-p-dioxin (TCDD) induces matrix metalloproteinase (MMP) expression and invasion in A2058 melanoma cells. *Toxicol. Appl. Pharmacol.* **210,** 212–224.

Vincenti, M. P., White, L. A., Schroen, D. J., Benbow, U., and Brinckerhoff, C. E. (1996). Regulating expression of the gene for matrix metalloproteinase-1 (MMP-1): Mechanisms that control enzyme activity, transcription and mRNA stability. *Crit. Rev. Eukaryot. Gene Expr.* **6,** 391–411.

Vogel, S., Gamble, M. V., and Blaner, W. S. (1999). Biosynthesis, absorption, metabolism and transport of retinoids. *In* "Handbook of Experimental Pharmacology. Retinoids. The Biochemical and Molecular Basis of Vitamin A and Retinoid Action" (H. Nau and W. S. Blaner, Eds.), pp. 31–95. Springer Verlag, Heidelberg, Germany.

Wang, S., and Hankinson, O. (2002). Functional involvement of the Brahma/SWI2-related gene 1 protein in cytochrome P4501A1 transcription mediated by the aryl hydrocarbon receptor complex. *J. Biol. Chem.* **277,** 11821–11827.

Wang, X., Sperkova, Z., and Napoli, J. L. (2001). Analysis of mouse retinal dehydrogenase type 2 promoter and expression. *Genomics* **74,** 245–250.

Wanner, R., Panteleyev, A., Henz, B. M., and Rosenbach, T. (1996). Retinoic acid affects the expression rate of the differentiation-related genes aryl hydrocarbon receptor, ARNT and keratin 4 in proliferative keratinocytes only. *Biochim. Biophys. Acta* **1317,** 105–111.

Watson, A. J., and Hankinson, O. (1992). Dioxin- and Ah receptor-dependent protein binding to xenobiotic responsive elements and G-rich DNA studied by *in vivo* footprinting. *J. Biol. Chem.* **267,** 6874–6878.

Watt, K., Jess, T. J., Kelly, S. M., Price, N. C., and McEwan, I. J. (2005). Induced alpha-helix structure in the aryl hydrocarbon receptor transactivation domain modulates protein-protein interactions. *Biochemistry* **44,** 734–743.

Wei, L. N. (2003). Retinoid receptors and their coregulators. *Annu. Rev. Pharmacol. Toxicol.* **43,** 47–72.

Weston, W. M., Nugent, P., and Greene, R. M. (1995). Inhibition of retinoic-acid-induced gene expression by 2,3,7,8-tetrachlorodibenzo-p-dioxin. *Biochem. Biophys. Res. Commun.* **207,** 690–694.

Widerak, M., Ghoneim, C., Dumontier, M. F., Quesne, M., Corvol, M. T., and Savouret, J. F. (2005). The aryl hydrocarbon receptor activates the retinoic acid receptoralpha through SMRT antagonism. *Biochimie* **88**(3–4), 387–397.

Yang, Y. M., Huang, D. Y., Liu, G. F., Zhong, J. C., Du, K., Li, Y. F., and Song, X. H. (2005). Effects of 2,3,7,8-tetrachlorodibenzo-p-dioxin on vitamin A metabolism in mice. *J. Biochem. Mol. Toxicol.* **19,** 327–335.

Yao, T. P., Ku, G., Zhou, N., Scully, R., and Livingston, D. M. (1996). The nuclear hormone receptor coactivator SRC-1 is a specific target of p300. *Proc. Natl. Acad. Sci. USA* **93,** 10626–10631.

Yin, H., Li, Y., and Sutter, T. R. (1994). Dioxin-enhanced expression of interleukin-1 beta in human epidermal keratinocytes: Potential role in the modulation of immune and inflammatory responses. *Exp. Clin. Immunogenet.* **11,** 128–135.

Yoh, S. M., and Privalsky, M. L. (2001). Transcriptional repression by thyroid hormone receptors. A role for receptor homodimers in the recruitment of SMRT corepressor. *J. Biol. Chem.* **276,** 16857–16867.

Yueh, M. F., Huang, Y. H., Hiller, A., Chen, S., Nguyen, N., and Tukey, R. H. (2003). Involvement of the xenobiotic response element (XRE) in Ah receptor-mediated induction of human UDP-glucuronosyltransferase 1A1. *J. Biol. Chem.* **278,** 15001–15006.

Yueh, M. F., Kawahara, M., and Raucy, J. (2005). Cell-based high-throughput bioassays to assess induction and inhibition of CYP1A enzymes. *Toxicol. In Vitro* **19,** 275–287.

Zhang, Q. Y., Dunbar, D., and Kaminsky, L. (2000). Human cytochrome P-450 metabolism of retinals to retinoic acids. *Drug Metab. Dispos.* **28,** 292–297.

Zile, M. H., Emerick, R. J., and DeLuca, H. F. (1967). Identification of 13-cis retinoic acid in tissue extracts and its biological activity in rats. *Biochim. Biophys. Acta* **141,** 639–641.

3

Role of Retinoic Acid in the Differentiation of Embryonal Carcinoma and Embryonic Stem Cells

Dianne Robert Soprano,*,† Bryan W. Teets,* and Kenneth J. Soprano†,‡

*Department of Biochemistry, Temple University School of Medicine, Philadelphia Pennsylvania 19140
†Fels Institute for Cancer Research and Molecular Biology, Temple University School of Medicine, Philadelphia, Pennsylvania 19140
‡Department of Microbiology and Immunology, Temple University School of Medicine, Philadelphia, Pennsylvania 19140

I. Introduction
II. Molecular Mechanism of Action of RA
III. Model Systems to Study Differentiation
  A. EC Cells
  B. ES Cells
IV. Role of RARs
V. RA-Regulated Genes
VI. Role of Specific RA-Regulated Genes
VII. Conclusions
  References

Retinoic acid (RA), the most potent natural form of vitamin A, plays an important role in many diverse biological processes such as embryogenesis and cellular differentiation. This chapter is a review of the mechanism of action of RA and the role of specific RA-regulated genes during the cellular differentiation of embryonal carcinoma (EC) and embryonic stem (ES) cells. RA acts by binding to its nuclear receptors and inducing transcription of specific target genes. The most studied mouse EC cell lines include F9 cells, which can be induced by RA to differentiate into primitive, parietal, and visceral endodermal cells; and P19 cells, which can differentiate to endodermal and neuronal cells upon RA treatment. ES cells can be induced to differentiate into a number of different cell types; many of which require RA treatment. Over the years, many RA-regulated genes have been discovered in EC and ES cells using a diverse set of techniques. Current research focuses on the elucidation how these genes affect differentiation in EC and ES cells using a variety of molecular biology approaches. However, the exact molecule events that lead from a pluripotent stem cell to a fully differentiated cell following RA treatment are yet to be determined. © 2007 Elsevier Inc.

I. INTRODUCTION

Retinoic acid (RA), the most potent natural form of vitamin A, plays an important role in mediating the growth and differentiation of both normal and transformed cells (Chambon, 1996; Soprano and Soprano, 2003). It is essential for many diverse biological functions including growth, vision, reproduction, embryonic development, differentiation of epithelial tissues, and immune responses.

The role of vitamin A during embryonic development was first recognized in the 1930s when maternal vitamin A deficiency was found to be associated with a number of defects (Hale, 1937; Mason, 1935). This was eventually termed the vitamin A-deficiency syndrome. Later, it was demonstrated that an excess of vitamin A caused a number of congenital abnormalities (Cohlan, 1953). Clearly the maintenance of retinoid homeostasis is critical during embryonic development. Following these initial observations, a large number of studies have examined the role of vitamin A and more specifically RA, retinoic acid receptors (RARs), and retinoid X receptors (RXRs) during embryonic development. This work has been extensively reviewed by a number of investigators (for reviews see Clagett-Dame and De Luca, 2002; Mark *et al.*, 2006; Ross *et al.*, 2000; Soprano and Soprano, 1995; Zile, 2001).

RA is also an important regulatory molecule for controlling cell growth and differentiation in both the adult and the embryo. It is critical for the maintenance of the differentiated state of all epithelial cells in the body and for hematopoietic cell differentiation (for reviews see De Luca *et al.*, 1995;

Oren et al., 2003). *In vitro*, RA induces differentiation of pluripotent embryonal carcinoma (EC) and embryonic stem (ES) cell lines into a number of specific cell types. ES cells are transiently present in the embryo and small numbers are also believed to be present in adult tissues. RA is a critical regulator of embryonic neurogenesis and has been shown to play an important role during adult neurogenesis *in vivo* (Jacobs et al., 2006; McCaffery and Drager, 2000). Therefore, the study of RA-induced differentiation *in vitro* is important to understand early embryonic development and differentiation of adult stem cells *in vivo*.

This chapter will summarize the current state of knowledge pertaining to the role of RA in the differentiation of EC and ES cells. Since the effects of RA during differentiation are due to the regulation of gene expression, we will begin by reviewing briefly the remarkable progress that has been made in understanding the mode of action of retinoids at the molecular level. We will then describe several model systems used to study RA-induced differentiation *in vitro*. Finally, we will review the current information pertaining to the role of retinoid nuclear receptors and RA-regulated genes during differentiation of EC and ES cells to a number of differentiated cell types.

II. MOLECULAR MECHANISM OF ACTION OF RA

RA functions by binding to ligand-inducible transcription factors (nuclear receptor proteins belonging to the steroid/thyroid hormone receptor superfamily) that activate or repress the transcription of downstream target genes (for review see Chambon, 1996; Soprano and Soprano, 2003). Six nuclear receptors (termed RARα, RARβ, RARγ, RXRα, RXRβ, and RXRγ), encoded by distinct genes, have been demonstrated to mediate the actions of RA. The natural metabolites all-*trans* RA (atRA) and 9-*cis* RA are high-affinity ligands for RARs, whereas 9-*cis* RA, phytanic acid, docosahexanoic acid, and unsaturated fatty acids, have been suggested to bind RXRs.

These proteins, as heterodimers (RAR/RXR) or homodimers (RXR/RXR), function to regulate transcription by binding to DNA sequences located within the promoter of target genes called retinoic acid response elements (RARE) or retinoid X response elements (RXRE), respectively (Fig. 1). RAREs consist of direct repeats of the consensus half-site sequence AGGTCA separated most commonly by five nucleotides (DR-5) while RXREs are typically direct repeats of AGGTCA with one nucleotide spacing (DR-1). The RAR/RXR heterodimer binds to the RARE with RXR occupying the 5′ upstream half-site and RAR occupying the 3′ downstream half-site.

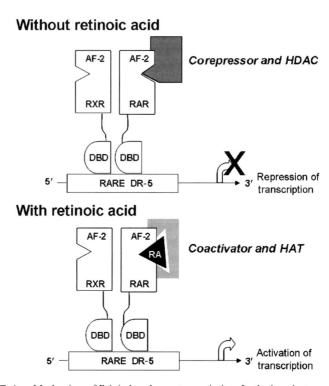

FIGURE 1. Mechanism of RA-induced gene transcription. In the inactivate promoter, the RAR and RXR exist as a heterodimer bound through its DNA-binding domain (DBD) to the RARE DR-5. Corepressors bind to RAR and recruit HDAC causing transcriptional repression. When RA is added, transcription is activated by RA binding to the RAR. The RAR bound to RA then recruits coactivators and HAT.

Over the last several years, a large number of proteins that interact with RARs have been demonstrated to play an important role in the ultimate control of their transcriptional activity (for review see Hart, 2002; Jepsen and Rosenfeld, 2002; Wei, 2003; Westin et al., 2000). In the absence of RA, the apo-receptor pair (RAR/RXR) binds to the RARE in the promoter of target genes and RAR recruits corepressors. These corepressors mediate their negative transcriptional effects by recruiting histone deacetylase complexes (HDACs). HDACs remove acetyl groups from histone proteins inducing a change in the chromatin structure causing DNA to be inaccessible to the transcriptional machinery. On the other hand, on RA binding (at physiological levels), there is a conformational change in the structure of the ligand-binding domain that results in the release of the corepressor and the recruitment of coactivators to the AF-2 region of the receptor. Some coactivators interact directly with the basal transcription machinery to enhance transcriptional activation, while others exhibit histone acetyltransferase (HAT) activity. HAT acetylates histone proteins causing the activation of transcription of the associated gene.

III. MODEL SYSTEMS TO STUDY DIFFERENTIATION

In order to study the role of RA during cellular differentiation *in vitro*, it is necessary to have available model systems that closely resemble development *in vivo*.

Mouse EC and ES cell lines have been well characterized and there are also several human lines available. These pluripotent cell lines can be maintained as undifferentiated cells and can be induced to differentiate *in vitro* to virtually any cell type. Furthermore, they are very amenable to genetic manipulations making them excellent model systems to address fundamental mechanistic questions during RA-induced differentiation.

A. EC CELLS

EC cells are the undifferentiated cells derived from teratocarcinomas, malignant multidifferentiated tumors that arise in the testes or ovaries when early mouse embryos are grafted into adult mice (for review see Smith, 2001). These cells can be propogated in culture continuously as undifferentiated cells; however, they retain the ability to differentiate to all three germ layers (ectoderm, mesoderm, and endoderm). Although EC cells are self-renewing and pluripotent in most cases, they lack the ability when reintroduced into the developing embryo to participate in embryogenesis and give rise to a wide variety of tissues. In particular, they are often aneuploid and therefore are not capable of proceeding through meiosis and producing mature sex cells. The most commonly used EC cell lines for the study of RA-dependent differentiation are F9 and P19 cells (Fig. 2).

F9 EC cells resemble the pluripotent stem cells of the inner cell mass of blastocysts. RA induces differentiation of F9 cells along a number of different pathways depending on the culture conditions (Gudas, 1991; Strickland and Mahdavi, 1978; Strickland *et al.*, 1980), mimicking the early commitment events that occur in 3- to 5-day blastocysts when the inner cell mass forms two endodermal layers. Treatment of F9 cells grown in monolayer culture with RA results in differentiation to primitive endoderm, while treatment with both RA and dibutyryl cyclic AMP causes differentiation to parietal endoderm. These parietal endoderm cells express high levels of plasminogen activator, laminin, and type IV collagen along with very low levels of alkaline phosphatase and lactate dehydrogenase (Strickland and Mahdavi, 1978) typical of parietal endoderm *in vivo*. On the other hand, treatment of F9 cells grown as aggregates (termed embryoid bodies) in bacterial dishes with

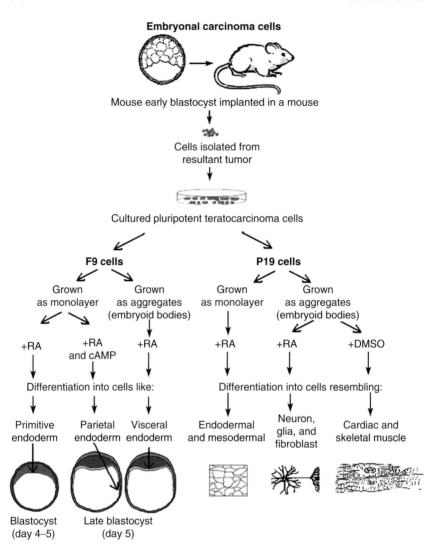

FIGURE 2. Terminal differentiation pathways of F9 and P19 EC cell lines. In F9 cells all pathways require RA, whereas in P19 the endodermal and neuronal pathways require RA while the myocardial pathway requires DMSO.

RA induces differentiation to visceral endoderm. Visceral endoderm is characterized by the expression of a number of proteins typical of visceral yolk sac including α-fetoprotein, albumin, retinol-binding protein, and transthyretin (Soprano et al., 1988; Young and Tilghman, 1984). Hence, the parietal endoderm and visceral endoderm obtained upon treatment of F9 cells with RA are indistinguishable from the parietal endodermal and visceral endodermal cells generated early in mouse embryogenesis.

P19 EC cells are also a pluripotent cell line that possesses the typical morphology of EC cells. Like F9 cells, P19 cells are a model system for the study of differentiation and early mammalian development. The P19 cell line was derived from a teratocarcinoma in C3H/HE mice, produced by grafting an embryo at 7 days of gestation to the testes of an adult male mouse (for review see Bain *et al.*, 1994; McBurney and Rogers, 1982). The cells contain a normal karyotype (McBurney and Rogers, 1982), which reduces the chance that the results obtained are due to a genetic abnormality. Depending on chemical treatment and growth conditions, P19 cells can be induced to differentiate into derivatives of all three germ layers. Dimethylsulfoxide treatment of P19 aggregates (embryoid bodies) produces cells with many of the characteristics of cardiac and skeletal muscle (McBurney *et al.*, 1982). Conversely, RA treatment of aggregates results in the formation of cells that resemble neurons, glia, and fibroblast-like cells (Jones-Villeneuve *et al.*, 1982). Treatment of P19 cells grown as a monolayer in the presence of RA results in the formation of endodermal and mesodermal derivatives (Mummery *et al.*, 1986).

B. ES CELLS

ES cells are derived from the inner cell mass of blastocysts (for review see Smith, 2001). These cells closely resemble EC cells in morphology, growth behavior, and marker expression. Furthermore, ES cells like EC cells are undifferentiated, immortal, and pluripotent cells that have the capacity to differentiate into cell types of all three primary germ layers. However, unlike EC cells, ES cells are diploid, participate in embryogenesis, and are able to differentiate into germ cells *in vivo*. Great care must be taken in the culturing of mouse ES cells to maintain the undifferentiated stem cell phenotype including the use of feeder cells or growth in the presence of the cytokine leukemia inhibitory factor (LIF). In addition to mice (Evans and Kaufman, 1981; Martin, 1981), ES cells have been prepared from fish (Hong *et al.*, 1996; Sun *et al.*, 1995), chicken (Pain *et al.*, 1996), rhesus monkey (Thomson and Marshall, 1998; Thomson *et al.*, 1995), marmoset (Thomson *et al.*, 1996), and humans (Shamblott *et al.*, 1998; Thomson *et al.*, 1998). A major application of mouse ES cells has been their use in the engineering of transgenic and knockout mice. ES cells also have great potential as a source of cells for transplantation in the treatment of numerous pathologies.

RA can induce differentiation of ES cells into a large number of different cell types including neurons, glial cells, adipocytes, chondrocytes, osteocytes, corneal epithelium, skeletal muscle, smooth muscle, and ventricular cardiomyocytes (Eiges and Benvenisty, 2002; Rohwedel *et al.*, 1999; Schuldiner *et al.*, 2000) (Fig. 3). In all cases, *in vitro* differentiation begins by removal of LIF from the culture medium and growth of ES cells as small aggregates termed embryoid bodies either in dishes containing a nonadhesive

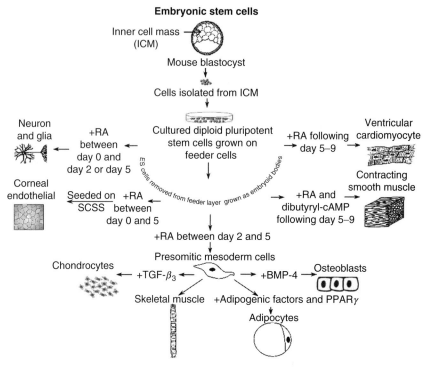

FIGURE 3. Terminal differentiation pathways in mouse ES cells treated with RA. Although RA is important to stimulate the cells to differentiate into the final cell types shown, other treatments and factors are generally necessary for terminal differentiation.

substratum or in hanging drop cultures. The specific cell types formed are dependent on both the timing of RA treatment and the addition of other factors. The timing of RA treatment can be at the time of formation (day 0) of embryoid bodies, 2 days following formation of embryoid bodies, or 5–9 days following formation of embryoid bodies.

Treatment of ES cell-derived embryoid bodies with RA from day 0 to either day 2 or day 5 following embryoid body formation results in differentiation of ES cells to neurons and glial cells (Bain et al., 1994; Fraichard et al., 1995; Glaser and Brustle, 2005; Gottlieb and Huettner, 1999; Strubing et al., 1995). These neurons and glial cells display key morphological, physiological, and biochemical properties of their normal counterparts. On the other hand, corneal epithelial cells can be derived following seeding of embryoid body cells treated with RA onto deepithelialized superficial corneoscleral slices (SCSS) (Wang et al., 2005).

Alternatively, treatment of ES cell-derived embryoid bodies with RA between day 2 and day 5 after embryoid body formation results in presomitic mesoderm. Removal of RA for 2 days (days 5–7 after embryoid body formation) followed by treatment with adipogenic factors (such as insulin, triidothyronine, or thiazolidinedione) and PPARγ results in the formation of adipocytes (Dani et al., 1997; Phillips et al., 2003), while treatment with BMP-4 or TGF-β3 results in the formation of osteoblasts and chondrocytes, respectively (Kawaguchi et al., 2005). The treatment of the embryoid bodies with RA is obligatory in the first phase for the commitment of these cells to presomitic mesoderm, while the subsequent treatment with the appropriate growth factors and culture milieu steers these precursor cells into distinct mesenchymal compartments. In addition, treatment of ES-derived embryoid bodies between day 2 and day 5 with RA alone can also result in the formation of skeletal muscle cells (Wobus et al., 1994).

Finally, treatment of ES cell-derived embryoid bodies following day 5–9 after formation with RA alone results in the formation of ventricular cardiomyocytes (Wobus et al., 1997), while treatment with both RA and dibutyryl cAMP results in the formation of contracting smooth muscle cells (Drab et al., 1997).

IV. ROLE OF RARs

There is an overwhelming amount of evidence that functional RARs and RXRs are obligatory in mediating RA-dependent differentiation of EC and ES cells. Studies with an RA-nonresponsive mutant line of P19 cells (termed RAC65) have demonstrated the importance of functional RARs in RA-dependent differentiation of P19 cells. RAC65 cells do not differentiate after treatment with RA (Jones-Villeneuve et al., 1983). These cells carry a rearrangement of one of the RARα genes resulting in the production of a truncated RARα protein that has lost its 70 C-terminal amino acids (Kruyt et al., 1991, 1992; Pratt et al., 1990). The truncated RARα acts as a dominant-negative repressor of transcription of RA-responsive target genes including RARβ2, Oct-3/4, Hox genes, and PBX genes (Kruyt et al., 1991; Pratt et al., 1993; Qin et al., 2004a; Schoorlemmer et al., 1995; Zwartkruis et al., 1993) with the resulting failure of RAC65 cells to differentiate upon treatment with RA. Similarly, transfection of expression vectors encoding truncated RAR receptors that interfere with the activity of endogenous RARs in a dominant-negative fashion also inhibits RA-induced differentiation of F9 EC cells (Espeseth et al., 1989).

To test the role of individual RAR isotypes and RXRα in mediating RA-induced differentiation, F9 cells lines containing functionally inactivated receptors have been created and studied. Another experimental approach has been to study the role of RAR isotype-selective and RXR-selective retinoids in the differentiation of F9 and P19 cells. Overall, these studies demonstrate that there are both specific and redundant functions of RARs and RXRs during RA-induced differentiation and that some artifactual redundancies are likely to be created as a result of the knockout of individual genes.

Comparison of RARα and RARγ has demonstrated that RARγ has a more prominent role in mediating RA-dependent differentiation and gene expression in F9 cells (Boylan et al., 1993, 1995; Taneja et al., 1996). Interestingly, RARγ is indispensable for the differentiation of F9 cells to primitive endoderm; however, both RARγ and RARα are required for efficient differentiation to parietal endoderm (Rochette-Egly et al., 2000; Taneja et al., 1995, 1997). Furthermore, RARα and RARγ play an important role in mediating RA metabolism in F9 cells since loss of RARγ results in a lower rate of production of polar metabolites of RA, while loss of RARα causes an increase in the metabolism of RA (Boylan et al., 1995). On the other hand, studies utilizing RAR-selective retinoids demonstrate that differentiation of P19 cells can be mediated by either RARα or RARγ (Taneja et al., 1996).

RARβ2 plays an important role in mediating both RA-induced growth arrest and late gene expression responses to RA in F9 cells (Faria et al., 1999). F9 cells that lack RARβ2 display an altered morphology and do not fully differentiate upon treatment with RA. The initial RA-dependent increase in early response gene expression was observed in the RARβ2$^{-/-}$ cells; however, their elevated expression was not sustained at later times following RA treatment along with a failure to increase the level of expression of at least one late RA-response gene (laminin B1). Taken together, this suggests that RARβ2 is not required for the initiation of RA-dependent differentiation in F9 cells; however, it is necessary for later events that culminate in the differentiated phenotype.

RXRα also plays a critical role in RA-induced differentiation of F9 cells to primitive and parietal endoderm but not to visceral endoderm, along with mediating growth arrest and apoptotic responses to RA (Chiba et al., 1997a; Clifford et al., 1996). Since differentiation to visceral endoderm is only delayed 1–2 days in cells lacking RXRα, it is likely that RXRβ and/or RXRγ can mediate visceral endoderm differentiation but not parietal and primitive endoderm differentiation in F9 cells. Finally, study of the loss of expression of RARα or RARγ along with RXRα demonstrates that the RAR/RXR heterodimer is the functional unit transducing the retinoid signal (Chiba et al., 1997b).

V. RA-REGULATED GENES

There has been great effort over the last 25 years by many investigators to elucidate the cascade of gene expression events that ultimately results in the differentiated phenotype displayed by EC and ES cells following RA treatment. The first step to achieve this goal has been to identify genes whose expression is altered by RA during differentiation along a number of different pathways. The approaches used to address this question have changed over the years taking advantage of the major technical developments in molecular and cellular biology. Among the technical approaches used include differential screening of cDNA libraries, subtractive cDNA libraries, differential display, microarray analysis, and proteomic analysis. An extremely long list of genes and expressed sequence tags (ESTs) that display altered expression upon treatment of EC and/or ES cells following RA exposure has resulted from these numerous studies. A listing of genes and ESTs differentially regulated in EC and ES cells induced by RA to differentiate along a number of different pathways can be obtained from a variety of sources including Gudas (1991), Gudas *et al.* (1994), Nishiguchi *et al.* (1994), Bouillet *et al.* (1995), Faria *et al.* (1998), Bain *et al.* (2000), Harris and Childs (2002), Wei *et al.* (2002), Sangster-Guity *et al.* (2004), and An *et al.* (2005). Several general trends are readily apparent on examination of these lists of genes and ESTs.

First, RA treatment of EC and ES cells results in both an increase and a decrease in the expression of particular genes. In some cases, the fold changes in expression are rather modest while in others the fold changes are very large. Furthermore, RA treatment causes changes in the expression of genes with a variety of different functions including transcription factors, RA metabolism and transport proteins, extracellular matrix proteins, protooncogenes, growth factors and their receptors, cytoskeletal proteins, proteins involved in cell metabolism, cell surface antigens, apoptosis-related proteins, cell-cycle control proteins, and proteins that mediate intracellular and extracellular signaling.

Second, the temporal pattern of gene expression during RA-induced differentiation has several phases. Regardless of the pathway of differentiation, there is a subset of genes whose expression is increased rapidly (within the first 12–16 h) upon RA treatment in the presence of cycloheximide indicating that they are primary response genes to RA. Promoter analysis of many of these genes demonstrates that they contain a RARE and that the RAR/RXR heterodimer mediates the increase in expression upon RA treatment. Finally, many of the same primary RA response genes are regulated in both EC and ES cells irrespective of the pathway of differentiation and are most often either transcription factors or proteins involved in RA metabolism and transport.

A much larger number of other genes display altered expression at later time points (1 or more days) following RA treatment and these changes in expression require new protein synthesis. Hence, these genes are secondary responders and are indirectly regulated by RA treatment. Furthermore, many of these genes are associated with a specific differentiation pathway. Differentiation of F9 cells to primitive endoderm is biphasic (early and late responding genes). A third phase occurs when these cells are induced to differentiate in the presence of RA and dibutyryl cAMP to parietal endoderm or to differentiate upon treatment of embryoid bodies with RA to visceral endoderm. Similarly P19 cells induced to differentiate to neuronal cells display three phases of gene expression changes, the initial primary response phase (0–16 h following RA treatment), the neural differentiation phase (16 h to 2 days following RA treatment), and the terminal differentiation phase (5–6 days following RA treatment).

Third, a final point to keep in mind is that the differentiation process often involves RA treatment along with additional treatments/factors such as dibutyryl cAMP or aggregation of cells (embryoid bodies). Therefore, some of the gene expression changes are mediated solely by RA, some by one of the additional treatments/factors, and a third group by a combination of RA and one or more of the additional treatments/factors. Some studies have attempted to address this issue, while others have not (Teramoto et al., 2005).

VI. ROLE OF SPECIFIC RA-REGULATED GENES

The availability of this large battery of RA-regulated genes associated with differentiation of EC and ES cells allows more hypothesis-driven experiments. These studies are focused on the determination of the specific role of individual genes during the differentiation process. The long-term goal of these studies is to understand the sequence of events at the molecular level during RA-induced differentiation in EC and ES cells. Many of these genes are expressed in early embryos and are likely to be important players in embryogenesis (Tables I and II).

RARβ2 and *RARα2*: The expression of two RARs, RARβ2 and RARα2, is rapidly increased following RA treatment of F9 cells and P19 cells (Hu and Gudas, 1990; Qin et al., 2004a; Shen et al., 1991). This increase in RARβ2 and RARα2 expression is mediated by RAREs located in the promoter of each of these genes (Shen et al., 1991).

The role of RARβ2 in mediating RA-dependent differentiation of F9 cells has been studied utilizing RARβ2$^{-/-}$ F9 cells (Faria et al., 1999). RARβ2$^{-/-}$ F9 cells have an altered morphology and fail to fully differentiate upon RA treatment. These cells fail to growth arrest after RA treatment suggesting that RARβ2 plays an important role in mediating growth. RARβ2 has been shown to be necessary for the RA-dependent increase in the protein level of p27 (an important cell cycle regulatory gene) by elevating p27 mRNA levels, rate of translation, and half-life of p27 (Li et al., 2004). Moreover, RARβ2 is required for the sustained expression of a number of other RA-responsive genes. A comparison of the gene expression profiles in wild-type and RARβ2$^{-/-}$ F9 cells demonstrates that RARβ2 regulates the expression of a wide variety of genes including transcription factors, cell surface-signaling

TABLE I. Mechanism of Action for Selected Genes During Endodermal Differentiation

Gene	Cell line	RA response	Mechanistic studies
RARγ	F9	Not RA regulated	Indispensable for formation of primitive endoderm and needed for efficient differentiation to parietal endoderm. Loss caused a decrease in the metabolism of RA
RXRα	F9	Not RA regulated	Indispensable for primitive and parietal endoderm and not required for visceral endoderm
RARα2	F9, P19	Primary	RARα needed for efficient differentiation to parietal endoderm. Loss of RARα caused an increase in the metabolism of RA
RARβ2	F9, P19	Primary	Required for sustained expression of RA-induced gene expression but not necessary for the initiation of differentiation. Also important for RA-dependent growth inhibition
Hoxa-1	F9, P19, ES	Primary	Not sufficient to induce differentiation without RA. Positive regulator of neuroectodermal and mesodermal differentiation, repressor of endodermal differentiation
Hoxb-2	F9, P19	Primary	Mechanistic studies not performed
Rex-1 (Zfp-42)	F9, ES	Secondary	Important for differentiation to primitive and visceral endoderm but not parietal endoderm
Pbx1	P19	Secondary	Necessary for endodermal differentiation
Disabled-2	F9	Unknown	Role in RA-dependent growth arrest

transduction molecules, and metabolic enzymes that are important during differentiation (Zhuang *et al.*, 2003).

Hoxa-1 and *Hoxb-1*: The expression of two homeobox domain genes, Hoxa-1 (Hox 1.6) and Hoxb-1 (Hox 2.9), are increased rapidly upon treatment of ES and EC cells with RA in the absence of new protein synthesis. Hoxa-1 was initially demonstrated by LaRosa and Gudas (1988) as an early response gene to RA treatment of F9 cells and was termed Ear-1. This RA-regulated increase in Hoxa-1 expression is mediated by a RARE located in the 3′ enhancer region of the Hoxa-1 gene (Langston and Gudas, 1992; Langston *et al.*, 1997) and RARγ (Boylan *et al.*, 1993). Hoxb-1 levels are also increased upon RA treatment (Simeone *et al.*, 1991). The Hoxb-1 promoter contains a RARE (Langston *et al.*, 1997; Ogura and Evans, 1995; Thompson *et al.*, 1998); however, Hoxa-1 also plays an important role in regulating its expression (Di Rocco *et al.*, 2001). ES cells that lack expression of Hoxa-1 display a reduced level of expression of Hoxb-1 upon RA treatment (Martinez-Ceballos *et al.*, 2005).

The role of Hoxa-1 in the RA-dependent differentiation of F9 cells has been studied utilizing F9 cells that stably express Hoxa-1 (Goliger and Gudas, 1992;

TABLE II. Mechanism of Action for Selected Genes During Neuronal Differentiation

Gene	Cell line	RA response	Mechanistic studies
CYP26A1	F9, P19	Primary	Hydroxylated retinoid products may play critical role in neurogenesis
Sox6	P19	Primary	Important in the regulation of cellular aggregation and neuronal differentiation.
Wnt-1	P19	Primary	Positive regulator of neurogenesis and inhibitor of gliogenesis
Mash-1	P19	Secondary	Not critical for neurogenesis
Ngn-1	P19	Secondary	Positive regulator of neurogenesis
NeuroD	P19	Secondary	Mechanistic studies not performed
N-cadherin	P19	Secondary	Positive regulator of neurogenesis.
Pbx1	P19	Secondary	Necessary for neuronal differentiation
CypA	P19	Not RA regulated	Positive regulator of neurogenesis

Shen and Gudas, 2000; Shen et al., 2000). Overexpression of Hoxa-1 caused an alteration in the morphology of the undifferentiated cells. However, the cells fail to differentiate in the absence of RA indicating that Hoxa-1 expression is not sufficient for differentiation. The Hoxa-1 overexpressing cells also are not inhibited from normal differentiation upon RA treatment. This suggests that Hoxa-1 plays an important role in the expression of genes that influence F9 cell morphology during RA-induced differentiation. Twenty-eight candidate genes that display altered expression in the Hoxa-1 overexpressing F9 cells have been identified. These genes include signaling molecules including BMP-4, superoxide dismutase, cadherin-6, transcription factors such as HMG-1, SAP18, the homeodomain proteins Gbx-2 and Evx-2, and cell cycle-regulated proteins including retinoblastoma-binding protein-2, and two novel clones termed clone 104 and HAIR-62.

ES cells isolated from Hoxa-1$^{-/-}$ mice have been studied to further elucidate the role of Hoxa-1 during RA-induced differentiation (Martinez-Ceballos et al., 2005). Wild-type and Hoxa-1$^{-/-}$ cells respond differently to both RA treatment and LIF removal. RA treatment of ES Hoxa-1$^{-/-}$ cells results in a reduced level of expression of a number of genes along the pathway to both neuroectoderm, including Fgf5, Nnat, Wnt3a, BDNF, and RhoB, and bone such as Postn, Col1A1, and BSP. In addition, the RA-treated Hoxa-1$^{-/-}$ cells have increased expression of a number of endodermal marker genes including Sox17, Gata6, Gata2, Dab2, and Lama1. It should be pointed out that it is not known where Hoxa-1 directly or indirectly regulates the expression of each of these genes or whether Hoxa-1 functions alone or in conjunction with other proteins. However, it is clear that Hoxa-1 functions as a positive regulator of neuroectodermal and mesodermal differentiation and a repressor of endoderm formation.

Rex-1 (Zfp-42): The expression of Rex-1 is reduced in F9 cells as early as 12 h following treatment with RA (Hosler et al., 1989). Rex-1 expression is also reduced in D3 ES cells induced to differentiate and is believed to be involved in trophoblast development and spermatogenesis (Rogers et al., 1991). Rex-1 encodes a protein containing four repeats of the zinc finger nucleic acid-binding motif and a potential acidic activator domain suggesting that it encodes a transcription factor; however, the specific function of this protein is not known. It is often used as a marker of stem cell character and pluripotency of cells.

While the RA-dependent reduction in expression of Rex-1 is at the transcriptional level, it is not a primary response gene to RA treatment. This decrease in Rex-1 expression is mediated by two DNA sequence motifs within its promoter, an octamer motif that binds Oct3/4 and a motif slightly upstream from the octamer motif that binds Rox-1 (Ben-Shushan et al., 1998; Hosler et al., 1993; Rosfjord and Rizzino, 1994). The expression level of both Oct3/4 and Rox-1 are greatly decreased upon treatment with RA. Therefore, RA causes a reduction in Oct3/4 and Rox-1 expression which in turn results in a reduction in Rex-1 expression.

The role of Rex-1 in RA-dependent differentiation of F9 cells has been studied using F9 cells with a targeted deletion of Rex-1 (F9 Rex-1$^{-/-}$ cells) (Thompson and Gudas, 2002). F9 Rex-1$^{-/-}$ cells failed to completely differentiate to primitive endoderm and visceral endoderm upon treatment with RA demonstrating a role for Rex-1 in these differentiation pathways. However, the F9 Rex-1$^{-/-}$ cells were able to differentiate to parietal endoderm in the presence of RA alone without dibutyryl cAMP. This suggests that primitive endoderm may not be a required precursor to parietal endoderm in the absence of Rex-1 protein.

Disabled-2: Disabled-2 expression is highly increased in F9 EC induced to differentiate to either parietal endoderm or visceral endoderm by RA (Cho et al., 1999). Mouse disabled-2 is highly homologous to the disabled-2 gene in *Drosophila* and its expression is highly correlated to growth suppression of F9 cells treated with RA. Transient expression of disabled-2 in F9 cells causes suppression of Elk-1 phosphorylation, c-Fos expression, and cell growth (Smith et al., 2001). Therefore, disabled-2 is likely to be an important mediator of RA-dependent growth inhibition during differentiation of F9 cells.

Pbx family of proteins: Pre-B-cell leukemia transcription factors (PBXs) are important cofactors for the transcriptional regulation mediated by a number of Hox proteins during embryonic development. PBX1, PBX2, and PBX3 mRNA and protein levels are elevated in P19 cells during endodermal and neuronal differentiation induced by RA (Knoepfler and Kamps, 1997; Qin et al., 2004a). The increases in PBX1 mRNA and PBX3 mRNA are secondary responses to RA treatment, while the increase of PBX2 mRNA appears to be a primary response. In addition, the half-lives of PBX1, PBX2, and PBX3 proteins are significantly extended by RA treatment. Study of the role of PBX proteins during differentiation of P19 cells has demonstrated that RA-dependent increase in PBX protein levels is an essential step in differentiation to both endodermal and neuronal cells (Qin et al., 2004b). Furthermore, PBX is necessary for the RA-dependent increase in BMP-4 and decorin during differentiation of P19 cells to endodermal cells.

CYP26a1: CYP26a1 is an RA-inducible cytochrome P450 enzyme that plays an important role in the catabolism of RA producing hydroxylated products. CYP26a1 mRNA levels are rapidly increased following RA treatment of EC and ES cells (Ray et al., 1997). The CYP26a1 promoter contains a canonical RARE in its 5′ promoter region that has been demonstrated to function in the RA-dependent regulation of transcription (Loudig et al., 2000). P19 cells stably expressing CYP26a1 undergo rapid neuronal differentiation in monolayer at low concentrations of RA suggesting that the hydroxylated products of RA may play a critical role in mediating neuronal differentiation (Pozzi et al., 2006; Sonneveld et al., 1999).

Sox6: The Sox (SRy-related HMG box) genes all contain a DNA-binding domain, the HMG box (high-mobility group) (Gubbay et al., 1990; Pevny and Lovell-Badge, 1997; Prior and Walter, 1996; Sinclair et al., 1990; Wegner, 1999). Sox genes are expressed in various phases of embryonic development and cell differentiation in a cell-specific manner. Sox1, Sox2, Sox3, Sox6, and Sox9 are expressed in the central nervous system during embryogenesis, and Sox5 and Sox6 are associated with spermatogenesis in the testis (Connor et al., 1995; Takamatsu et al., 1995).

It was demonstrated that RA-treated P19 cell aggregates express Sox6 during the neural differentiation phase. Constitutive overexpression of Sox6 caused marked cellular aggregation and the cells differentiated into microtubule-associated protein 2-expressing neuronal cells in the absence of RA (Hamada-Kanazawa et al., 2004a). On the other hand, suppression of Sox6 expression resulted in nearly totally blocking neuronal differentiation and causing apoptosis upon RA treatment of P19 cells (Hamada-Kanazawa et al., 2004b). Furthermore, Sox6 overexpression leads to an increase in N-cadherin, Mash-1, and Wnt-1 levels suggesting that Sox6 is involved upstream in the Wnt-1-signaling pathway in the early stage of neuronal differentiation (Hamada-Kanazawa et al., 2004a). These Sox6 overexpressing cells contained higher levels of E-cadherin and N-cadherin than those of wild-type P19 cells, although the N-cadherin levels were lower than that of aggregated wild-type P19 cells treated with RA (Hamada-Kanazawa et al., 2004a). This suggests that Sox6 may also be involved in the N-cadherin-signaling pathway. Taken together, these studies demonstrate that Sox6 is important in regulating cellular aggregation and neuronal differentiation.

Wnt-1: Wnt-1 is also expressed during vertebrate embryogenesis in the central nervous system (McMahon *et al.*, 1992; Nusse and Varmus, 1982), and is the vertebrate counterpart of the *Drosophila wingless* gene (Rijsewijk *et al.*, 1987). It is a secreted factor that is low in abundance in undifferentiated P19 cells, rapidly increases during the neural differentiation phase (16 h to 2 days following RA treatment), and finally decreases in level during the terminal differentiation phase (Papkoff, 1994; Schuuring *et al.*, 1989; St-Arnaud *et al.*, 1989).

Interestingly, two groups have studied the effects of overexpression of Wnt-1 in P19 cells and have obtained differing results. The first group overexpressed Wnt-1 and showed that the levels of Wnt-4 and Wnt-6 were elevated, indicating that the initiation of differentiation had occurred. However, these cells did not differentiate into neuronal cells in the absence of RA (Smolich and Papkoff, 1994). The second group found that overexpression of Wnt-1 caused the cells to differentiate solely into neurons instead of the normal mix of neurons and glia cells (Tang *et al.*, 2002). The one difference between the protocols used by these two groups was how the cells were plated. Smolich and Papkoff plated their cells as monolayer, whereas Tang *et al.* plated their cells as aggregates. This suggests that while Wnt-1 appears to be important for neuronal differentiation, its overexpression alone is not sufficient to cause the cells to differentiate in the absence of RA. Additional factor(s) associated with cellular aggregation are also required in the presence of Wnt-1 for the cells to differentiate to neurons. Interestingly, although Wnt-1 is not sufficient for neural differentiation, it appears to be important in determining cell fate because gliogenesis is abolished in the overexpressing aggregate cells. Mash-1 and Ngn-1 is upregulated during the neural differentiation phase of the Wnt-1 overexpressing aggregate cells indicating that Wnt-1 promotes neural fate and inhibits glial fate in P19 cells through activation of neural basic helix-loop-helix (bHLH) gene expression (Tang *et al.*, 2002).

Neurogenin-1 and *NeuroD*: Neurogenin-1 (Ngn-1) is a vertebrate determination factor that activates a cascade of downstream bHLH genes including NeuroD during neurogenesis (Ma *et al.*, 1996). bHLH proteins are expressed in multiple stages in neural lineages and specify cell fates (Jan and Jan, 1993). Ngn-1 can also actively inhibit gliogenesis in a manner that is independent of its ability to promote NeuroD expression (Sun *et al.*, 2001). Ngn-1 and Mash-1 are expressed in dividing cells in the complementary regions of the developing central nervous system and in distinct sublineages of the peripheral nervous system (Ma *et al.*, 1997). RA treatment induces the expression of many bHLH genes including Mash-1 (Johnson *et al.*, 1992) and NeuroD2 (Farah *et al.*, 2000; Oda *et al.*, 2000). Ngn-1 is one of the RA-inducible genes in P19 cells that are expressed during the neural differentiation phase, while NeuroD is an RA-inducible gene that is expressed during the terminal differentiation phase of these cells.

A study demonstrated that overexpression of Ngn-1 was able to direct neuronal differentiation when P19 cells are grown as aggregates in the absence of RA; however, RA was required to achieve the maximum number of differentiated cells (Kim *et al.*, 2004). Analysis of the promoter of the NeuroD gene demonstrated that Ngn-1 increases its rate of transcription, suggesting that RA regulates the bHLH cascade involved in neuronal differentiation of P19 cells. However, since these cells were grown as aggregates, it is possible that other factors regulated by aggregation are needed to complement Ngn-1 during differentiation of P19 cells to neurons.

N-Cadherin: N-cadherin is a cell adhesion molecule and is considered to play an important role in the development of the central nervous system (Geiger and Ayalon, 1992; Takeichi, 1995). *In vitro* N-cadherin has been shown to be essential for neurite outgrowth (Bixby and Zhang, 1990). N-cadherin is one of the RA-inducible genes that is expressed during the neural differentiation phase of P19 cells. Overexpression of N-cadherin in P19 cells was found to be sufficient to initiate neuronal differentiation, suggesting that its expression may normally operate *in vivo* to bias cells to a neuronal fate (Gao *et al.*, 2001). In addition, N-cadherin was found to induce the RA-signaling pathway, but was not able to substitute for the role of aggregation of P19 cells during neuronal differentiation (Gao *et al.*, 2001). These cells also showed an elevated level of Wnt-1 that could be due to interaction of N-cadherin with either β-catenin (Miller and Randell, 1996) or the fibroblast growth factor receptors (Umbhauer *et al.*, 2000).

CypA: Cyclophilin A (CypA) was first identified and purified from bovine spleen (Handschumacher *et al.*, 1984) and has been characterized as a housekeeping gene that belongs to the immunophilin protein family (Steele *et al.*, 2002; Thellin *et al.*, 1999; Zhong and Simons, 1999). It has also been found that CypA promotes proper subcellular localization of Zpr1p, a zinc finger-containing protein (Ansari *et al.*, 2002). CypA expression is found in numerous tissues, but its expression is highest in neuronal cells (Goldner *et al.*, 1996), and particularly the nuclei of ganglia sensor neurons (Chiu *et al.*, 2003). P19 cells grown as aggregates and treated with RA did not differentiate when CypA expression was knocked down by siRNA (Song *et al.*, 2004), suggesting that CypA is important for neuronal differentiation. Additionally, CypA has been reported to enhance transcription of the activity of a RARE containing reporter vector. Therefore, the role of CypA in neuronal differentiation of P19 cells appears to be in the regulation of transcription of RA-responsive genes.

VII. CONCLUSIONS

In vitro differentiation of EC and ES cells by RA mimics events that occur during early development. Differentiation by RA requires functional RARs and RXRs. The expression of a large battery of genes is modulated upon RA treatment of EC and ES along several differentiation pathways. Current studies are addressing the role of these genes in mediating RA-dependent differentiation. However, the exact molecule events that lead from a pluripotent cell to a fully differentiated cell following RA treatment are yet to be determined. Future work will strive to elucidate these RA-dependent differentiation pathways at the molecular level.

ACKNOWLEDGMENTS

The support of National Institutes of Health grant DK070650 is gratefully acknowledged.

REFERENCES

An, J., Yuan, Q., Wang, C., Liu, L., Tang, K., Tian, H. Y., Jing, N. H., and Zhao, F. K. (2005). Differential display of proteins involved in the neural differentiation of mouse embryonic carcinoma P19 cells by comparative proteomic analysis. *Proteomics* **5,** 1656–1668.

Ansari, H., Greco, G., and Luban, J. (2002). Cyclophilin A peptidyl-prolyl isomerase activity promotes ZPR1 nuclear export. *Mol. Cell. Biol.* **22,** 6993–7003.

Bain, G., Ray, W. J., Yao, M., and Gottlieb, D. I. (1994). From embryonal carcinoma cells to neurons: The P19 pathway. *BioEssays* **16,** 343–348.

Bain, G., Mansergh, F. C., Wride, M. A., Hance, J. E., Isogawa, A., Rancourt, S. L., Ray, W. J., Yoshimura, Y., Tsuzuki, T., Gottlieb, D. I., and Rancourt, D. E. (2000). ES cell neural differentiation reveals a substantial number of novel ESTs. *Funct. Integr. Genomics* **1,** 127–139.

Ben-Shushan, E., Thompson, J. R., Gudas, L. J., and Bergman, Y. (1998). Rex-1, a gene encoding a transcription factor expressed in the early embryo, is regulated via Oct-3/4 and Oct-6 binding to an octamer site and a novel protein, Rox-1, binding to an adjacent site. *Mol. Cell. Biol.* **18,** 1866–1878.

Bixby, J. L., and Zhang, R. (1990). Purified N-cadherin is a potent substrate for the rapid induction of neurite outgrowth. *J. Cell Biol.* **110,** 1253–1260.

Bouillet, P., Oulad-Abdelghani, M., Vicaire, S., Garnier, J.-M., Schuhbaur, B., Dolle, P., and Chambon, P. (1995). Efficient cloning of cDNAs of retinoic acid-responsive genes in P19 embryonal carcinoma cells and characterization of a novel muse gene, stra1 (Mouse LERK-2/Eplg2). *Dev. Biol.* **170,** 420–433.

Boylan, J. F., Lohnes, D., Taneja, R., Chambon, P., and Gudas, L. J. (1993). Loss of retinoic acid receptor γ function in F9 cells by gene disruption results in aberrant *Hoxa-1* expression and differentiation upon retinoic acid treatment. *Proc. Natl. Acad. Sci. USA* **90,** 9601–9605.

Boylan, J. F., Lufkin, D., Achkar, C. C., Taneja, R., Chambon, P., and Gudas, L. J. (1995). Targeted disruption of retinoic acid receptor α (RARα) and RARγ results in receptor-specific alterations in retinoic acid-mediated differentiation and retinoic acid metabolism. *Mol. Cell. Biol.* **15,** 843–851.

Chambon, P. (1996). A decade of molecular biology of retinoic acid receptors. *FASEB J.* **10,** 940–954.

Chiba, H., Clifford, J., Metzger, D., and Chambon, P. (1997a). Distinct retinoid X receptors-retinoic acid receptor heterodimers are differentially involved in the control of expression of retinoid target genes in F9 embryonal carcinoma cells. *Mol. Cell. Biol.* **17,** 3013–3020.

Chiba, H., Clifford, J., Metzger, D., and Chambon, P. (1997b). Specific and redundant function of retinoid X receptor/retinoic acid receptor heterodimers in differentiation, proliferation, and apoptosis of F9 embryonal carcinoma cells. *J. Cell Biol.* **139**, 735–747.

Chiu, R., Rey, O., Zheng, J. Q., Twiss, J. L., Song, J., Pang, S., and Yokoyama, K. K. (2003). Effects of altered expression and localization of cyclophilin A on differentiation of p19 embryonic carcinoma cells. *Cell. Mol. Neurobiol.* **23**, 929–943.

Cho, S. Y., Cho, S. Y., Lee, S. H., and Park, S. S. (1999). Differential expression of mouse Disabled 2 gene in retinoic acid-treated F9 embryonal carcinoma cells and early mouse embryos. *Mol. Cells* **30**, 179–184.

Clifford, J., Chiba, H., Sobieszczuk, D., Metzger, D., and Chambon, P. (1996). RXRα–null F9 embryonal carcinoma cells are resistant to the differentiation, anti-proliferative and apoptotic effects of retinoids. *EMBO J.* **15**, 4142–4155.

Clagett-Dame, M., and De Luca, H. F. (2002). The role of vitamin A in mammalian reproduction and embryonic development. *Annu. Rev. Nutr.* **22**, 347–381.

Cohlan, S. Q. (1953). Excessive intakes of vitamin A as a cause of congenital anomalies in the rat. *Science* **117**, 535–536.

Connor, F., Wright, E., Denny, P., Koopman, P., and Ashworth, A. (1995). The Sry-related HMG box-containing gene Sox6 is expressed in the adult testis and developing nervous system of the mouse. *Nucleic Acids Res.* **17**, 3365–3372.

Dani, C., Smith, A. G., Dessolin, S., Leroy, P., Staccini, L., Villageois, P., Darimont, C., and Ailhaud, G. (1997). Differentiation of embryonic stem cells into adipocytes *in vitro*. *J. Cell Sci.* **110**, 1279–1285.

De Luca, L. M., Darwiche, N., Jones, C. S., and Scita, G. (1995). Retinoids in differentiation and neoplasia. *Sci. Med.* **2**, 28–37.

Di Rocco, G., Gavalas, A., Popperl, H., Krumlauf, R., Mavilio, F., and Zappavigna, V. (2001). The recruitment of SOX/OCT complexes and the differential activity of HOXA1 and HOXB1 modulate the hox-b1 auto-regulatory enhancer function. *J. Biol. Chem.* **276**, 20506–20515.

Drab, M., Haller, H., Bychkov, R., Erdmann, B., Lindschau, C., Haase, H., Morano, I., Luft, F. C., and Wobus, A. M. (1997). From totipotent embryonic stem cells to spontaneously contracting smooth muscle cells: A retinoic acid and db-cAMP *in vitro* differentiation model. *FASEB J.* **11**, 905–915.

Eiges, R., and Benvenisty, N. (2002). A molecular view on pluripotent stem cells. *FEBS Lett.* **529**, 135–141.

Espeseth, A. S., Murphy, S. P., and Linney, E. (1989). Retinoic acid receptor expression vector inhibits differentiation of F9 embryonal carcinoma cells. *Genes Dev.* **3**, 1647–1656.

Evans, M. J., and Kaufman, M. (1981). Establishment in culture of pluripotential cells from mouse embryos. *Nature* **292**, 154–156.

Farah, M. H., Olson, J. M., Sucic, H. B., Hume, R. I., Tapscott, S. J., and Turner, D. L. (2000). Generation of neurons by transient expression of neural bHLH proteins in mammalian cells. *Development* **127**, 693–702.

Faria, T. N., LaRosa, G. J., Wilen, E., Liao, J., and Gudas, L. J. (1998). Characterization of genes which exhibit reduced expression during the retinoic acid-induced differentiation of F9 teratocarcinoma cells: Involvement of cyclin D3 in RA-mediated growth arrest. *Mol. Cell. Endocrinol.* **143**, 155–166.

Faria, T. N., Mendelsohn, C., Chambon, P., and Gudas, L. J. (1999). The targeted disruption of both alleles of RARβ2 in F9 cells results in the loss of retinoic acid-associated growth arrest. *J. Biol. Chem.* **274**, 26783–26788.

Fraichard, A., Chassande, O., Bilbaut, G., Dehay, C., Savatier, P., and Samarut, J. (1995). In vitro differentiation of embryonic stem cells into glial cells and functional neurons. *J. Cell. Sci.* **108**, 3181–3188.

Gao, X., Bian, W., Yang, J., Tang, K., Kitani, H., Atsumi, T., and Jing, N. (2001). A role of N-cadherin in neuronal differentiation of embryonic carcinoma P19 cells. *Biochem. Biophys. Res. Commun.* **284**, 1098–1103.

Geiger, B., and Ayalon, O. (1992). Cadherin. *Annu. Rev. Cell Biol.* **8,** 307–320.
Glaser, T., and Brustle, O. (2005). Retinoic acid induction of ES-cell-derived neurons: The radial glia connection. *Trends Neurosci.* **28,** 397–400.
Goliger, J. A., and Gudas, L. J. (1992). Mouse F9 teratocarcinoma stem cells expressing the stably transfected homeobox gene Hox 1.6 exhibit an altered morphology. *Gene Expr.* **2,** 147–160.
Gottlieb, D. I., and Huettner, J. E. (1999). An *in vitro* pathway from embryonic stem cells to neurons and glia. *Cells Tissues Organs* **165,** 165–172.
Gubbay, J., Collignon, J., Koopman, P., Capel, B., Economou, A., and Munsterberg, A. (1990). A gene mapping to the sex-determining region of the mouse Y chromosome is a member of a novel family of embryonically expressed genes. *Nature* **346,** 245–250.
Gudas, L. J. (1991). Retinoic acid and teratocarcinoma stem cells. *Sem. Dev. Biol.* **2,** 171–179.
Gudas, L. J., Sporn, M. B., and Roberts, A. B. (1994). Cellular biology and biochemistry of retinoids. *In* "The Retinoids: Biology, Chemistry and Medicine" (M. B. Sporn, A. B. Roberts, and D. S. Goodman, Eds.), 2nd ed., pp. 443–520. New York, Raven Press.
Hale, F. (1937). Relation of maternal vitamin A deficiency to microphthalmia in pigs. *Texas State J. Med.* **33,** 228–232.
Hamada-Kanazawa, M., Ishikawa, K., Nomoto, K., Uozumi, T., Kawai, Y., Narahara, M., and Miyake, M. (2004a). Sox6 expression causes cellular aggregation and the neuronal differentiation of P19 embryonic carcinoma cells in the absence of retinoic acid. *FEBS Lett.* **560,** 192–198.
Hamada-Kanazawa, M., Ishikawa, K., Ogawa, D., Kanai, M., Kawai, Y., Narahara, M., and Miyake, M. (2004b). Suppression of Sox6 in P19 cells leads to failure of neuronal differentiation by retinoic acid and induces retinoic acid-dependent apoptosis. *FEBS Lett.* **577,** 60–66.
Handschumacher, R. E., Harding, M. W., Rice, J., Drugge, R. J., and Speicher, D. W. (1984). Cyclophilin: A specific cytosolic binding protein for cyclosporin A. *Science* **226,** 44–47.
Harris, T. M., and Childs, G. (2002). Global gene expression patterns during differentiation of F9 embryonal carcinoma cells into parietal endoderm. *Funct. Integr. Genomics* **2,** 105–119.
Hart, S. M. (2002). Modulation of nuclear receptor dependent transcription. *Biol. Res.* **35,** 295–303.
Hong, Y., Winkler, C., and Schartl, M. (1996). Pluripotentency and differentiation of embryonic stem cell lines from the medakafish (*Oryzias latipes*). *Mech. Dev.* **60,** 33–44.
Hosler, B. A., LaRosa, G. J., Grippo, J. F., and Gudas, L. J. (1989). Expression of *REX-1*, a gene containing zinc finger motifs, is rapidly reduced by retinoic acid in F9 teratocarcinoma cells. *Mol. Cell. Biol.* **9,** 5623–5629.
Hosler, B. A., Rogers, M. B., Kozak, C. A., and Gudas, L. J. (1993). An octamer motif contributes to the expression of the retinoic acid-regulated zinc finger gene Rex-1 (Zfp-42) in F9 teratocarcinoma cells. *Mol. Cell. Biol.* **13,** 2919–2928.
Hu, L., and Gudas, L. J. (1990). Cyclic AMP analogs and retinoic acid influence the expression of retinoic acid receptor $\alpha\beta$ and γmRNAs in F9 teratocarcinoma cells. *Mol. Cell. Biol.* **10,** 391–396.
Jacobs, S., Lie, D. C., DeCicco, K. L., Shi, Y., DeLuca, L. M., Gage, F. H., and Evans, R. M. (2006). Retinoic acid is required early during adult neurogenesis in the dentate gyrus. *Proc. Natl. Acad. Sci. USA* **103,** 3902–3907.
Jan, Y. N., and Jan, L. Y. (1993). HLH proteins, fly neurogenesis, and vertebrate myogenesis. *Cell* **75,** 827–830.
Jepsen, K., and Rosenfeld, M. G. (2002). Biological roles and mechanistic actions of corepressor complexes. *J. Cell Sci.* **115,** 689–698.
Johnson, J. E., Zimmerman, K., Saito, T., and Anderson, D. J. (1992). Induction and repression of mammalian achaete-scute homologue (MASH) gene expression during neuronal differentiation of P19 embryonal carcinoma cells. *Development* **114,** 75–87.

Jones-Villeneuve, E. M. V., McBurney, M. W., Rogers, K. A., and Kalnins, V. I. (1982). Retinoic acid induces embryonal carcinoma cells to differentiate into neurons and glial cells. *J. Cell Biol.* **94,** 253–262.

Jones-Villeneuve, E. M. V., Rudnicki, M. A., Harris, J. F., and McBurney, M. W. (1983). Retinoic acid-induced neural differentiation of embryonal carcinoma cells. *Mol. Cell. Biol.* **3,** 2271–2279.

Kawaguchi, J., Mee, P. J., and Smith, A. G. (2005). Osteogenic and chondrogenic differentiation of embryonic stem cells in response to specific growth factors. *Bone* **36,** 758–769.

Kim, S., Yoon, Y. S., Kim, J. W., Jung, M., Kim, S. U., Lee, Y. D., and Suh-Kim, H. (2004). Neurogenin1 is sufficient to induce neuronal differentiation of embryonal carcinoma P19 cells in the absence of retinoic acid. *Cell. Mol. Neurobiol.* **24,** 343–356.

Knoepfler, P. S., and Kamps, M. P. (1997). The Pbx family of proteins is strongly upregulated by a posttranscriptional mechanism during retinoic acid-induced differentiation of P19 embryonal carcinoma cells. *Mech. Dev.* **63,** 5–14.

Kruyt, F. A., van den Brink, C. E., Defize, L. H., Donath, M. J., Kastner, P., Kruijer, W., Chambon, P., and van der Saag, P. T. (1991). Transcriptional regulation of retinoic acid receptor beta in retinoic acid-sensitive and -resistant P19 embryonalcarcinoma cells. *Mech. Dev.* **33,** 171–178.

Kruyt, F. A., van der Veer, L. J., Mader, S., van den Brink, C. E., Feijen, A., Jonk, L. J., Kruijer, W., and van der Saag, P. T. (1992). Retinoic acid resistance of the variant embryonal carcinoma cell line RAC65 is caused by expression of a truncated RAR alpha. *Differentiation* **49,** 27–37.

Langston, A. W., and Gudas, L. J. (1992). Identification of a retinoic acid responsive enhancer 3' of the murine homeobox gene Hox1-6. *Mech. Dev.* **38,** 217–228.

Langston, A. W., Thompson, J. R., and Gudas, L. J. (1997). Retinoic acid-responsive enahncers located 3' of the HoxA and HoxB homeobox gene clusters. Functional analysis. *J. Biol. Chem.* **272,** 2167–2175.

LaRosa, G. J., and Gudas, L. J. (1988). An early effect of retinoic acid: Cloning of an mRNA (Ear-1) exhibiting rapid and protein synthesis-independent induction during teratocarcinoma stem cell differentiation. *Proc. Natl. Acad. Sci. USA* **85,** 329–333.

Li, R., Faria, T. N., Boehm, M., Nabel, E. G., and Gudas, L. J. (2004). Retinoic acid causes cell growth arrest and an increase in p27 in F9 wild type but not in F9 retinoic acid receptor β_2 knockout cells. *Exp. Cell Res.* **294,** 290–300.

Loudig, O., Babichuk, C., White, J., Abu-Abed, S., Mueller, C., and Petkovich, M. (2000). Cytochrome P450RAI(CYP26) promoter, a distinct composite retinoic acid response element underlies the complex regulation of retinoic acid metabolism. *Mol. Endocrinol.* **14,** 1483–1497.

Ma, Q., Kintner, C., and Anderson, D. J. (1996). Identification of neurogenin, a vertebrate neuronal determination gene. *Cell* **87,** 43–52.

Ma, Q., Sommer, L., Cserjesi, P., and Anderson, D. J. (1997). Mash1 and neurogenin1 expression patterns define complementary domains of neuroepithelium in the developing CNS and are correlated with regions expressing notch ligands. *J. Neurosci.* **17,** 3644–3652.

Mark, M., Ghyselinck, N. B., and Chambon, P. (2006). Function of retinoid nuclear receptors: Lessons from genetic and pharamacological dissections of the retinoic acid signaling pathway during mouse embryogenesis. *Ann. Rev. Pharmacol. Toxicol.* **46,** 451–480.

Martin, G. R. (1981). Isolation of a pluripotent cell line from early mouse embryos cultured in medium conditioned by teratocarcinoma stem cells. *Proc. Natl. Acad. Sci. USA* **78,** 7634–7638.

Martinez-Ceballos, E., Chambon, P., and Gudas, L. J. (2005). Differences in gene expression between wild type and hoax-1 knockout embryonic stem cells after retinoic acid treatment or leukemia inhibitory factor (LIF) removal. *J. Biol. Chem.* **280,** 16484–16498.

Mason, K. E. (1935). Foetal death, prolonged gestation, and difficult parturition in the rat as a result of vitamin A deficiency. *Am. J. Anat.* **57**, 303–349.
McBurney, M. W., and Rogers, B. J. (1982). Isolation of male embryonal carcinoma cells and their chromosome replication patterns. *Dev. Biol.* **89**, 503–508.
McBurney, M. W., Jones-Villeneuve, E. M. V., Edwards, M. K. S., and Anderson, P. J. (1982). Control of muscle and neuronal differentiation in a cultured embryonal carcinoma cell line. *Nature* **299**, 165–167.
McCaffery, P. M., and Drager, U. C. (2000). Regulation of retinoic acid signaling in the embryonic nervous system: A master differentiation factor. *Cytokine Growth Factor Rev.* **11**, 233–249.
McMahon, A. P., Joyner, A. L., Bradley, A., and McMahon, J. A. (1992). The midbrain–hindbrain phenotype of $Wnt\text{-}1^-/Wnt\text{-}1^-$ mice results from stepwise deletion of engrailed-expressing cells by 9.5 days postcoitum. *Cell* **69**, 581–595.
Miller, J. R., and Randell, T. M. (1996). Signal transduction through β-catenin and specification of cell fate during embryogenesis. *Genes Dev.* **10**, 2527–2539.
Mummery, C. L., Feijen, A., Molenaar, W. H., van den Brink, C. E., and De Laat, S. W. (1986). Establishment of a differentiated mesodermal line form P19 EC cells expressing functional PDGF and EGF receptors. *Exp. Cell Res.* **165**, 229–242.
Nishiguchi, S., Joh, T., Horie, K., Zou, Z., Yasunaga, T., and Shimada, K. (1994). A survey of genes expressed in undifferentiated mouse embryonal carcinoma F9 cells: Characterization of low-abundance mRNAs. *J. Biochem.* **116**, 128–139.
Nusse, R., and Varmus, H. E. (1982). Many tumors induced by the mouse mammary tumor virus contain a provirus integrated in the same region of the host genome. *Cell* **31**, 99–109.
Oda, H., Iwata, I., Yasunami, M., and Ohkubo, H. (2000). Structure of the mouse NDRF gene and its regulation during neuronal differentiation of P19 cells. *Brain Res. Mol. Brain Res.* **77**, 37–46.
Ogura, T., and Evans, R. M. (1995). Evidence for two distinct retinoic acid response pathways for HOXB1 gene regulation. *Proc. Natl. Acad. Sci. USA* **92**, 392–396.
Oren, T., Sher, J. A., and Evans, T. (2003). Hematopoiesis and retinoids: Development and disease. *Leuk. Lymphoma* **44**, 1881–1891.
Pain, B., Clark, M. E., Shen, M., Nakazawa, H., Sakurai, M., Samarut, J., and Etches, R. J. (1996). Long-term *in vitro* cultures and characterization of avian embryonic stem cells with multiple morphogenetic potentialities. *Development* **122**, 2339–2348.
Papkoff, J. (1994). Identification and biochemical characterization of secreted Wnt-1 protein from P19 embryonal carcinoma cells induced to differentiate along the neuroectodermal lineage. *Oncogene* **9**, 313–317.
Pevny, L. H., and Lovell-Badge, R. (1997). Sox genes find their feet. *Curr. Opin. Genet. Dev.* **7**, 338–344.
Phillips, B. W., Vernochet, C., and Dani, C. (2003). Differentiation of embryonic stem cells for pharmacological studies on adipose cells. *Pharm. Res.* **47**, 264–268.
Pozzi, S., Rossetti, S., Bistulfi, G., and Sacchi, N. (2006). RAR-mediated epigenetic control of the cytochrome P450 Cyp26a1 in embryocarcinoma cells. *Oncogene* **25**, 1400–1407.
Pratt, M. A., Kralova, J., and McBurney, M. W. (1990). A dominant negative mutation of the alpha retinoic acid receptor gene in a retinoic acid-nonresponsive embryonal carcinoma cell. *Mol. Cell. Biol.* **10**, 6445–6553.
Pratt, M. A., Langston, A. W., Gudas, L. J., and McBurney, M. W. (1993). Retinoic acid fails to induce expression of Hox genes in differentiation-defective murine embryonal carcinoma cells carrying a mutant gene for alpha retinoic acid receptor. *Differentiation* **53**, 105–113.
Prior, H. M., and Walter, M. A. (1996). SOX genes: Architects of development. *Mol. Med.* **2**, 405–412.
Qin, P., Haberbusch, J. M., Soprano, K. J., and Soprano, D. R. (2004a). Retinoic acid regulates the expression of PBX1, PBX2 and PBX3 in P19 cell both transcriptionally and post-transcriptionally. *J. Cell. Biochem.* **92**, 147–163.

Qin, P., Haberbusch, J. M., Zhang, Z., Soprano, K. J., and Soprano, D. R. (2004b). Pre-B cell leukemia transcription factor (PBX) proteins are important mediators for retinoic acid-dependent endodermal and neuronal differentiation of mouse embryonal carcinoma P19 cells. *J. Biol. Chem.* **279,** 16263–16271.

Ray, W. J., Bain, G., Yao, M., and Gottlieb, D. I. (1997). CYP26, a novel mammalian cytochrome P450, is induced by retinoic acid and defines a new family. *J. Biol. Chem.* **272,** 18702–18708.

Rijsewijk, F., Schuermann, M., Wagenaar, E., Parren, P., Weigel, D., and Nusse, R. (1987). The *Drosophila* homolog of the mouse mammary oncogene int-1 is identical to the segment polarity gene wingless. *Cell* **50,** 649–657.

Rochette-Egly, C., Plassat, J. L., Taneja, R., and Chambon, P. (2000). The AF-1 and AF-2 activating domains of retinoic acid receptor (RARalpha) and their phosphorylation are differently involved in parietal endodermal differentiation of F9 cells and retinoid-induced expression of target genes. *Mol. Endo.* **14,** 1398–1410.

Rogers, M. B., Hosler, B. A., and Gudas, L. J. (1991). Specific expression of a retinoic acid-regulated, zinc-finger gene, Rex-1, in preimplantation embryos, trophoblast and spermatocytes. *Development* **113,** 815–824.

Rohwedel, J., Guan, K., and Wobus, A. M. (1999). Induction of cellular differentiation by retinoic acid in vitro. *Cells Tissues Organs* **165,** 190–202.

Rosfjord, E., and Rizzino, A. (1994). The octamer motif present in the Rex-1 promoter binds Oct-1 and Oct-3 expressed by EC and ES cells. *Biochem. Biophys. Res. Commun.* **203,** 1795–1802.

Ross, S. A., McCaffrey, P. J., Drager, U. C., and De Luca, L. M. (2000). Retinoids in embryonal development. *Physiol. Rev.* **80,** 1021–1054.

Sangster-Guity, N., Yu, L. M., and McCormick, P. (2004). Molecular profiling of embryonal carcinoma cells following retinoic acid or histone deacetylase inhibitor treatment. *Cancer Biol. Ther.* **3,** 1109–1120.

Schoorlemmer, J., Jonk, L., Sanbing, S., van Puijenbroek, A., Feijen, A., and Kruijer, W. (1995). Regulation of Oct-4 gene expression during differentiation of EC cells. *Mol. Biol. Rep.* **21,** 129–140.

Schuldiner, M., Yanuka, O., Itskovitz-Eldor, J., Melton, D. A., and Benvenisty, N. (2000). Effects of eight growth factors on the differentiation of cells derived from human embryonic stem cells. *Proc. Natl. Acad. Sci. USA* **97,** 11307–11312.

Schuuring, E., Deemter, L., van Roelink, H., and Nusse, R. (1989). Transient expression of the proto-oncogene *int-1* during differentiation of P19 embryonal carcinoma cells. *Mol. Cell. Biol.* **9,** 1357–1361.

Shamblott, M., Axelman, J., Wang, S., Buggs, E., Littlefield, J., Donovan, P., Blumenthal, P., Huggins, G., and Gearhart, J. (1998). Derivation of pluripotent stem cells from cultured human primordial germ cells. *Proc. Natl. Acad. Sci. USA* **95,** 13726–13731.

Shen, J., Wu, H., and Gudas, L. J. (2000). Molecular cloning and analysis of a group of genes differentially expressed in cells which overexpress the Hoxa-1 homeobox gene. *Exp. Cell Res.* **259,** 274–283.

Shen, S., and Gudas, L. J. (2000). Molecular cloning of a novel retinoic acid-responsive gene, HAIR-62, which is also up-regulated in Hoxa-1-overexpressing cells. *Cell Growth Differ.* **11,** 11–17.

Shen, S., Kruyt, F. A., den Hertoz, J., van der Saag, P. T., and Kruijer, W. (1991). Mouse and human retinoic acid receptor beta 2 promoters: Sequence comparison and localization of retinoic acid responsiveness. *DNA Seq.* **2,** 111–119.

Simeone, A., Acampora, D., Nigro, V., Faiella, A., D'Esposito, M., Stornauiolo, A., Mavilio, F., and Boncinelli, E. (1991). Differential regulation by retinoic acid of the homeobox genes of the four Hox loci in human embryonal carcinoma cells. *Mech. Dev.* **33,** 215–228.

Sinclair, A., Berta, P., Palmer, M., Hawkins, J., Griffiths, B., and Smith, M. (1990). A gene from the human sex-determining region encodes a protein with homology to a conserved DNA-binding motif. *Nature* **346**, 240–244.

Smith, A. G. (2001). Embryo-derived stem cells: Of mice and men. *Annu. Rev. Cell Dev. Biol.* **17**, 435–462.

Smith, E. R., Capo-chichi, C. D., He, J., Smedberg, J. L., Yang, D. H., Prowse, A. H., Goodwin, A. K., Hamilton, T. C., and Xu, X. X. (2001). Disabled-2 mediates c-Fos suppression and the cell growth regulatory activity of retinoic acid in embryonic carcinoma cells. *J. Biol. Chem.* **276**, 47303–47310.

Smolich, B. D., and Papkoff, J. (1994). Regulated expression of Wnt family members during neuroectodermal differentiation of P19 embryonal carcinoma cell: Overexpression of *Wnt-1* perturbs normal differentiation-specific properties. *Dev. Biol.* **166**, 300–310.

Sonneveld, E., van den Brink, C. E., Tertoolen, L. G., van der Burg, B., and van der Saag, P. T. (1999). Retinoic acid hydroxylase (CYP26) is a key enzyme in neuronal differentiation of embryonal carcinoma cells. *Dev. Biol.* **21**, 390–404.

Song, J., Lu, Y. C., Yokoyama, K., Rossi, J., and Chiu, R. (2004). Cyclophilin A is required for retinoic acid-induced neuronal differentiation in p19 cells. *J. Biol. Chem.* **279**, 24414–24419.

Soprano, D. R., and Soprano, K. J. (1995). Retinoids as teratogens. *Annu. Rev. Nutr.* **15**, 111–132.

Soprano, D. R., and Soprano, K. J. (2003). Role of RARs and RXRs in mediating the molecular mechanism of action of Vitamin A. *In* "Molecular Nutrition" (J. Zempleni and H. Daniel, Eds.), pp. 135–149. CABI, Cambridge, MA.

Soprano, D. R., Soprano, K. J., Wyatt, M. L., and Goodman, D. S. (1988). Induction of the expression of retinol-binding protein and transthyretin in F9 embryonal carcinoma cells differentiated to embryoid bodies. *J. Biol. Chem.* **263**, 17897–17900.

St.-Arnaud, R., Craig, J., McBurney, M. W., and Papkoff, J. (1989). The *int-1* proto-oncogene is transcriptionally activated during neuroectodermal differentiation of P19 mouse embryonal carcinoma cells. *Oncogene* **4**, 1077–1080.

Steele, B. K., Meyers, C., and Ozbun, M. A. (2002). Variable expression of some "housekeeping" genes during human keratinocyte differentiation. *Anal. Biochem.* **307**, 341–347.

Strickland, S., and Mahdavi, V. (1978). The induction of differentiation in teratocarcinoma stem cells by retinoic acid. *Cell* **15**, 393–403.

Strickland, S., Smith, K. K., and Marotti, K. R. (1980). Hormonal induction of differentiation in teratocarcinoma stem cells: Generation of parietal endoderm by retinoic acid and dibutyryl cAMP. *Cell* **21**, 347–355.

Strubing, C., Ahnert-Hilger, G., Shan, J., Wiedenmann, B., Hescheler, J., and Wobus, A. M. (1995). Differentiation of pluripotent embryonic stem cells into the neuronal lineage *in vitro* gives rise to mature inhibitory and excitatory neurons. *Mech. Dev.* **53**, 275–287.

Sun, L., Bradford, C. S., Ghosh, C., Collodi, P., and Barnes, D. W. (1995). ES-like cell cultures derived from early zebrafish embryos. *Mol. Mar. Biol. Biotechnol.* **4**, 193–199.

Sun, Y., Nadal-Vicens, M., Misono, S., Lin, M. Z., Zubiaga, A., Hua, X., Fan, G., and Greenberg, M. E. (2001). Neurogenin promotes neurogenesis and inhibits glial differentiation by independent mechanisms. *Cell* **104**, 365–376.

Takamatsu, N., Kanda, H., Tsuchiya, I., Yamada, S., Ito, M., Kabeno, S., Shiba, T., and Yamashita, S. (1995). A gene that is related to SRY and is expressed in the testes encodes a leucine zipper-containing protein. *Mol. Cell. Biol.* **15**, 3759–3766.

Takeichi, M. (1995). Morphogenetic roles of classic cadherins. *Curr. Opin. Cell Biol.* **7**, 619–627.

Taneja, R., Bouillet, P., Boylan, J. F., Gaub, M. P., Roy, B., Gudas, L. J., and Chambon, P. (1995). Reexpression of retinoic acid receptor (RAR) gamma or overexpression of RAR alpha or RAR beta in RAR gamma-null F9 cells reveals a partial functional redundancy between the three RAR types. *Proc. Natl. Acad. Sci. USA* **92**, 7854–7858.

Taneja, R., Roy, B., Plassat, J.-L., Zusi, C. F., Ostrowski, J., Reczek, P. R., and Chambon, P. (1996). Cell-type and promoter-context dependent retinoic acid receptor (RAR) redundancies for RARβ2 and *Hoxa-1* activation in F9 and P19 cells can be artefactually generated by gene knockouts. *Proc. Natl. Acad. Sci. USA* **93**, 6197–6202.

Taneja, R., Rochette-Egly, C., Plassat, J. L., Penna, L., Gaub, M. P., and Chambon, P. (1997). Phosphorylation of activation functions AF-1 and AF-2 of RARα and RARγ is indispensable for differentiation of F9 cells upon retinoic acid and cAMP treatment. *EMBO J.* **16**, 6452–6465.

Tang, K., Yang, J., Gao, X., Wang, C., Liu, L., Kitani, H., Atsumi, T., and Jing, N. (2002). Wnt-1 promotes neuronal differentiation and inhibits gliogenesis in P19 cells. *Biochem. Biophys. Res. Commun.* **293**, 167–173.

Teramoto, S., Kihara-Negishi, F., Sakurai, T., Yamada, T., Hashimoto-Tamaoki, T., Tamura, S., Kohno, S., and Oikawa, T. (2005). Classification of neural differentiation-associated genes in P19 embryonal carcinoma cells by their expression patterns induced after cell aggregation and/or retinoic acid treatment. *Oncol. Rep.* **14**, 1231–1238.

Thellin, O., Zorzi, W., Lakaye, B., De Borman, B., Coumans, B., Hennen, G., Grisar, T., Igout, A., and Heinen, E. (1999). Housekeeping genes as internal standards: Use and limits. *J. Biotechnol.* **75**, 291–295.

Thomson, J. A., and Marshall, V. S. (1998). Primate embryonic stem cells. *Curr. Top. Dev. Biol.* **38**, 133–165.

Thomson, J. A., Kalishman, J., Golos, T. G., Durning, M., Harris, C. P., Becker, R. A., and Hearn, J. P. (1995). Isolation of a primate embryonic stem cell line. *Proc. Natl. Acad. Sci. USA* **92**, 7844–7848.

Thomson, J. A., Kalishman, J., Golos, T. G., Durning, M., Harris, C. P., and Hearn, J. P. (1996). Pluripotent cell lines derived from common marmoset (*Callithix jacchus*) blastocysts. *Biol. Reprod.* **55**, 254–259.

Thomson, J. A., Itskovitz-Eldor, J., Shapiro, S., Waknitz, M., Swiergiel, J., Marshall, V., and Jones, J. (1998). Embryonic stem cell lines derived from human blastocytes. *Science* **282**, 1145–1147.

Thompson, J. R., and Gudas, L. J. (2002). Retinoic acid induces parietal endoderm but not primitive endoderm and visceral endoderm differentiation in F9 teratocarcinoma stem cells with a targeted deletion of the Rex-1 (Zfp-42) gene. *Mol. Cell. Endocrinol.* **195**, 119–133.

Thompson, J. R., Huang, D. Y., and Gudas, L. J. (1998). The murine Hoxb1 3'RAIDR enhancer contains multiple regulatory elements. *Cell Growth Differ.* **9**, 969–981.

Umbhauer, M., Penzo-Mendez, A., Clavilier, L., Boucaut, J., and Riou, J. (2000). Signaling specificities of fibroblast growth factor receptors in early Xenopus embryo. *J. Cell Sci.* **113**, 2865–2875.

Wang, Z., Ge, J., Huang, B., Gao, Q., Liu, B., Wang, L., Yu, L., Fan, Z., Lu, Z., and Liu, J. (2005). Differentiation of embryonic stem cells into corneal epithelium. *Sci. China Life Sci.* **48**, 471–480.

Wegner, M. (1999). From head to toes: The multiple facets of Sox proteins. *Nucleic Acids Res.* **27**, 1409–1420.

Wei, L.-N. (2003). Retinoid receptors and their coregulators. *Ann. Rev. Pharmacol. Toxicol.* **43**, 47–72.

Wei, Y., Harris, T., and Childs, G. (2002). Global gene expression patterns during neural differentiation of P19 embryonic carcinoma cells. *Differentiation* **70**, 204–219.

Westin, S., Rosenfeld, M. G., and Glass, C. K. (2000). Nuclear receptor coactivators. *Adv. Pharmacol.* **47**, 89–112.

Wobus, A. M., Rohwedel, J., Maltsev, V., and Hescheler, J. (1994). *In vitro* differentiation of embryonic stem cells into cardiomyocytes or skeletal muscles cells is specifically modulated by retinoic acid. *Rouxs Arch. Dev. Biol.* **204**, 36–45.

Wobus, A. M., Kaomei, G., Shan, J., Wellner, M. C., Rohwedel, J., Ji., G., Fleischmann, B., Katus, H. A., Hescheler, J., and Franz, W. J. (1997). Retinoic acid accelerates embryonic stem cell-derived cardiac differentiation and enhances development of ventricular cardiomyocytes. *J. Mol. Cell. Cardiol.* **29,** 1525–1539.

Young, P. R., and Tilghman, S. M. (1984). Induction of alpha-fetoprotein synthesis in differentiating F9 teratocarcinoma cells is accompanied by a genome-wide loss of DNA methylation. *Mol. Cell. Biol.* **4,** 898–907.

Zhong, H., and Simons, J. W. (1999). Direct comparison of GAPDH, beta-actin, cyclophilin, and 28S rRNA as internal standards for quantifying RNA levels under hypoxia. *Biochem. Biophys. Res. Commun.* **259,** 523–526.

Zhuang, Y., Faria, T. N., Chambon, P., and Gudas, L. J. (2003). Identification and characterization of retinoic acid receptors beta2 target genes in F9 teratocarcinoma cells. *Mol. Cancer Res.* **1,** 619–630.

Zile, M. (2001). Function of vitamin A in vertebrate embryonic development. *J. Nutr.* **131,** 705–708.

Zwartkruis, F., Kruyt, F., van der Saag, P. T., and Meijlink, F. (1993). Induction of HOX-2 genes in P19 embryocarcinoma cells is dependent on retinoic acid receptor alpha. *Exp. Cell Res.* **205,** 422–425.

4

Metabolism of Retinol During Mammalian Placental and Embryonic Development

Geoffroy Marceau,*,†,1 Denis Gallot,*,‡,1 Didier Lemery,*,‡ and Vincent Sapin*,†

*Université d'Auvergne, JE 2447, ARDEMO, F-63000, Clermont-Ferrand, France
†INSERM, U.384, Laboratoire de Biochimie, Faculté de Médecine
F-63000, Clermont-Ferrand, France
‡CHU Clermont-Ferrand, Maternité, Hôtel-Dieu, F-63000
Clermont-Ferrand, France

I. General Aspects of Retinol Transport and Metabolism in Mammalian Species
II. Placental Transport and Metabolism of Retinol During Mammalian Development
III. Embryonic Metabolism of Retinol During Mammalian Development
References

Retinol (vitamin A) is a fat-soluble nutrient indispensable for a harmonious mammalian gestation. The absence or excess of retinol and its active derivatives [i.e., the retinoic acids (RAs)] can lead to abnormal development of embryonic and extraembryonic (placental) structures. The embryo is unable to synthesize the retinol and is strongly dependent on the maternal delivery of retinol itself or precursors: retinyl

[1]Both authors contribute equally to this work.

esters or carotenoids. Before reaching the embryonic tissue, the retinol or the precursors have to pass through the placental structures. During this placental step, a simple diffusion of retinol can occur between maternal and fetal compartments; but retinol can also be used *in situ* after its activation into RA[1] or stored as retinyl esters. Using retinol-binding protein knockout model, an alternative way of embryonic retinol supply was described using retinyl esters incorporated into maternal chylomicrons. In the embryo, the principal metabolic event occurring for retinol is its conversion into RAs, the active molecules implicated on the molecular control of embryonic morphogenesis and organogenesis. All these placental and embryonic events of retinol transport and metabolism are highly regulated. Nevertheless, some genetic and/or environmental abnormalities in the transport and/or metabolism of retinol can be related to developmental pathologies during mammalian development. © 2007 Elsevier Inc.

I. GENERAL ASPECTS OF RETINOL TRANSPORT AND METABOLISM IN MAMMALIAN SPECIES

The retinol (vitamin A) belongs to the "retinoids" family including both the compounds possessing one of the biological activities of the retinol (ROH) and the many synthetic analogues related structurally to the retinol, with or without a biological activity. Provitamin A is the dietary source of retinol and is supplied as carotenoids (mainly β-carotene) in vegetables and preformed retinyl esters (long-chain fatty acid esters of retinol: palmitate, oleate, stearate, and linoleate) in animal meat (Blomhoff, 1994). Retinol plays a central role in many essential biological processes such as vision, immunity, reproduction, growth, development, control of cellular proliferation, and differentiation (Chambon, 1996). The main active forms of retinol are retinoic acids (RAs), except for reproduction and vision, where retinol and retinal also play important roles.

Two vehicles are described for mammalian blood transport of retinoids. First, the retinyl esters and carotenoids can be incorporated in intact or remnant chylomicrons or very low-density lipoproteins (Debier and Larondelle, 2005). Second, the main form of retinol blood transport (1 μmol/l) is the association with a specific binding protein (RBP), which is itself

---

[1]Abbreviations: ADH, alcohol dehydrogenases; CRABP, cellular retinoic acid-binding protein; CRBP, cellular retinol-binding protein; dpc, days post coïtum; LRAT, lecithin retinol acyltransferase; RA, retinoic acid; Ral, retinaldehyde; RALDH, retinaldehyde dehydrogenase; RBP, retinol-binding protein; RDH, retinol dehydrogenase; ROH, retinol.

complexed with transthyretin (ratio 1 mol/1mol). The constitution of this ternary complex prevents the glomerular filtration of the small RBP-retinol form (21 kDa) and increases the affinity of RBP for retinol (Bellovino et al., 2003). A binding to other plasma proteins, such as albumin or lipocalins, is also described for retinol. Albumin could serve as a transporter for RA, which circulates in very small levels in the blood. The transfer of retinol to target cells involves a specific membrane-bound RBP receptor (Sivaprasadarao et al., 1998). To date, the debate still remains concerning the molecular mechanisms of the cellular retinol penetration: endocytosis, dissociation of RBP-retinol complex, and intracellular degradation of RBP or extracellular dissociation of RBP-retinol complex and delivery of retinol via transmembrane pore. The uptake of remnant chylomicrons and very low-density lipoproteins (containing retinyl esters and carotenoids) is realized by target tissues using, respectively, the lipoprotein lipase and low-density lipoproteins receptor pathways. Bound to albumin, RA can be transferred into the tissues by passive diffusion, with an efficiency of transfer, which is cell type and tissue specific.

To be biologically active (Fig. 1), retinol must first be oxidized to retinaldehyde and then to RA. A large number of enzymes catalyze the reversible oxidation of retinol to retinaldehyde: the alcohol dehydrogenases (ADH), the retinol dehydrogenase (RDH) of the microsomal fraction, and some members of the cytochrome P450 family. Several enzymes are able to catalyze irreversibly the oxidation of retinaldehyde to RA: the retinal dehydrogenases (RALDH1, 2, 3, and 4) and also members of the cytochrome P450 family (Liden and Eriksson, 2006). These enzymatic reactions could be antagonized and/or stopped by several toxic molecules, namely ethanol, citral, nitrofen, or bisdiamine, leading to an exogenous alteration of RA production. Specific isomerization reactions are also likely to occur within the cells, since there are at least two RA stereoisomers *in vivo* (all-*trans* and 9-*cis* RA) exhibiting distinct biochemical activities. The catabolism of all-*trans* and 9-*cis* RA is also an important mechanism for controlling RA levels in cell and tissues and is carried out by three specific members of cytochrome P450s, CYP26A1, B1, and C1 (19). RA is catabolized to products such as 4-oxo-RA, 4-hydroxy-RA, 18-hydroxy-RA, and 5,18-epoxy-RA, which are finally excreted. These compounds can also undergo glucuronidation (Marill et al., 2003). An alternative metabolic pathway was present for intracellular retinol: the formation and storage as retinyl esters. Indeed, retinol may be esterified by two enzymes (lecithin retinol acyltransferase and diacylglycerol *O*-acyltransferase) into mostly long-chain retinyl esters such as retinyl palmitate, stearate, oleate, and linoleate. These esters are then stored in cytosolic lipid droplets. The mobilization of these retinyl esters and the release of retinol esters are realized by a retinyl ester hydrolase.

Since retinol, retinaldehyde, and RA are lipids, they lack appreciable water solubility and consequently must be bound to proteins within cells. Several intracellular-binding proteins for retinol, retinaldehyde, and RA

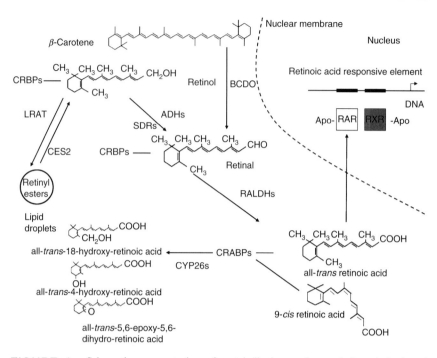

FIGURE 1. Schematic representation of metabolic (generation and degradation) and molecular-signaling pathway of RAs. Retinol is bound to its blood-binding protein (RBP) in a ternary complex (with TTR). It gets into the cell using a receptor (p63) mechanism and is linked to the CRBPs. Retinol could be stored as retinyl esters in lipid droplets or converted in RA bound to their cytoplasmic-binding proteins (CRABPs). RA enters into the nucleus to activate the nuclear receptors (RARS and RXRs) that are able to regulate transcription of target genes and is finally degraded by cytochrome P450 enzymes in active forms. Abbreviations: ADH, alcohol dehydrogenase; BCDO, β-carotene dioxygenase; CES2, carboxylesterase type 2; CRABP, cellular retinoic acid-binding protein; CRBP, cellular retinol-binding protein; LRAT, lecithin retinol acyltransferase; RA, retinoic acid; RALDH, retinaldehyde dehydrogenase; RAR, retinoic acid receptor; RARE, RAR responsive element; Ral, retinaldehyde; ROH, retinol; RBP, retinol-binding protein; RXR, retinoid X receptor; TTR, transthyretin.

have been identified and extensively characterized. They include cellular retinol-binding proteins type 1 and 2 (CRBP1 and 2) and cellular RA-binding proteins type 1 and 2 (CRABP1 and 2). The CRBP1 is a key protein to regulate the metabolism of retinol by orientating to storage, export of retinol, or conversion into RA (Ghyselinck et al., 1999). Both the CRABPs bind RA controlling the intracellular levels of retinoids, acting as cofactors for RA-metabolizing enzymes, and/or participating in the cytoplasmic-nuclear transport of RA (Napoli, 1999; Ong, 1994).

II. PLACENTAL TRANSPORT AND METABOLISM OF RETINOL DURING MAMMALIAN DEVELOPMENT

The placenta regulates the transport and metabolism of maternal nutrients transferred to the fetus. Abnormalities in these placental functions may have deleterious consequences for fetal development (Miller et al., 1993; Rossant and Cross, 2001). Since there is no de novo fetal synthesis of retinol, the developing mammalian embryo is dependent on the maternal circulation for its vitamin A supply. The presence of measurable hepatic vitamin A stored at birth is indicative of the functionality of placental transport during gestation (Ross and Gardner, 1994; Satre et al., 1992). A number of studies have investigated the ability of retinoids to pass through the placental barrier in mice or rabbits (Collins et al., 1994; Creech Kraft et al., 1989, 1991; Kochhar et al., 1988; Sass et al., 1999; Ward and Morriss-Kay, 1995). It is well established that each retinoid (e.g., retinol, 13-cis, 9-cis, all-trans RA, and their glycuronoconjugates) presents a specific rate of transfer. It has been proposed that this peculiarity could account for the variability in teratological effects of comparable amounts of different maternally absorbed retinoids (Nau et al., 1996).

The different intracellular-binding proteins for retinol, retinaldehyde and RA, are expressed in the mouse (Sapin et al., 1997), rat (Bavik et al., 1997), porcine (Johansson et al., 2001), and human placenta (Blanchon et al., 2002). CRBP1 is detected in the mouse visceral endoderm of the yolk sac, the mouse trophoblastic layer of the placental labyrinth closest to the fetal endothelium, the porcine areolar trophoblasts (Johansson et al., 1997, 2001), and the human villous trophoblastic cells (Blanchon et al., 2002). The CRBP2 is also described to be expressed in the mouse yolk sac and trophoblastic giant cells (Xueping et al., 2002). Moreover, it was shown that both fetal as well as maternal CRBP2 are required to ensure adequate delivery of vitamin A to the developing fetus when dietary vitamin A is limiting (Xueping et al., 2002). In addition, Johansson et al. (1999) precise that retinol metabolism may occur in the CRBP1 positive villous stromal cells and decidual cells of the basal plate in human placenta (Johansson et al., 1999). For RA transport, cellular RA-binding proteins are also described in placenta (Green and Ford, 1986; Levin et al., 1987). CRABP1 was found in hamster, human, and porcine placenta (Johansson et al., 2001; Okuno et al., 1987; Willhite et al., 1992), as well as CRABP2 in human placenta (Astrom et al., 1992). Throughout mouse placentation, the expression patterns of the *CRABP1* and *2* genes partly overlap in the decidual tissue and the vacuolar zones of the decidua, suggesting a role for these binding proteins in sequestering free RA from maternal blood, thus regulating its availability to the embryo.

Vitamin A is provided to the fetus through a limited and tightly controlled placental transfer (Bates, 1983; Moore, 1971). The amount of retinol provided to the fetus is usually maintained constant until maternal stores are depleted (Ismadi and Olson, 1982; Pasatiempo and Ross, 1990; Ross and Gardner, 1994). The first step of retinol transfer from maternal blood to embryo implicates the RBP. The exact mechanism of transfer remains discussed but it seems to involve RBP receptor. Indeed, a receptor for RBP has been characterized in the human placenta (Sivaprasadarao and Findlay, 1994). It is clearly established that maternal RBP does not cross the placental barrier and does not enter the developing embryo. Similar studies show convincingly that RBP of fetal origin is unable to cross the placenta and enter the maternal circulation. Thus, for retinol bound to RBP in the maternal circulation to be transferred to the fetus, it must be dissociated from maternal RBP after the binding to its receptor at the maternal face of placental barrier. Bound to the CRBPs, the retinol passes through the cytoplasm of the trophoblastic cells and enters the fetal circulation, where a new complex is formed using transthyretin and RBP of fetal origin (Quadro et al., 2004).

However, some studies have suggested that RBP might be dispensable for retinol placental transfer, as homozygous RBP null mutant mice are viable and fertile (Clagett-Dame and DeLuca, 2002). Accumulation of hepatic retinoids stores is not impaired in $RBP^{-/-}$ embryo. Indeed, the knockout mice accumulate retinol and retinyl ester in the liver at a higher rate compared with wild-type animals (Quadro et al., 1999). The normal sizes observed for litters from RBP-deficient dams and the usual good health of their pups indicate that the retinol bound to RBP is not the only source for retinoids reaching the embryo. Results demonstrate that retinyl esters in lipoproteins particles can be a significant source for retinoids (present postprandially in maternal blood) and can be used by the fetus to support embryogenesis (Quadro et al., 2005). These data are consistent with several previous works establishing that very low-density lipoproteins and low-density lipoproteins can be taken up by the placental cells (Bonet et al., 1995). This postprandial pathway may be also the delivery pathway of carotenoids (precursors of vitamin A) to the fetus through the placenta.

If the mammalian placenta is able to regulate the transfer of the retinoids from mother to the fetus, it also expresses several proteins with metabolic activities related to retinoids. In human placenta, a large number of enzymes catalyzing the oxidation of retinol into retinaldehyde are identified; among them are the nonspecific ADH of class I (Estonius et al., 1996) or class III (Sharma et al., 1989). In the guinea pig, a low-ADH activity is detected in placenta throughout gestation (Card et al., 1989), like during late pregnancy in the rat (Zorzano and Herrera, 1989) or in the ewe (Clarke et al., 1989). Specific enzymes like RDH of the microsomal fraction are able to oxidize retinol in retinaldehyde in the human placenta (personal unpublished data).

Among them, the more specific RDH, catalyzing the oxidation of 9-*cis* but not all-*trans* retinol, is expressed in human and mouse placenta (Gamble *et al.*, 1999). Other enzymes with an RDH activity are described in placenta: 17β-hydroxysteroid and 11β-hydroxysteroid dehydrogenases (Brown *et al.*, 2003; Lin *et al.*, 2006; Persson *et al.*, 1991). Moreover, the human type 1 isoforms of 3β-hydroxysteroid dehydrogenase/isomerase are expressed in the placenta (Thomas *et al.*, 2004), with a potentiality to oxidize the retinol in retinal. An RDH is also found in the yolk sac of rat embryos (Bavik *et al.*, 1997).

Secondarily, retinaldehyde had to be oxidized to RA by retinal dehydrogenase: RALDH1, 2, 3, and 4 (Liden and Eriksson, 2006). These enzymes were localized in the yolk sac of rat embryos (Bavik *et al.*, 1997). They were also detected in human choriocarcinoma (JEG-3 cell line) and placental cells. The retinol conversion into all-*trans* RA was demonstrated using high-performance liquid chromatography (HPLC) experiments (Blanchon *et al.*, 2002). The presence of 9-*cis* ROH conversion to 9-*cis* RA is also detected in human placenta. Two other mammalian placentas have also been shown to produce RA from retinol: the porcine (Parrow *et al.*, 1998) and the mouse (yolk sac) placenta (Bavik *et al.*, 1997). This RA generation was experimentally blocked by the presence of ethanol. This point may be a possible linkage between the nutrient supply of retinol to the placenta, the generation of strong developmental morphogene, and placental gene regulation and physiology. During pregnancy, placental cells may be exposed to deleterious maternal conditions, including alcohol abuse. Links have been established between alcohol abuse, fetal malformations, and alterations of retinoid metabolism (Leo and Lieber, 1999). The interferences of alcohol on synthesis of functional retinoids from retinol are clearly demonstrated (Wang, 2005). In this way, the alterations of placental retinoids metabolism by maternal ethanol ingestion may provide a novel and additional explanation for the genesis of fetal alcoholic syndrome and highlight the placental roles in this pathology.

The catabolism of all-*trans* and 9-*cis* RA is also an important mechanism for controlling RA levels in placental cell and tissues. Creech Kraft *et al.* (1989) have demonstrated that the early human placenta is able to metabolize 13-*cis* RA. In mouse, placenta's cytochrome also allows the transformation of all-*trans* RA in polar metabolites (Eckhoff *et al.*, 1989). It has been established that the three specific members of cytochrome P450s (CYP26A1, B1, and C1) are expressed at a high level in placenta (Ray *et al.*, 1997; Taimi *et al.*, 2004; Trofimova-Griffin and Juchau, 1998). Nevertheless, their protective activities are limited, when high levels of RAs are present in maternal blood. Indeed, they are unable to protect the fetus against teratogenic maternal blood levels of 13-*cis* RA as demonstrated by the large spectrum of fetal malformations occurring when mothers were treated with Roaccutane®.

The human term placental tissues (and more precisely, the villous mesenchymal fibroblasts) are able to esterify retinol (Sapin *et al.*, 2000). Nevertheless, the retinyl esters are never detected in human umbilical blood at delivery (Sapin *et al.*, 2000). It is well accepted that the placenta can be considered as a transitory, primitive functional liver during the first stages of mammalian development. During this period, the placenta stores retinol, waiting for liver maturity and functionality marked by the capacity to secrete RBP. This hypothesis is supported by results concerning the switch in retinoids content of embryonic and placental compartments during the development of the mouse conceptus (Satre *et al.*, 1992). During early organogenesis, the retinyl ester content of the placenta is nearly eightfold higher than the embryonic content. At the end of gestation, the embryonic retinyl ester content is nearly fourfold greater than placental one. Abnormalities concerning this switch between placental and embryonic retinyl esters stores are associated with intrauterine growth retardations (Sapin *et al.*, 2004). Little is known about enzymes implicated in the metabolism (anabolism and catabolism) of placental retinyl esters. The diacylglycerol acyltransferase (DGAT1) showing a nonspecific enzymatic property to esterify the retinol (Orland *et al.*, 2005) is expressed in the human primordial placenta (Gimes and Toth, 1993) and in the amniotic epithelium of amniotic membranes (personal unpublished data). At the opposite, the lecithin retinol acyltransferase (LRAT) esterifying more specifically the retinol (O'Byrne *et al.*, 2005) seems to be not expressed in placenta and amniotic membranes (personal unpublished data). The enzymes involved into the release of retinol from retinyl ester, that is, retinyl ester hydrolase (Linke *et al.*, 2005) like carboxylesterase, are active in rat placenta (Lassiter *et al.*, 1999). The carboxylesterase-2 isoform is also expressed in human amniotic epithelium (Zhang *et al.*, 2002), and more particularly into microsomal fraction (Yan *et al.*, 1999). Moreover, the lipoxygenase is able to oxidize all-*trans* retinol acetate (one form of retinyl ester) in human term placenta (Datta and Kulkarni, 1996).

The maternal and fetal blood levels of the β-carotene (provitamin A) strongly suggest that it may be used as a precursor of retinol in placenta (Dimenstein *et al.*, 1996). Note that we also detected the expression of the enzymes β-carotene-15,15'-dioxygenase (BCDO isoform 2) catalyzing the key step of retinol's cleaving β-carotene into two molecules of retinal (Kiefer *et al.*, 2001) in the amniotic part of human term fetal membranes, but not in the placenta (personal unpublished data). In conclusion, all these data reveal the complex properties of the mammalian placenta in term of retinoids metabolism: the production of active retinoids from retinol, the *cis/trans* isomerization and degradation of several RAs, the retinyl ester formation and hydrolysis, and the cleavage of β-carotene into retinal.

III. EMBRYONIC METABOLISM OF RETINOL DURING MAMMALIAN DEVELOPMENT

Vitamin A deficiency and excess have profound effects on the development of the vertebrate embryo (Lammer *et al.*, 1985; Wilson *et al.*, 1953). A molecular basis for these phenomena was proposed when it was found that vitamin A active forms, RAs, act through ligand-activated transcription factors and that RA is able to change the expression pattern of homeobox genes clusters (Conlon, 1995; Gudas, 1994). RA is indispensable for patterning the anteroposterior body axis, for morphogenesis and organogenesis (Mark and Chambon, 2003). Almost every organ or tissue can be affected by RAs if the embryo is treated with them at a critical time in development (Shenefelt, 1972). It illustrates the crucial role of RAs in the regulation of distinct developmental events (Ross *et al.*, 2000; Zile, 1998). The biogeneration of RA in the embryo appears to be the first developmental step in the initiation of RA-regulated signaling pathways (Zile, 2001). Tissue distribution of RA results from the balancing activities of RA-synthesizing enzymes (including retinaldehyde dehydrogenases RALDH1–4), and RA-catabolizing cytochrome P450 hydroxylases (CYP26A1, B1, and C1) (Mark *et al.*, 2004). Regulation of retinoid synthesis and catabolism can both be viewed as important ways in which distinct spatiotemporal patterns of active retinoids are maintained in the developing mammalian embryo. Therefore, the function of vitamin A is inseparable from its metabolism.

All of the physiologically important vitamin A metabolites and enzyme systems regulating vitamin A metabolism have been demonstrated in embryos. RA has been detected very early during vertebrate development. Using a transgenic mouse line carrying a β-galactosidase (*lacZ*) reporter gene under the regulation of three copies of the RARE from the RARβ2 gene, Rossant *et al.* (1991) found that prior to implantation sporadic staining was present in the inner cell mass of the blastocyst. RA has also been detected in the preimplantation porcine blastocyst (Parrow *et al.*, 1998). After implantation (egg cylinder stage or preprimitive streak stage) all-*trans* retinaldehyde (20 fmol/embryo), but not all-*trans* RA, was identified in mouse embryos (Ulven *et al.*, 2000). Reporter mice showed strong transgene expression in the posterior half of the embryo at the primitive streak stage. This observation was consistent with previous studies demonstrating greater RA synthesis, concentration, and activity in the posterior part of the early vertebrate embryo (Nieuwkoop, 1952). As somite formation and neural tube closure began, there was a sharp anterior boundary of expression corresponding to the preotic sulcus, which forms the border between presumptive rhombomeres 2 and 3 in the developing hindbrain. At later stages of development, transgene expression was noted in the somites, developing heart, lens and neural retina, the endoderm layer of the developing gut, the mesenchyme at the base of the developing limb buds, and the cervical and lumbar regions of the developing spinal cord (Colbert *et al.*, 1993; Moss *et al.*, 1998; Reynolds *et al.*, 1991). At even later times, the expression was noted in ectoderm between the mandible and maxilla and in the nasal placode, developing ear, skin, and somite-derived tissues, a number of internal organs (stomach, metanephric kidneys, and lung), eye, and developing limbs (Rossant *et al.*, 1991; Vermot *et al.*, 2003). Using HPLC, all-*trans* ROH and all-*trans* RA were identified as the primary retinoids in whole mouse embryos from 9 to 14 days post coïtum (dpc) (Horton and Maden, 1995), mouse limb buds (Satre and Kochhar, 1989), and human embryonic tissues (Creech Kraft *et al.*, 1993), with all-*trans* ROH representing the most abundant retinoid. Further study of individual tissues of the mouse embryo at 10.5 and 13 dpc revealed that all tissues contained at least some detectable RA (Horton and Maden, 1995). Spinal cord contained the highest amount of all-*trans* RA, which was enriched 15-fold over forebrain levels. 13-*cis*-RA has also been observed in the limb buds of E11 mouse embryos (Satre and Kochhar, 1989). It does not bind directly to the nuclear retinoids receptors and probably requires isomerization to all-*trans* RA before it acts (Repa *et al.*, 1993). Conversely, 9-*cis* RA is able to bind to and to activate the nuclear retinoids receptors, but this metabolite was never demonstrated in murine limb buds nor in whole embryos (Horton and Maden, 1995; Scott *et al.*, 1994). 4-oxo-all-*trans* RA and 4-oxo-all-*trans* retinol have been detected, but these metabolites do not seem to play an essential role in normal mammalian embryonic development and 4-oxo-all-*trans* RA is thought to represent an early step in the degradation of all-*trans* RA (Frolik *et al.*, 1979). Another all-*trans* ROH metabolite, 2-hydroxymethyl-3-methyl-5-(2'-oxopropyl)-2,5 dihydrothiophene, was identified in rat conceptus between 9.5 and 10 dpc, but it is unknown whether this metabolite plays a functional role during embryogenesis (Wellik and DeLuca, 1996).

As previously presented, the ADH family consists of numerous enzymes able to catalyze the reversible oxidation of a wide variety of substrates (ethanol, retinol) to the corresponding aldehydes. Three forms (ADH1, 3, and 4) are highly conserved in vertebrates and mammals. In humans, ADH4 demonstrated higher ROH dehydrogenase activity than ADH1, whereas ADH3 had insignificant ROH dehydrogenase activity (Deltour *et al.*, 1999). The mRNA for cytosolic ADH4 has been detected by polymerase chain reaction analysis in the egg-cylinder stage mouse embryo (Ulven *et al.*, 2000). The expression of ADH4 mRNA corresponds well both spatially and temporally with the presence of RA-like activity (Ang *et al.*, 1996). However, the enzyme is not absolutely essential for embryogenesis, as homozygous mutant mice null for ADH4 are viable and fertile as are ADH1 null mutant mice (Deltour *et al.*, 1999a,b). The synthesis of RA from ROH may be competitively inhibited by ethanol leading to RA deficiency. Maternal ethanol consumption during rat gestation modifies the retinyl ester and RA contents in developing fetal organs. Taken together, these experiments show definite interaction between ethanol and vitamin A, contributing to explain the mechanisms of prenatal ethanol consumption embryopathy (Zachman and Grummer, 1998).

Three members of the RALDH family (RALDH1, 2, and 3) can account for all of the all-*trans* RA generated in the early embryo (McCaffery and Dräger, 1997). RALDH2 plays a crucial role in the synthesis of RA and RALDH2 null mutant mice die *in utero* before 10.5 dpc (Niederreither*et al.*, 1999). Nevertheless, the ability to rescue the development of null mutant embryos by providing mothers with all-*trans* RA suggests that precisely located regions of all-*trans* RA synthesis are not essential, at least for some early all-*trans* RA-dependent morphogenetic events (Clagett-Dame *et al.*, 2002). In the mouse embryo RALDH2 appears first and is expressed in the mesoderm adjacent to the node and primitive streak but not within the node itself during gastrulation (Niederreither *et al.*, 1997). At later stages of development, RALDH2 expression localizes to undifferentiated somites, mesenchyme surrounding the neural tube, developing gut, heart, lung, kidney, eye, differentiating limbs, and specific regions of the head (Batourina *et al.*, 2001; Malpel *et al.*, 2000; Moss *et al.*, 1998; Niederreither *et al.*, 1997; Wagner *et al.*, 2000). RALDH1 and 3 appear later in development. Between 9 and 10 dpc, RALDH1 protein is found in the ventral mesencephalon of the mouse embryo, the dorsal retina, the thymic primordia, and the medial aspect of the otic vesicles (Haselbeck *et al.*, 1999). Later it is expressed in the mesonephros (McCaffery *et al.*, 1991). RALDH3 activity is first detected in the rostral head at 8.5–8.75 dpc and it is expressed in the surface ectoderm overlying the prospective eye field at 9 dpc. At a later stage, it localizes to the ventral retina, dorsal pigment epithelium, lateral ganglionic eminence, dorsal margin of the otic vesicle, and olfactory neuroepithelium (Mic *et al.*, 2000). RALDH4 expression is detected in fetal liver at E14.5 but not earlier (Lin *et al.*, 2003).

Numerous cytochrome P450 enzymes are believed to play a role in embryonic all-*trans* RA oxidation. Both Cyp26A1 and B1 mRNAs are induced by all-*trans* RA whereas Cyp26C1 mRNA is downregulated by all-*trans* RA (Reijntjes *et al.*, 2005). They are expressed in the early embryo as well as later in development and may play a critical role in regulating the access of ligand to the nuclear retinoids receptors in specific regions of the developing embryo. Disruption of the murine Cyp26A1 gene is embryolethal (Abu-Abed *et al.*, 2001). Expression of Cyp26A1 begins at the same time as RALDH2. Its mRNA is detected as early as 6 dpc in mouse and one day later is found in embryonic endoderm, mesoderm, and primitive streak. At 7.5 dpc, expression in posterior domains is diminished, and the anterior regions of all three embryonic germ layers show expression. Between 8.5 and 10.5 dpc, the mRNA is expressed in prospective rhombomere 2, neural crest cells involved in the formation of cranial ganglia V, VII/VIII, and IX/X, the caudal neural plate, the tailbud mesoderm, and the hindgut (Fujii et al., 1997). The mRNA has been shown to be particularly abundant in human cephalic tissues during the late embryonic early fetal period of development (Trofimova-Griffin et al., 2000). Cyp26B1 mRNA shows a dynamic pattern of expression in the developing hindbain and is found between the somites, in the dorsal and ventral aspects of the limb buds, and in the node region of presomitic rat embryos (McLean et al., 2001). Cyp26C1 mRNA is expressed in the hindbrain, inner ear, first branchial arch, and tooth buds during murine development (Tahayato et al., 2003).

Concerning the other metabolic activities related to the ROH or its precursors during mammalian development, the presence of carotene-15,15′ dioxygenase mRNA has been detected in maternal tissue at the site of embryo implantation during early stages of mouse embryogenesis (7.5 and 8.5 dpc) (Paik et al., 2001). It suggests that this enzyme may be acting to provide needed retinoid to the embryo. A weak signal of the carotene-15,15′ dioxygenase mRNA can also be detected in embryonic tissues still 15 dpc, but the functionality of β-carotene cleavage remains still discussed (Redmond et al., 2001). Vitamin A is mainly stored in the stellate cells of the liver as retinyl esters in lipid droplets but may also be found during embryonic life in lung (Chytil, 1996; Zachman and Valceschini, 1998). Fetal CRBP2 is expressed transiently in the mouse yolk sac, lung, and liver during development. Both loss of maternal and loss of fetal CRBP2 contribute to increased neonatal mortality, when dietary vitamin A is reduced to marginal levels. Nevertheless, the role of CRBP2 for retinoids metabolism seems to be limited for the embryonic part. Indeed, the CRBP2 plays a specific role in ensuring adequate transport of vitamin A to the developing fetus, particularly when maternal vitamin A is limited (Xueping et al., 2002). Similar role can be played by lecithin:retinol acyltransferase, during mammalian development (Liu and Gudas, 2005). Due to its high expression during mouse development, CRBP1 seemed to be strongly important for the regulation of this retinoids storage. Indeed, CRBP1 (and not CRBP2) is specifically expressed in several tissues including spinal cord, lung, and liver (Dolle et al., 1990; Gustafson et al., 1993). Nevertheless, CRBP1 mutant embryos from mothers fed with a vitamin A-enriched diet are healthy. They do not present any of the congenital abnormalities related to RA deficiency. During development, ROH and retinyl ester levels are decreased in CRBP1 deficient embryos and fetuses by 50% and 80%, respectively (Ghyselinck et al., 1999). The CRBP1 deficiency does not alter the expression patterns of RA-responding genes during development. Therefore, CRBP1 is required in prenatal life to maintain normal amounts of ROH and to ensure its efficient storage as retinyl esters but seems of secondary importance for RA synthesis, under conditions of maternal vitamin A sufficiency (Matt et al., 2005).

ACKNOWLEDGMENTS

Grant support: "ARDEMO" team was supported by the Minister of Research and Technology (JE 2447). GM and VS were supported by an INSERM grant, respectively Poste Accueil and Contrat d'Interface. DG was supported by a grant from the Société Française de Médecine Périnatale and from the Collège National des Gynécologues et Obstétriciens Fançais.

REFERENCES

Abu-Abed, S., Dollé, P., Metzgze, D., Beckett, B., Chambon, P., and Petkovich, M. (2001). The retinoic acid-metabolizing enzyme, CYP26A1, is essential for normal hindbrain patterning, vertebral identity, and development of posterior structures. *Genes Dev.* **15**, 226–240.

Ang, H. L., Deltour, L., Hayamizu, T. F., Zgombic-Knight, M., and Duester, G. (1996). Retinoic acid synthesis in mouse embryos during gastrulation and craniofacial development linked to class IV alcohol dehydrogenase gene expression. *J. Biol. Chem.* **271**, 9526–9534.

Astrom, A., Pettersson, U., and Voorhees, J. J. (1992). Structure of the human cellular retinoic acid-binding protein II gene. Early transcriptional regulation by retinoic acid. *J. Biol. Chem.* **267**, 25251–25255.

Bates, C. J. (1983). Vitamin A in pregnancy and lactation. *Proc. Nutr. Soc.* **42**, 65–79.

Batourina, E., Gim, S., Bello, N., Shy, M., Clagett-Dame, M., Srinivas, S., Costantini, F., and Mendelsohn, C. (2001). Vitamin A controls epithelia/mesenchymal interactions through Ret expression. *Nat. Genet.* **27**, 74–78.

Bavik, C., Ward, S. J., and Ong, D. E. (1997). Identification of a mechanism to localize generation of retinoic acid in rat embryos. *Mech. Dev.* **69**, 155–167.

Bellovino, D., Apreda, M., Gragnoli, S., Massimi, M., and Gaetani, S. (2003). Vitamin A transport: *In vitro* models for the study of RBP secretion. *Mol. Aspects Med.* **24**, 411–420.

Blanchon, L., Sauvant, P., Bavik, C., Gallot, D., Charbonne, F., Alexandre-Gouabau, M. C., Lemery, D., Jacquetin, B., Dastugue, B., Ward, S., and Sapin, V. (2002). Human choriocarcinoma cell line JEG-3 produces and secretes active retinoids from retinol. *Mol. Hum. Reprod.* **8**, 485–493.

Blomhoff, R. (1994). Transport and metabolism of vitamin A. *Nutr. Rev.* **52**, 13–23.

Bonet, B., Chait, A., Gown, A. M., and Knopp, R. H. (1995). Metabolism of modified LDL by cultured human placental cells. *Atherosclerosis* **112**, 125–136.

Brown, W. M., Metzger, L. E., Barlow, J. P., Hunsaker, L. A., Deck, L. M., Royer, R. E., and Vander Jagt, D. L. (2003). 17-Beta-Hydroxysteroid dehydrogenase type 1: Computational design of active site inhibitors targeted to the Rossmann fold. *Chem. Biol. Interact.* **143–144**, 481–491.

Card, S. E., Tompkins, S. F., and Brien, J. F. (1989). Ontogeny of the activity of alcohol dehydrogenase and aldehyde dehydrogenases in the liver and placenta of the guinea pig. *Biochem. Pharmacol.* **38**, 2535–2541.

Chambon, P. (1996). A decade of molecular biology of retinoic acid receptors. *FASEB J.* **10**, 940–954.

Chytil, F. (1996). Retinoids in lung development. *FASEB J.* **10**, 986–992.

Clagett-Dame, M., and DeLuca, H. F. (2002). The role of vitamin A in mammalian reproduction and embryonic development. *Annu. Rev. Nutr.* **22**, 347–381.

Clarke, D. W., Smith, G. N., Patrick, J., Richardson, B., and Brien, J. F. (1989). Activity of alcohol dehydrogenase and aldehyde dehydrogenase in maternal liver, fetal liver and placenta of the near-term pregnant ewe. *Dev. Pharmacol. Ther.* **12**, 35–41.

Colbert, M. C., Linney, E., and LaMantia, A. S. (1993). Local sources of retinoic acid coincide with retinoid-mediated transgene activity during embryonic development. *Proc. Natl. Acad. Sci. USA* **90**, 6572–6576.

Collins, M. D., Tzimas, G., Hummler, H., Burgin, H., and Nau, H. (1994). Comparative teratology and transplacental pharmacokinetics of all-trans-retinoic acid, 13-cis-retinoic acid, and retinyl palmitate following daily administrations in rats. *Toxicol. Appl. Pharmacol.* **127**, 132–144.

Conlon, R. A. (1995). Retinoic acid and pattern formation in vertebrates. *Trends Genet.* **11**, 314–319.

Creech Kraft, J., Lofberg, B., Chahoud, I., Bochert, G., and Nau, H. (1989). Teratogenicity and placental transfer of all-trans-, 13-cis-, 4-oxo-all-trans-, and 4-oxo-13-cis-retinoic acid after administration of a low oral dose during organogenesis in mice. *Toxicol. Appl. Pharmacol.* **100**, 162–176.

Creech Kraft, J., Eckhoff, C., Kochhar, D. M., Bochert, G., Chahoud, I., and Nau, H. (1991). Isotretinoin (13-cis-retinoic acid) metabolism, cis-trans isomerization, glucuronidation, and transfer to the mouse embryo: Consequences for teratogenicity. *Teratog. Carcinog. Mutagen.* **11**, 21–30.

Creech Kraft, J., Shepard, T., and Juchau, M. R. (1993). Tissue levels of retinoids in human embryos/fetuses. *Reprod. Toxicol.* **7**, 11–15.

Datta, K., and Kulkarni, A. P. (1996). Co-oxidation of all-trans retinol acetate by human term placental lipoxygenase and soybean lipoxygenase. *Reprod. Toxicol.* **10**, 105–112.

Debier, C., and Larondelle, Y. (2005). Vitamins A and E: Metabolism, roles and transfer to offspring. *Br. J. Nutr.* **93**, 153–174.

Deltour, L., Foglio, M. H., and Duester, G. (1999a). Impaired retinol utilization in *Adh4* alcohol dehydrogenase mutant mice. *Dev. Genet.* **25**, 1–10.

Deltour, L., Foglio, M. H., and Duester, G. (1999b). Metabolic deficiencies in alcohol dehydrogenase *Adh1*, *Adh3*, and *Adh4* null mutant mice. Overlapping roles of *Adh1* and *Adh4* in ethanol clearance and metabolism of retinol to retinoic acid. *J. Biol. Chem.* **274**, 16796–16801.

Dimenstein, R., Trugo, N. M., Donangelo, C. M., Trugo, L. C., and Anastacio, A. S. (1996). Effect of subadequate maternal vitamin-A status on placental transfer of retinol and beta-carotene to the human fetus. *Biol. Neonate* **69**, 230–234.

Dolle, P., Ruberte, E., Leroy, P., Morriss-Kay, G., and Chambon, P. (1990). Retinoic acid receptors and cellular retinoid binding proteins I. A systematic study of their differential pattern of transcription during mouse organogenesis. *Development* **110**, 1133–1151.

Eckhoff, C., Lofberg, B., Chahoud, I., Bochert, G., and Nau, H. (1989). Transplacental pharmacokinetics and teratogenicity of a single dose of retinol (vitamin A) during organogenesis in the mouse. *Toxicol. Lett.* **48**, 171–184.

Estonius, M., Svensson, S., and Hoog, J. O. (1996). Alcohol dehydrogenase in human tissues: Localisation of transcripts coding for five classes of the enzyme. *FEBS Lett.* **397**, 338–342.

Frolik, C. A., Roberts, A. B., Tavela, T. E., Roller, P. P., Newton, D. L., and Sporn, M. B. (1979). Isolation and identification of 4-hydroxy- and 4-oxoretinoic acid. *In vitro* metabolites of all-trans-retinoic acid in hamster trachea and liver. *Biochemistry* **18**, 2092–2097.

Fujii, H., Sato, T., Kaneko, S., Gotoh, O., Fujii-Kuriyama, Y., Osawa, K., Kato, S., and Hamada, H. (1997). Metabolic inactivation of retinoic acid by a novel P450 differentially expressed in developing mouse embryos. *EMBO J.* **16**, 4163–4173.

Gamble, M. V., Shang, E., Zott, R. P., Mertz, J. R., Wolgemuth, D. J., and Blaner, W. S. (1999). Biochemical properties, tissue expression, and gene structure of a short chain dehydrogenase/reductase able to catalyze cis-retinol oxidation. *J. Lipid Res.* **40**, 2279–2292.

Ghyselinck, N. B., Bavik, C., Sapin, V., Mark, M., Bonnier, D., Hindelang, C., Dierich, A., Nilsson, C. B., Hakansson, H., Sauvant, P., Azais-Braesco, V., Frasson, M., *et al.* (1999). Cellular retinol-binding protein I is essential for vitamin A homeostasis. *EMBO J.* **18**, 4903–4914.

Gimes, G., and Toth, M. (1993). Low concentration of Triton X-100 inhibits diacylglycerol acyltransferase without measurable effect on phosphatidate phosphohydrolase in the human primordial placenta. *Acta Physiol. Hung.* **81**, 101–108.

Green, T., and Ford, H. C. (1986). Intracellular binding proteins for retinol and retinoic acid in early and term human placentas. *Br. J. Obstet. Gynaecol.* **93**, 833–838.

Gudas, L. J. (1994). Retinoids and vertebrate development. *J. Biol. Chem.* **269**, 15399–15402.

Gustafson, A. L., Dencker, L., and Eriksson, U. (1993). Non-overlapping expression of CRBP I and CRABP I during pattern formation of limbs and craniofacial structures in the early mouse embryo. *Development* **117**, 451–460.

Haselbeck, R. J., Hoffmann, I., and Duester, G. (1999). Distinct functions for *Aldh1* and *Raldh2* in the control of ligand production for embryonic retinoid signalling pathways. *Dev. Genet.* **25**, 353–364.

Horton, C., and Maden, M. (1995). Endogenous distribution of retinoids during normal development and teratogenesis in the mouse embryo. *Dev. Dyn.* **202**, 312–323.

Ismadi, S. D., and Olson, J. A. (1982). Dynamics of the fetal distribution and transfer of Vitamin A between rat fetuses and their mother. *Int. J. Vitam. Nutr. Res.* **52**, 112–119.

Johansson, S., Gustafson, A. L., Donovan, M., Romert, A., Eriksson, U., and Dencker, L. (1997). Retinoid binding proteins in mouse yolk sac and chorio-allantoic placentas. *Anat. Embryol. (Berl.)* **195**, 483–490.

Johansson, S., Gustafson, A. L., Donovan, M., Eriksson, U., and Dencker, L. (1999). Retinoid binding proteins-expression patterns in the human placenta. *Placenta* **20**, 459–465.

Johansson, S., Dencker, L., and Dantzer, V. (2001). Immunohistochemical localization of retinoid binding proteins at the materno-fetal interface of the porcine epitheliochorial placenta. *Biol. Reprod.* **64**, 60–68.

Kiefer, C., Hessel, S., Lampert, J. M., Vogt, K., Lederer, M. O., Breithaupt, D. E., and von Lintig, J. (2001). Identification and characterization of a mammalian enzyme catalyzing the asymmetric oxidative cleavage of provitamin A. *J. Biol. Chem.* **276**, 14110–14116.

Kochhar, D. M., Penner, J. D., and Satre, M. A. (1988). Derivation of retinoic acid and metabolites from a teratogenic dose of retinol (vitamin A) in mice. *Toxicol. Appl. Pharmacol.* **96**, 429–441.

Lammer, E. J., Chen, D. T., Hoar, R. M., Agnish, N. D., Benke, P. J., Braun, J. T., Curry, C. J., Fernhoff, P. M., Grix, A. W., Jr., Lott, I. T., Macash, R. G., Nada, G. R., et al. (1985). Retinoic acid embryopathy. *N. Engl. J. Med.* **313**, 837–841.

Lassiter, T. L., Barone, S., Jr., Moser, V. C., and Padilla, S. (1999). Gestational exposure to chlorpyrifos: Dose response profiles for cholinesterase and carboxylesterase activity. *Toxicol. Sci.* **52**, 92–100.

Leo, M. A., and Lieber, C. S. (1999). Alcohol, vitamin A, and beta-carotene: Adverse interactions, including hepatotoxicity and carcinogenicity. *Am. J. Clin. Nutr.* **69**, 1071–1085.

Levin, M. S., Li, E., Ong, D. E., and Gordon, J. I. (1987). Comparison of the tissue-specific expression and developmental regulation of two closely linked rodent genes encoding cytosolic retinol-binding proteins. *J. Biol. Chem.* **262**, 7118–7124.

Liden, M., and Eriksson, U. (2006). Understanding retinol metabolism—structure and function of retinol dehydrogenases. *J. Biol. Chem.* **281**(19), 13001–13004.

Lin, M., Zhang, M., Abraham, M., Smith, S. M., and Napoli, J. L. (2003). Mouse retinal dehydrogenase 4 (RALDH4), molecular cloning, cellular expression, and activity in 9-cis-retinoic acid biosynthesis in intact cells. *J. Biol. Chem.* **278**, 9856–9861.

Lin, S. X., Shi, R., Qiu, W., Azzi, A., Zhu, D. W., Dabbagh, H. A., and Zhou, M. (2006). Structural basis of the multispecificity demonstrated by 17beta-hydroxysteroid dehydrogenase types 1 and 5. *Mol. Cell. Endocrinol.* **248**, 38–46.

Linke, T., Dawson, H., and Harrison, E. H. (2005). Isolation and characterization of a microsomal acid retinyl ester hydrolase. *J. Biol. Chem.* **280**, 23287–23294.

Liu, L., and Gudas, L. J. (2005). Disruption of the lecithin:retinol acyltransferase gene makes mice more susceptible to vitamin A deficiency. *J. Biol. Chem.* **280**, 40226–40234.

Malpel, S., Mendelsohn, C., and Cardoso, W. V. (2000). Regulation of retinoic acid signalling during lung morphogenesis. *Development* **127**, 3057–3067.

Marill, J., Idres, N., Capron, C. C., Nguyen, E., and Chabot, G. G. (2003). Retinoic acid metabolism and mechanism of action: A review. *Curr. Drug. Metab.* **4**, 1–10.

Mark, M., and Chambon, P. (2003). Functions of RARs and RXRs *in vivo*: Genetic dissection of the retinoid signalling pathway. *Pure Appl. Chem.* **75**, 1709–1732.

Mark, M., Ghyselinck, N. B., and Chambon, P. (2004). Retinoic acid signalling in the development of branchial arches. *Curr. Opin. Genet. Dev.* **14**, 591–598.

Matt, N., Schmidt, C. K., Dupe, V., Dennefeld, C., Nau, H., Chambon, P., Mark, M., and Ghyselinck, N. B. (2005). Contribution of cellular retinol-binding protein type 1 to retinol metabolism during mouse development. *Dev. Dyn.* **233**, 167–176.

McCaffery, P., and Dräger, U. C. (1997). A sensitive bioassay for enzymes that synthesize retinoic acid. *Brain Res. Brain Res. Protoc.* **3**, 232–236.

McCaffery, P., Tempst, P., Lara, G., and Dräger, U. C. (1991). Aldehyde deshydrogenase is a positional marker in the retina. *Development* **112**, 693–702.

McLean, G., Abu-Abed, S., Dollé, P., Tahayato, A., Chambon, P., and Petkovich, M. (2001). Cloning of a novel retinoic-acid metabolizing cytochrome P450, Cyp26B1, and comparative expression analysis with Cyp26A1 during early murine development. *Mech. Dev.* **107**, 195–201.

Mic, F. A., Molotkov, A., Fan, X., Cuenca, A. E., and Duester, G. (2000). RALDH3, a retinaldehyde dehydrogenase that generates retinoic acid, is expressed in the ventral retina, otic vesicle and olfactory pit during mouse development. *Mech. Dev.* **97**, 227–230.

Miller, R. K., Faber, W., Asai, M., D'Gregorio, R. P., Ng, W. W., Shah, Y., and Neth-Jessee, L. (1993). The role of the human placenta in embryonic nutrition. Impact of environmental and social factors. *Ann. NY Acad. Sci.* **678**, 92–107.

Moore, T. (1971). Vitamin A transfer from mother to offspring in mice and rats. *Int. J. Vitam. Nutr. Res.* **41**, 301–306.

Moss, J. B., Xavier-Neto, J., Shapiro, M. D., Nayeem, S. M., McCaffery, P., Drager, U. C., and Rosenthal, N. (1998). Dynamic patterns of retinoic acid synthesis and response in the developing mammalian heart. *Dev. Biol.* **199**, 55–71.

Napoli, J. L. (1999). Interactions of retinoid binding proteins and enzymes in retinoid metabolism. *Biochim. Biophys. Acta* **1440**, 139–162.

Nau, H., Elmazar, M. M., Ruhl, R., Thiel, R., and Sass, J. O. (1996). All-trans-retinoyl-beta-glucuronide is a potent teratogen in the mouse because of extensive metabolism to all-trans-retinoic acid. *Teratology* **54**, 150–156.

Niederreither, K., McCaffery, P., Dräger, U. C., Chambon, P., and Dolle, P. (1997). Restricted expression and retinoic acid-induced downregulation of the retinaldehyde dehydrogenase type 2 (RALDH-2) gene during mouse development. *Mech. Dev.* **62**, 67–78.

Niederreither, K., Subbarayan, V., Dollé, P., and Chambon, P. (1999). Embryonic retinoic acid synthesis is essential for early mouse post-implantation development. *Nat. Genet.* **21**, 444–448.

Nieuwkoop, P. D. (1952). Activation and organization of the central nervous system in amphibians I. Induction and activation. *J. Exp. Zool.* **120**, 83–108.

O'Byrne, S. M., Wongsiriroj, N., Libien, J., Vogel, S., Goldberg, I. J., Baehr, W., Palczewski, K., and Blaner, W. S. (2005). Retinoid absorption and storage is impaired in mice lacking lecithin: retinol acyltransferase (LRAT). *J. Biol. Chem.* **280**, 35647–35657.

Ong, D. E. (1994). Cellular transport and metabolism of vitamin A: Roles of the cellular retinoid-binding proteins. *Nutr. Rev.* **52**, 24–31.

Orland, M. D., Anwar, K., Cromley, D., Chu, C. H., Chen, L., Billheimer, J. T., Hussain, M. M., and Cheng, D. (2005). Acyl coenzyme A dependent retinol esterification by acyl coenzyme A: Diacylglycerol acyltransferase 1. *Biochim. Biophys. Acta* **1737**, 76–82.

Okuno, M., Kato, M., Moriwaki, H., Kanai, M., and Muto, Y. (1987). Purification and partial characterization of cellular retinoic acid-binding protein from human placenta. *Biochim. Biophys. Acta* **923**, 116–124.

Paik, J., During, A., Harrison, E. H., Mendelsohn, C. L., Lai, K., and Blaner, N. S. (2001). Expression and characterization of a murine enzyme able to cleave beta-carotene. The formation of retinoids. *J. Biol. Chem.* **276**(34), 32160–32168.

Parrow, V., Horton, C., Maden, M., Laurie, S., and Notarianni, E. (1998). Retinoids are endogenous to the porcine blastocyst and secreted by trophectoderm cells at functionally-active levels. *Int. J. Dev. Biol.* **42**, 629–632.

Pasatiempo, A. M., and Ross, A. C. (1990). Effects of food or nutrient restriction on milk vitamin A transfer and neonatal vitamin A stores in the rat. *Br. J. Nutr.* **63**, 351–362.

Persson, B., Krook, M., and Jornvall, H. (1991). Characteristics of short-chain alcohol dehydrogenases and related enzymes. *Eur. J. Biochem.* **200**, 537–543.

Quadro, L., Blaner, W. S., Salchow, D. J., Vogel, S., Piantedosi, R., Gouras, P., Freeman, S., Cosma, M. P., Colantuoni, V., and Gottesman, M. E. (1999). Impaired retinal function and vitamin A availability in mice lacking retinol-binding protein. *EMBO J.* **18**, 4633–4644.

Quadro, L., Hamberger, L., Gottesman, M. E., Colantuoni, V., Ramakrishnan, R., and Blaner, W. S. (2004). Transplacental delivery of retinoid: The role of retinol-binding protein and lipoprotein retinyl ester. *Am. J. Physiol. Endocrinol. Metab.* **286**, 844–851.

Quadro, L., Hamberger, L., Gottesman, M. E., Wang, F., Colantuoni, V., Blaner, W. S., and Mendelsohn, C. L. (2005). Pathways of vitamin A delivery to the embryo: Insights from a new tunable model of embryonic vitamin A deficiency. *Endocrinology* **146**, 4479–4490.

Ray, W. J., Bain, G., Yao, M., and Gottlieb, D. I. (1997). CYP26, a novel mammalian cytochrome P450, is induced by retinoic acid and defines a new family. *J. Biol. Chem.* **272**, 18702–18708.

Redmond, T. M., Gentleman, S., Duncan, T., Yu, S., Wiggert, B., Gantt, E., and Cunningham, F. X., Jr. (2001). Identification, expression, and substrate specificity of a mammalian beta-carotene 15,15′-dioxygenase. *J. Biol. Chem.* **276**, 6560–6565.

Reijntjes, S., Blentic, A., Gale, E., and Maden, M. (2005). The control of morphogen signalling: Regulation of the synthesis and catabolism of retinoic acid in the developing embryo. *Dev. Biol.* **285**, 224–237.

Repa, J. J., Hanson, K. K., and Clagett-Dame, M. (1993). All-trans-retinol is a ligand for the retinoic acid receptors. *Proc. Natl. Acad. Sci. USA* **90**, 7293–7297.

Reynolds, K., Mezey, E., and Zimmer, A. (1991). Activity of the β-retinoic acid receptor promoter in transgenic mice. *Mech. Dev.* **36**, 15–29.

Ross, A. C., and Gardner, E. M. (1994). The function of vitamin A in cellular growth and differentiation, and its roles during pregnancy and lactation. *Adv. Exp. Med. Biol.* **352**, 187–200.

Ross, S. A., McCaffery, P. J., Dräger, U. C., and De Luca, L. M. (2000). Retinoids in embryonal development. *Physiol. Rev.* **80**, 1021–1054.

Rossant, J., and Cross, J. C. (2001). Placental development: Lessons from mouse mutants. *Nat. Rev. Genet.* **2**, 538–548.

Rossant, J., Zirngibl, R., Cado, D., Shago, M., and Giguère, V. (1991). Expression of a retinoic acid response element-hsplacZ transgene defines specific domains of transcriptional activity during mouse embryogenesis. *Genes Dev.* **5**, 1333–1344.

Sapin, V., Ward, S. J., Bronner, S., Chambon, P., and Dolle, P. (1997). Differential expression of transcripts encoding retinoid binding proteins and retinoic acid receptors during placentation of the mouse. *Dev. Dyn.* **208**, 199–210.

Sapin, V., Chaib, S., Blanchon, L., Alexandre-Gouabau, M. C., Lemery, D., Charbonne, F., Gallot, D., Jacquetin, B., Dastugue, B., and Azais-Braesco, V. (2000). Esterification of vitamin A by the human placenta involves villous mesenchymal fibroblasts. *Pediatr. Res.* **48**, 565–572.

Sapin, V., Gallot, D., Marceau, G., Dastugue, B., and Lemery, D. (2004). Implications trophoblastiques des rétinoïdes: Aspects fondamentaux et hypothèses physiopathologiques. *Reprod. Hum. Horm.* **17**, 155–158.

Sass, J. O., Tzimas, G., Elmazar, M. M., and Nau, H. (1999). Metabolism of retinaldehyde isomers in pregnant rats: 13-Cis- and all-trans-retinaldehyde, but not 9-cis-retinaldehyde, yield very similar patterns of retinoid metabolites. *Drug Metab. Dispos.* **27**, 317–321.

Satre, M. A., and Kochhar, D. M. (1989). Elevations in the endogenous levels of the putative morphogen retinoic acid in embryonic mouse limb-buds associated with limb dysmorphogenesis. *Dev. Biol.* **133**, 529–536.

Satre, M. A., Ugen, K. E., and Kochhar, D. M. (1992). Developmental changes in endogenous retinoids during pregnancy and embryogenesis in the mouse. *Biol. Reprod.* **46**, 802–810.

Scott, W. J., Jr., Walter, R., Tzimas, G., Sass, J. O., Nau, H., and Collins, M. D. (1994). Endogenous status in retinoids and their cytosolic binding proteins in limb buds of chick vs mouse embryos. *Dev. Biol.* **165**, 397–409.

Sharma, C. P., Fox, E. A., Holmquist, B., Jornvall, H., and Vallee, B. L. (1989). cDNA sequence of human class III alcohol dehydrogenase. *Biochem. Biophys. Res. Commun.* **164**, 631–637.

Shenefelt, R. E. (1972). Morphogenesis of malformation in hamsters caused by retinoic acid: Relation to dose and stage at treatment. *Teratology* **5**, 103–118.

Sivaprasadarao, A., and Findlay, J. B. (1994). Structure-function studies on human retinol-binding protein using site-directed mutagenesis. *Biochem. J.* **300**, 437–442.

Sivaprasadarao, A., Sundaram, M., and Findlay, J. B. (1998). Interactions of retinol-binding protein with transthyretin and its receptor. *Methods Mol. Biol.* **89**, 155–163.

Tahayato, A., Dolle, P., and Petkovich, M. (2003). Cyp26C1 encodes a novel retinoic acid-metabolizing enzyme expressed in the hindbrain, inner ear, first branchial arch and tooth buds during murine development. *Gene Expr. Patterns* **3**, 449–454.

Taimi, M., Helvig, C., Wisniewski, J., Ramshaw, H., White, J., Amad, M., Korczak, B., and Petkovich, M. (2004). A novel human cytochrome P450, CYP26C1, involved in metabolism of 9-cis and all-trans isomers of retinoic acid. *J. Biol. Chem.* **279**, 77–85.

Thomas, J. L., Duax, W. L., Addlagatta, A., Kacsoh, B., Brandt, S. E., and Norris, W. B. (2004). Structure/function aspects of human 3beta-hydroxysteroid dehydrogenase. *Mol. Cell. Endocrinol.* **215**, 73–82.

Trofimova-Griffin, M. E., and Juchau, M. R. (1998). Expression of cytochrome P450RAI (CYP26) in human fetal hepatic and cephalic tissues. *Biochem. Biophys. Res. Commun.* **252**, 487–491.

Trofimova-Griffin, M. E., Brzezinski, M. R., and Juchau, M. R. (2000). Patterns of CYP26 expression in human prenatal cephalic and hepatic tissues indicate an important role during early brain development. *Brain Res. Dev. Brain Res.* **120**, 7–16.

Ulven, S. M., Gundersen, T. E., Weedon, M. S., Landaas, V. O., Sakhi, A. K., Fromm, S. H., Geronimo, B. A., Moskaug, J. O., and Blomhoff, R. (2000). Identification of endogenous retinoids, enzymes, binding proteins, and receptors during early postimplantation development in mouse: Important role of retinal dehydrogenase type 2 in synthesis of all-trans retinoic acid. *Dev. Biol.* **220**, 379–391.

Vermot, J., Niederreither, K., Garnier, J. M., Chambon, P., and Dolle, P. (2003). Decreased embryonic retinoic acid synthesis results in a DiGeorge syndrome phenotype in newborn mice. *Proc. Natl. Acad. Sci. USA* **100**, 1763–1768.

Wang, X. D. (2005). Alcohol, vitamin A, and cancer. *Alcohol* **35**, 251–258.

Wagner, E., McCaffery, P., and Drager, U. C. (2000). Retinoic acid in the formation of the dorsoventral retina and its central projections. *Dev. Biol.* **222**, 460–470.

Ward, S. J., and Morriss-Kay, G. M. (1995). Distribution of all-trans-, 13-cis- and 9-cis-retinoic acid to whole rat embryos and maternal serum following oral administration of a teratogenic dose of all-trans-retinoic acid. *Pharmacol. Toxicol.* **76**, 196–201.

Wellik, D. M., and DeLuca, H. F. (1996). Metabolites of all-trans-retinol in day 10 conceptuses of vitamin A-deficient rats. *Arch. Biochem. Biophys.* **330**, 355–362.

Willhite, C. C., Jurek, A., Sharma, R. P., and Dawson, M. I. (1992). Structure-affinity relationships of retinoids with embryonic cellular retinoic acid-binding protein. *Toxicol. Appl. Pharmacol.* **112,** 144–153.

Wilson, J. G., Roth, C. B., and Warkany, J. (1953). An analysis of the syndrome of malformations induced by maternal vitamin A deficiency. Effects of restoration of vitamin A at various times during gestation. *Am. J. Anat.* **92,** 189–217.

Xueping, E., Zhang, L., Lu, J., Tso, P., Blaner, W. S., Levin, M. S., and Li, E. (2002). Increased neonatal mortality in mice lacking cellular retinol-binding protein II. *J. Biol. Chem.* **277,** 36617–36623.

Yan, B., Matoney, L., and Yang, D. (1999). Human carboxylesterases in term placentae: Enzymatic characterization, molecular cloning and evidence for the existence of multiple forms. *Placenta* **20,** 599–607.

Zachman, R. D., and Grummer, M. A. (1998). The interaction of ethanol and vitamin A as a potential mechanism for the pathogenesis of Fetal Alcohol syndrome. *Alcohol. Clin. Exp. Res.* **22,** 1544–1556.

Zhang, W., Xu, G., and McLeod, H. L. (2002). Comprehensive evaluation of carboxylesterase-2 expression in normal human tissues using tissue array analysis. *Appl. Immunohistochem. Mol. Morphol.* **10,** 374–380.

Zile, M. H. (1998). Vitamin A and embryonic development: An overview. *J. Nutr.* **128,** 455S–458S.

Zile, M. H. (2001). Function of vitamin A in vertebrate embryonic development. *J. Nutr.* **131,** 705–708.

Zorzano, A., and Herrera, E. (1989). Disposition of ethanol and acetaldehyde in late pregnant rats and their fetuses. *Pediatr. Res.* **25,** 102–106.

5

CONVERSION OF β-CAROTENE TO RETINAL PIGMENT

HANS K. BIESALSKI,* GURUNADH R. CHICHILI,[†]
JÜRGEN FRANK,* JOHANNES VON LINTIG,[‡] AND
DONATUS NOHR*

*Department of Biological Chemistry and Nutrition, University of Hohenheim
Stuttgart, Germany
[†]Institute of Biology I, Animal Physiology and Neurobiology, University of Freiburg
Freiburg, Germany
[‡]Molecular Immunogenetics Program, Oklahoma Medical Research Foundation
Oklahoma City, Oklahoma 73104

I. General Aspects of Vitamin A Metabolism
II. Conversion of β-Carotene to Vitamin A
   A. Historical Background
   B. Cloning of the Enzymes from Drosophila, Chicken, Mouse, and Human
   C. An Enzyme for Eccentric Cleavage of β-Carotene
   D. Cleavage Mechanisms: Eccentric Versus Central; Monooxygenase Versus Dioxygenase
   E. Regulatory Mechanisms
   F. Carotenoid Cleavage Enzymes During Development and in Various Tissues of the Adult

III. β-Carotene as Provitamin A in Retinal Pigment
     Epithelial Cells
IV. Alternative Routes of Vitamin A Supply
    References

Vitamin A and its active metabolite retinoic acid (RA)[1] play a major role in development, differentiation, and support of various tissues and organs of numerous species. To assure the supply of target tissues with vitamin A, long-lasting stores are built in the liver from which retinol can be transported by a specific protein to the peripheral tissues to be metabolized to either RA or reesterified to form intracellular stores. Vitamin A cannot be synthesized *de novo* by animals and thus has to be taken up from animal food sources or as provitamin A carotenoids, the latter being converted by central cleavage of the molecule to retinal in the intestine. The recent demonstration that the responsible β-carotene cleaving enzyme β,β-carotene 15,15′-monooxygenase (Bcmo1) is also present in other tissues led to numerous investigations on the molecular structure and function of this enzyme in several species, including the fruit fly, chicken, mouse, and also human. Also a second enzyme, β,β-carotene-9′,10′-monooxygenase (Bcmo2), which cleaves β-carotene eccentrically to apo-carotenals has been described. Retinal pigment epithelial cells were shown to contain Bcmo1 and to be able to cleave β-carotene into retinal *in vitro*, offering a new pathway for vitamin A production in another tissue than the intestine, possibly explaining the more mild vitamin A deficiency symptoms of two human siblings lacking the retinol-binding protein for the transport of hepatic vitamin A to the target tissues. In addition, alternative ways to combat vitamin A deficiency of specific targets by the supplementation with β-carotene or even molecular therapies seem to be the future. © 2007 Elsevier Inc.

I. GENERAL ASPECTS OF VITAMIN A METABOLISM

Numerous experimental and clinical studies have shown that vitamin A (retinol) and its biologically active metabolite RA exert important effects on vertebrate development, cellular growth, and differentiation. These pleiotropic

[1]Abbreviations: ARAT, acyl-CoA-retinol acyltransferase; Bcmo1, β,β-carotene 15,15′-monooxygenase; Bcmo2, β,β-carotene-9′,10′-monooxygenase; CRBP, cellular retinol-binding protein; CRABP, cellular retinoic acid-binding protein; EST, expressed sequence tag; LPL, lipoprotein lipase; LRAT, retinol acyltransferase; PPAR, peroxysome proliferators-activated receptor; R, retinol; RA, retinoic acid; RALDH, retinal dehydrogenase; RAR, retinoic acid receptor; RBP, retinol-binding protein; RE, retinyl ester; REH, retinyl ester hydrolase; RolDH, retinol dehydrogenase; RXR, retinoic X receptor; TTR, transthyretin.

effects of RA are exerted through retinoic acid receptors (RAR). To fulfill these basic functions, a continuous supply of target tissues independent from acute dietary intake is necessary. This is achieved through long-lasting liver stores. The dietary vitamin A, absorbed as retinol along with other lipids and fat-soluble vitamins, is packaged as retinyl esters (REs) in nascent chylomicrons and released into the bloodstream. Most retinyl palmitate remains in the remnants which are removed from the circulation by the liver where REs are stored. The supply of extrahepatic target tissues is maintained by retinol, formed after hydrolysis of the hepatic REs and secretion into the bloodstream. The uptake of retinol into various vitamin A-dependent target tissues is either mediated through specific membrane receptors or intracellular apoCRBP (Blaner et al., 1994).

Following uptake into the cells, retinol is metabolized to either RA or reesterified to form intracellular REs. It is generally accepted that circulating retinol bound to RBP is the predominant source of vitamin A for the target cells. To ensure sufficient and continuous supply, the retinol/RBP plasma level is homeostatically controlled. Even in cases of a high dietary vitamin A intake, the plasma retinol/RBP level remains constant as long as liver stores contain REs.

Animals in general are unable to synthesize vitamin A *de novo*. All retinoids in the diet are either in the form of vitamin A from animal food sources or as provitamin A carotenoids usually from plant sources (Dowling and Wald, 1960). Of all known carotenoids, β-carotene is believed to be the most important for animal and human nutrition because of its provitamin A activity. For humans, up to 80% of the daily vitamin A supply is derived from provitamin A carotenoids derived from fruits and vegetables. For strict herbivores, dietary carotenoids are the sole source for vitamin A formation (Sommer, 1997). The benefits of β-carotene supplementation to combat vitamin A deficiency have been demonstrated in several studies (Carlier et al., 1993; Sommer, 1989; West et al., 1997), including the reduction of the incidence of maternal night blindness (Christian et al., 1998).

On the basis of different conversion factors for β-carotene cleavage, ranging from 1:6 up to 1:36 (1:12 commonly accepted), it may wonder whether a mean intake of β-carotene of 1–2 mg in western diets will substantially contribute to vitamin A supply (RDA 1 mg/day). The cleavage calculation is based on classical bioavailability studies which measure the formation of retinol in blood following intake of β-carotene. The recent demonstration that the Bcmo1 appears in different tissues allows to speculate that the most important conversion into vitamin A appears in selected tissues and not only in the intestine. This assumption would allow to understand why despite a very low conversion rate of β-carotene to vitamin A the mean intake of β-carotene may indeed sufficiently cover the need of different vitamin A-dependent tissues such as the retinal pigment epithelium (RPE).

II. CONVERSION OF β-CAROTENE TO VITAMIN A

A. HISTORICAL BACKGROUND

As early as in the 1930s, Moore provided first evidence that a plant-derived carotenoid is the direct precursor for vitamin A in animals describing a conversion of β,β-carotene to vitamin A. Karrer et al. (1930) proposed a central cleavage mechanism at the 15,15′ carbon double bond for this conversion. Glover and Redfearn (1954) proposed an eccentric cleavage reaction and stepwise β-oxidation-like process, leading to a controversial debate about the importance of one or the other cleavage pathway, which not ended before the beginning of the new century when the molecular identification of both enzymes was achieved.

In the 1960s, several groups described independently the β,β-carotene 15,15′-monooxygenase in cell free systems, claiming a central cleavage of β,β-carotene, especially, as retinal was the only product found in these experiments. Due to the requirement of molecular oxygen, the enzyme was named β,β-carotene 15,15′-dioxygenase (EC 1.13.11.21). Leuenberger and coworkers isotope labeling provided strong evidence that the enzyme catalyzes a monooxygenase mechanism. Therefore, in the following this enzyme is named Bcmo1.

B. CLONING OF THE ENZYMES FROM *DROSOPHILA*, CHICKEN, MOUSE, AND HUMAN

In 2000, two research groups independently succeeded in cloning the key enzyme in vitamin A formation (von Lintig and Vogt, 2000; Wyss et al., 2000). The approach by von Lintig and Vogt relied on sequence conservation of Bcmo1 to the plant enzyme VP14, which catalyzes the first step in the pathway leading to the plant growth factor abscisic acid (for recent review see Moise et al., 2005). Using an expression cloning strategy in an *Escherichia coli* strain, they identified a Bcmo1 from the fruit fly *Drosophila melanogaster*. Direct genetic evidence that this enzyme catalyzes the key step in vitamin A formation was provided by von Lintig et al. (2001). They showed that mutations in the gene encoding Bcmo1 causes visual chromophore deficiency and thus blindness in *Drosophila* mutant *ninaB* (von Lintig et al., 2001). Wyss et al. (2000) cloned the first Bcmo1 from a vertebrate, that is, from chicken intestinal mucosa. The recombinant protein was tested for cleavage activity and incubated with β-carotene *in vitro*. Retinal was the only product found and no other β-apo-carotenals were detected. In the following years, the mouse and also the human homologue were cloned and characterized (for recent review see von Lintig et al., 2005). Bcmo1 is a cytosolic enzyme that does not appear in the nucleus or other organelles.

Kinetic studies with recombinant human Bcmo1 with β-carotene as a substrate led to a K_m of 7.1 ± 1.8 μM and V_{max} of 10.4 ± 3.3 nmol retinal/mg × min. While β-cryptoxanthin as a substrate revealed a K_m of 30 ± 3.8 μM and a V_{max} of 0.9 ± 0.2 nmol retinal/mg × min, neither zeaxanthin nor lycopene was accepted as substrates. This supports the hypothesis that the carotenoid substrate molecule must have at least at one site a nonsubstituted β-ionone ring.

C. AN ENZYME FOR ECCENTRIC CLEAVAGE OF β-CAROTENE

Besides Bcmo1 another protein with significant sequence identity, RPE65 was described in vertebrates, not in *Drosophila*. It is exclusively expressed in the RPE (Hamel *et al.*, 1993). RPE65 has been shown to be an all-*trans* to 11-*cis* retinoid isomerase in the visual cycle (Redmond *et al.*, 2005). Searches in expressed sequence tag (EST) databases led to the identification of an EST fragment from mouse highly similar, but neither identical with Bcmo1 nor with RPE65, thus representing a candidate cDNA for an enzyme catalyzing the eccentric cleavage of carotenoids (Kiefer *et al.*, 2001). Subsequent investigations (primer design, RACE-PCR, sequence analysis) revealed that the cDNA encoded a protein of 532 amino acids with an ~40% sequence identity with the mouse Bcmo1. Also the human and zebrafish homologue could be subsequently cloned. Thus, in mammals besides Bcmo1 and RPE65, a third type, Bcmo2, of this family of nonheme iron oxygenases exists (Fig. 1).

Enzymatic activity tests revealed that no retinoids were formed, but a compound was detected with a UV/vis spectrum resembling those of long-chained apo-carotenals. Using the *E. coli* test system, it could be found that the β-carotene content of that strain expressing Bcmo2 was largely reduced compared to a control strain. HPLC analysis revealed significant amounts of β-apo-10'-carotenal, obviously formed from β-carotene. In addition, another cleavage product was found, β-ionone. Therefore, as the enzyme catalyzes the eccentric cleavage of β-carotene at the 9',10' carbon double bond, it was called Bcmo2. A Bcmo2 from ferrets was cloned and characterized. This analysis revealed that Bcmo2 cleaves β-carotene and with lower turnover rates the acyclic carotenoid lycopene (Hu *et al.*, 2006).

D. CLEAVAGE MECHANISMS: ECCENTRIC VERSUS CENTRAL; MONOOXYGENASE VERSUS DIOXYGENASE

As described above, the initial debate about central versus eccentric cleavage of β-carotene had not been settled before the proof that both pathways are most likely possible in vertebrates. Depending on the specific tissue, each pathway may be used preferentially in β,β-carotene metabolism

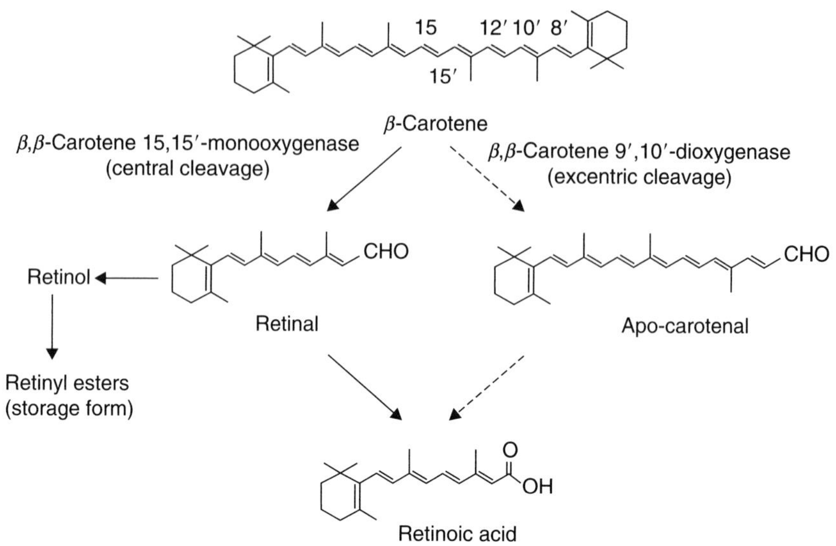

FIGURE 1. Cleavage pathways for β-carotene leading to either two molecules of retinal (central cleavage) or to one molecule of β-apo-carotenal and one molecule of β-ionone. Both products represent intermediates, as retinal is reduced to retinol to be subsequently bound to proteins or stored as retinyl esters in the liver. β-Apo-carotenals can be shortened to RA in a process similar to β-oxidation.

or the interaction of both enzymes may modify carotenoids to yield certain apo-carotenoid products. The debate about monooxygenase versus dioxygenase was solved by Woggon and coworkers (Leuenberger et al., 2001), who clearly could show by labeling experiments with ^{17}O and ^{18}O that the enzyme catalyzes the oxidative cleavage in a monooxygenase reaction type. Therefore, the gene symbol was changed from Bcdo1 to Bcmo1 for β-carotene 15,15′-monooxygenase 1 and the EC number was changed to 1.14.99.36.

E. REGULATORY MECHANISMS

Vitamin A deficiency increases β-carotene cleavage activity whereas vitamin A supplementation downregulates the monooxygenase activity. Lutein, lycopene, 15,15′-dehydro-β-carotene, and astaxanthine inhibit the cleavage activity of Bcmo1, in which lycopene and 15,15′-dehydro-β-carotene acted in a competitive manner, while lutein and astaxanthine activity also included a noncompetitive component. Also the nutritional status—especially the vitamin A level—influences the conversion of β-carotene to retinal. Rats with low vitamin A plasma level showed higher Bcmo1 activity, while those with a protein restriction had a lower Bcmo1 activity, both compared to control levels (Parvin and Sivakumar, 2000). In 2003, the first detailed promoter analysis was published for the mouse Bcmo1 gene. Redmond and coworkers identified an AP2, a bHLH, and a peroxysome proliferators-activated receptor (PPAR) response element besides the core promoter elements (TATA and CACA boxes; Boulanger et al., 2003). Potentially interesting among those promoter sites is the PPRE, as it was shown that PPARγ and retinoic X receptor (RXR)α bind as heterodimer to this response element and luciferase expression is driven in cell lines that normally express Bcmo1 (TC7, PF11, and monkey RPE). The functionality and specificity of this PPRE was also proven by electromobility shift assay (EMSA) and by using the PPARγ-specific agonist ciglitazone as well as the RXRα-specific agonist 9-*cis* RA.

So far known, there is only one other gene in vitamin A metabolism to contain PPRE, the cellular retinol-binding protein II (CRBP-II). As Bcmo1, mRNA levels of this gene are also upregulated by PPAR ligands as well as by 9-*cis* RA (Scharff et al., 2001). CRBP-II is expressed in huge amounts in the small intestine, the major site of vitamin A synthesis, and may act downstream of Bcmo1 by binding the cleavage product retinal. Thus, common mechanisms in the regulation of the genes involved in the vitamin A biosynthetic pathway may contribute to vitamin A homeostasis, and the involvement of PPARs may interlink vitamin A formation to the regulation of overall lipid metabolism.

F. CAROTENOID CLEAVAGE ENZYMES DURING DEVELOPMENT AND IN VARIOUS TISSUES OF THE ADULT

Retinoids play an essential role in vertebrate development and the present model assumes that the retinoids needed by the embryo are mainly derived from preformed maternal vitamin A. Since molecular data were missing, it could not be elucidated whether the embryo itself can take also advantage of provitamins. von Lintig and his group (Lampert et al., 2003) used the zebrafish (*Danio rerio*) as a vertebrate model system to study the expression patterns of Bcmo1 during development. The zebrafish Bcmo1 homologue was expressed with the beginning of segmentation stages and was detected in several embryonic structures including the eye and the neural crest. A targeted gene knockdown of Bcmo1 by morpholinos led to specific malformations in the architecture of the branchial arch skeleton and the eye, underlining the importance of this enzyme. Even though only small amounts of β,β-carotene (besides huge amounts of retinoids) can be found in zebrafish yolk, these studies reveal that at least certain tissues may rely on a local vitamin A synthesis from the provitamin for retinoid signaling during development.

In mouse, the situation remains more unclear, as transient high levels of Bcmo1 mRNA expression were described at embryonic day 7 (E7) which decreased to minimal levels at days E11–E15 by Northern blot analysis. *In situ* hybridization studies in contrast detected Bcmo1 mRNA only in maternal tissues surrounding the embryo, not in the embryo itself at E7.5 and 8.5. Thus, the situation in the mouse needs to be further elucidated, especially concerning the temporal and local distribution of Bcmo1 during mouse development.

In adult mammals, most of the vitamin A is synthesized in the intestinal epithelia from where it is transported for storage to the liver. However, on cloning Bcmo1, its tissue-specific mRNA distribution pattern could be shown in several tissues in several species, including human (Lindqvist and Andersson, 2004). These include the small intestine, the liver, kidney, testes, uterus, skin, skeletal muscle, and in a huge amount in the RPE of monkey and human (Yan *et al.*, 2001), in the latter its function by converting β-carotene to vitamin A was described *in vitro* by Chichili *et al.* (2005; see Section IV).

Taken together, these results indicate that besides an external vitamin A supply via the circulation, provitamin A may tissue specifically impact retinoid metabolism and functions in various cell types and tissues.

A major question for the future will be to explain a possible physiological role of eccentric cleavage products like β-ionone or β-apo-carotenals in vertebrates. The elucidation of functions or physiological interactions of both central and eccentric cleavage enzymes is of a major interest and thus, suitable animal models are needed.

III. β-CAROTENE AS PROVITAMIN A IN RETINAL PIGMENT EPITHELIAL CELLS

Several epidemiological studies have shown that the consumption of large quantities of carotenoid-rich fruits decreases the risk of ocular disease conditions such as age-related macular degeneration (AMD; Mayne, 1996; Snodderly, 1995), which is the leading cause of blindness in the elderly people of the developed world. In an interventional study, the supplementation with β-carotene along with other antioxidants resulted in a significant decrease of the development of advanced AMD (Snodderly, 1995). Also in rats and monkey, protective effects of β-carotene on light-induced retinal damage have been described. One major cause for the development of AMD is oxidative processes mainly in the macular region of the retina, which is almost permanently exposed to focused radiant energy in a highly oxygenated environment, such giving a high potential for oxygen free radicals and singlet oxygen to be generated (Chichili *et al.*, 2006). As β-carotene is a potent quencher of singlet oxygen, this function may be responsible for the protective effects on light-induced damage, as it is already known for other tissues. Indeed, studies in our laboratory (Chichili et al., 2006) could show that β-carotene either given as pure substance or as an extract from β-carotene-rich tomatoes to retinal pigment epithelial cells (lines ARPE19 or D407) in vitro was not only taken up by the cells as revealed by HPLC but also was able to reduce signs of oxidative and nitrosative stress. After treating the β-carotene-free cells with 1-mM H_2O_2, cells could be positively labeled with an antibody against nitrotyrosine indicating nitrosative stress, the intensity of the fluorescence signal correlating with the degree of the stress. This intensity was strongly decreased in cells that had been pretreated with β-carotene.

We found comparable results looking for protein carbonylation and lipid peroxidation, in both cases β-carotene was able to reduce the effects induced by oxidative stress.

Besides its function as protective agent against oxidative stress, β-carotene also plays a role as provitamin A; however, it remained unknown, whether it may also play a role as provitamin A in retinal pigment epithelial cells, that is, are these cells able to take up β-carotene and convert it into vitamin A.

The key step in vitamin A formation is the oxidative cleavage of provitamin A carotenoids by Bcmo1. As described above, the gene for Bcmo1 was cloned from several species and it was shown that it is widely distributed throughout a plethora of vertebrate tissues, including the intestine, liver, kidney, prostate, testis, ovary, and skeletal muscle (Lindqvist and Andersson, 2002). It can also be found in monkey retina and human retinal pigment epithelial cells (Bhatti et al., 2003).

In a study of our group, we investigated whether Bcmo1 has any role in the conversion of provitamin A in the RPE which plays an essential role in the retinoic metabolism of the eye (Chichili et al., 2006).

Cultured retinal pigment epithelial cells (D407) expressed Bcmo1 mRNA as well as Bcmo1 on the protein level, detected by real-time PCR and Western blot, respectively. When a water-soluble beadlet formulation of β-carotene (gift from BASF, Ludwigshafen) was applied to the cells, it was taken up by them and could be converted to retinol. In addition, a dose- and time-dependent upregulation of Bcmo1 mRNA occurred, which in turn resulted in enzymatically active Bcmo1 protein. If one of the possible metabolic end-products, RA, was applied to the cells, this led to an upregulation of Bcmo1 mRNA at lower concentrations and a downregulation at higher concentrations, indicating a demand-oriented regulation of Bcmo1 via retinoic acid receptors (RAR and RXR), as an incubation with an RAR-specific antagonist led to a time-dependent upregulation of Bcmo1 mRNA and protein. Both receptor types are present in D407 cells and could be upregulated via the application of RA as well as β-carotene. To complete the study, also the expression of CYP26A1 was studied, an enzyme that cleaves RA into polar metabolites which can be discarded by the cells. CYP26A1 was upregulated up to tenfold by β-carotene, indicating that from β-carotene RA is synthesized which subsequently induces CYP26A1.

In summary, the retinal pigment epithelial cells expresses Bcmo1 which can be induced by β-carotene. In addition, β-carotene can be converted into vitamin A in a regulated manner, thus providing functional evidence for an eye-specific provitamin A metabolism in humans. This fits well to the findings in the siblings with a mutation in the RBP gene described above, who are not able to recruit vitamin A from the liver but exhibit only mild clinical symptoms such as reduced ability to see during dawn and modest retinal dystrophy. Here, the alternative way to produce vitamin A from β-carotene in the eye might be the way to circumvent vitamin A deficiency, at least in the eyes.

IV. ALTERNATIVE ROUTES OF VITAMIN A SUPPLY

In 1999, we described a novel case of two sisters aged 14 and 17 years with very low plasma retinol (0.19 μmol/L) and RBP concentrations below the limit of detection (<0.6 μmol/L) (Biesalski *et al.*, 1999). The levels of plasma REs, all-*trans* RA, and 13-*cis* RA were within normal ranges. The affected sisters came to attention because of ocular manifestations of vitamin A deficiency (night blindness). Eye examination (Seeliger *et al.*, 1999) showed no signs of xerophthalmia but dark adaption was raised and visual acuity was reduced with atrophy of the RPE. No other abnormalities were detected. Growth and physiological functions were all normal and other clinical signs of severe vitamin A deficiency were lacking. A noninvasive liver store test revealed no depletion of vitamin A storage. A fat absorption test demonstrated normal fat and vitamin A absorption. Plasma levels of other fat-soluble vitamins (β-carotene, α-tocopherol) were normal so that a nutritional vitamin A deficiency could be excluded. We detected two point mutations in the RBP gene, one on each of the two alleles that result in single amino acid substitutions. The girl's virtual lack of RBP was caused by the genetic defect of the RBP gene. RBP is mainly synthesized in the liver on instruction of a single gene. Surprisingly, a mild vitamin A deficiency was solely seen in the eye. This implies that retinol–RBP may be of particular importance in maintaining the health and function of the eye by delivering vitamin A to the retina. But how did the girls tissues obtain their retinol since none of the other tissues or their function were in any way impaired? The case of these two sisters demonstrates that other forms of vitamin A can be utilized effectively by most other tissues to provide functional compensation for the loss of retinol–RBP. All retinoids present in the body originate from the diet either as preformed vitamin A from animal sources (primarily as retinol and RE) or as provitamin A carotenoid from plant sources (usually as β-carotene, α-carotene or β-cryptoxanthin). Within the small intestine, a portion of provitamin A carotenoids is cleaved into retinal by the Bcmo1, following enzymatic conversion to RE. REs together with unconverted carotenoids are packed into nascent chylomicrons and taken to the blood via lymphatic fluid. The majority of chylomicron retinoid and carotenoid is cleared from the circulation by the liver. However, a significant percentage, 25–30%, of chylomicron retinoid is cleared from the circulation by extrahepatic tissues (Goodman et al., 1965). Extrahepatic tissue must be able to take up, metabolize, and store vitamin A carried in chylomicrons. Intracellular vitamin A stores can be mobilized as needed and are independent from the retinol–RBP supply. In addition, 5–10% of the vitamin A in chylomicrons isolated from intestinal lymph is present as unesterified retinol, which can be transferred to other plasma lipoproteins or cell membranes and thus become widely distributed (Harrison et al., 1995). A small proportion of vitamin A absorbed into enterocytes as retinol or β-carotene leaves the intestine as RA via the portal circulation. RA bound to albumin is present in the fasting and postprandial circulations in low concentrations and turns over rapidly. Numerous organs besides the intestine can convert retinol to RA and plasma RA can equilibrate with intracellular RA. As seen by the two sisters, an RBP-independent pathway of vitamin A distribution to organs exists (Fig. 2).

Recapitulating, after consumption of a vitamin A-rich meal, the circulation contains relatively large levels of RE in chylomicrons and their remnants (Fig. 2). Although the liver takes up the majority of the dietary vitamin A, many tissues take up some dietary vitamin A from chylomicrons. RA bound to albumin is present in the fasting and postprandial circulations. Provitamin A carotenoids are found in the circulation both postprandially in chylomicrons and under fasting conditions in lipoproteins. In many tissues, provitamin A can be converted into retinal and subsequently oxidized to retinol (Section II.A). In accordance to the two siblings introduced by Biesalski et al. (1999), only mild symptoms (impaired vision at the time of weaning) associated with quite low vitamin A plasma levels were detected in mice lacking RBP (Quadro et al., 1999). Adult RBP null mice are viable, breed, and have normal vision when maintained on a vitamin A-sufficient diet. This implies the presence of an RBP-independent pathway of vitamin A distribution for most tissues and a preference of retinol bound to RBP (retinol–RBP) for the eye (Vogel et al., 2002). The role of provitamin A carotenoids in vitamin A supply is open but may be examined by means of $RBP^{(-/-)}/Bcmo1^{(-/-)}$ double knockout mice. Such detailed studies would improve our understanding of vitamin A metabolism in extrahepatic tissues.

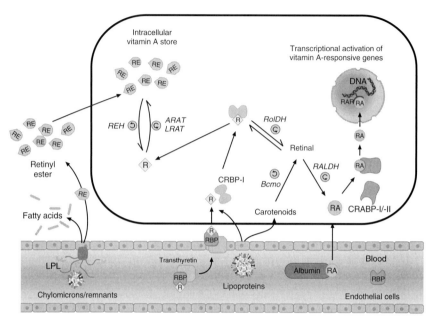

FIGURE 2. Vitamin A supply of extrahepatic tissue from the blood. Compensation of RBP deficiency by REs, RA, and carotenoids. Normally, vitamin A is delivered to the cells by the RBP–TTR–retinol complex. After binding to a receptor, retinol can be released from the complex originally formed in the liver. After absorption, there is either intracellular binding to CRBP-I, oxidation to RA, or reesterification through either ARAT or LRAT. The resulting REs constitute an intracellular pool that can be assessed through hydrolysis by REH. REs can also be derived directly from lipid metabolism. During degradation of chylomicrons to remnants, LPL releases not only fatty acids but also REs, all of which are absorbed by cells, enabling vitamin A supply to target cells independently of the controlled hepatic release of the RBP–TTR–retinol complex. In addition, retinol and carotenoids from plasma lipoproteins may enter the cell. Retinol is directly bound to CRBP-I, provitamin A carotenoids may partially cleaved to retinal by intracellular BCO and subsequently oxidized to RA by an RALDH. Small amounts of RA (bound to albumin) in the blood can pass the cell membrane and attach to specific cellular retinoic acid-binding proteins (CRABP-I and II) within the cell. RA can act within the nucleus of a cell to regulate transcription of vitamin A-responsive genes by binding to specific transcription factors. (From Biesalski and Grimm, 2005.) (See Color Insert.)

REFERENCES

Bhatti, R. A., Yu, S., Boulanger, A., Fariss, R. N., Guo, Y., Bernstein, S. L., Gentleman, S., and Redmond, T. M. (2003). Expression of beta-carotene 15,15′ monooxygenase in retina and RPE-choroid. *Invest. Ophthalmol. Vis. Sci.* **44,** 44–49.

Biesalski, H. K., and Grimm, P. (2005). "Pocket Atlas of Nutrition." Thieme Verlag, Stuttgart.

Biesalski, H. K., Frank, J., Beck, S. C., Heinrich, F., Reifen, R., Gollnick, H., Seeliger, M., Wissinger, M., and Zrenner, E. (1999). Biochemical but not clinical vitamin A deficiency due to mutations in the gene for retinol-binding protein. *Am. J. Clin. Nutr.* **69**(5), 931–936.

Blaner, W. S., Wei, S., Kurlandsky, S. B., and Episkopou, V. (1994). Retinoid transport in rodents. In "Retinoids: From Basic Science to Clinical Applications" (M. A. Livrea and G. Vidali, Eds.), pp. 53–78. Birkhäuser Verlag, Basel.
Boulanger, A., McLemore, P., Copeland, N. G., Gilbert, D. J., Jenkins, N. A., Yu, S. S., Gentleman, S., and Redmond, T. M. (2003). Identification of beta-carotene 15,15′-monooxygenase as a peroxisome proliferator-activated receptor target gene. FASEB J. 17, 1304–1306.
Carlier, C., Coste, J., Etchepare, M., Periquet, B., and Amedee-Manesme, O. (1993). A randomised controlled trial to test equivalence between retinyl palmitate and beta carotene for vitamin A deficiency. Brit. Med. J. 307, 1106–1110.
Chichili, G. R., Nohr, D., Schäffer, M., von Lintig, J., and Biesalski, H. K. (2005). ß.carotene conversion into vitamin A in human retinal pigment epithelial cells. Invest. Ophthalmol. Vis. Sci. 46(10), 3562–3569.
Chichili, G. R., Nohr, D., Frank, J., Flaccus, A., Fraser, P. D., Enfissi, E. M. A., and Biesalski, H. K. (2006). Protective effect of tomato with elevated beta-carotene levels on oxidative stress in ARPE-19 cells. Br. J. Nutr. 96(4), 643–649.
Christian, P., West, K. P., Jr., Khatry, S. K., Katz, J., LeClerq, S., Pradhan, E. K., and Shrestha, S. R. (1998). Vitamin A or beta-carotene supplementation reduces but does not eliminate maternal night blindness in Nepal. J. Nutr. 128, 1458–1463.
Dowling, J. E., and Wald, G. (1960). The biological function of vitamin A acid. Proc. Natl. Acad. Sci. USA 46, 587–608.
Glover, J., and Redfearn, E. R. (1954). The mechanism of transformation of β-carotene into vitamin A in vivo. Biochem. J. 58, 15–16.
Goodman, D. S., Huang, H. S., and Shiratori, T. (1965). Mechanism of the biosynthesis of vitamin A from β-carotene. J. Lipid Res. 6, 390–396.
Hamel, C. P., Tsilou, E., Harris, E., Pfeffer, B. A., Hooks, J. J., Detrick, B., and Redmond, T. M. (1993). A developmentally regulated microsomal protein specific for the pigment epithelium of the vertebrate retina. J. Neurosci. Res. 34, 414–425.
Harrison, E. H., Gad, M. Z., and Ross, A. C. (1995). Hepatic uptake and metabolism of chylomicron retinyl esters: Probable role of plasma membrane/endosomal retinyl ester hydrolases. J. Lipid Res. 36, 1498–1506.
Hu, K. Q., Liu, C., Ernst, H., Krinsky, N. I., Russell, R. M., and Wang, X. D. (2006). The biochemical characterization of ferret carotene-9′,10′-monooxygenase catalyzing cleavage of carotenoids in vitro and in vivo. J. Biol. Chem. 281, 19327–19338.
Karrer, P., Helfenstein, A., Wehri, H., and Wettstein, A. (1930). Pflanzenfarbstoffe. XXV. Leber die Konstitution des Lycopins und Carotins Acta 14, 154–162.
Kiefer, C., Hessel, S., Lampert, J. M., Vogt, K., Lederer, M. O., Breithaupt, D. E., and von Lintig, J. (2001). Identification and characterization of a mammalian enzyme catalyzing the asymmetric oxidative cleavage of provitamin A. J. Biol. Chem. 276, 14110–14116.
Lampert, J. M., Holzschuh, J., Hessel, S., Driever, W., Vogt, K., and von Lintig, J. (2003). Provitamin A conversion to retinal via the beta, beta-carotene 15,15′-oxygenase (bcox) is essential for pattern formation and differentiation during zebrafish embryogenesis. Development 130, 2173–2186.
Leuenberger, M. G., Engeloch-Jarret, C., and Woggon, W. D. (2001). The reaction mechanism of the enzyme-catalyzed central cleavage of beta-carotene to retinal. Angew. Chem. Int. Ed. Engl. 40, 2613–2617.
Lindqvist, A., and Andersson, S. (2002). Biochemical properties of purified recombinant human beta carotene 15,15′-monooxygenase. J. Biol. Chem. 277, 23942–23948.
Lindqvist, A., and Andersson, S. (2004). Cell type-specific expression of beta carotene 15,15′-monooxygenase in human tissues. J. Histochem. Cytochem. 52, 491–499.
Mayne, S. T. (1996). Beta-carotene, carotenoids, and disease prevention in humans. FASEB J. 10, 690–701.

Moise, A. R., von Lintig, J., and Palczewski, K. (2005). Related enzymes solve evolutionarily recurrent problems in the metabolism of carotenoids. *Trends Plant Sci.* **10**, 178–186.

Moore, T. (1930). Vitamin A and carotene VI. The conversion of carotene to vitamin A *in vivo*. *Biochem. J.* **24**, 692–702.

Parvin, S. G., and Sivakumar, B. (2000). Nutritional status affects intestinal carotene cleavage activity and carotene conversion to vitamin A in rats. *J. Nutr.* **130**, 573–577.

Quadro, L., Blaner, W. S., Salchow, D. J., Vogel, S., Piantedosi, R., Gouras, P., Freeman, S., Cosma, M. P., Colantuoni, V., and Gottesman, M. E. (1999). Binding protein. *EMBO J.* **18**, 4633–4644.

Redmond, T. M., Poliakov, E., Yu, S., Tsai, J. Y., Lu, Z., and Gentleman, S. (2005). Mutation of key residues of RPE65 abolishes its enzymatic role as isomerohydrolase in the visual cycle. *Proc. Natl. Acad. Sci. USA* **102**, 13658–13663.

Scharff, E. I., Koepke, J., Fritzsch, G., Lucke, C., and Ruterjans, H. (2001). Crystal structure of diisopropylfluorophosphatase from Loligo vulgaris. *Structure (Camb.)* **9**, 493–502.

Seeliger, M. W., Biesalski, H. K., Wissinger, B., Gollnick, G., Gielen, S., Frank, J., Beck, S. C., and Zrenner, E. (1999). Phenotype in retinol deficiency due to a hereditary defect in retinol binding protein (RBP) synthesis. *Invest. Ophtalmol. Vis. Sci.* **40**(1), 3–11.

Snodderly, D. M. (1995). Evidence for protection against age-related macular degeneration by carotenoids and antioxidant vitamins. *Am. J. Clin. Nutr.* **62**, 1448S–1461S.

Sommer, A. (1989). New imperatives for an old vitamin (A). *J. Nutr.* **119**, 96–100.

Sommer, A. (1997). Vitamin A deficiency, child health, and survival. *Nutrition* **13**, 484–485.

Vogel, S., Piantedosi, R., O'Byrne, S. M., Kako, Y., Quadro, L., Gottesman, M. E., Goldberg, I. J., and Blazer, W. S. (2002). Retinol-binding protein deficient mice: Biochemical basis for impaired vision. *Biochemistry* **41**, 15360–15368.

von Lintig, J., and Vogt, K. (2000). Filling the gap in vitamin A research. Molecular identification of an enzyme cleaving beta-carotene to retinal. *J. Biol. Chem.* **275**, 11915–11920.

von Lintig, J., Dreher, A., Kiefer, C., Wernet, M. F., and Vogt, K. (2001). Analysis of the blind Drosophila mutant *ninaB* identifies the gene encoding the key enzyme for vitamin A formation *in vivo*. *Proc. Natl. Acad. Sci. USA* **98**, 1130–1135.

von Lintig, J., Hessel, S., Isken, A., Kiefer, C., Lampert, J. M., Voolstra, O., and Vogt, K. (2005). Towards a better understanding of carotenoid metabolism in animals. *Biochim. Biophys. Acta* **1740**, 122–131.

West, K. P., LeClerq, S. C., Shrestha, S. R., Wu, L. S., Pradhan, E. K., Khatry, S. K., Katz, J., Adhikari, R., and Sommer, A. (1997). Effects of vitamin A on growth of vitamin A-deficient children: Field studies in Nepal. *J. Nutr.* **127**, 1957–1965.

Wyss, A., Wirtz, G., Woggon, W., Brugger, R., Wyss, M., Friedlein, A., Bachmann, H., and Hunziker, W. (2000). Cloning and expression of beta,beta-carotene 15,15′-dioxygenase. *Biochem. Biophys. Res. Commun.* **271**, 334–336.

Yan, W., Jang, G. F., Haeseleer, F., Esumi, N., Chang, J., Kerrigan, M., Campochiaro, M., Campochiaro, P., Palczewski, K., and Zack, D. J. (2001). Cloning and characterization of a human beta,beta-carotene-15,15′-dioxygenase that is highly expressed in the retinal pigment epithelium. *Genomics* **72**, 193–202.

6

Vitamin A-Storing Cells (Stellate Cells)

Haruki Senoo, Naosuke Kojima, and Mitsuru Sato

Department of Cell Biology and Histology, Akita University School of Medicine 1-1-1 Hondo, Akita 010-8543, Japan

I. Introduction
II. Morphology of HSCs
III. Regulation of Vitamin A Homeostasis by HSCs
IV. HSCs in Arctic Animals
V. Roles of HSCs During Liver Regeneration
VI. Production and Degradation of ECM Components by HSCs
VII. Reversible Regulation of Morphology, Proliferation, and Function of the HSCs by 3D Structure of ECM
VIII. Stimulation of Proliferation of HSCs and Tissue Formation of the Liver by a Long-Acting Vitamin C Derivative
IX. Extrahepatic Stellate Cells
X. Conclusions
References

Hepatic stellate cells (HSCs; also called as vitamin A-storing cells, lipocytes, interstitial cells, fat-storing cells, Ito cells) exist in the space between parenchymal cells and sinusoidal endothelial cells of the hepatic lobule, and store 80% of vitamin A in the whole body as retinyl palmitate in lipid droplets in the cytoplasm. In physiological conditions, these cells play pivotal roles in the regulation of vitamin A homeostasis; they express specific receptors for retinol-binding protein (RBP), a binding protein specific for retinol, on their cell surface, and take up the complex of retinol and RBP by receptor-mediated endocytosis. HSCs in Arctic animals such as polar bears and Arctic foxes store 20–100 times the levels of vitamin A found in human or rat. HSCs play an important role in the liver regeneration. A gradient of vitamin A-storage capacity exists among the SCs in a hepatic lobule. The gradient was expressed as a symmetrical biphasic distribution starting at the periportal zone, peaking at the middle zone, and sloping down toward the central zone in the hepatic lobule. In pathological conditions such as liver fibrosis, HSCs lose vitamin A and synthesize a large amount of extracellular matrix (ECM) components including collagen, proteoglycan, and adhesive glycoproteins. Morphology of these cells also changes from the star-shaped SCs to that of fibroblasts or myofibroblasts. The three-dimensional structure of ECM components was found to regulate reversibly the morphology, proliferation, and functions of the HSCs. Molecular mechanisms in the reversible regulation of the SCs by ECM imply cell surface integrin-binding to ECM components followed by signal transduction processes and then cytoskeleton assembly. SCs also exist in extrahepatic organs such as pancreas, lung, kidney, and intestine. Hepatic and extrahepatic SCs form the SC system. © 2007 Elsevier Inc.

I. INTRODUCTION

The hepatic lobule consists of parenchymal cells (PCs) and non-parenchymal cells associated with the sinusoids: endothelial cells (ECs), Kupffer cells, pit cells, dendritic cells, and SCs (Bloom and Fawcett, 1994; Wake, 1971, 1980) (Fig. 1). Sinusoidal endothelial cells (SECs) express lymphocyte costimulatory molecules (Kojima *et al.*, 2001) and form the greater part of the extremely thin lining of the sinusoids, which are larger than ordinary capillaries and more irregular in shape. Kupffer cells are tissue macrophages and components of the diffuse mononuclear phagocyte system. They are usually situated on the endothelium with cellular processes extending between the underlying ECs. The greater part of their irregular cell surface is exposed to the blood in the lumen of the sinusoid. Pit cells are natural killer cells. Dendritic cells, located in the portal triad in human (Prickett *et al.*, 1988), and in periportal and central areas in rat (Steiniger *et al.*, 1984) that capture and process antigens, migrate to lymphoid organs and secrete cytokines to initiate immune responses (Banchereau and Steinman, 1998). The hepatic stellate cells (HSCs) (Blomhoff and Wake, 1991; Bloom and Fawcett, 1994;

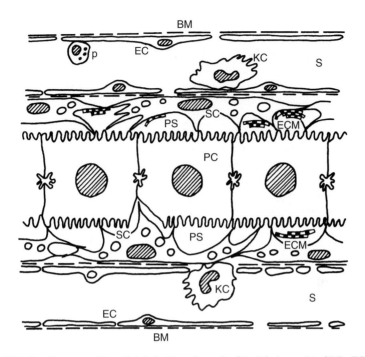

FIGURE 1. Structure of hepatic lobule. Hepatic cords of the lobule consist of PCs. ECs form the thin lining of the sinusoids (S). Kupffer cells (KC) are tissue macrophages and belong to the monocyte–macrophage cell lineage. Stellate cells (HSC) lie in the space between PCs and ECs, and store 80% of vitamin A of the whole body as retinyl palmitate in the lipid droplets in the cytoplasm.

Sato *et al.*, 2003; Senoo, 2004; Senoo *et al.*, 1997; Wake, 1971, 1980) that lie in the space between SECs and PCs are considered to be derived from mesenchymal origin. Both ECs and SCs are derived from mesenchymal tissue, namely, septum transversum. Kupffer cells are from monocyte–macrophage system. SCs that store vitamin A in their cytoplasm have been found in extrahepatic organs (kidney, intestine, lung, pancreas, and so on) and characterized (Matano *et al.*, 1999; Nagy *et al.*, 1997; Wake, 1980). The purpose of this chapter is to survey recent progress in studies of structure and function of the HSCs (vitamin A-storing cells).

II. MORPHOLOGY OF HSCs

HSCs [Fig. 1; fine structure of the HSCs is thoroughly described in the review of Wake (1980)] distribute regularly within hepatic lobules. The cell consists of a spindle-shaped or angular cell body and long and branching cytoplasmic processes which encompass the endothelial tubes of sinusoids (Wake, 1995, 1998). Some processes penetrate the hepatic cell plates (plate-like structures formed by hepatic PCs) to reach the neighboring sinusoids to taper off to several subendothelial processes. Accordingly, a single SC wraps two or three, sometimes four, sinusoids with long processes. The total length of sinusoids surrounded by a single SC is 60–140 μm in the rat liver.

The subendothelial processes of the SCs are flat and have three cell surfaces; inner, outer, and lateral. The inner one is smooth and adheres to the adluminal (basal) surface of ECs. Between the two cells, namely between ECs and SCs, the basement membrane components such as type IV collagen and laminin are intercalated. The outer surface, facing to the perisinusoidal space (space of Disse), is decorated with short microvillous protrusions. The lateral edges of the subendothelial processes are characteristically studded with numerous spikelike microprojections whose tips make contacts with the microvillous facets of the hepatic PCs. The SCs adhere to ECs through basement membrane components, and, on the other hand, make spotty contacts with PCs.

The HSCs have been demonstrated at molecular and morphological levels to adhere each other by adherens junctions (Hiagashi *et al.*, 2004) and gap junctions (Greenwel *et al.*, 1993).

III. REGULATION OF VITAMIN A HOMEOSTASIS BY HSCs

Vitamin A (Fig. 2) is known to regulate diverse cellular activities such as cell proliferation, differentiation, morphogenesis, and tumorigenesis (Blomhoff, 1994; Chawla *et al.*, 2001). In physiological conditions, HSCs store 80% of the total vitamin A in the whole body as retinyl palmitate in lipid droplets in the cytoplasm, and regulate both transport and storage of vitamin A.

The concentration of vitamin A in the bloodstream is regulated within the physiological range by these HSCs. By receptor-mediated endocytosis, the cells take up retinol from the blood, where it circulates as a complex of retinol and a specific binding protein called retinol-binding protein (RBP) (Blomhoff, 1994) (Fig. 3). Once inside the cell, free retinol has several fates, one of which is reformation of the complex with RBP and returns to the bloodstream (Blomhoff, 1992a,b,c; Senoo, 2000). Thus, the HSCs are important for the regulation of homeostasis of vitamin A.

When [^{3}H]retinol was injected via portal vein, the largest amount of the labeled retinol was taken up by the liver within 90 min after injection, although the labeled material was detected in all organs examined (Senoo *et al.*, 1984). The radioactivity of the retinol in the liver did not change until

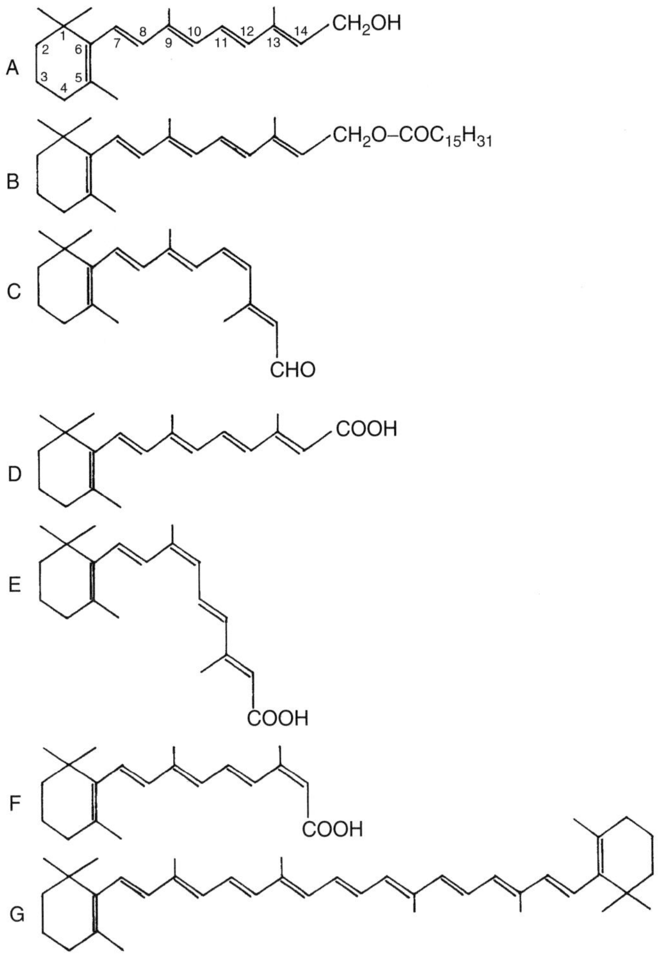

FIGURE 2. Structural formulas of some naturally occurring vitamin A and β-carotene. Vitamin A circulates in the plasma as retinol (A) that binds to a specific carrier protein, RBP. Retinoid is stored as retinyl palmitate (B) in the HSCs. 11-*Cis* retinal (C) exists in the retina with rhodopsin. All-*trans* retinoic acid (D) binds to nuclear retinoic acid receptors (RARα, RARβ, RARγ), and 9-*cis* retinoic acid (E) binds to nuclear retinoid X receptors (RXRα, RXRβ, RXRγ). These nuclear receptors regulate transcription of various genes. 13-*Cis* retinoic acid (F) can bind to RARs. β-Carotene (G) forms two retinals and finally two retinols.

6 days after the injection. These results were consistent with the reports that main storage site of vitamin A in mammals is the liver (Wake, 1971, 1980).

To examine the distribution of vitamin A in the liver, radioactivity per cell was determined after cell fractionation (Senoo and Hata, 1993a; Senoo *et al.*, 1984, 1991). Specific activity of [^{3}H]retinol (per cell) was the highest in the HSC fraction, both 90 min and 6 days after injection. These results strongly

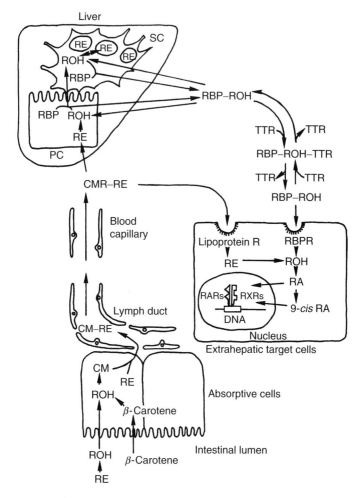

FIGURE 3. Major pathway for vitamin A transport in the body. Dietary retinyl esters (RE) are hydrolyzed to retinol (ROH) in the intestinal lumen before absorption by enterocytes, and carotenoids are absorbed and then partially converted to retinol in the enterocytes. In the enterocytes, retinol reacts with fatty acid to form esters before incorporation into chylomicrons (CM). Chylomicrons then reach the general circulation by way of the intestinal lymph, and chylomicron remnants (CMR) are formed in blood capillaries. Chylomicron remnants, which contain almost all the absorbed retinol, are mainly cleared by the liver PCs and to some extent also by cells in other organs. In liver PCs, retinyl esters are rapidly hydrolyzed to retinol, which then binds to RBP. A complex of retinol–RBP is secreted and transported to HSCs. SCs store vitamin A mainly as retinyl palmitate and secrete retinol–RBP directly into the blood. Most retinol–RBP in the bloodstream is reversibly complexed with transthyretin (TTR). The uncomplexed retinol–RBP is presumably taken up in a variety of cells by cell surface receptors specific for RBP.

support earlier morphological observations (Wake, 1971, 1980) that the SC is the storage site of vitamin A in the liver, and are not inconsistent with reports on the retinol transfer from PCs to SCs (Andersen *et al.*, 1992; Blomhoff *et al.*, 1990; Gyøen *et al.*, 1987; Malaba *et al.*, 1996; Senoo *et al.*, 1990, 1993).

Immunoelectronmicroscopic studies suggest that RBP mediates the paracrine transfer of retinol from hepatic PCs to the SCs and that SCs bind and internalize RBP by receptor-mediated endocytosis (Malaba *et al.*, 1996; Senoo *et al.*, 1993). RBP receptor was cloned and characterized (Båvik *et al.*, 1991, 1992, 1993; Smeland *et al.*, 1995). The SCs may have pivotal roles in type 2 diabetes, because RBP was reported to contribute to insulin resistance in obesity and type 2 diabetes (Yang *et al.*, 2005).

IV. HSCs IN ARCTIC ANIMALS

More than 50 years ago, Rodahl reported that animals (polar bears and seals) in the Arctic area were able to store a large amount of vitamin A in the liver (Rodahl, 1949a,b; Rodahl and Moore, 1943). To investigate the cellular and molecular mechanisms in transport and storage of vitamin A in these Arctic animals, we performed a study in the Svalbard archipelago (situated at 80°N, 15°E) (Higashi and Senoo, 2003; Senoo *et al.*, 1999). After getting permission to hunt the animals from the district governor of Svalbard, 11 Arctic foxes (*Alopex lagopus*), 14 bearded seals (*Erignathus barbatus*), 22 glaucous gulls (*Larus hyperboreus*), 5 fulmars (*Fulmarus glacialis*), 4 Brünnich's guillemots (*Uria lomvia*), 6 ringed seals (*Phoca hispida*), 5 hooded seals (*Cystophora cristata*), 6 puffins (*Fratercula arctica*), 5 Svalbard ptarmigans (*Lagopus mutus hyperboreus*), and 7 Svalbard reindeers (*Rangifer tarandus platyrhynchus*) were caught in the period from August 1996 to September 2001. Three polar bears (*Ursus maritimus*) were shot in self-defense at Svalbard February and August 1998 in Ny Ålesund and Hornsund. We also obtained 13 brown bears (*Ursus arctos*) from Jämtland, Gävleborg, and Dalarna, 4 red foxes (*Vulpes vulpes*) from Västergötaland, and 8 gray gulls (*Larus argentatus*) from Skåne, Sweden.

Fresh organs, namely, the liver, kidney, spleen, lung, and jejunum, were examined by morphological methods and high-performance liquid chromatography (HPLC). Serum from each animal was analyzed with HPLC.

The Arctic animals stored vitamin A in HSCs (Figs. 4–6). Only a small amount of vitamin A existed within other organs such as kidney, spleen, lung, and jejunum. Top predators among Arctic animals stored 6- to 23-μmol retinyl ester per gram liver which is 20–100 times the levels normally found in other animals including humans, rats, and mice. These results indicate that the HSCs in these animals have high ability for uptake and enough capacity for storage of vitamin A.

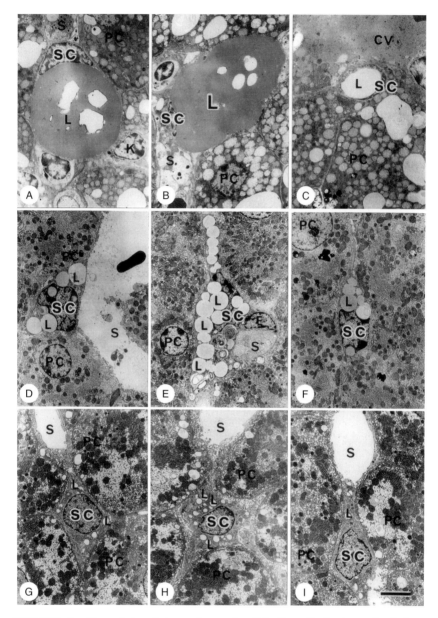

FIGURE 4. Transmission electron micrographs of the liver of polar bears, Arctic foxes, and rats. Electron micrographs of the liver of polar bears (A–C), Arctic foxes (D–F), and rats (G–I) were taken in portal area (A, D, and G), intermediate area (B, E, and H), and central area (C, F, and I) of the hepatic lobule.

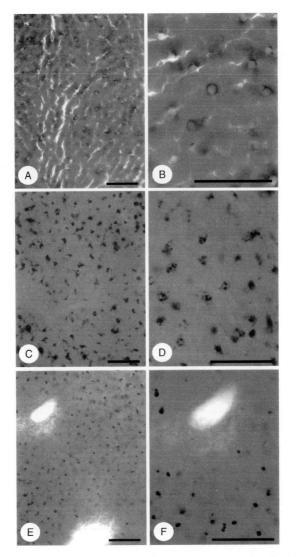

FIGURE 5. Gold chloride staining specifically demonstrating black-stained HSCs of polar bears (A and B), Arctic foxes (C and D), and rats (E and F). Scale bars indicate 100 μm. (See Color Insert.)

The existence of a gradient of vitamin A-storing capacity in the liver was reported and it was found to be independent on the vitamin A amount in the organ (Figs. 7 and 8; Higashi and Senoo, 2003). This gradient was expressed as a symmetrical biphasic distribution starting at the periportal zone, peaking at the middle zone, and sloping down toward the central zone in the liver lobule.

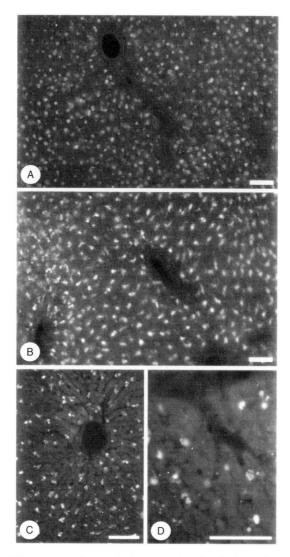

FIGURE 6. Fluorescence micrographs demonstrating vitamin A autofluorescence in HSCs of polar bears (A), Arctic foxes (B), and rats (C and D). Scale bars indicate 100 μm. (See Color Insert.)

Xenobiotics (such as PCBs and dioxins) may reduce the threshold of vitamin A toxicity (Nilsson *et al.*, 1999) and both vitamin A and fat-soluble xenobiotics have a tendency to accumulate in the food chain (Barrie *et al.*, 1992; Dewailly *et al.*, 1989; Holden, 1998; Jarman *et al.*, 1992; Muir *et al.*, 1998; Skaare *et al.*, 2001; Wiig *et al.*, 1998). Kidney total vitamin A, which may be used as a biomarker for retinoid-related toxicity or excess, in polar

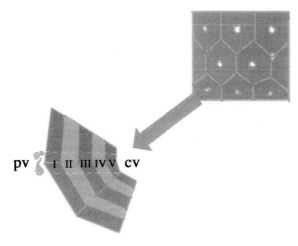

FIGURE 7. Zonal division of the liver lobule. To make a zonal morphometric analysis, the liver lobule was divided histologically into five zonal areas (zones I–V) of equal widths from the portal vein (pv) to the central vein (cv). (See Color Insert.)

bear and bearded seal was below 1% of their liver value, which is in the normal range for most animals (Senoo *et al.*, 2004). Arctic fox and glaucous gull, however, had kidney levels of about 9% and 42% of the liver values, respectively. This increased kidney concentrations and decreased capacity for storage in HSCs of total vitamin A in Arctic fox and glaucous gull are most likely signs of vitamin A toxicity that deserve attention. Nuclear deviation has been reported in PCs on sinusoidal surface in Arctic animals (Sato *et al.*, 2001a). These data are alarming and have not been observed previously in free-living animals.

V. ROLES OF HSCs DURING LIVER REGENERATION

It is well known that liver cells including PCs and SCs show a remarkable growth capacity after partial hepatectomy (PHx). Following 70% PHx in rodents, liver mass is almost completely restored after 14 days. PC proliferation starts after ∼24 h, in the areas surrounding portal tracts and proceeds to the pericentral areas by 36–38 h. As a result of the early PC proliferation, avascular clusters of PCs are observed from 3 days after PHx. Nonparenchymal cells enter DNA synthesis ∼24 h after PCs, with peak activity at 48 h or later. Not only proliferation of PCs but also activation of sinusoidal liver cells including HSCs are involved in the regeneration process through cell–cell interaction and cytokine networks (Mabuchi *et al.*, 2004). PCs and

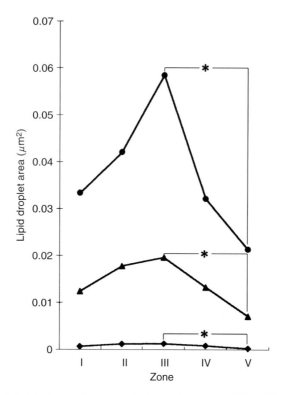

FIGURE 8. Intralobular zonal gradient of vitamin A storage in HSCs. The zonal gradient of vitamin A storage is expressed as a symmetric biphasic profile with a peak at zone III and a downward slope toward zone V. Characteristically, the area of lipoid droplets in zone V is slightly smaller than that in zone I. Graphs were plotted with the mean value, depicting three animal species: polar bear (●), Arctic fox (▲), and rat (◆).

SCs interaction and the SC activation at different time points after 70% PHx were investigated in the rat (Mabuchi *et al.*, 2004). At 3 and 7 days after 70% PHx, the hepatic microcirculation was studied using intravital fluorescence microscopy. In separate groups, SCs and PCs were isolated and liver tissue was processed for histology and immunohistochemistry using anti-desmin (a marker of quiescent SCs) antibody and anti-alpha-smooth muscle actin (α-SMA; a marker of "activated" SCs) antibody. In the isolated PCs and SCs, double immunostaining was used to establish the presence of activated SCs [desmin and bromodeoxyuridine (BrdU); desmin and α-SMA].

PC clusters were often seen *in vivo* at 3 days after 70% PHx. The distance between SCs fell from 61.7 ± 2.1 μm in controls to 36.1 ± 1.4 μm ($p < 0.001$) while the SCs:PCs ratio rose [0.71 ± 0.01 to 1.08 ± 0.03 ($p < 0.001$)]. In >80% of *in vivo* microscopic fields at 3 days after 70% PHx, clusters of HSCs were observed especially near PC clusters. At 1 and 3 days after PHx, >20% of cells in the PC-enriched fraction were SCs which adhered to PCs. At 3 days after PHx, in addition to desmin staining, isolated SCs were also positive for BrdU and α-SMA, and formed clusters suggesting that these HSCs were activated. At 3 days after PHx, SCs in the SC fraction were only positive for desmin, which indicated that adherence to PCs is required for HSC activation. Thus, these data suggest that HSCs are activated by adhering to PCs during the early phase of hepatic regeneration.

As mentioned earlier, under physiological conditions, HSCs within liver lobules store about 80% of the total body vitamin A in lipid droplets in their cytoplasm, and these cells show zonal heterogeneity in terms of vitamin A-storing capacity (Figs. 7 and 8). The status of vitamin A storage in SCs in the liver regeneration was examined (Higashi et al., 2005b). Morphometry at the electron microscopic level, fluorescence microscopy for vitamin A autofluorescence, and immunofluorescence microscopy for desmin and α-SMA were performed on sections of liver from rats at various times after the animal had been subjected to 70% PHx.

Under the electron microscope, the mean area of vitamin A-storing lipid droplets per HSC gradually decreased toward 3 days after PHx, and then returned to normal within 14 days after it. However, the heterogeneity of vitamin A-storing lipid droplet area per HSC within the hepatic lobule disappeared after PHx and did not return to normal by 14 days thereafter, even though the liver volume had returned to normal.

These results suggest that HSCs alter their vitamin A-storing capacity during liver regeneration and that the recovery of vitamin A homeostasis requires a much longer time than that for liver volume.

VI. PRODUCTION AND DEGRADATION OF ECM COMPONENTS BY HSCs

In pathological conditions such as liver cirrhosis, the HSCs lose vitamin A, proliferate vigorously, and synthesize and secrete a large amount of extracellular matrix (ECM) components such as collagen, proteoglycan, and glycoprotein. The structure of the cells also changes from star-shaped SCs to that of fibroblast-like cells or myofibroblasts (Majno, 1979) with well-developed rough-surfaced endoplasmic reticulum and Golgi apparatus (Fig. 9) (Blomhoff and Wake, 1991; Sato et al., 2003; Senoo and Wake, 1985; Senoo et al., 1997).

In order to elucidate cell type or types responsible for collagen metabolism among non-parenchymal cells in the liver, collagen production by SCs, Kupffer cells, and SECs was analyzed (Senoo et al., 1984). SCs were found to produce collagen on day 8 in primary culture, although collagen production was not induced at an earlier stage of culture (day 2). Capability of collagen production by cells was retained in the secondary culture, suggesting that the

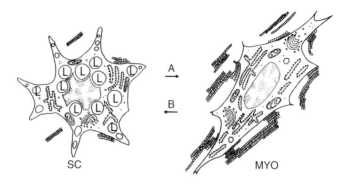

FIGURE 9. Phenotypic changes of the HSCs. In physiological conditions, the HSCs store 80% of vitamin A in the whole body as retinyl palmitate in the lipid droplets (L) in the cytoplasm. Whereas, in pathological conditions, such as liver cirrhosis, these cells lose vitamin A, and synthesize a large amount of ECM. Morphology of the cells also changes from the star-shaped SCs to that of the fibroblasts or myofibroblasts (MYO) (passage of A). Inductive conditions to passage B imply the reversibility of hepatic fibrosis.

SC is a candidate cell responsible for collagen production. Kupffer cells and SECs produced little collagen either on day 2 or day 8 in primary culture under the conditions employed.

Types of collagen produced by SCs in secondary culture were analyzed by fluorography after SDS-polyacrylamide slab gel electrophoresis under nonreducing and reducing conditions (Fig. 10) (Senoo et al., 1984). Type I collagen is the major component synthesized (Fig. 10, lane A). Minor components include type III collagen which remained at the γ-region under nonreducing condition (Fig. 10, lane A), but migrated to the α1-region after reduction (Fig. 10, lane B), and type IV collagen which remained slightly below the origin under nonreducing condition (Fig. 10, lane A), but migrated to α-region slightly lower than the β-region after reduction (Fig. 10, lane B). All these bands were susceptible to purified bacterial collagenase (Fig. 10, lanes C and D). Quantitation of these collagen bands by densitometry indicated that the percentage of type I, type III, and type IV collagens was 88.2%, 10.4%, and 1.4%, which is consistent with an observation on collagen types in human alcoholic liver cirrhosis. The dysregulation of collagen gene expression in HSCs is a central pathogenetic step during the development of hepatic fibrosis (Davis et al., 1996). The causes of the dysregulation have been studied by proteome analysis and several candidates such as cytoglobin/STAP or galectin-1 and galectin-3 were reported (Kawada et al., 2001; Maeda et al., 2003).

Long-Evans cinnamon (LEC)-like colored rats spontaneously develop hepatocellular carcinoma with cholangiofibrosis after chronic hepatitis. To investigate the role of HSCs in induction and suppression of hepatic fibrosis, the liver of LEC rats was morphologically examined (Imai et al., 2000a).

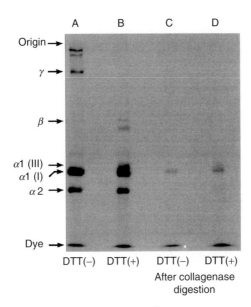

FIGURE 10. Fluorescence autoradiograms of [^{3}H]proline-labeled proteins. SCs at confluency in the secondary passage were cultured in a medium containing [^{3}H]proline for 18 h. Collagenous proteins were partially purified from the cell layer and medium with pepsin digestion after precipitation with ammonium sulfate. SDS-polyacrylamide slab gel electrophoresis was performed before (A and B) or after (C and D) the treatment with purified bacterial collagenase, and processed for fluorescence autoradiography. (B and D) Electrophoresed after reduction with dithiothreitol. Arrows indicate the migration positions of carrier rat collagen chains.

The liver of LEC rats 1.5 years of age showed cholangiofibrosis and subcellular injury of hepatic PCs. However, no diffuse hepatic fibrosis was observed in the liver, and HSCs around the regions of cholangiofibrosis were negative for α-SMA, an indicator of "activated" HSCs as mentioned earlier. Under the electron microscope, the area of lipid droplets of an SC in the liver of LEC rats was 1.6–1.8 times as large as that of normal Wistar rats. The HSCs did not participate in the accumulation of collagen fibers around themselves when the cells contained a large amount of vitamin A-lipid droplets, even though the development of hepatic lesions was in progress. Antagonistic relationship between the storage of vitamin A and production of collagen in the SCs (Senoo and Wake, 1985) was strongly supported.

Matrix metalloproteinases (MMPs) and tissue inhibitor of metalloproteinases (TIMP) were reported to be synthesized by hepatic PCs and SCs (Arthur, 2002; Benyon and Arthur, 2001; Friedman, 2000; Friedman and Arthur, 2002; Iredale et al., 1998; Lindsay and Thorgeirsson, 1995; Montfort et al., 1990; Pinzani and Marra, 2001; Poynard et al., 2002). Reports indicate that a differential expression of MMP activity, hence the remodeling of

ECM components, is dependent on the substratum used for the culture of HSCs (Li *et al.*, 1999; Wang *et al.*, 2003). Three-dimensional (3D) structure of ECM can regulate reversibly morphology, proliferation, and functions of HSCs (Senoo and Hata, 1994a,b; Senoo *et al.*, 1996, 1998). SCs can take up collagen fibrils by endocytosis (Mousavi *et al.*, 2005) and phagocytosis (Higashi *et al.*, 2005a). Thus, the HSCs play pivotal roles in blood flow in sinusoid, homeostasis of vitamin A in the whole body, remodeling, fibrosis, and cirrhosis of the liver.

VII. REVERSIBLE REGULATION OF MORPHOLOGY, PROLIFERATION, AND FUNCTION OF THE HSCs BY 3D STRUCTURE OF ECM

Tissues are not composed solely of cells. A substantial part of their volume is intercellular space that is largely filled by an intricate network of macromolecules constituting ECM. This matrix comprises a variety of polysaccharides and proteins that are secreted locally and assembled into an organized meshwork (Alberts *et al.*, 2002). ECM was considered to serve mainly as a relatively inactive scaffolding to stabilize the physical structure of tissues until recently. But now it is clear that ECM plays a far more active complex role in regulating the behavior of the cells that contact, influencing their morphology, development, migration, proliferation, and functions (Hata and Senoo, 1992; Senoo and Hata, 1993b, 1994b; Senoo *et al.*, 1996).

We reported that HSCs proliferated better and synthesized more collagen on type I collagen-coated culture dishes than on polystyrene dishes (Senoo and Hata, 1994a). We also demonstrated that the SCs formed a mesh-like structure, proliferated slowly, and synthesized a small amount of collagen on a basement membrane gel (Senoo and Hata, 1994a) prepared from murine Engelbreth-Holm-Swarm (EHS) tumor, a gel consisting largely of laminin, type IV collagen, heparan sulfate proteoglycan, and nidogen (Kleinman *et al.*, 1986).

Other reports also indicate that ECM can regulate the functions of SCs. The regulation of the retinol esterification activities, the central process in vitamin A storage, was evaluated in SCs cultured on type I or type IV collagen-coated dishes in the presence of [^{3}H]retinol (Davis and Vucic, 1989). Uptake of [^{3}H]retinol into the cells and esterification into retinyl palmitate were enhanced when the cells were cultured on type IV collagen-coated dishes. The basement membrane gel was reported to be able to maintain the differentiated phenotype such as storage of lipids in cultured SCs (Friedman *et al.*, 1989, 1990). Types of collagen synthesized by SCs were also modulated by ECM (Davis *et al.*, 1987). The cells synthesized mainly type IV collagen on type I collagen-coated culture dishes, and synthesized equal amounts of type I and type IV collagen on type IV collagen-coated dishes. The total amount of collagen synthesized by SCs was more on type I collagen than on type IV collagen. Thus, these recent studies support the idea that ECM regulates phenotypes of the HSCs such as collagen metabolism and storage of vitamin A in lipid droplets in the cytoplasm. Responses of the HSCs to cytokines are also modulated by ECM (Davis, 1988). When transforming growth factor-β (TGF-β) was applied to HSCs cultured on type I or type IV collagen-coated culture dishes, collagen synthesis of the cells inoculated on type I collagen-coated dishes was stimulated. On the other hand, there was no response to TGF-β in terms of collagen synthesis by the HSCs inoculated on type IV collagen. Thus, reactions of the HSCs to cytokines are modulated by ECM.

Our studies and other works clearly show that ECM can regulate morphology, proliferation, and functions of HSCs.

We reported that morphology, proliferation, and collagen synthesis of the SCs were reversibly regulated by 3D structure of ECM (Imai and Senoo, 1998, 2000; Imai et al., 2000b; Kojima et al., 1998, 1999). The cellular processes of the HSCs were demonstrated to be extended and retracted according to the ECM, and speculated to have important functions in transport and storage of vitamin A and transport of metalloproteinases (Sato et al., 1997, 1998, 1999, 2001b, 2003, 2004). These data also indicate that the SCs are not static, but dynamic in the changeable 3D structure of ECM in the space between PCs and SECs.

The dynamic movement of cultured SCs was analyzed with a video-enhanced optical microscopy (Miura et al., 1997). When cultured on polystyrene surface, the SCs spread well, flattened with extensive stress fibers. The cell surface ruffling activity of filopodia and lamellipodia was prominent (Fig. 11), reflecting weak adhesion to the substratum. All filopodia remained dynamic throughout the 4-h recording, and extended and retracted repeatedly (Figs. 12 and 13). Within 1 h after inoculating in or on type I collagen gel, the SCs began to extend cellular processes (Fig. 12A–D), and the cellular processes appeared to adhere to and extend along type I collagen fibers. After repeated extension and retraction of cellular processes, HSCs displayed a number of long cellular processes with distal fine branches by 4-h culture on type I collagen gel (Fig. 12D). The cellular processes also extended in or on type III collagen gel, but not in type IV collagen-coated dishes or on Matrigel comprising from the basement membrane components.

The role of microtubule organization in maintenance of the cellular process structure was demonstrated by video recording of the SC culture after addition of colchicine (Fig. 13). In the presence of 1-μM colchicine, once extended cellular processes after overnight culture on type I collagen gel were time-dependently retracted (Fig. 13). The effects of colchicine were also dose-dependent at a concentration of 0.1–1.0 μM, and almost all cells changed to round shapes within a few hours in the presence of

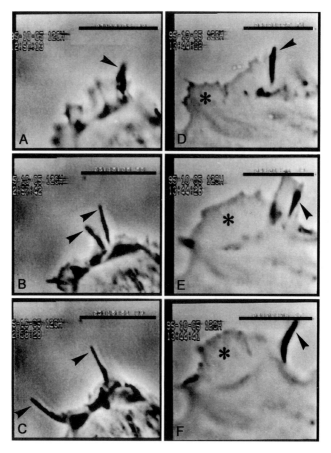

FIGURE 11. Motility of filopodia and lamellipodia in rat HSCs inoculated on polystyrene surface. The cells were monitored by phase-contrast video-enhanced microscopy. Photographs were taken at 29 s (B) and 55 s (C) after taking a picture of (A), and at 24 s (E) and 39 s (F) after taking a picture of (D), respectively. Arrowheads indicate filopodia and asterisks indicate lamellipodia. Scale bars indicate 20 μm.

1-μM colchicine. Virtually no effects were seen after treatment with 1-μM γ-lumicolchicine as a control. The cold treatment at 4 °C, which is known to induce the degradation of cold-labile form of microtubules, also induced the retraction of elongated cellular processes within 3 h. The once extended cellular processes were also partly retracted 1 h after treatment with 4-μg/ml cytochalasin B, as seen after treatment with 0.5-μM colchicine. The effects of cytochalasin B at the concentration of 4 μg/ml appeared to be weaker than that of 1-μM colchicine and a part of cellular processes still remained. However, almost all cells were changed to round shapes after overnight treatment with cytochalasin B.

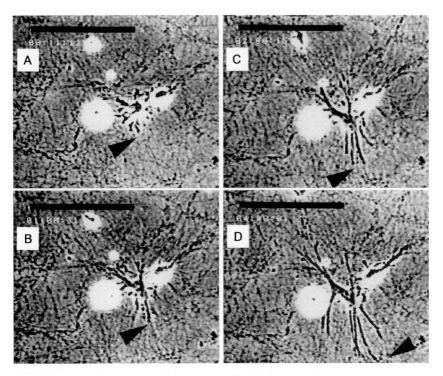

FIGURE 12. Extension of cellular processes in cultured rat stellate cells on type I collagen gel. The cells were cultured on type I collagen gel, and monitored by phase-contrast time-lapse video microscopy for up to 4 h. Photographs were taken 71 min (A), 2 h (B), 3 h (C), and 4 h (D) after inoculation. Arrowheads indicate the front of elongating cellular processes. Scale bars indicate 200 μm.

Fibroblasts were also reported to change the phenotype according to 3D structure of collagen (Cukierman et al., 2001; Grinnell et al., 2003; Kojima et al., 1999), and the integrin α2β1 was demonstrated to recognize 3D structure of triple helical collagen peptide (Emsley et al., 2000).

VIII. STIMULATION OF PROLIFERATION OF HSCs AND TISSUE FORMATION OF THE LIVER BY A LONG-ACTING VITAMIN C DERIVATIVE

A long-acting vitamin C derivative, l-ascorbic acid 2-phosphate (Asc 2-P), was found to stimulate cell proliferation, collagen accumulation, and tissue formation (Hata and Senoo, 1989; Kurata et al., 1993). On the basis of this discovery, Asc 2-P was added to the medium in which HSCs were cultured

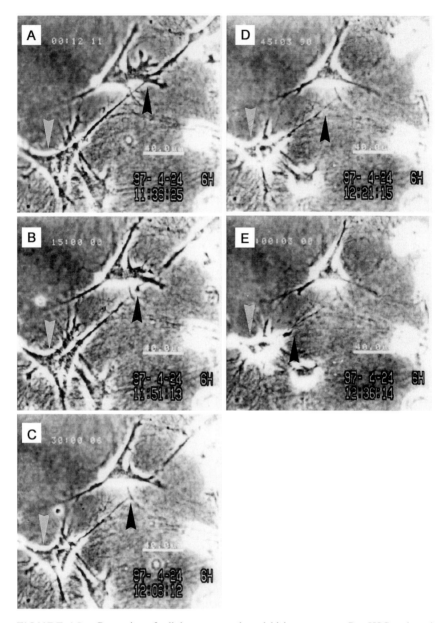

FIGURE 13. Retraction of cellular processes by colchicine treatment. Rat HSCs cultured overnight on type I collagen gel were monitored for 1 h after the addition of 1-μM colchicine. Pictures show phase-contrast video-enhanced micrographs 12 s (A), 15 min (B), 30 min (C), 45 min (D), and 60 min (E) after the addition of colchicine. Black arrowheads indicate the front of cellular process of the cell marked with white arrowheads. Scale bars indicate 100 μm.

(Senoo and Hata, 1994a). The cells in the medium supplemented with Asc 2-P stretched better than the cells in the control medium. Asc 2-P stimulated cell proliferation and collagen synthesis of the HSCs, and formation of the liver tissue-like structure in coculture of PCs and fibroblasts (Senoo *et al.*, 1989).

IX. EXTRAHEPATIC STELLATE CELLS

Previous studies using fluorescence microscopy, transmission electron microscopy, and electron microscopic autoradiography showed that cells that stored vitamin A distributed in extrahepatic organs, namely, lung, digestive tract, spleen, adrenal gland, testis, uterus, lymph node, thymus, bone marrow, adventitia of the aorta, lamina propria of the trachea, oral mucosa, and tonsil (Matano *et al.*, 1999; Nagy *et al.*, 1997; Wake, 1980). Morphology of these cells was similar to that of fibroblasts. These cells emanate autofluorescence of vitamin A and contain lipid droplets in the cytoplasm. These cells and HSCs form the SC system that regulates homeostasis of vitamin A in the whole body. Extrahepatic SCs also can synthesize and secrete ECM components.

Vitamin A distribution and content in tissues of a lamprey (*Lampetra japonica*) (Wold *et al.*, 2004) (Fig. 14) and an arrowtooth halibut (*Atheresthes evermanni*) (Fig. 15) were analyzed. HSCs showed an abundance of vitamin A stored in lipid droplets in their cytoplasm. Similar cells storing vitamin A were present in the intestine, kidney, gill, and heart in both female and male lampreys. Morphological data obtained by gold chloride staining method, fluorescence microscopy, and transmission electron microscopy and HPLC quantification of vitamin A were consistent. The highest level of total vitamin A measured by HPLC was found in the intestine. The second and third highest concentrations of vitamin A were found in the liver and the kidney, respectively. These vitamin A-storing cells were not epithelial cells, but mesoderm-derived cells. Similar cells were distributed in the arrowtooth halibut. We propose as a hypothesis that these cells belong to the SC system (family) that stores vitamin A and regulates homeostasis of the vitamin in the whole body in these animals. Fibroblastic cells in the skin and somatic muscle stored little vitamin A. These results indicate that there is difference in the vitamin A-storing capacity between the splanchnic and intermediate mesoderm-derived cells (SCs) and somatic and dorsal mesoderm-derived cells (fibroblasts) in these animals. SCs derived from the splanchnic and intermediate mesoderm have high capacity and fibroblasts derived from the somatic and dorsal mesoderm have low capacity for the storage of vitamin A in these animals.

Pancreatic SCs, one sort of extrahepatic SCs, are now considered to be responsible for the induction of chronic pancreatitis and pancreatic fibrosis

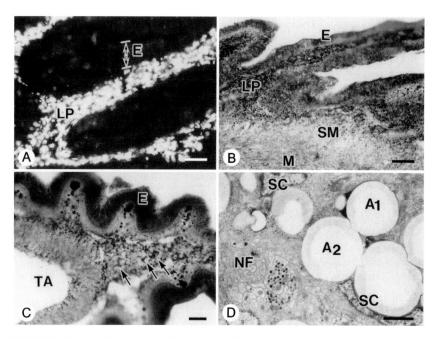

FIGURE 14. Stellate cells in lamina propria of the intestine in lamprey (*Lampetra japonica*). Vitamin A-autofluorescence is detected in cells of the lamina propria (LP) (A). Gold chloride-reacted cells were distributed in the lamina propria (B and C). An electron micrograph showing membrane-bound (type I, A1) and nonmembrane-bound (type II, A2) lipid droplets (Wake, 1974) in a cell (SC) in the lamina propria of the intestine. E, epithelium; M, muscle; NF, unmyelinated nerve fiber; SM, submucosa; TA, typhlosolar artery. Scale bars indicate 5 μm (A–C) and 1 μm (D).

(Apte *et al.*, 1998; Bachem *et al.*, 1998, 2002a,b; Masamune *et al.*, 2002; Wells and Crawford, 1998). These extrahepatic SCs are now to be targets of the treatment of inflammation and organ fibrosis.

X. CONCLUSIONS

HSCs that lie in the space between PCs and SECs play pivotal roles in the regulation of homeostasis of vitamin A in the whole body. HSCs in top predators of Arctic animals store vitamin A which is 20–100 times the levels normally found in other animals, including humans. The existence of a gradient of vitamin A-storing capacity in the liver was reported and it is independent on the vitamin A amount in the organ. This gradient was expressed as a symmetrical biphasic distribution starting at the periportal zone, peaking at the middle zone, and sloping down toward the central zone

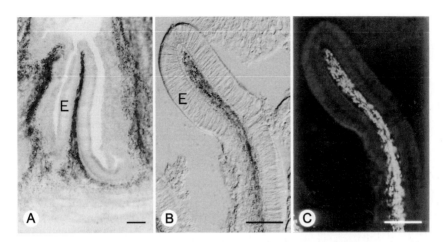

FIGURE 15. Stellate cells in the pyloric cecum. The pyloric cecum of the arrowtooth halibut (*Atheresthes evermanni* Jordan et Starks) was observed by Sudan III staining (A), differential interference microscopy (B), and fluorescence microscopy (C) for detecting autofluorescence of vitamin A. Scale bars indicate 100 mm. (See Color Insert.)

in the liver lobule. In pathological conditions such as liver cirrhosis, their phenotype changes from that of star-shaped SCs to that of fibroblasts or myofibroblasts. The 3D structure of the ECM can reversibly regulate the morphology, proliferation, and functions of the SCs. Molecular mechanisms in the regulation of the SCs by 3D structure of the ECM imply cell surface integrin binding to the matrix components followed by signal transduction processes and cytoskeleton assembly. HSCs play key roles in the regeneration of the liver. The SC system consists of hepatic and extrahepatic SCs and regulates vitamin A homeostasis of the whole body.

ACKNOWLEDGMENTS

The authors thank Mitsutaka Miura (Akita University School of Medicine) for his technical assistance.

REFERENCES

Alberts, B., Johnson, A., Lewis, J., Raff, M., Roberts, K., and Walter, P. (2002). "Molecular Biology of the Cell," 4th ed., pp. 1090–1118. Garland Science, New York.

Andersen, K. B., Nilsson, A., Blomhoff, R., Øyen, T. B., Gabrielsen, O. S. D., Norum, K. R., and Blomhoff, R. (1992). Direct mobilization of retinol from hepatic perisinusoidal stellate cells to plasma. *J. Biol. Chem.* **267**, 1340–1344.

Apte, M. V., Haber, P. S., Applegate, T. L., Norton, I. D., McCaughan, G. W., Korsten, M. A., Pirola, R. C., and Wilson, J. S. (1998). Periacinar stellate shaped cells in rat pancreas: Identification, isolation, and culture. *Gut* **43**, 128–133.

Arthur, M. J. P. (2002). Reversibility of liver fibrosis and cirrhosis following treatment for hepatitis C. *Gastroenterology* **122**, 1525–1528.

Bachem, M. G., Schneider, E., Gross, H., Weidenbach, H., Schmid, R. M., Menke, A., Siech, M., Beger, H., Grnert, A., and Adler, G. (1998). Identification, culture, and characterization of pancreatic stellate cells in rats and humans. *Gastroenterology* **115**, 421–432.

Bachem, M. G., Schmid-Kotsas, A., Schuenemann, M., Fundel, M., Adler, G., Menke, A., Siech, M., Buck, A., and Gruenert, A. (2002a). The increased deposition of connective tissue in pancreas carcinoma is the result of a paracrine stimulation of pancreatic stellate cells by cancer cells. *Mol. Biol. Cell* **13**, 4a.

Bachem, M. G., Schmid-Kotsas, A., Buck, A., Siech, M., and Gruenert, A. (2002b). An TGFβ1-antisense oligonucleotide reduces autocrine stimulated TGFβ1-, fibronectin- and collagen type I-synthesis of cultured pancreatic stellate cells. *Mol. Biol. Cell* **13**, 345a.

Banchereau, J., and Steinman, R. M. (1998). Dendritic cells and the control of immunity. *Nature* **392**, 245–252.

Barrie, L. A., Gregor, D., Hargrave, B., Lake, R., Muir, D., Shearer, R., Tracey, N., and Bidleman, T. (1992). Arctic contaminants: Sources, occurrence and pathways. *Sci. Total Environ.* **122**, 1–74.

Båvik, C. O., Eriksson, U., Allen, R. A., and Peterson, P. A. (1991). Identification and partial characterization of a retinal pigment epithelial membrane receptor for plasma retinol-binding protein. *J. Biol. Chem.* **266**, 14978–14985.

Båvik, C. O., Busch, C., and Eriksson, U. (1992). Characterization of a plasma retinol-binding protein membrane receptor expressed in the retinal pigment epithelium. *J. Biol. Chem.* **267**, 23035–23042.

Båvik, C. O., Lévy, F., Hellman, U., Wernstedt, C., and Eriksson, U. (1993). The retinal pigment epithelial membrane receptor for plasma retinol-binding protein. *J. Biol. Chem.* **268**, 20540–20546.

Benyon, R. C., and Arthur, M. J. P. (2001). Extracellular matrix degradation and the role of hepatic stellate cells. *Semin. Liver Dis.* **21**, 373–384.

Blomhoff, R. (1994). "Vitamin A in Health and Disease." Marcel Dekker Inc., New York.

Blomhoff, R., and Wake, K. (1991). Perisinusoidal stellate cells of the liver: Important roles in retinol metabolism and fibrosis. *FASEB J.* **5**, 271–277.

Blomhoff, R., Green, M. H., Berg, T., and Norum, K. R. (1990). Transport and storage of vitamin A. *Science* **250**, 399–404.

Blomhoff, R., Green, M. H., Green, J. B., Berg, T., and Norum, K. R. (1992a). Vitamin A metabolism: New perspectives on absorption, transport, and storage. *Physiol. Rev.* **71**, 951–990.

Blomhoff, R., Green, M. H., and Norum, K. R. (1992b). Vitamin A: Physiological and biochemical processing. *Annu. Rev. Nutr.* **12**, 37–57.

Blomhoff, R., Senoo, H., Smeland, S., Bjerknes, T., and Norum, K. R. (1992c). Cellular uptake of vitamin A. *J. Nutr. Sci. Vitaminol.* **38**, 327–330.

Bloom, W., and Fawcett, D. W. (1994). "A Textbook of Histology," 12th ed., pp. 652–668. Chapman & Hall, New York.

Chawla, A., Repa, J. J., Evans, R. M., and Mangelsdorf, D. J. (2001). Nuclear receptors and lipid physiology: Opening the X-files. *Science* **294**, 1866–1870.

Cukierman, E., Pankov, R., Stevens, D. R., and Yamada, K. M. (2001). Taking cell-matrix adhesion to the third dimension. *Science* **294**, 1708–1712.

Davis, B. H. (1988). Transforming growth factor β responsiveness is modulated by extracellular collagen matrix during hepatic Ito cell culture. *J. Cell. Physiol.* **136**, 547–553.

Davis, B. H., and Vucic, A. (1989). Modulation of vitamin A metabolism during hepatic and intestinal culture. *Biochem. Biophys. Acta* **1010,** 318–324.
Davis, B. H., Pratt, B. M., and Madri, J. A. (1987). Retinol and extracellular collagen matrices modulate hepatic Ito cell collagen phenotype and cellular retinol binding protein levels. *J. Biol. Chem.* **262,** 10280–10286.
Davis, B. H., Chen, A., and Beno, D. W. A. (1996). Raf and mitogen-activated protein kinase regulate stellate cell collagen gene expression. *J. Biol. Chem.* **271,** 11039–11042.
Dewailly, E., Nantel, A., Weber, J.-P., and Meyer, F. (1989). High levels of PCBs in breast milk of Inuit women from arctic Quebec. *Bull. Environ. Contam. Toxicol.* **43,** 641–646.
Emsley, J., Knight, C. G., Farndale, R. W., Barnes, M. J., and Liddington, R. C. (2000). Structural basis of collagen recognition by integrin α2β1. *Cell* **101,** 47–56.
Friedman, S. L. (2000). Molecular regulation of hepatic fibrosis, an integrated cellular response to tissue injury. *J. Biol. Chem.* **275,** 2247–2250.
Friedman, S. L., and Arthur, M. J. P. (2002). Reversing hepatic fibrosis. *Sci. Med.* **8,** 194–205.
Friedman, S. L., Roll, F. J., Boyles, J., Arenson, D. M., and Bissell, D. M. (1989). Maintenance of differentiated phenotype of cultured rat hepatic lipocytes by basement membrane matrix. *J. Biol. Chem.* **264,** 10756–10762.
Friedman, S. L., Roll, F. J., Boyles, J., and Bissell, D. M. (1990). Autocrine regulation of lipocyte matrix production: Activation by extracellular matrix. *Cells Hepatic Sinusoid* **2,** 61–63.
Greenwel, P., Rubin, J., Schwartz, M., Hertzberg, E. L., and Rojkind, N. (1993). Liver fat-storing cell clones obtained from a CCl$_4$-cirrhotic rat are heterogeneous with regard to proliferation, expression of extracelllar matrix components, interleukin-6, and connexin 43. *Lab. Invest.* **69,** 210–216.
Grinnell, F., Ho, C.-H., Tamariz, E., Lee, D. J., and Skuta, G. (2003). Dendritic fibroblasts in three-dimensional collagen matrices. *Mol. Biol. Cell* **14,** 384–395.
Gyøen, T., Bjerkelund, T., Blomhoff, H. K., Norum, K. R., Berg, T., and Blomhoff, R. (1987). Liver takes up retinol-binding protein from plasma. *J. Biol. Chem.* **262,** 10926–10930.
Hata, R., and Senoo, H. (1989). L-Ascorbic acid 2-phosphate stimulates collagen accumulation, cell proliferation, and formation of a three-dimensional tissuelike substance by skin fibroblast. *J. Cell. Physiol.* **138,** 8–16.
Hata, R., and Senoo, H. (1992). Extracellular matrix system regulates cell growth, tissue formation, and cellular functions. *Tissue Cult. Res. Commun.* **11,** 337–343.
Higashi, N., and Senoo, H. (2003). Distribution of vitamin A-storing lipid droplets in hepatic stellate cells in liver lobules-A comparative study. *Anat. Rec. Part A* **271,** 240–248.
Hiagashi, N., Kojima, N., Miura, M., Imai, K., Sato, M., and Senoo, H. (2004). Cell-cell junctions between mammalian (human and rat) hepatic stellate cells. *Cell Tissue Res.* **317,** 35–43.
Higashi, N., Wake, K., Sato, M., Kojima, N., Imai, K., and Senoo, H. (2005a). Degradation of extracellular matrix by extrahepatic stellate cells in the intestine of the lamprey, *Lampetra japonica*. *Anat. Rec. Part A* **285A,** 668–675.
Higashi, N., Sato, M., Kojima, N., Irie, T., Kawamura, K., Mabuchi, A., and Senoo, H. (2005b). Vitamin A storage in hepatic stellate cells in the regenerating rat liver: With special reference to zonal heterogeneity. *Anat. Rec.* **286A,** 899–907.
Holden, C. (1998). Polar bears and PCBs. *Science* **280,** 2053.
Imai, K., and Senoo, H. (1998). Morphology of sites of adhesion between hepatic stellate cells (vitamin A-storing cells) and a three-dimensional extracellular matrix. *Anat. Rec.* **250,** 430–437.
Imai, K., and Senoo, H. (2000). Morphology of sites of adhesion between exracellular matrix and hepatic stellate cells. *Connect. Tissue* **32,** 395–400.
Imai, K., Sato, M., Kojima, N., Miura, M., Sato, T., Sugiyama, T., Enomoto, K., and Senoo, H. (2000a). Storage of lipid droplets in and production of extracellular matrix by hepatic stellate

cells (vitamin A-storing cells) in Long-Evans cinnamon-like colored (LEC) rats. *Anat. Rec.* **258**, 338–348.
Imai, K., Sato, T., and Senoo, H. (2000b). Adhesion between cells and exracellular matrix with special reference to hepatic stellate cell adhesion to three-dimensional collagen fibers. *Cell Struct. Funct.* **25**, 329–336.
Iredale, J. P., Benyon, R. C., Pickering, J., McCullen, M., Northrop, M., Pawley, S., Hovell, C., and Arthur, M. J. P. (1998). Mechanisms of spontaneous resolution of rat liver fibrosis. Hepatic stellate cell apoptosis and reduced hepatic expression of metalloproteinase inhibitors. *J. Clin. Invest.* **102**, 538–549.
Jarman, W. M., Simon, M., Norstrom, R. J., Burns, S. A., Bacon, C. A., Simonelt, B. R. T., and Risenbrough, R. W. (1992). Global distribution of Tris(4-chlorophenyl)methanol in high trophic level birds and mammals. *Environ. Sci. Technol.* **26**, 1770–1774.
Kawada, N., Kristensen, D. B., Asahina, K., Nakatani, K., Minamiyama, Y., Seki, S., and Yoshizato, K. (2001). Characterization of a stellate cell activation-associated protein (STAP) with peroxidase activity found in rat hepatic stellate cells. *J. Biol. Chem.* **276**, 25318–25323.
Kleinman, H. K., McGarvey, M. L., Hassell, J. R., Star, V. L., Cannon, F. B., Laurie, G. W., and Martin, G. R. (1986). Basement membrane complexes with biological activity. *Biochemistry* **25**, 312–318.
Kojima, N., Sato, M., Imai, K., Miura, M., Matano, Y., and Senoo, H. (1998). Hepatic stellate cells (vitamin A-storing cells) change their cytoskeleton structure by extracellular matrix components through a signal transduction system. *Histochem. Cell Biol.* **110**, 121–128.
Kojima, N., Sato, M., Miura, M., Imai, K., and Senoo, H. (1999). Alteration in distribution of focal adhesion components by signaling inhibitors in hepatic stellate cells and fibroblasts cultured on type I collagen gel. *Cells Hepatic Sinusoid* **7**, 24–25.
Kojima, N., Sato, M., Suzuki, A., Sato, T., Satoh, S., Kato, T., and Senoo, H. (2001). Enhanced expression of B7-1, B7-2, and intercellular adhesion molecule 1 in sinusoidal endothelial cells by warm ischemia/reperfusion injury in rat liver. *Hepatology* **34**, 751–757.
Kurata, S., Senoo, H., and Hata, R. (1993). Transcriptional activation of type I collagen genes by ascorbic acid 2-phosphate in human skin fibroblasts and its failure in cells from a patient with α2(I)-chain-defective Ehlers-Danlos syndrome. *Exp. Cell Res.* **206**, 63–71.
Li, Y.-L., Sato, M., Kojima, N., Miura, M., and Senoo, H. (1999). Regulatory role of extracellular matrix components in expression of matrix metalloproteases in cultured hepatic stellate cells. *Cell Struct. Funct.* **24**, 255–261.
Lindsay, C. K., and Thorgeirsson, U. P. (1995). Localization of messenger RNA for tissue inhibitor of metalloproteinases-1 and type IV collagenases/gelatinases in monkey hepatocellular carcinomas. *Clin. Exp. Metastasis* **13**, 381–388.
Mabuchi, A., Mullaney, I., Sheard, P., Hessian, O., Zimmermann, A., Senoo, H., and Wheatley, T. M. (2004). Role of hepatic stellate cells in the early phase of liver regenration in rat: Formation of tight adhesions to parenchymal cells. *J. Hepatol.* **40**, 910–916.
Maeda, N., Kawada, N., Seki, S., Arakawa, T., Ikeda, K., Iwao, H., Okuyama, H., Hirabayashi, J., Kasai, K., and Yoshizato, K. (2003). Stimulation of proliferation of rat hepatic stellate cells by galectin-1 and galectin-4 through different intracellular signaling pathways. *J. Biol. Chem.* **278**, 18938–18944.
Majno, G. (1979). The story of the myofibroblasts. *Am. J. Surg. Pathol.* **6**, 535–542.
Malaba, L., Smeland, S., Senoo, H., Norum, K. R., Berg, T., Blomhoff, R., and Kindberg, G. M. (1996). Retinol-binding protein and asialo-orosomucoid are taken up by different pathways in liver cells. *J. Biol. Chem.* **270**, 15686–15692.
Masamune, A., Kikuta, K., Satoh, M., Sakai, Y., Satoh, A., and Shimosegawa, T. (2002). Ligands of peroxisome proliferator-activated receptor-γ block activation of pancreatic stellate cells. *J. Biol. Chem.* **277**, 141–147.

Matano, Y., Miura, M., Kojima, N., Sato, M., Imai, K., and Senoo, H. (1999). Hepatic stellate cells and extrahepatic stellate cells (extrahepatic vitamin A-storing cells). *Cells Hepatic Sinusoid* **7**, 26–27.

Miura, M., Sato, M., Toyoshima, I., and Senoo, H. (1997). Extension of long cellular processes of hepatic stellate cells cultured on extracellular type I collagen gel by microtubule assembly: Observation utilizing time-lapse video-microscopy. *Cell Struct. Funct.* **22**, 487–492.

Montfort, I., Perez-Tamayo, R., Alvizouli, A. M., and Tello, E. (1990). Collagenase of hepatocytes and sinusoidal liver cells in the reversibility of experimental cirrhosis of the liver. *Virchows Arch. B Cell Pathol.* **59**, 281–289.

Mousavi, S. A., Sato, M., Spørstol, M., Smedsrød, B., Berg, T., and Senoo, H. (2005). Uptake of collagen into hepatic stellate cells: Evidence of the involvement of urokinase plasminogen activator associated protein/Endo 180. *Biochem. J.* **387**, 39–46.

Muir, D. C. G., Wagemann, R., Hargrave, B. T., Thomas, D. J., Peakall, D. B., and Norstrom, R. J. (1998). Arctic marine ecosystem contamination. *Sci. Total Environ.* **122**, 75–134.

Nagy, N. E., Holven, K. B., Roos, N., Senoo, H., Kojima, N., Norum, K. R., and Blomhoff, R. (1997). Storage of vitamin A in extrahepatic stellate cells in normal rats. *J. Lipid Res.* **38**, 645–658.

Nilsson, C. B., Trossvik, C., Manzoor, E., Azais-Baesco, V., Blaner, W. S., Fex, G., Harrison, E. H., Bennekum, A. M. V., and Håkansson, H. (1999). Coinciding time- and dose-related effects of 2,3,7,8-tetrachlodibenzo-*p*-dioxin on renal retinol esterification and serum retinoic acid levels in the rat. In a doctoral thesis by Charlotte Nilsson "Studies on the effects of 2,3,7,8-tetrachlodibenzo-*p*-dioxin on vitamin A homeostasis." Stockholm.

Pinzani, M., and Marra, F. (2001). Cytokine receptors and signaling in hepatic stellate cells. *Semin. Liver Dis.* **21**, 397–416.

Poynard, T., Mchutchison, J., Manns, M., Trepo, C., Lindsay, K., Goodman, Z., Ling, M.-H., and Albrecht, J. (2002). Impact of pegylated interferon alfa-2b and ribavirin on liver fibrosis in patients with chronic hepatitis C. *Gastroenterology* **122**, 1303–1313.

Prickett, T. C. R., McKenzie, J. L., and Hart, D. N. J. (1988). Characterization of interstitial dendritic cells in human liver. *Transplantation* **46**, 754–761.

Rodahl, K. (1949a). Toxicity of polar bear liver. *Nature (London)* **164**, 530–531.

Rodahl, K. (1949b). Vitamin sources in Arctic regions. *Norsk polarinstitutt skrifter* **91**, 1–64.

Rodahl, K., and Moore, T. (1943). The vitamin A content and toxicity of bear and seal liver. *Biochem. J.* **37**, 166–168.

Sato, M., Imai, K., Kojima, N., Miura, M., and Senoo, H. (1997). Long cellular processes of hepatic stellate cells cultured on or in type I collagen gel. *Cells Hepatic Sinusoid* **6**, 85–89.

Sato, M., Sato, T., Kojima, N., Miura, M., Imai, K., and Senoo, H. (1998). Induction of cellular processes containing collagenase and retinoid by integrin-binding to interstitial collagen in hepatic stellate cell culture. *Cell Biol. Int.* **22**, 115–125.

Sato, M., Kojima, N., Miura, M., Imai, K., and Senoo, H. (1999). Intracellular signaling for process elongation in cultured hepatic stellate cells on type I collagen gel. *Cells Hepatic Sinusoid* **7**, 32–33.

Sato, M., Miura, M., Kojima, N., Higashi, N., Imai, K., Sato, T., Wold, H. L., Moskaug, J. Ø., Blomhoff, R., Wake, K., Roos, N., Berg, T., *et al.* (2001a). Nuclear deviation in hepatic parenchymal cells on sinusoidal surface in arctic animals. *Cell Struct. Funct.* **26**, 71–77.

Sato, M., Sato, T., Kojima, N., Miura, M., Imai, K., and Senoo, H. (2001b). Induction of cellular process elongation mediated by microtubule-associated protein 2 in hepatic stellate cells cultured on type I collagen gel. *Cells Hepatic Sinusoid* **8**, 205–206.

Sato, M., Suzuki, S., and Senoo, H. (2003). Hepatic stellate cells: Unique characteristics in cell biology and phenotype. *Cell Struct. Funct.* **28**, 105–112.

Sato, M., Sato, T., Kojima, N., Imai, K., Higashi, N., Wang, D.-R., and Senoo, H. (2004). Three-dimensional structure of extracellular matrix regulates gene expression in cultured stellate cells to induce process elongation. *Comp. Hepatol.* **3**(Suppl. 1), S4.

Senoo, H. (2000). Digestion, metabolism. *In* "The Digital Handbook of Experimental Laboratory Animals: The Rat" (G. J. Krinke, Ed.), pp. 359–383. Academic Press, London.

Senoo, H. (2004). Structure and function of hepatic stellate cells. *Med. Electron Miocrosc.* **37,** 3–15.

Senoo, H., and Hata, R. (1993a). Isolation of perisinusoidal stellate cells (vitamin A-storing cells, fat-storing cells) of the liver. *Connect. Tissue* **25,** 129–137.

Senoo, H., and Hata, R. (1993b). Tissue formation and extracellular matrix system—cellular devices for adhesion to extracellular matrix. *Tissue Cult. Res. Commun.* **12,** 237–245.

Senoo, H., and Hata, R. (1994a). Extracellular matrix regulates and L-ascorbic acid 2-phosphate further modulates morphology, proliferation, and collagen synthesis of the perisinusoidal stellate cells. *Biochem. Biophys. Res. Commun.* **200,** 999–1006.

Senoo, H., and Hata, R. (1994b). Extracellular matrix regulates cell morphology, proliferation, and tissue formation. *Acta Anat. Nippon.* **69,** 719–733.

Senoo, H., and Wake, K. (1985). Suppression of experimental hepatic fibrosis by administration of vitamin A. *Lab. Invest.* **52,** 182–194.

Senoo, H., Hata, R., Nagai, Y., and Wake, K. (1984). Stellate cells (vitamin A-storing cells) are the primary site of collagen synthesis in non-parenchymal cells in the liver. *Biomed. Res.* **5,** 451–458.

Senoo, H., Tsukada, Y., Sato, T., and Hata, R. (1989). Co-culture of fibroblasts and hepatic parenchymal cells induces metabolic changes and formation of a three-dimensional structure. *Cell Biol. Int. Rep.* **13,** 197–206.

Senoo, H., Stang, E., Nilsson, A., Kindberg, G. M., Berg, T., Roos, N., Norum, K. R., and Blomhoff, R. (1990). Internalization of retinol-binding protein in parenchymal and stellate cells of rat liver. *J. Lipid Res.* **31,** 1229–1239.

Senoo, H., Hata, R., Wake, K., and Nagai, Y. (1991). Isolation and serum free culture of stellate cells. *Cells Hepatic Sinusoid* **3,** 259–262.

Senoo, H., Smeland, S., Malaba, L., Bjerknes, T., Stang, E., Roos, N., Berg, T., Norum, K. R., and Blomhoff, R. (1993). Transfer of retinol-binding protein from HepG2 human hepatoma cells to cocultured rat stellate cells. *Proc. Natl. Acad. Sci. USA* **90,** 3616–3620.

Senoo, H., Imai, K., Sato, M., Kojima, N., Miura, M., and Hata, R. (1996). Three-dimensional structure of extracellular matrix reversibly regulates morphology, proliferation and collagen metabolism of perisinusoidal stellate cells (vitamin A-storing cells). *Cell Biol. Int.* **20,** 501–512.

Senoo, H., Sato, M., and Imai, K. (1997). Hepatic stellate cells—From the viewpoint of retinoid handling and function of the extracellular matrix. *Acta Anat. Nippon.* **72,** 79–94.

Senoo, H., Imai, K., Matano, Y., and Sato, M. (1998). Molecular mechanisms in the reversible regulation of morphology, proliferation and collagen metabolism in hepatic stellate cells by the three-dimensional structure of the extracellular matrix. *J. Gastroenterol. Hepatol.* **13**(Suppl.), S19–S32.

Senoo, H., Imai, K., Wake, K., Wold, H. L., Moskaug, J. O., Kojima, N., Matano, Y., Miura, M., Sato, M., Roos, N., Langvatn, R., Norum, K. R., *et al.* (1999). Vitamin A-storing system in mammals and birds in Arctic area: A study in the Svalbard archipelago. *Cells Hepatic Sinusoid* **7,** 34–35.

Senoo, H., Wake, K., Wold, H. L., Higashi, N., Imai, K., Moskaug, J. Ø., Kojima, N., Miura, M., Sato, T., Sato, M., Roos, N., Berg, T., *et al.* (2004). Decreased capacity for vitamin A storage in hepatic stellate cells in arctic animals. *Comp. Hepatol.* **3**(Suppl. 1), S18.

Skaare, J. U., Bernhoft, A., Wiig, Ø., Norum, K. R., Haug, E., Eide, D. M., and Derocher, A. E. (2001). Relationship between plasma levels of organochlorines, retinol and thyroid hormones from polar bears (*Ursus maritimus*) at Svalbard. *J. Toxicol. Environ. Health* **62,** 227–241.

Smeland, S., Bjerknes, T., Malaba, L., Eskild, W., Norum, K. R., and Blomhoff, R. (1995). Tissue distribution of the receptor for plasma retinol-binding protein. *Biochem. J.* **305,** 419–424.

Steiniger, B., Klempnauer, J., and Wonigeit, K. (1984). Phenotype and histological distribution of interstitial dendritic cells in the rat pancreas, liver, heart, and kidney. *Transplantation* **38**, 169–175.

Wake, K. (1971). "Sternzellen" in the liver: Perisinusoidal cells with special reference to storage of vitamin A. *Am. J. Anat.* **132**, 429–462.

Wake, K. (1974). Development of vitamin A-rich lipid droplets in multivesicular bodies of rat liver stellate cells. *J. Cell Biol.* **63**, 683–691.

Wake, K. (1980). Perisinusoidal stellate cells (fat-storing cells, interstitial cells, lipocytes), their related structure in and around the liver sinusoids, and vitamin A-storing cells in extrahepatic organs. *Int. Rev. Cytol.* **66**, 303–353.

Wake, K. (1995). Structure of the sinusoidal wall in the liver. *Cells Hepatic Sinusoid* **5**, 241–246.

Wake, K. (1998). Hepatic stellate cells. *Connect. Tissue* **30**, 245–246.

Wang, D.-R., Sato, M., Li, L.-L., Miura, M., Kojima, N., and Senoo, H. (2003). Stimulation of pro-MMP-2 production and activation by native form of extracellular type I collagen in cultured hepatic stellate cells. *Cell Struct. Funct.* **28**, 505–513.

Wells, R. G., and Crawford, J. M. (1998). Pancreatic stellate cells. The new stars of chronic pancreatitis? *Gastroenterology* **115**, 491–493.

Wiig, Ø., Derocher, A. E., Cronin, M. M., and Skaare, J. U. (1998). Female pseudohermaphrodite polar bears at Svalbard. *J. Wildl. Dis.* **34**, 792–796.

Wold, H. L., Wake, K., Higashi, N., Wang, D., Kojima, N., Imai, K., Blomhoff, R., and Senoo, H. (2004). Vitamin A distribution and content in tissues of the lamprey (*Lampetra japonica*). *Anat. Rec.* **276A**, 134–142.

Yang, Q., Graham, T. E., Mody, N., Preitner, F., Peroni, O. D., Zabolotny, J. M., Kotani, K., Quadro, L., and Kahn, B. B. (2005). Serum retinol binding protein 4 contributes to insulin resistance in obesity and type 2 diabetes. *Nature* **436**, 356–362.

7

USE OF MODEL-BASED COMPARTMENTAL ANALYSIS TO STUDY VITAMIN A KINETICS AND METABOLISM

Christopher J. Cifelli, Joanne Balmer Green, and Michael H. Green

Department of Nutritional Sciences, The Pennsylvania State University University Park, Pennsylvania 16801

I. Introduction
II. Highlights of Whole-Body Vitamin A Metabolism
III. Early Kinetic Studies of Vitamin A Metabolism
IV. Overview of Compartmental Analysis
V. Use of Model-Based Compartmental Analysis to Study Vitamin A Kinetics
   A. Whole-Body Models
   B. Effects of Vitamin A Status on Vitamin A Kinetics
   C. Vitamin A Kinetics in Specific Organs
   D. Exogenous Factors That Affect Vitamin A Metabolism
   E. Physiological Interpretation of Three- and Four-Compartment Models
   F. Vitamin A Kinetic Studies in Humans
VI. Conclusions
   References

We discuss the use of mathematical modeling, and specifically model-based compartmental analysis, to analyze vitamin A kinetic data obtained in rat and human studies over the past 25 years. Following an overview of whole-body vitamin A metabolism, a review of early kinetic studies, and an introduction to the approach and terminology of compartmental analysis, we summarize studies done in this laboratory to develop models of whole-body vitamin A metabolism in rats at varying levels of vitamin A status. Highlights of the results of these studies include the extensive recycling of vitamin A among plasma and tissues before irreversible utilization and the existence of significant extrahepatic pools of the vitamin. Our studies also document important differences in vitamin A kinetics as a function of vitamin A status and the importance of plasma retinol pool size in vitamin A utilization rate. Later we describe vitamin A kinetics and models developed for specific organs including the liver, eyes, kidneys, small intestine, lungs, testes, adrenals, and remaining carcass, and we discuss the effects of various exogenous factors (e.g., 4-HPR, dioxin, iron deficiency, dietary retinoic acid, and inflammation) on vitamin A dynamics. We also briefly review the retrospective application of model-based compartmental analysis to human vitamin A kinetic data. Overall, we conclude that the application of model-based compartmental analysis to vitamin A kinetic data provides unique insights into both quantitative and descriptive aspects of vitamin A metabolism and homeostasis in the intact animal. © 2007 Elsevier Inc.

I. INTRODUCTION

If PubMed is used to search the key word "vitamin A" over the past several decades, an astounding number of scientific papers (35,000 since 1950) are identified. Even before that time, vitamin A was considered an intriguing nutrient: it was known to have roles in the visual cycle, reproduction, tissue health and differentiation, and immune function. When it was discovered in the 1980s that retinoids affect gene transcription, there was an explosion of interest in the molecular and cellular actions of vitamin A that continues to generate exciting results today.

Along with interest in the molecular and cellular aspects of vitamin A action (for reviews see Gudas *et al.*, 1994; Soprano and Soprano, 2003), much has also been learned in the last 25 years about the whole-body metabolism of vitamin A in intact animals. Based largely on data from kinetic studies in the rat, it has become clear that this system is much more complex and interesting than previously thought. Since vitamin A metabolism is believed to be similar in rats and humans, and since that assumption is supported by several experiments that have been carried out in humans (see later), kinetic studies done in the past several decades have deepened our knowledge about the vitamin A system in both species.

Here we will review the contributions of mathematical modeling and kinetic studies to our understanding of vitamin A metabolism in the intact animal. We will highlight early work which laid the foundation for subsequent studies in this laboratory, then give some background on model-based compartmental analysis, the method we have used to analyze kinetic data collected in intact rats and humans. Next we will describe whole-body and organ-level models of vitamin A metabolism, discuss the effects of vitamin A status on retinol kinetics, present work on several exogenous factors that alter vitamin A kinetics, and review information on retinol kinetics in humans. We conclude with a summary of the contributions of kinetic studies and mathematical modeling to current understanding of vitamin A homeostasis and metabolism.

II. HIGHLIGHTS OF WHOLE-BODY VITAMIN A METABOLISM

Many excellent reviews of vitamin A metabolism are available (Blomhoff et al., 1991; Ross, 2003a; Ross and Zolfaghari, 2004). As summarized by Green and Green (2005a) and illustrated in Fig. 1, ingested preformed vitamin A and provitamin A carotenoids are processed in the intestine and absorbed into plasma via the lymph as a component of chylomicrons, the triglyceride-rich absorptive lipoproteins that carry dietary fat into the body. This vitamin A, primarily in the form of retinyl esters, is mainly cleared with chylomicron remnants into liver hepatocytes. There, retinyl esters are hydrolyzed to retinol and processed either for secretion into plasma in complex with the vitamin A transport protein, retinol-binding protein (RBP) or for storage; some of the retinol in hepatocytes may be directly transferred to vitamin A-storing liver perisinusoidal stellate cells, a main body storage site for the vitamin. Secreted retinol:RBP complexes in plasma with transthyretin, protecting the complex from filtration by the kidneys. It is likely that retinol bound to RBP leaves the circulation and delivers retinol to target cells for storage as retinyl esters or oxidation to active forms, including retinoic acid (RA) in nonocular tissues and retinaldehyde in the retina of the eye. In cells, RA regulates gene expression through its interaction with RA receptors and RA response elements on nuclear DNA. Specific intracellular carrier proteins and enzymes regulate key aspects of vitamin A metabolism at the cellular level.

As will become evident in subsequent sections, whole-body vitamin A metabolism is much more complex than this simple overview would suggest. Our review of kinetic studies of vitamin A will illustrate how the application of mathematical modeling to isotope data collected over time has added

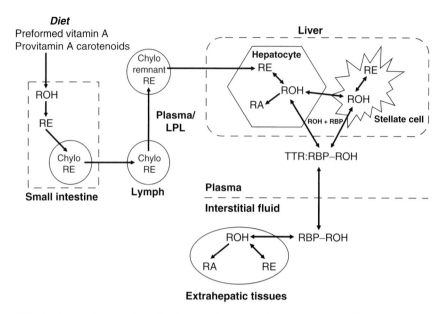

FIGURE 1. Simple schematic of whole-body vitamin A metabolism. Illustrated are the digestion, absorption, and initial handling of dietary vitamin A, the processing of vitamin A in the liver, and the transport and recycling of retinol to extrahepatic tissues. See text for more details. ROH, retinol; RE, retinyl esters; chylo, chylomicrons; LPL, lipoprotein lipase; RBP, retinol-binding protein; TTR, transthyretin; RA, retinoic acid.

to our understanding of vitamin A kinetics, dynamics, and homeostasis. See Green and Green (1996) for another discussion of these topics.

III. EARLY KINETIC STUDIES OF VITAMIN A METABOLISM

A number of papers published in the 1960s and 1970s used isotopic techniques and kinetic methods to study various aspects of vitamin A metabolism. Several examples that were important to the development of subsequent kinetic work will be briefly discussed here as background to more recent experiments.

A seminal report by Sauberlich *et al.* (1974) laid the groundwork for estimating vitamin A requirements in humans. Eight male subjects received intravenous or oral doses of [^{14}C]-labeled retinyl acetate, and plasma retinol concentrations and specific activity, as well as radioactivity lost in breath, urine, and feces, were determined during vitamin A depletion (up to 771 days) and repletion (up to 372 days). The authors determined that it took about 26 days for the labeled dose to equilibrate with the total body vitamin A pool and that body pool sizes ranged from 1.10 to 3.07 mmol. They found that vitamin A utilization rate decreased during depletion, providing early support for the idea that the vitamin is conserved in the face of low vitamin A intake. Adding to this idea, Hicks et al. (1984) later reported that excretion of labeled vitamin A metabolites into bile of rats fed increasing levels of vitamin A and dosed with [^{3}H]retinyl acetate was constant when liver vitamin A levels were low (up to 112 nmol/g) and then increased rapidly (by eightfold) to a plateau at 490 nmol/g.

Autopsy studies in the late 1960s and early 1970s (reviewed by Sauberlich et al., 1974) documented the importance of the liver in vitamin A storage, and thus indirect methods for estimating liver vitamin A were sought. Rietz et al. (1973, 1974) developed an isotopic method for estimating the body vitamin A pool in rats. Rats were given an intravenous dose of [^{3}H]vitamin A, and specific radioactivity (cpm/IU vitamin A) in plasma, and radioactivity in liver, was determined. Assuming that the dose had equilibrated with body vitamin A pools, liver vitamin A was calculated as radioactivity in liver divided by plasma vitamin A specific activity. Estimates compared well with fluorometric determinations of liver vitamin A. Over the years, isotope dilution methods for estimating vitamin A stores have been further developed. Results from modern isotope dilution studies, in which plasma data are collected following an oral dose of stable isotope-labeled vitamin A, are providing important information about vitamin A status in various populations and the efficacy of vitamin A interventions in improving vitamin A status (for review see Furr et al., 2005).

The work of Rietz et al. (1973) corroborated the conclusions of Sewell et al. (1967), which indicated that vitamin A stores are in a dynamic state. In Sewell et al.'s study, rats received an oral dose of tritium-labeled retinyl acetate. Feces and urine were collected for 5 days, and blood, liver, and kidneys were obtained at the time of killing (5–45 days after dosing). When the log of liver radioactivity was plotted against time, the decline followed a single exponential with a turnover time of 82 days. Since liver vitamin A concentrations remained constant while radioactivity decreased, the authors suggested that there is a dynamic exchange of vitamin A among blood, liver, and vitamin A-requiring tissues. As discussed subsequently, this is a feature of the vitamin A system that has been documented and quantified by modeling.

Sundaresan (1977) studied the rate of metabolism of retinol in vitamin A-deficient rats dosed intermittently with RA. [^{14}C]Retinol was injected intraperitoneally, and radioactivity in urine and feces was monitored for 10 weeks. Blood and livers were also collected at the time of killing. A plot of the log of daily urinary radioactivity versus time after dosing showed three phases which Sundaresan interpreted to indicate (1) initial rapid metabolism and excretion of newly absorbed vitamin A, (2) normal physiological metabolism, and (3) vitamin A metabolism after liver reserves were exhausted. In conjunction with work by Carney *et al.* (1976) in vitamin A-depleted rats, this study provides early support for the existence of two distinct pools of vitamin A in the liver: one includes newly absorbed vitamin A and is small and rapidly turning-over, the other contains stored vitamin A (Anonymous, 1977).

In the early 1980s, our laboratory collaborated with Underwood's group in a study designed to use kinetics to estimate vitamin A disposal rate (utilization rate) in rats (Lewis *et al.*, 1981). Control and vitamin A-deficient rats received a dose of [^{3}H]retinol-labeled plasma, and tritium kinetics in plasma were monitored for 48 h. Data on plasma tracer concentration versus time after dosing were plotted semi-logarithmically but did not follow the expected single exponential (i.e., a straight line), indicating either that plasma retinol was kinetically heterogeneous or that retinol was recycling to plasma before irreversible utilization. This work led both to the realization that vitamin A metabolism is more complex than had been previously believed and to further studies in our laboratory which used kinetic methods and compartmental analysis to describe and quantitate whole-body and organ-level vitamin A dynamics. Before reviewing those studies, we will present some background on compartmental analysis.

IV. OVERVIEW OF COMPARTMENTAL ANALYSIS

Mathematical models are mathematically formalized representations of a system that allow for the study of complex processes that are occurring simultaneously. In different disciplines of biology, mathematical modeling has been used to gain a deeper understanding of physiological systems and processes. For example, mathematical modeling has been used to study the rate of uptake of endogenous and exogenous compounds, to calculate enzyme kinetics, to predict pharmacological responses to drugs, and to calculate nutrient intake. For general discussions of the applications of mathematical modeling in the life sciences, see Robertson (1983), Hargrove (1998), Wastney *et al.* (1999), and Novotny *et al.* (2003).

Compartmental analysis (Foster and Boston, 1983; Green and Green, 1990a,b; Wastney *et al.*, 1999), and in particular model-based compartmental analysis, is the form of mathematical modeling that we will focus on here. Another form of compartmental analysis, empirical compartmental analysis, has also been used in the vitamin A field; see Green and Green (1990a,b) for theoretical background and Green *et al.* (1987) for an application. In nutrition, model-based compartmental analysis has been used to model mineral uptake and utilization (Birge *et al.*, 1969; Pinna *et al.*, 2001; Wastney *et al.*, 1996), lipoprotein metabolism (Adiels *et al.*, 2005; Berman *et al.*, 1982), glucose homeostasis (Malmendier *et al.*, 1974), digestion and absorption (Moore-Colyer *et al.*, 2003), and vitamin kinetics (Coburn *et al.*, 2003; Green *et al.*, references cited herein); see Green and Green (1990a) for other examples.

Compartmental modeling involves the representation of a system by a finite number of homogenous states and lumped processes, called compartments, which interact by means of material exchange (DiStefano and Landaw, 1984; Green and Green, 1990a; Wastney *et al.*, 1999, pp. 7–9) (Fig. 2). Compartmental analysis assumes that the system under study exhibits deterministic behavior, meaning that the future state of the system may be predicted based on its current state and future input; no probabilistic effects are included (Carson *et al.*, 1983, p. 56). Compartmental modeling provides both quantitative and predictive information about the system of interest, as well as unique insights into underlying mechanisms and metabolic processes that govern the system's kinetic behavior. The approach is unique in that it allows the researcher to investigate aspects of a system that might be difficult to study experimentally; model predictions may also provide unexpected insights into the metabolism of the compound of interest, and they may lead to the generation of new hypotheses and/or experiments (Green, 1992; Green and Green, 1990a).

The overall goal of model-based compartmental analysis is to describe and quantify the kinetics and often the dynamics of a particular system of interest (Green and Green, 1990a). Note that, although all of the work discussed here will be based on *in vivo* studies, model-based compartmental analysis can also be fruitfully applied to data collected from *in vitro* systems (see Blomhoff *et al.*, 1989 for an example). To use model-based compartmental analysis, the investigator begins by formalizing a conceptual

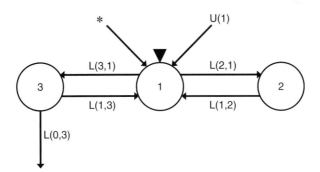

FIGURE 2. A three-compartment mammillary model. Compartments are shown as circles; movement between compartments is represented by arrows and quantified by fractional transfer coefficients [L(I,J)s or the fraction of compartment J's retinol transferred to compartment I per unit time]. U(1) represents input of newly absorbed dietary retinol, the asterisk shows the site of introduction of the tracer (typically plasma), and the triangle indicates that this compartment is a site of sampling.

compartmental model based on what is known and theorized about the system under study (Fig. 2); the model includes not just the compartmental structure but also estimates of the fractional transfer coefficients (see later) that quantify movement between compartments and out of the system. Then an appropriate *in vivo* experiment is designed and data are collected. In the case of retinol, a stable- or radioisotopic tracer in a physiological form (i.e., as part of the plasma transport complex or in chylomicrons) is administered intravenously; alternately, the tracer may be solubilized in oil and given orally. After dose administration, tracer concentration in plasma (and perhaps in organs and excreta) is followed over time (Fig. 3). Plasma tracer data (as fraction of administered dose), along with other relevant information (e.g., initial conditions, tracee mass, and sites of input), are analyzed using appropriate software. Here we will concentrate on the Simulation, Analysis and Modeling computer program (SAAM).

First introduced in 1962, SAAM mathematically compares the proposed model to the data and provides statistical information about model solutions (Berman and Weiss, 1978; Wastney *et al.*, 1999, pp. 95–138; www.WinSAAM.com; Stefanovski *et al.*, 2003); thus the modeler is able to evaluate the closeness of fit for each solution. Solution results are evaluated by comparing the observed and calculated data both graphically and numerically, with adjustments being made to the model until a close fit between the observed and calculated values is obtained. During model development, the known or suspected physiology and biochemistry of the system are kept in mind so that a physiologically reasonable model is developed (Green and Green, 1990a,b).

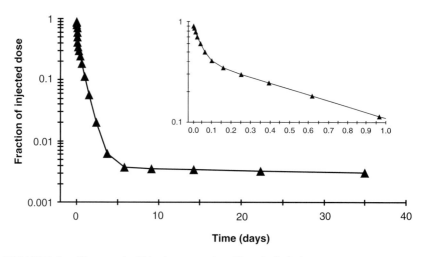

FIGURE 3. Plasma retinol kinetics versus time. Hypothetical plasma tracer response curve after intravenous administration of [^{3}H]retinol-labeled plasma to a rat with high liver vitamin A stores. Symbols represent observed data and the line is a model simulation. The inset shows the first day's data on an expanded scale.

To determine the simplest model that will provide an adequate fit to a particular data set, multiple models are tested, with model complexity being increased only when it results in a significant improvement in the weighted sum of squares (Landaw and DiStefano, 1984). Once a satisfactory fit is obtained, weighted nonlinear regression analysis is applied using the SAAM program to obtain best fit values for the fractional transfer coefficients describing movement between compartments (Fig. 2) as well as their statistical uncertainty (i.e., fractional standard deviation); then other kinetic parameters can be calculated (Table I). Typically, the modeling analysis assumes that the system is in a steady state, although the SAAM program can accommodate more complex situations (see later). Overall, the structure of the compartmental model provides a visual picture of how the compartments are linked, whereas the model-based parameters provide information about transfer between compartments, recycling, movement out of the system, and compartment masses.

In this laboratory, we have used model-based compartmental analysis to delineate the underlying mechanisms that control vitamin A homeostasis. Specifically, we have used modeling to study vitamin A kinetics in plasma and tissues under different conditions, including various levels of vitamin A status and after treatment with different exogenous factors, to better understand vitamin A utilization and whole-body vitamin A metabolism. The subsequent sections will review these studies and integrate the results to provide a better understanding of the kinetic behavior of the vitamin A system.

V. USE OF MODEL-BASED COMPARTMENTAL ANALYSIS TO STUDY VITAMIN A KINETICS

A. WHOLE-BODY MODELS

Following up on the implications of the collaborative study with Underwood's group (Lewis *et al.*, 1981) (see above), we designed an experiment to model vitamin A metabolism in rats with marginal liver vitamin A levels (<350 nmol) (Green *et al.*, 1985). Rats received an intravenous dose of plasma containing [^{3}H]retinol/RBP/transthyretin prepared *in vivo* using vitamin A-deficient donor rats (Green and Green, 1990b), and plasma tracer concentration was monitored from 3 min to 35 days after dosing; rats were killed at 1, 2, or 15 days (2/time) or at 35 days ($n = 5$); liver, kidneys, adrenals, small intestine, eyes, skin, and remaining carcass were removed for analysis.

When fraction of the dose remaining in plasma versus time was plotted on a semi-log scale for rats killed at 35 days (Green *et al.*, 1985), a curve like that shown in Fig. 3 was obtained; similar patterns have been found in a number

TABLE I. Model-Based Kinetic Parameters[a]

Parameter	Abbreviation; unit	Definition
Fractional transfer coefficient	L(I,J); d^{-1}	Portion of vitamin A in compartment J that is transferred to compartment I each day
Mean transit time	t(I); h	Average length of time that a molecule of retinol which reaches compartment I remains there before leaving that compartment reversibly or irreversibly
Mean residence time	T(I,J); day	Average of the distribution of times that a molecule of retinol spends in compartment I before irreversibly leaving compartment I after entering the system via compartment J
System residence time	T(sys); day	Average amount of time a molecule of vitamin A spends in the system
Traced mass	M(I); nmol	The amount of retinol in each compartment during a steady state
Transfer rate	R(I,J); nmol/day	The amount of vitamin A transferred from compartment J to compartment I each day; includes the system disposal rate
Fractional catabolic rate	FCR(I,J); day^{-1}	The fraction of retinol in compartment I that leaves irreversibly each day after entering the system via compartment J
System fractional catabolic rate	FCR(sys); day^{-1}	The fraction of the total traced mass irreversibly lost each day
Recycling number	v(I)	The average number of times a molecule of retinol recycles to compartment I before leaving compartment I irreversibly
Recycling time	tt(1); day	The average length of time it takes a molecule of retinol to return to compartment 1 after leaving compartment 1 reversibly

[a]For a more detailed discussion of these and other kinetic parameters, see Green and Green (1990a).

of other studies that will be discussed here. The initial rapid decrease in plasma tracer concentration between 0 and 2.5 h after dosing indicates that vitamin A is leaving plasma for other pools. As the tracer mixes into those pools, it begins to recycle to plasma, slowing down the disappearance of tracer from plasma. Once mixing is complete (about 8 days in Fig. 3), the plasma tracer response profile comes into a terminal slope that is a function of the system fractional catabolic rate for vitamin A (Table I).

Data on fraction of dose remaining in plasma versus time, and in tissues at the time of killing, were analyzed by model-based compartmental analysis using the SAAM software and the "super rat" approach (Landaw and DiStefano, 1984). In this technique, data from different subjects are collected at various times and then modeled together in one data set to develop a composite model. This method is extremely useful in that it allows the modeling of tissue kinetics over time. Using the combined data set from the experiment of Green et al. (1985), a model with eight compartments was developed for data from plasma, liver, kidneys, and rest of carcass. On the basis of the sites of sampling, the model included one compartment for plasma retinol (and a minor one for retinyl esters in plasma lipoproteins), two compartments (one more rapidly turning-over than the other) in liver, two in kidneys, and two in the remaining carcass; irreversible loss of tracer was assumed to occur from the fast turning-over compartments in liver, kidney, and carcass. The model predicted that there was extensive recycling of retinol among plasma, liver, and other organs before irreversible utilization of the vitamin. In fact, the model predicted that the plasma retinol turnover rate was 13 times the disposal rate (24 nmol/day). The model also predicted that 48% of plasma retinol recycled to the liver, and 52% was transferred to nonhepatic tissues. We hypothesized that vitamin A turnover was a "high-response" system in which changes in recycling could permit rapid adjustment in vitamin A distribution in response to physiological, nutritional, or metabolic state. This prediction had been suggested by the study of Lewis et al. (1981) and thus challenged the conventional wisdom that retinol secreted by the liver is transported directly to target tissues for uptake and utilization.

The 1985 model also predicted that there were significant amounts of vitamin A in extrahepatic tissues, in contrast to the idea prevalent at the time that more than 90% of whole-body vitamin A is found in the liver, irrespective of vitamin A status. In these rats with marginal liver stores, 44% of whole-body vitamin A was predicted to be present in extrahepatic tissues. In a subsequent modeling study with vitamin A-deficient rats (Lewis et al., 1990), it was predicted that 93% of body vitamin A was extrahepatic.

These findings related to retinol recycling and the distribution of body vitamin A pools illustrate the sort of "extra" information that can be derived from modeling. In the case of recycling, a kineticist might hypothesize that retinol recycles among tissues based on the shape of the plasma tracer response curve, but compartmental analysis provides a tool to estimate the extent of recycling compared to the utilization rate. In the case of vitamin A distribution, while it would be possible to analytically measure vitamin A levels in all body tissues, modeling provides an indirect but reliable way to estimate this distribution.

In a subsequent experiment, modeling was used to describe and quantitate retinol metabolism in rats with low vitamin A status (liver vitamin A, ~3.5 nmol) (Lewis et al., 1990). The design and methods used were similar to the 1985 study, but the number of animals and sampling times were larger and data were collected from additional sites (including urine and feces). A compartmental model was developed to fit the data for plasma, liver, kidneys, urine, feces, and carcass. As in the earlier study, the model predicted that plasma retinol turnover rate (70 nmol/day) was about 12 times the disposal rate. Vitamin A recycling through kidneys, liver, and carcass was extensive. The results indicated that, even though plasma retinol levels and liver vitamin A stores were low in these rats (0.35 μmol/L and 3.5 nmol, respectively), vitamin A recycling was high. It was suggested that a high-response system in the face of low vitamin A status might be a mechanism to lessen the effects of vitamin A depletion. Furthermore, although vitamin A intake was very low, plasma retinol concentrations were about 1/5 normal, and liver vitamin A stores were essentially depleted, these rats grew normally and appeared healthy; that is, they seemed to adapt to the chronic condition of low vitamin A intake. Of interest, the model predicted that "carcass" contained 39 times more vitamin A than the liver in these animals.

Comparing the studies done on rats with marginal (Green et al., 1985) versus low vitamin A levels (Lewis et al., 1990), vitamin A intake was estimated to be $\sim$32 versus $\sim$7 nmol/day, and disposal rates were predicted to be 24 versus 5.8 nmol/day, respectively. Both groups of rats were in negative liver vitamin A balance during the kinetic studies (8.8 and 0.16 nmol/day, respectively), so the true disposal rates were closer to 33 and 6 nmol/day, respectively.

A third study has been done in rats with very adequate vitamin A stores (liver vitamin A, $\sim$4550 nmol) (Green and Green, 1987); because of the extensive liver stores, the experiment was carried out for 115 days. While the data have not yet been fully analyzed due to challenging complexities in the model, we developed a theoretical model to validate the approach being used (Green and Green, 2005b). Using a simulated data set for vitamin A tracer and tracee values over time in plasma and organs (liver, small intestine, kidneys, lungs, testes, adrenals, eyes, and remaining carcass), we tested the forcing function feature of SAAM (Wastney et al., 1999, pp. 123–126) which uncouples the system so that the kinetics of individual organs with respect to plasma can be modeled (see later). Once individual organs have been fit, the plasma forcing function is removed and the full data set is modeled simultaneously. On the basis of both the preliminary and theoretical analyses, it is clear that vitamin A recycling to plasma from all of the organs will be a prominent feature of the model for vitamin A-adequate rats. Of note is that the theoretical model (Green and Green, 2005b) predicted that <0.1% of plasma retinol turnover went to the eyes or adrenals, <1% went to the lungs or testes, 4% went to the liver or small intestine, 30% went to the remaining carcass, and 60% went to the kidneys. The irreversible utilization rate was predicted to be 36 nmol/day, which is only 10% of the plasma retinol turnover rate (378 nmol/day).

As mentioned earlier, an important aspect of model development and interpretation is that the process often leads to the generation of new hypotheses. For example, we have emphasized the extensive recycling of retinol between plasma and tissues prior to irreversible loss that was predicted by the models developed for rats with low, marginal, and high vitamin A stores. In the case of the kidneys, the models predicted rapid exchange of retinol between the fast turning-over kidney compartment and plasma, with numerous cycles before loss (Green et al., 1985; Lewis et al., 1990). The rapidity of retinol turnover, coupled with the amount of recycling, led us to hypothesize that (1) the retinol:RBP complex is filtered by the kidney, (2) the majority of the filtered retinol is immediately reabsorbed by the proximal tubule epithelial cells, and (3) the retinol is resecreted as retinol:RBP. At the time of our initial study (Green et al., 1985), we assumed that RBP taken up by the kidneys would be degraded, and we speculated that lipoproteins might be the vehicle for renal retinol recycling. However, the amount of retinol involved could not be accounted for by retinyl esters in lipoproteins. Timely work by Soprano et al. (1986) indicated that certain extrahepatic tissues, including the kidneys, have the potential to synthesize and secrete RBP (i.e., they contain RBP mRNA). In more recent years, other researchers have studied the molecular mechanisms underlying the reabsorption and resecretion of retinol by the kidneys (Christensen et al., 1999; Raila et al., 2005). These experiments have shown that megalin, a membrane-bound protein expressed in the renal proximal tubule cells, is responsible for the uptake of filtered retinol and that this process is essential for the maintenance of whole-body vitamin A homeostasis. In the case of renal vitamin A metabolism, we suggest that development of the whole-body models expanded our understanding of vitamin A kinetics and turnover at the time the studies were done and also generated hypotheses that could be tested in later experiments.

As is evident in the preceding paragraphs, development of a vitamin A compartmental model that includes tissues and excreta requires long-term sampling of the organs of interest as well as plasma. As will be illustrated below, useful information on whole-body vitamin A metabolism may also be obtained using several other modeling approaches that do not involve extensive tissue sampling.

B. EFFECTS OF VITAMIN A STATUS ON VITAMIN A KINETICS

As the data from the three studies presented above were analyzed, it became evident that there were interesting differences in vitamin A kinetics as a function of vitamin A status. Thus, in a 1987 paper (Green et al., 1987), we compared plasma tracer kinetics at three levels of vitamin A status using empirical compartmental modeling (Green and Green, 1990a,b). In this approach, observed plasma tracer data are fit to a multiexponential equation, and kinetic parameters such as vitamin A disposal rate, plasma fractional catabolic rate, transit and residence times, and recycling number are computed from the curves' exponential constants and coefficients. As is evident in Fig. 4, there are dramatic differences in the steepness of the bend

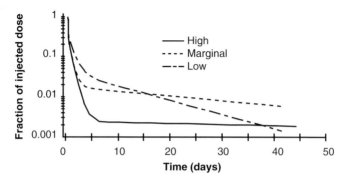

FIGURE 4. Plasma retinol kinetics in rats with various liver vitamin A stores. Model-simulated plasma tracer response curves after intravenous administration of [^{3}H]retinol-labeled plasma to rats with low (---; liver vitamin A, 8 nmol), marginal (-----; liver vitamin A, 150 nmol), or high vitamin A status (—; liver vitamin A, 3450 nmol). Based on data from Green et al. (1987).

in the tracer response curves as well as in the final slope for rats with high, marginal, or low liver vitamin A levels. These features of the curves are related to the size of liver stores, the recycling of labeled vitamin A, and the fractional catabolic rate. As predicted by the three whole-body models presented above, the 1987 analysis revealed significant differences in vitamin A disposal rate among the three groups. Disposal rate ranged from 4.2 nmol/day in the rats with low vitamin A status to 41.3 nmol/day in those with high liver reserves; also the total time an average retinol molecule spent in plasma (mean residence time; Table I) was significantly lower in rats with high vitamin A status. Statistical analysis indicated that 90% of the variance in vitamin A disposal rate could be accounted for by variation in the plasma retinol pool size. We also found that there was a significant negative correlation between the fraction of the labeled dose in plasma at 5 days and the natural log of total liver vitamin A. This observation was extended in subsequent studies (Adams and Green, 1994; Duncan et al., 1993) that became useful for later work on the application of isotope dilution analysis to the prediction of vitamin A status in humans (Furr et al., 2005).

In two later studies, we further investigated the determinants of vitamin A utilization in rats (Green and Green, 1994a; Kelley and Green, 1998), following our observation in the 1987 paper that there is a significant effect of plasma retinol pool size on vitamin A utilization. In the two subsequent studies, data were collected on plasma retinol kinetics versus time after administration of [^{3}H]retinol-labeled plasma as described earlier. Then model-based compartmental analysis was applied to describe whole-body vitamin A kinetics as viewed from the plasma space. In this approach, processes with similar kinetics are lumped into the same compartment, as opposed to the whole-body modeling approach described earlier in which data are obtained for various anatomical compartments. Here, the resulting model is simpler than the whole-body models. However, the same kinetic parameters may be estimated, making this method extremely useful. In the 1994 work (Green and Green, 1994a), rats were fed different levels of dietary vitamin A to affect vitamin A stores and balance. After a 41-day kinetic study, plasma tracer data were fit to a three-compartment, mammillary model (Fig. 2). The central plasma compartment, which is the site of dietary and tracer input, exchanges retinol with two extravascular compartments. One of these is a slower turning-over pool that includes retinyl ester stores, and the other is a faster turning-over pool of (presumably) retinol. After data for each rat in each of the four dietary groups were modeled, data from each group were analyzed using the multiple studies feature in SAAM (Lyne et al., 1992). With this tool, information from different subjects or animals that have been treated similarly is analyzed as one data set. Using the multiple studies feature, mean model parameters were estimated as a function of dietary intake. The results showed that the number of recyclings of retinol to plasma (recycling number; Table I) was not affected by vitamin A status and averaged 12–13 in all groups. In rats with low or marginal vitamin A status, vitamin A intake, vitamin A reserves, and plasma retinol concentration all influenced vitamin A kinetics. In rats with marginal vitamin A status (liver vitamin A, ~500 nmol), only 40% of the slow turning-over pool (compartment 3) could be accounted for by liver vitamin A. In the group with depleted liver vitamin A stores (<10 nmol), liver contained 3% of the vitamin A in the slow turning-over pool (275 nmol). See subsequent discussion about the slowly turning-over extrahepatic pool of vitamin A.

In the other study (Kelley and Green, 1998), a similar approach was used to investigate factors that influence vitamin A utilization rate in vitamin A-adequate rats under conditions of low vitamin A intake. After compartmental analysis, multiple linear regression analysis was used to examine the impact of plasma retinol pool size, vitamin A intake, and liver vitamin A levels on vitamin A utilization. We found that, if liver stores are adequate, vitamin A disposal is not decreased to compensate for low vitamin A intake as long as plasma retinol concentration is normal. We also found that there are appreciable extrahepatic pools of vitamin A in rats, especially when liver levels are low (Green and Green, 1996) (Fig. 5). This extrahepatic pool appears to deplete even more slowly than the liver in response to lowered dietary vitamin A input. In contrast, when liver vitamin A levels are very high, the model-predicted total traced mass underestimates measured liver vitamin A levels (Fig. 5). This has implications when isotope dilution techniques (Furr et al., 2005) are used to assess very high vitamin A stores that will not be traced.

Overall, our modeling studies lead us to hypothesize that both hepatic and extrahepatic pools of vitamin A help maintain normal plasma retinol

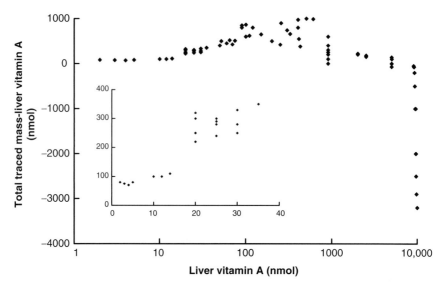

FIGURE 5. Relationship between liver vitamin A and model-predicted total traced mass. The inset shows the data on an expanded scale from 0 to 40 nmol liver vitamin A. When the *y*-axis is positive, the value provides a measure of extrahepatic vitamin A that has been traced; when it is negative, the value estimates liver stores of vitamin A that do not equilibrate with tracer (i.e., they turn over extremely slowly). Data are from Green and Green (1996).

levels even when dietary vitamin A intake is low, and we have identified plasma retinol as the main determinant of irreversible utilization of vitamin A. In the study by Kelley and Green (1998), plasma retinol pool size accounted for 92% of the variability in utilization.

C. VITAMIN A KINETICS IN SPECIFIC ORGANS

1. Liver

As noted earlier, the liver plays a central role in vitamin A homeostasis: it processes chylomicron remnants that contain newly absorbed dietary vitamin A, it stores vitamin A as retinyl esters, it synthesizes and secretes the plasma vitamin A transport protein (RBP), and it mobilizes retinol for delivery to extrahepatic tissues (Ross, 2003b). Several cell types, principally hepatocytes and perisinusoidal stellate cells, are known to be involved in liver vitamin A metabolism. In view of the liver's importance, we have been particularly interested in modeling the kinetics and turnover of hepatic vitamin A under different conditions and in using this approach to better understand vitamin A metabolism at the cellular level. As a result of these

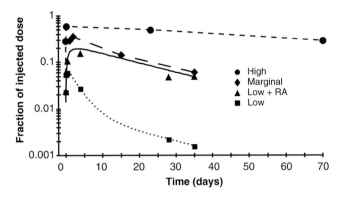

FIGURE 6. Liver vitamin A kinetics versus time. Observed data (symbols) and model-predicted tracer response curves (lines) in liver after intravenous administration of [^{3}H]retinol-labeled plasma to rats with low (■; liver vitamin A, 1.8 nmol), moderate (♦; liver vitamin A, 256 nmol), or high vitamin A status (●; liver vitamin A, 4750 nmol), or low vitamin A status under chronic dietary supplementation with RA (▲; liver vitamin A, 6 nmol; RA intake, 800 nmol/day).

investigations, a more complete understanding of the kinetic and dynamic behavior of liver vitamin A has been obtained.

Some interesting information is obtained by comparing models developed for rats with low, marginal, or high vitamin A status (see above). Despite dramatic differences in vitamin A nutriture and kinetic responses (Fig. 6), two compartments were sufficient to fit the liver data at all three levels of vitamin A status (Cifelli *et al.*, 2005; Green and Green, 2005b; Green *et al.*, 1985; Lewis *et al.*, 1990). In each case, the model indicates that plasma retinol exchanges with a small, fast turning-over liver compartment that in turn exchanges with a slowly turning-over compartment which presumably corresponds to the retinyl ester storage pool in liver stellate cells. The models for all three groups predicted extensive recycling of retinol between liver and plasma before irreversible utilization of vitamin A.

Despite these similarities in model structure, significant differences were found in model-derived kinetic parameters as a function of vitamin A status. Specifically, the mean residence time (Table I) for an average retinol molecule in liver was 0.358 days in rats with low vitamin A status (Cifelli *et al.*, 2005), 10.6 days in those with marginal liver vitamin A levels (Green *et al.*, 1985), and 132 days in the high status group (Green and Green, 2005b). The longer tissue residence time when stores were higher is directly related to both the larger body pools of vitamin A with which the radiolabeled retinol could equilibrate and a lower fractional catabolic rate (Table I). The models predicted liver vitamin A levels of 1.8, 256, and 4750 nmol, respectively, in the three groups. Similarly, the model-predicted mean transit time for a molecule of retinol in liver (Table I) was ~16 times longer in rats with marginal vitamin A status as compared to those with low vitamin A stores. Taken together, these results demonstrate that hepatic uptake, storage, and utilization of vitamin A are affected by vitamin A status.

Given the complex molecular and physiological processes that are involved in liver retinol metabolism, it might seem surprising that only two kinetically distinct compartments were needed to fit the data across a broad range of vitamin A levels. The models and the model-derived kinetic parameters suggest that the first, fast turning-over compartment contains "transient" retinol. That is, retinol in the first compartment, which presumably includes vitamin A in both hepatocytes and stellate cells, is rapidly either esterified for long-term storage, enzymatically converted to RA (utilized), or secreted into plasma bound to RBP. In contrast, the second, more slowly turning-over compartment presumably contains mostly retinyl esters and represents long-term storage in stellate cells.

These hypotheses were directly explored in a study published in 1993 (Green *et al.*, 1993). In that work, model-based compartmental analysis was used to study liver vitamin A metabolism of retinol and retinyl esters in whole liver and isolated hepatocytes from rats that had received a vitamin A tracer in either the plasma RBP/transthyretin transport complex or chylomicrons; tracer and tracee levels in stellate cell retinol and retinyl esters were estimated from values in total liver minus hepatocytes. Using currently accepted ideas about hepatic vitamin A metabolism along with information from a previously developed model of whole-body vitamin A metabolism (Green *et al.*, 1985), the model developed to fit these data described hepatic pathways for metabolism of diet-derived vitamin A as well as recycling retinol in hepatocytes and stellate cells. A complex model, with multiple pools of retinol and retinyl esters in each cell type, was required to fit the data. The model confirmed that (1) both hepatocytes and stellate cells can take up recycling retinol from plasma; (2) hepatocytes and stellate cells appear capable of retinol secretion, presumably as retinol bound to RBP; and (3) there is a transfer of vitamin A between hepatocytes and stellate cells, including a direct movement of chylomicron-derived vitamin A from hepatocytes to stellate cells for storage. Interestingly, the model led us to hypothesize and expand on other ideas about hepatic vitamin A metabolism by suggesting that extracellular apoRBP may be taken up by and/or interacts with the surface of stellate cells for movement of retinol from cellular retinol-binding protein I (CRBP I) to apoRBP for resecretion or movement of holoRBP off the cell. Conversely, the opposite is hypothesized to occur for storage of vitamin A in stellate cells. That is, holoRBP equilibrates with apoCRBP I in stellate cells. Overall, the model predicted that liver perisinusoidal stellate cells are more important than previously thought in terms of hepatic secretion of retinol into plasma.

2. Other Organs

As noted earlier, it has long been known (Wolbach and Howe, 1925) that vitamin A is involved in numerous metabolic and physiological processes throughout the body, including sustaining the visual cycle of the eye (Wald, 1968), supporting reproduction in the testes (Livera *et al.*, 2002), and maintaining epithelial barriers in the small intestine (Olson *et al.*, 1981; Zile *et al.*, 1977) and lung (Biesalski and Nohr, 2003). In view of these vastly different tissue-specific actions of vitamin A, we have been interested in examining retinol dynamics and kinetics in specific organs. Thus, we have used model-based compartmental analysis to develop compartmental models describing vitamin A kinetics in various organs in rats at different levels of vitamin A status (Cifelli *et al.*, 2005; Green and Green, 2005b; Green *et al.*, 1992). In each of the studies, plasma tracer data obtained after intravenous administration of [^{3}H]retinol-labeled plasma were fit to a multiexponential equation using SAAM. Then the "forcing function" option was used to develop a compartmental model for the tracer data from each organ subsystem based on the mathematical description of the plasma and individual organ tracer response profiles (Foster *et al.*, 1979; Green *et al.*, 1992; Wastney *et al.*, 1999, pp. 123–126). This approach makes use of the fact that each of the sampled organs exchanges vitamin A solely with plasma. Thus, we could uncouple each organ from the whole system and model each individually, allowing us to develop initial models for each organ before working with all organs simultaneously.

Two compartment models were sufficient to fit the data for adrenals, kidneys, lungs, small intestine, and testes in rats with low, marginal, and high liver vitamin A stores (Cifelli *et al.*, 2005; Green *et al.*, 1985, 1992; Lewis *et al.*, 1990). As in the case of liver, retinol kinetics were adequately described by one compartment that rapidly exchanged vitamin A with plasma and a second, more slowly turning-over compartment that included tissue stores of the vitamin. In contrast, only one slowly turning-over compartment was required to fit tracer data from eyes of rats in the three groups (Green *et al.*, 1992). Not surprisingly, kinetic parameters differed among organs within each dietary group. For example, in the rats that had the highest level of liver vitamin A stores, the fractional turnover of vitamin A in the eyes was lower than in the other organs examined (i.e., the transit time for vitamin A was longer in the eyes than in other organs), and the small intestine showed a higher rate of plasma vitamin A turnover than other organs.

Although the organs studied could be described using similar model structures in the three groups, differences in model-derived kinetic parameters and in tracer response curves indicated significant differences in vitamin A turnover in individual organs as a function of vitamin A status. For instance, the models predicted that the vitamin A traced mass (Table I) was higher in the adrenals (23 times), eyes (1.6 times), kidneys (24 times), lungs (164 times), small intestine (30 times), and testes (13 times) of rats with high compared to low vitamin A status (Green and Green, 2005b; Green et al., 1992). In addition, the model-predicted mean transit times (Table I) for vitamin A in each organ were significantly higher in the low vitamin A status rats as compared to those with high vitamin A status, with the most dramatic difference observed for the eyes (12.5 days in rats with high vitamin A status versus 75 days in those with low vitamin A status) (Green et al., 1992). Similar differences in model-predicted total traced mass and transit times were observed between the marginal and high vitamin A status rats. Interestingly, the differences in vitamin A kinetics between the rats with marginal and high status occurred despite the fact that plasma retinol concentrations were not statistically different between the two groups. This suggests that, as is true for plasma (Fig. 4) and liver (Fig. 6), tracer response curves for other organs are sensitive to liver stores of vitamin A.

D. EXOGENOUS FACTORS THAT AFFECT VITAMIN A METABOLISM

The vitamin A kinetic and modeling studies described above were done under the assumption that the system was in a (quasi) steady state. It is also relevant and interesting to study vitamin A metabolism when the system has been perturbed (i.e., is not in a steady state). Models of such states provide insight into which parameters have been affected by the perturbation, and they help target physiological processes that should be further explored.

We have used two different experimental approaches to model vitamin A metabolism under perturbation by exogenous factors. One approach is to compare two groups of animals—one group is under control conditions that are close to the steady state and the other has been perturbed and may or may not be close to a new steady state. The researcher develops the control model and then looks for minimal changes in model parameters or structure that explain the perturbation. As illustrated in subsequent sections, we have used this approach to study effects of the anticancer drug 4-HPR, the dioxin TCDD, iron deficiency, and dietary RA on vitamin A metabolism.

A second approach may be used under certain circumstances to study effects of specific perturbations on vitamin A metabolism, as illustrated later in the discussions about the dioxin TCDD and inflammation. In this method, a tracer kinetic study is begun with the animals under control conditions: rats are given the tracer, and plasma tracer data are collected over a long enough period of time (e.g., 3 weeks) so that whole-body vitamin A kinetics can be modeled. Then the perturbing agent is administered either acutely or chronically, and additional plasma tracer data are collected until the predetermined end of the study. In this design, the animal acts as its own control: the control model is developed based on the first data collection period, and then the minimal changes in fractional transfer coefficients that will predict and explain the perturbation are identified. Both of these approaches provide useful insights into the impact of various exogenous agents or conditions on the vitamin A system.

1. 4-HPR

One exogenous agent that is known to affect vitamin A metabolism and has been studied using compartmental analysis is the synthetic anticancer retinoid N-(4-hydroxyphenyl)retinamide (4-HPR; fenretinide). Although it is a useful therapy in human breast cancer, 4-HPR is associated with reductions in plasma concentrations of retinol and RBP and with reduced dark adaptation. To gain insight into the mechanisms behind these effects, we compared plasma retinol kinetics for 35 days after administration of [^{3}H]retinol-labeled plasma to control rats and those treated with 4-HPR (Adams et al., 1995). Data fit a three-compartment model similar to that shown in Fig. 2. Results showed that 4-HPR treatment was associated with a reduced rate of plasma retinol turnover, delayed and reduced retinol recycling to plasma, and significant reductions in vitamin A utilization. Taken together with other information, our data indicate that 4-HPR may block access and binding of retinol to RBP, thus leading to vitamin A accumulation in certain cells.

2. TCDD

Another exogenous agent we have studied is the persistent environmental toxin 2,3,7,8-tetrachlorodibenzo-p-dioxin (TCDD). It is known that exposure to TCDD has profound effects on vitamin A homeostasis in both experimental animals and humans (for review see Zile, 1992 and references therein). In rats, TCDD exposure is associated with a dramatic fall in liver vitamin A levels, an increase in kidney vitamin A, and often an increase in plasma retinol concentrations. In order to gain a better understanding of the impact of TCDD on vitamin A metabolism in the whole animal, we used model-based compartmental analysis to investigate which kinetic processes are affected by dioxin in the rat (Kelley et al., 1998). In a study that was similar in design to those described earlier, data were collected on vitamin A kinetics in plasma, tissues, and excreta of normal rats versus those exposed to weekly oral doses of TCDD. In a concurrent experiment, tracer was administered in triglyceride-rich absorptive lipoproteins (chylomicrons) so that the initial metabolism of newly absorbed retinyl esters could be modeled. Plasma tracer data for the first study were fit to a four-compartment model (Fig. 7) in which retinol in the central plasma compartment exchanged with vitamin A in both a small, rapidly turning-over extravascular pool (compartment 2) and a larger, slower turning-over compartment (compartment 3). This latter compartment also exchanged with another extravascular pool (compartment 4) which we hypothesized contained the slowest turning-over pools of body vitamin A in liver perisinusoidal stellate cells and

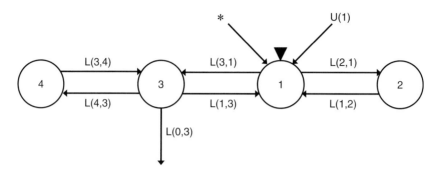

FIGURE 7. A four-compartment model. Compartments are shown as circles; movement between compartments is represented by arrows and quantified by fractional transfer coefficients [L(I,J)s or the fraction of compartment J's retinol transferred to compartment I per unit time]. U(1) represents input of newly absorbed dietary retinol, the asterisk shows the site of introduction of the tracer (typically plasma), and the triangle indicates that this compartment is a site of sampling.

carcass. The model included vitamin A loss into urine and feces from compartment 3. The model developed for the second study was more complex at the front end to accommodate the metabolism of dietary vitamin A: plasma metabolism of chylomicrons, hepatic uptake of chylomicron remnants, and liver processing of labeled retinyl esters. Considering the modeling results from both experiments, the data indicate that repeated exposure to TCDD causes more rapid movement of vitamin A out of storage pools in the liver and possibly other tissues, presumably due to increases in hydrolysis of retinyl esters to retinol, as well as an increase in vitamin A utilization. The models predict that, as liver vitamin A levels fell in the TCDD-treated animals, a larger fraction of the remaining pool was mobilized to maintain a constant rate of retinol secretion into plasma so that plasma retinol levels remained constant (albeit slightly higher than in controls). Regarding effects of TCDD on liver vitamin A levels, our models predict that vitamin A is transferred into storage pools normally but also mobilized out more quickly and degraded more rapidly.

To gain a deeper understanding of the TCDD-related increase in vitamin A degradation, we designed a subsequent study in which [^{3}H]retinol-labeled plasma was administered to normal rats, and plasma tracer responses were followed for 21 days until the kinetics had reached a final terminal slope (Kelley et al., 2000). Then TCDD was administered weekly over the next 21 days, and resultant changes in plasma tracer response were measured (Fig. 8). We hypothesized that, if the initial effect of TCDD was to mobilize vitamin A from liver stores, this would be reflected by an increase in plasma tracer concentrations. Alternately, if increased degradation was the initial consequence of TCDD administration, then plasma tracer concentrations

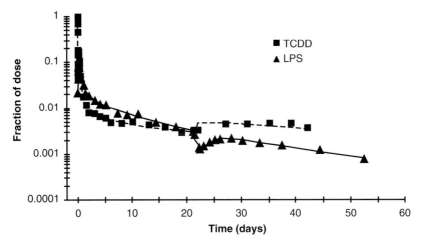

FIGURE 8. Plasma retinol kinetics before, during, and after perturbations. Shown are plasma tracer response curves after intravenous administration of [^{3}H]retinol-labeled plasma to a representative rat given the dioxin TCDD 21 days later (■) or after oral administration of [^{3}H]retinol in oil to rats who then received the inflammatory agent lipopolysaccharide (LPS; ▲) 21 days later. Symbols represent observed data and the lines are model simulations. Note the increase in plasma tracer concentration after TCDD administration and the decrease following LPS treatment. Data from the TCDD experiment are from Kelley et al. (2000), and data from the LPS study are from Gieng (2006).

would fall after TCDD treatment. The results showed that TCDD administration was associated with an increase in plasma tracer concentrations within 6 days (Fig. 8), indicating that TCDD caused an increased mobilization of tracer into plasma. By applying model-based compartmental analysis to the data collected after the TCDD perturbation, we found that by changing one model parameter, the fractional transfer coefficient describing movement from the very slowly turning-over liver pool (compartment 4; Fig. 7) to the slow turning-over hepatic compartment (compartment 3), we were able to fit the changes in plasma tracer response due to TCDD treatment. The adjustment needed corresponded to more than a fourfold increase in mobilization of tracer from compartment 4 to compartment 3. In physiological terms, this likely indicates that a major impact of TCDD on vitamin A dynamics is to increase mobilization of vitamin A from storage sites, most likely from perisinusoidal stellate cells. This might occur via an effect on either a nonspecific carboxyl ester hydrolase or a member of the bile-salt-dependent retinyl ester hydrolase family (Harrison and Gad, 1989), which is located in both stellate cells and hepatocytes (Blaner et al., 1985; Blomhoff et al., 1985). We concluded that increased degradation of the vitamin was a secondary effect. Overall, this study illustrates the usefulness of model-based compartmental analysis in discriminating between possible mechanisms by which a system perturbation affects isotope kinetics.

3. Iron Deficiency

Modeling has also been used to ascertain the most likely mechanism by which iron deficiency alters vitamin A kinetics (Jang *et al.*, 2000). Iron and vitamin A deficiencies often coexist in human populations, and in animal models, iron deficiency is associated with lowered plasma retinol concentrations and increased liver vitamin A levels. We used model-based compartmental analysis to study the effects of iron deficiency on vitamin A metabolism. Kinetic data were compared in control and iron-deficient rats following administration of [^{3}H]retinol-labeled plasma, and the approach applied by Kelley and Green (1998) was used to model the results. As in our previous studies, either a three- or four-compartment model was sufficient to fit plasma tracer data obtained during 48 days after administration of label. Visual inspection of the curves indicated that iron deficiency decreased retinol recycling from the slow turning-over pool(s) to plasma and that fractional irreversible utilization of vitamin A was lower in the iron-deficient animals. The model predicted a decrease in irreversible utilization of vitamin A and a decrease in vitamin A absorption efficiency in iron-deficient rats. The results suggested that liver vitamin A accumulation in iron deficiency might be due to impaired release of the vitamin into plasma and imply that decreased vitamin A mobilization from liver might account for the lower plasma retinol pool size. It is possible that the same enzyme that causes the increased mobilization of retinol in TCDD-treated rats is inhibited by iron deficiency.

Data from the TCDD and iron studies have also been analyzed using nonsteady state compartmental analysis (Green and Green, 2003) to reflect the fact that, in both cases, the vitamin A system was not actually in a steady state. Starting from the steady state solution, we developed a parallel model for tracee and set the model parameters [L(I,J)s; Table I] equal in the tracer and tracee models. Then we looked for minimal changes in the steady state model that would accurately describe the changes in liver vitamin A levels in iron-deficient or TCDD-treated rats. In both cases, we found that, by making time variant the one model parameter related to mobilization of liver vitamin A stores, we could account for the changes in liver vitamin A. The rate of retinol mobilization was held constant at a different value in the two models in order to maintain homeostasis of the plasma retinol pool and the two other non-liver pools of vitamin A. In iron deficiency, liver vitamin A levels increased with time whereas with TCDD treatment, levels decreased with time. These analyses emphasize the added information about both tracer and tracee that can be gained through a nonsteady state model.

4. Dietary RA

We have also studied the effects of RA on whole-body vitamin A metabolism (Lewis, 1987). All-*trans*-RA is an active metabolite of vitamin A that regulates numerous physiological processes in normal cells. Before the discovery of the molecular mechanisms of RA action, it was shown that dietary RA could partially substitute for vitamin A, influence hepatic vitamin A levels, and spare whole-body vitamin A stores (Lamb et al., 1974). In view of these observations, we were interested in the effects of chronic RA administration on vitamin A kinetics in rats with low vitamin A status. Using data collected in an *in vivo* kinetic study done at the same time as that on rats with low vitamin A status (Lewis et al., 1990), parallel models were developed for rats with low vitamin A status with or without RA supplementation (Cifelli et al., 2005). To develop models for individual organs, the "forcing function" option in WinSAAM was applied (Wastney et al., 1999, pp. 123–126) (see earlier discussion). Once all organs were satisfactorily fit for each group, the forcing function was removed and the entire data set was modeled together, allowing for the determination of various kinetic parameters. The final model is shown in Fig. 9. Despite its apparent complexity, the model indicates that two compartments were needed to fit data for most of the organs examined in this study. That is, each organ could be characterized kinetically by a fast turning-over compartment that exchanges retinol with both plasma and the second, more slowly turning-over compartment in that organ. Therefore, despite differences in the physiological and molecular

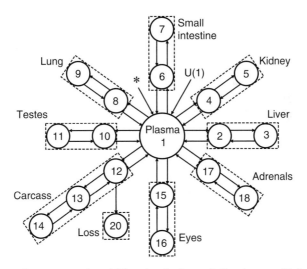

FIGURE 9. Compartmental model for vitamin A metabolism in rats with low vitamin A status with or without supplementation with dietary RA. Compartments are shown as circles and sampled tissues are indicated within rectangles; movement between compartments is represented by arrows and quantified by fractional transfer coefficients [L(I,J)s or the fraction of compartment J's retinol transferred to compartment I per unit time]. U(1) represents input of newly absorbed dietary retinol, and the asterisk shows the site of introduction of the tracer (plasma). Irreversible loss from the system was modeled from carcass compartment 12 to a sink (compartment 20).

processes involved in retinol metabolism in various organs, the compartmental structures for the different organs are kinetically alike.

The tracer response curves for the liver (Fig. 6), kidneys, small intestine, and lungs of RA-treated rats were visually different from the unsupplemented rats as early as 2 h after administration of label. These differences were reflected in the model-predicted kinetic parameters. Specifically, the tissue residence times for vitamin A in the liver, kidneys, small intestine, and lungs were 14, 3.5, 5, and 75 times greater, respectively, in the RA-treated rats as compared to the unsupplemented ones. Similarly, the total traced mass of vitamin A in liver, kidneys, small intestine, and lungs was 11, 3, 5, and 31 times greater, respectively, in the RA-treated rats. The differences in the observed and model-predicted fractions of injected dose were a result of increased fractional input and decreased fractional output of vitamin A in the liver, kidneys, small intestine, and lungs of the RA-treated rats. For the other organs studied (eyes, testes, adrenals, and remaining carcass), there were no differences in vitamin A kinetics between the groups.

The differences in individual organ vitamin A kinetics were paralleled by differences in whole-body vitamin A kinetics in RA-supplemented versus untreated rats. For instance, the model-predicted vitamin A disposal rate was 20% lower, and the system fractional catabolic rate was 50% lower, in RA-treated rats. Together, the lower disposal and system catabolic rates contributed to a greater system residence time and total traced mass in the RA-treated rats, as evidenced by the differences in the tracer response curves for liver between the groups (Fig. 6). It is interesting that the liver tracer response curve for rats with low vitamin A status supplemented with RA was kinetically similar to the curve for rats with marginal vitamin A stores. These results suggest that retinol kinetics in liver are directly affected by hepatic vitamin A levels (Fig. 6). Overall, we concluded that vitamin A recycling, uptake, and mass were affected in a tissue-specific manner in rats with low vitamin A status during chronic RA administration. This resulted in retinol sparing and a positive vitamin A balance as was observed in earlier studies (Dowling and Wald, 1960; Lamb *et al.*, 1974).

5. Inflammation

The final exogenous factor we have modeled is the effect of inflammation on vitamin A kinetics. It is known that, like vitamin A deficiency, inflammation is also associated with low plasma retinol levels. That is, hyporetinolemia is often observed during inflammation and the acute-phase response to infection. In developing countries, interpretation of low plasma retinol levels may be complicated due to the presence of both inadequate dietary sources of vitamin A and chronic/acute infection. In recent work (Gieng, 2006; Gieng *et al.*, 2005), the effects of prolonged inflammation and hyporetinolemia on vitamin A kinetics were studied in rats by applying model-based compartmental analysis to *in vivo* kinetic data. Rats were orally dosed with

[³H]retinol in oil, and serial blood samples were collected beginning at 2 h. On day 21, after tracer had equilibrated with tracee pools of vitamin A, osmotic minipumps were implanted subcutaneously. Pumps delivered either saline solution or one of two inflammatory agents [lipopolysaccharide (LPS) or interleukin-6 (IL-6)] for 3 (LPS; Fig. 8) or 7 days (IL-6). Both agents caused a decrease in plasma retinol and tracer concentrations; plasma retinol and tracer concentrations returned to control levels 6 (LPS) or 16 days (IL-6) after treatment.

During modeling, four hypotheses were tested to explain the hyporetinolemia of inflammation. We postulated that the decrease in plasma retinol was caused by (1) an increased urinary excretion of retinol, (2) a redistribution of retinol out of the vascular bed, (3) an increase in degradation of retinol, or (4) a decrease in mobilization of retinol from liver stores. To account for the observed negative liver vitamin A balance during the kinetic study, it was necessary to develop a tracee model in parallel with the tracer model. In order to fit tracer and tracee data, a fourth compartment (Fig. 10) was needed in the model presented earlier (Fig. 2). Compartment 4 in Fig. 10 acted kinetically like compartment 3 in Fig. 2; it was needed to prevent plasma retinol levels from dropping more than was observed experimentally. The model that best fits the data (Fig. 10) predicted that inflammation-induced hyporetinolemia was caused by a rapid and transient but significant reduction in the mobilization of vitamin A into plasma. The specific activity

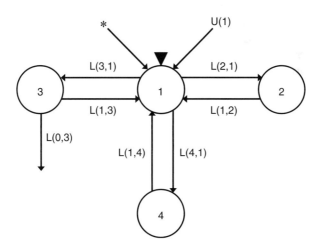

FIGURE 10. A four-compartment mammillary model. Compartments are shown as circles; movement between compartments is represented by arrows and quantified by fractional transfer coefficients [L(I,J)s or the fraction of compartment J's retinol transferred to compartment I per unit time]. U(1) represents input of newly absorbed dietary retinol, the asterisk shows the site of introduction of the tracer (typically plasma), and the triangle indicates that this compartment is a site of sampling.

of plasma retinol (i.e., [^{3}H]retinol/retinol) decreased during inflammation, indicating that the mobilization of retinol from liver stores was inhibited while there was still some apoRBP being synthesized in hepatocytes to allow for mobilization of unlabeled, diet-derived vitamin A. It was hypothesized that delivery of vitamin A to tissues, such as the eyes, lungs, or kidneys, might be impaired with chronic low plasma retinol levels such as occurs during chronic inflammatory disease. This has important public health implications. Specifically, administering vitamin A during inflammation may not ameliorate low plasma retinol concentrations, although it might increase the delivery of chylomicron vitamin A to extrahepatic tissues. Normally, the liver parenchymal cells secrete diet-derived vitamin A into plasma quite quickly (von Reinersdorff *et al.*, 1998). It may be that the chronic administration of large vitamin A supplements during inflammation, when RBP synthesis is depressed, may block the normal secretion of diet-derived retinol from liver parenchymal cells and lead to a toxic accumulation of vitamin A, causing liver damage.

E. PHYSIOLOGICAL INTERPRETATION OF THREE- AND FOUR-COMPARTMENT MODELS

Although large, complex models such as the ones developed by Green *et al.* (1985) and Lewis *et al.* (1990) are important to our understanding of whole-body vitamin A metabolism, three- and four-compartment models that view the vitamin A system from the plasma space can be developed from more manageable experiments and with less extensive modeling. Of course, with the latter approach, the investigator is not able to postulate definitive correspondence between anatomical sites and the model's compartments as is possible in a large model developed after sampling multiple tissues as well as plasma. However, this limitation does not negate the usefulness of the more straightforward approach.

In the studies discussed here which use this approach, either a three- (Fig. 2) or four-compartment model (Fig. 10) has provided a good fit to vitamin A kinetic data. Compartment 1 is clearly the plasma pool of retinol bound to RBP and transthyretin. It is worth emphasizing that plasma retinol acts as one kinetically homogeneous pool with a mean transit time (Table I) of about 2 h. In the postabsorptive state after a vitamin A-rich meal, there would be a transient increase in plasma retinyl esters. These would be in a kinetically distinguishable pool from compartment 1. On the basis of kinetics, we hypothesize that compartment 2 represents retinol that has either entered interstitial fluid or been filtered by the kidneys as holoRBP; it also likely includes a relatively rapidly turning-over intracellular pool of retinol in some of the organs shown in Fig. 9 (e.g., carcass).

When liver vitamin A is not changing substantially during the course of a kinetic study, then plasma tracer data can often be fit to a three-compartment model (Fig. 2), with the majority of compartment 3 corresponding to liver retinyl ester stores. However, when liver vitamin A balance is negative, such as when TCDD was administered (Green and Green, 2003) or when liver vitamin A balance was positive as in the case of iron deficiency (Jang et al., 2000), or in the LPS-induced inflammation study (Gieng, 2006), then a fourth compartment is required to fit the tracer and tracee data (liver vitamin A) (Figs. 7 and 10). The kinetic behavior of vitamin A in compartment 3 in the three-compartment model versus compartment 4 in the four-compartment model is very similar; that is, all of the slowest turning-over vitamin A lumps into one compartment (compartment 3) in a steady state condition. When liver vitamin A is changing substantially over time (i.e., is not in a steady state), then compartment 4 is needed as liver vitamin A decreases or increases compartment 3; compartment 4 contains sufficient tracer to feed the plasma retinol tracer compartment as liver tracee is depleted or expands. This was especially important in the inflammation study, in which the model predicted that during inflammation, mobilization of retinol from liver retinyl ester stores was inhibited, likely due to unavailability of apoRBP. The recycling of retinol from compartment 4 and to a lesser extent from compartment 2 prevented plasma retinol concentration from dropping even further into a vitamin A-deficient-like state. Thus, compartment 4 in Fig. 10 plays an important role in plasma retinol homeostasis. It will be interesting in future kinetic studies to determine the location of compartment 4 vitamin A. Likely candidates for this extrahepatic pool of kinetically active vitamin A that plays an important role in whole-body vitamin A metabolism are adipose tissue, small intestine, and skin. Future research will be needed to identify the source(s) of apoRBP involved in vitamin A turnover from this pool (plasma versus the organs themselves).

F. VITAMIN A KINETIC STUDIES IN HUMANS

As noted elsewhere (Green and Green, 2005a), there are as yet few published studies on vitamin A kinetics in humans which were designed to be analyzed by compartmental analysis. However, in two cases, model-based compartmental analysis was applied retrospectively to human data, and some very interesting results were obtained. In the first instance, Green and Green (1994b) analyzed data collected by Goodman in 1965–1966 on the long-term kinetics of plasma retinol in three human subjects who were given an intravenous dose of autologous plasma labeled *in vitro* with [^{14}C]retinol. Data for each subject were fit to a three-compartment model (Fig. 2) and then modeled as one data set to arrive at a working model for vitamin A metabolism. Average parameters were determined using the SAAM program. This analysis confirmed the suspected similarities in many aspects of vitamin A metabolism in humans compared with rats: extensive recycling among plasma and tissues, similar sources of plasma retinol input, and vitamin A stores similar to what had been published from autopsy studies of liver.

In a second retrospective application, von Reinersdorff et al. (1998) used model-based compartmental analysis to develop a model for retinol kinetics in a human subject after oral administration of 105 μmol of [^{13}C]retinyl palmitate. From a modeler's point of view, there were several nonideal factors in this experiment, including the mass of retinol in the oral dose (it was too large to be considered a tracer), the short duration of the study (7 days), and the fact that the dose was administered orally. Oral administration introduces significant constraints: in order to model the rapid absorptive phase, an adequate number of early samples must be collected to characterize the absorption and initial metabolism of the dose. However, oral dosing is safer and easier to use in humans. Note that some interesting work has been done to develop artificial chylomicrons for the intravenous administration of labeled fat-soluble substances in humans (Redgrave et al., 1993), and this possibility is worth pursuing if vitamin A kinetic studies in humans are planned. The von Reinersdorff data from the postabsorptive phase were fit to a three-compartment model that was similar to the one developed for the Goodman data (Green and Green, 1994b), and the kinetic parameters were similar. For example, the model predicted that the transit time for plasma retinol was ~2 h, that ~50 μmol of retinol passed through the plasma each day, and that the estimated vitamin A utilization rate was 4 μmol/day. These latter two estimates again indicate the extensive recycling of retinol through the plasma compartment.

In a third human study (Tang et al., 2003), plasma retinol kinetics were followed for 52 days after oral administration of octadeuterated retinyl acetate, and data from one subject were modeled using model-based compartmental analysis (Furr et al., 2005). Following absorption and hepatic secretion of retinol into plasma, the three-compartment model shown in Fig. 2 was compatible with the data; estimates of model parameters were similar to those calculated in the other two studies. The model indicated that about 0.02 pools of the body's vitamin A were used irreversibly each day, and the vitamin A disposal rate was estimated to be 9 μmol/day.

In addition to the limitation mentioned here related to administering the labeled dose orally in human studies and the resultant needs for extensive and early sampling, there are several other factors that would need to be considered in designing human vitamin A kinetic studies (Green and Green, 2005a). The need for a large number of samples and a long study duration are obvious constraints; there are also technical problems in the mass versus detectability of the administered dose. In spite of these challenges to applying model-based compartmental analysis to human kinetic data, such studies would provide interesting and important information on vitamin A metabolism in humans.

VI. CONCLUSIONS

On the basis of the work reviewed here, it is evident that much has been learned about whole-body vitamin A metabolism since the 1980s. At that time (Green and Green, 1996), conventional wisdom indicated that newly absorbed vitamin A was transported to the liver to be either stored there or resecreted bound to RBP and that circulating holoRBP delivered retinol to target tissues for uptake and utilization. Building on the work of others, we have applied mathematical modeling to *in vivo* kinetic data collected over the past several decades. These studies describe and quantitate whole-body and organ-level vitamin A metabolism under different conditions and have contributed to the realization that this system is complex and intriguingly regulated. For example, modeling has shed light on the role of retinol recycling in plasma retinol homeostasis; it has revealed that the liver is not the only source of input of retinol into plasma, quantified the role of various tissues in whole-body vitamin A dynamics, and provided hypotheses about hepatic vitamin A metabolism. Modeling has also demonstrated the importance of plasma retinol pool size in vitamin A utilization and indicated the contribution of extrahepatic tissues to overall vitamin A storage and plasma retinol homeostasis.

As discussed earlier, both large complex models and more simple three- and four-compartment models make unique and useful contributions to our understanding of whole-body and organ-level vitamin A metabolism. To build a large model requires extensive long-term sampling of both plasma and organs (and in some cases, excreta), and the resulting model quantitates vitamin A kinetics in each of the sites sampled. On the other hand, when vitamin A metabolism is viewed from the plasma space, processes with similar kinetics are lumped into the same compartment. It is interesting that both large and simpler models for vitamin A metabolism indicate that, in addition to the plasma retinol pool, the remaining body vitamin A pools are either about the size of the plasma retinol pool or are larger and slowly turning-over. Those latter pools presumably map onto vitamin A storage sites in the liver and many other tissues. While the small pools are metabolically very active, the larger pools do turn over and are important in overall whole-body vitamin A kinetics.

Although the ability to apply model-based compartmental analysis to the study of vitamin A metabolism requires an expertise that is not typically part of a modern biologist's skill set, the payoffs in understanding aspects of the system which are difficult to study directly justify the effort required to learn and use this approach. Besides the wealth of unique information that may be learned through modeling biological systems, the approach provides continuing challenges and learning opportunities for the investigator.

REFERENCES

Adams, W. R., and Green, M. H. (1994). Prediction of liver vitamin A in rats by an oral isotope dilution technique. *J. Nutr.* **124,** 1265–1270.
Adams, W. R., Smith, J. E., and Green, M. H. (1995). Effects of N-(4-hydroxyphenyl) retinamide on vitamin A metabolism in rats. *Proc. Soc. Exp. Biol. Med.* **208,** 178–185.
Adiels, M., Packard, C., Caslake, M. J., Stewart, P., Soro, A., Westerbacka, J., Wennberg, B., Olofsson, S. O., Taskinen, M. R., and Boren, J. (2005). A new combined multicompartmental model for apolipoprotein B-100 and triglyceride metabolism in VLDL subfractions. *J. Lipid Res.* **46,** 58–67.
Anonymous (1977). "Turnover" of vitamin A. *Nutr. Rev.* **35,** 310–313.
Berman, M., and Weiss, M. F. (1978). "SAAM 27." DHEW Publication No. (NIH) 78-180. Washington, DC.
Berman, M., Grundy, S. M., and Howard, B. V. (Eds.) (1982). "Lipoprotein Kinetics and Modeling." Academic Press, New York.
Biesalski, H. K., and Nohr, D. (2003). Importance of vitamin-A for lung function and development. *Mol. Aspects Med.* **24,** 431–440.
Birge, S. J., Peck, W. A., Berman, M., and Whedon, G. D. (1969). Study of calcium absorption in man: A kinetic analysis and physiologic model. *J. Clin. Invest.* **48,** 1705–1713.
Blaner, W. S., Smith, J. E., Dell, R. B., and Goodman, D. S. (1985). Spatial distribution of retinol-binding protein and retinyl palmitate hydrolase activity in normal and vitamin A-deficient rat liver. *J. Nutr.* **115,** 856–864.
Blomhoff, R., Rasmussen, M., Nilsson, A., Norum, K. R., Berg, T., Blaner, W. S., Kato, M., Mertz, J. R., Goodman, D. S., Ericksson, U., and Peterson, P. A. (1985). Hepatic retinol metabolism. Distribution of retinoids, enzymes, and binding proteins in isolated rat liver cells. *J. Biol. Chem.* **260,** 13560–13565.
Blomhoff, R., Nenseter, M. S., Green, M. H., and Berg, T. (1989). A multicompartmental model of fluid-phase endocytosis in rabbit liver parenchymal cells. *Biochem. J.* **262,** 605–610.
Blomhoff, R., Green, M. H., Green, J. B., Berg, T., and Norum, K. R. (1991). Vitamin A metabolism: New perspectives on absorption, transport, and storage. *Physiol. Rev.* **71,** 951–990.
Carney, S. M., Underwood, B. A., and Loerch, J. D. (1976). Effects of zinc and vitamin A deficient diets on the hepatic mobilization and urinary excretion of vitamin A in rats. *J. Nutr.* **106,** 1773–1781.
Carson, E. R., Cobelli, C., and Finkelstein, L. (1983). "The Mathematical Modeling of Metabolic and Endocrine Systems: Model Formulation, Identification, and Validation." John Wiley and Sons, New York.
Christensen, E. I., Moskaug, J. O., Vorum, H., Jacobsen, C., Gundersen, T. E., Nykaer, A., Blomhoff, R., Wilnow, T. E., and Muestrup, S. K. (1999). Evidence for an essential role of megalin in transepithelial transport of retinol. *J. Am. Soc. Nephrol.* **10,** 685–695.
Cifelli, C. J., Green, J. B., and Green, M. H. (2005). Dietary retinoic acid alters vitamin A kinetics in both the whole body and in specific organs of rats with low vitamin A status. *J. Nutr.* **135,** 746–752.
Coburn, S. P., Townsend, D. W., Ericson, K. L., Reynolds, R. D., Ziegler, P. J., Costill, D. L., Mahuren, J. D., Schaltenbrand, W. E., Pauly, T. A., Wang, Y., Fink, W. J., Pearson, D. R., *et al.* (2003). Modeling short (7 hour)- and long (6 week)-term kinetics of vitamin B-6 metabolism with stable isotopes in humans. *In* "Mathematical Modeling in Nutrition and the Health Sciences" (J. A. Novotny, M. H. Green, and R. C. Boston, Eds.), pp. 173–192. Kluwer Academic/Plenum Publishers, New York.
DiStefano, J. J., III, and Landaw, E. M. (1984). Multiexponential, multicompartmental, and noncompartmental modeling. I. Methodological limitations and physiological interpretations. *Am. J. Physiol.* **246,** R651–R664.

Dowling, J. E., and Wald, G. (1960). The biological function of vitamin A acid. *Proc. Natl. Acad. Sci. USA* **46**, 587–608.

Duncan, T. E., Green, J. B., and Green, M. H. (1993). Liver vitamin A levels in rats are predicted by a modified isotope dilution technique. *J. Nutr.* **123**, 933–939.

Foster, D. M., and Boston, R. C. (1983). The use of computers in compartmental analysis: The SAAM and CONSAM programs. *In* "Compartmental Distribution of Radiotracers" (J. S. Robertson, Ed.), pp. 73–142. CRC Press, Boca Raton.

Foster, D. M., Aamodt, R. L., Henkin, R. I., and Berman, M. (1979). Zinc metabolism in humans: A kinetic model. *Am. J. Physiol.* **237**, R340–R349.

Furr, H. C., Green, M. H., Haskell, M., Mokhtar, N., Nestel, P., Newton, S., Ribaya-Mercado, J. D., Tang, G., Tanumihardjo, S., and Wasantwisut, E. (2005). Stable isotope dilution techniques for assessing vitamin A status and bioefficacy of provitamin A carotenoids in humans. *Public Health Nutr.* **8**, 596–607.

Gieng, S. H. (2006). Kinetic analysis and dynamics of inflammation-induced hyporetinolemia in the rat. The Pennsylvania State University, University Park, PA. Doctoral dissertation.

Gieng, S. H., Patel, D., Green, M. H., Green, J. B., and Rosales, F. J. (2005). Model-based compartmental analysis indicates a reduced mobilization of hepatic vitamin A during inflammation in rats. *FASEB J.* **19**, A442 (abs. 279.9).

Green, M. H. (1992). Introduction to modeling. *J. Nutr.* **122**, 690–694.

Green, M. H., and Green, J. B. (1987). Multicompartmental analysis of whole body retinol dynamics in vitamin A-sufficient rats. *Fed. Proc.* **46**, 1011 (abs. 4047).

Green, M. H., and Green, J. B. (1990a). The application of compartmental analysis to research in nutrition. *Annu. Rev. Nutr.* **10**, 41–61.

Green, M. H., and Green, J. B. (1990b). Experimental and kinetic methods for studying vitamin A dynamics *in vivo*. *Meth. Enzymol.* **190**, 304–317.

Green, M. H., and Green, J. B. (1994a). Vitamin A intake and status influence retinol balance, utilization and dynamics in rats. *J. Nutr.* **124**, 2477–2485.

Green, M. H., and Green, J. B. (1994b). Dynamics and control of plasma retinol. *In* "Vitamin A in Health and Disease" (R. Blomhoff, Ed.), pp. 119–133. Marcel Dekker, Inc., New York.

Green, M. H., and Green, J. B. (1996). Quantitative and conceptual contributions of mathematical modeling to current views on vitamin A metabolism, biochemistry, and nutrition. *Adv. Food Nutr. Res.* **40**, 3–24.

Green, M. H., and Green, J. B. (2003). The use of model-based compartmental analysis to study vitamin A metabolism in a non-steady state. *In* "Mathematical Modeling in Nutrition and the Health Sciences" (J. A. Novotny, M. H. Green, and R. C. Boston, Eds.), pp. 159–172. Kluwer Academic/Plenum Publishers, New York.

Green, M. H., and Green, J. B. (2005a). Contributions of mathematical modeling to understanding whole-body vitamin A metabolism and to the assessment of vitamin A status. *Sight Life Newsletter* **2**, 4–10.

Green, M. H., and Green, J. B. (2005b). The application of model-based compartmental analysis to the study of the kinetic behavior of vitamin A in vitamin A-sufficient rats: Validation of a theoretical model. *In* "Mathematical Modeling in Nutrition and Toxicology" (J. L. Hargrove and C. D. Berdanier, Eds.), pp. 175–187. Mathematical Biology Press, Athens.

Green, M. H., Uhl, L., and Green, J. B. (1985). A multicompartmental model of vitamin A kinetics in rats with marginal liver vitamin A stores. *J. Lipid Res.* **26**, 806–818.

Green, M. H., Green, J. B., and Lewis, K. C. (1987). Variation in retinol utilization rate with vitamin A status in the rat. *J. Nutr.* **117**, 694–703.

Green, M. H., Green, J. B., and Lewis, K. C. (1992). Model-based compartmental analysis of retinol kinetics in organs of rats at different levels of vitamin A status. *In* "Retinoids: Progress in Research and Clinical Applications" (M. Livrea and L. Packer, Eds.), pp. 185–204. Marcel Dekker, New York.

Green, M. H., Green, J. B., Berg, T., Nourm, K. R., and Blomhoff, R. (1993). Vitamin A metabolism in rat liver: A kinetic model. *Am. J. Physiol.* **264**, G509–G521.

Gudas, L., Sporn, M. B., and Roberts, A. B. (1994). Cellular biology and biochemistry of the retinoids. *In* "The Retinoids" (M. B. Sporn, A. B. Roberts, and D. S. Goodman, Eds.), 2nd ed., pp. 444–520. Raven Press, New York.

Hargrove, J. L. (1998). "Dynamic Modeling in the Health Sciences." Springer-Verlag, New York.

Harrison, E. H., and Gad, M. Z. (1989). Hydrolysis of retinyl palmitate by enzymes of rat pancreas and liver. Differentiation of bile salt-dependent and bile salt-independent, neutral retinyl ester hydrolases in rat liver. *J. Biol. Chem.* **264**, 17142–17147.

Hicks, V. A., Gunning, D. B., and Olson, J. A. (1984). Metabolism, plasma transport and biliary excretion of radioactive vitamin A and its metabolites as a function of liver reserves of vitamin A in the rat. *J. Nutr.* **114**, 1327–1333.

Jang, J.-T., Green, J. B., Beard, J. L., and Green, M. H. (2000). Kinetic analysis shows that iron deficiency decreases liver vitamin A mobilization in rats. *J. Nutr.* **130**, 1291–1296.

Kelley, S. K., and Green, M. H. (1998). Plasma retinol is a major determinant of vitamin A utilization in rats. *J. Nutr.* **128**, 1767–1773.

Kelley, S. K., Nilsson, C. B., Green, M. H., Green, J. B., and Håkansson, H. (1998). Use of model-based compartmental analysis to study effects of 2,3,7,8-tetrachlorodibenzo-*p*-dioxin on vitamin A kinetics in rats. *Toxicol. Sci.* **44**, 1–13.

Kelley, S. K., Nilsson, C. B., Green, M. H., Green, J. B., and Håkansson, H. (2000). Mobilization of vitamin A stores in rats after administration of 2,3,7,8-tetrachlorodibenzo-*p*-dioxin: A kinetic analysis. *Toxicol. Sci.* **55**, 478–484.

Lamb, A. J., Apiwatanaporn, P., and Olson, J. A. (1974). Induction of rapid, synchronous vitamin A deficiency in the rat. *J. Nutr.* **104**, 1140–1148.

Landaw, E. M., and DiStefano, J. J., III (1984). Multiexponential, multicompartmental, and noncompartmental modeling. II. Data analysis and statistical considerations. *Am. J. Physiol.* **246**, R665–R677.

Lewis, K. C. (1987). Multicompartmental tracer kinetic analysis of retinol metabolism in rats with low vitamin A status. The Pennsylvania State University, University Park, PA. Doctoral dissertation.

Lewis, K. C., Green, M. H., and Underwood, B. A. (1981). Vitamin A turnover in rats as influenced by vitamin A status. *J. Nutr.* **111**, 1135–1144.

Lewis, K. C., Green, M. H., Green, J. B., and Zech, L. A. (1990). Retinol metabolism in rats with low vitamin A status: A compartmental model. *J. Lipid Res.* **31**, 1535–1548.

Livera, G., Rouiller-Fabre, V., Pairault, C., Levacher, C., and Habert, R. (2002). Regulation and perturbation of testicular functions by vitamin A. *Reproduction* **124**, 173–180.

Lyne, A., Boston, R., Pettigrew, K., and Zech, L. (1992). EMSA: A SAAM service for the estimation of population parameters based on model fits to identically replicated experiments. *Comput. Methods Program Biomed.* **38**, 117–151.

Malmendier, C. L., Delcroix, C., and Berman, M. (1974). Interrelations in the oxidative metabolism of free fatty acids, glucose, and glycerol in normal and hyperlipemic patients. A compartmental model. *J. Clin. Invest.* **54**, 461–476.

Moore-Colyer, M. J., Morrow, H. J., and Longland, A. C. (2003). Mathematical modelling of digesta passage rate, mean retention time and *in vivo* apparent digestibility of two different lengths of hay and big-bale grass silage in ponies. *Br. J. Nutr.* **90**, 109–118.

Novotny, J. A., Green, M. H., and Boston, R. C. (Eds.) (2003). "Mathematical Modeling in Nutrition and the Health Sciences." Kluwer Academic/Plenum Publishers, New York.

Olson, J. A., Rojanapo, W., and Lamb, A. J. (1981). The effect of vitamin A status on the differentiation and function of goblet cells in the rat intestine. *Ann. N Y Acad. Sci.* **359**, 181–191.

Pinna, K., Woodhouse, L. R., Sutherland, B., Shames, D. M., and King, J. C. (2001). Exchangeable zinc pool masses and turnover are maintained in healthy men with low zinc intakes. *J. Nutr.* **131**, 2288–2294.

Raila, J., Willnow, T. E., and Schweigert, F. J. (2005). Megalin-mediated reuptake of retinol in the kidneys of mice is essential for vitamin A homeostasis. *J. Nutr.* **135,** 2512–2516.

Redgrave, T. G., Ly, H. L., Quintao, E. C. R., Ramberg, C. F., and Boston, R. C. (1993). Clearance from plasma of triacylglycerol and cholesteryl ester after intravenous injection of chylomicron-like lipid emulsions in rats and man. *Biochem. J.* **290,** 843–847.

Rietz, P., Vuilleumier, J. P., Weber, F., and Wiss, O. (1973). Determination of the vitamin A bodypool of rats by an isotopic dilution method. *Experientia* **29,** 168–170.

Rietz, P., Wiss, O., and Weber, F. (1974). Metabolism of vitamin A and the determination of vitamin A status. *Vitam. Horm.* **32,** 237–249.

Robertson, J. S. (Ed.) (1983). "Compartmental Distribution of Radiotracers." CRC Press, Boca Raton.

Ross, A. C. (2003a). Retinoid production and catabolism: Role of diet in regulating retinol esterification and retinoic acid oxidation. *J. Nutr.* **133,** 291S–296S.

Ross, A. C. (2003b). Hepatic metabolism of vitamin A. *In* "Hepatology: A Textbook of Liver Disease" (D. Zakim and T. Boyer, Eds.), pp. 149–168. Saunders, Philadelphia.

Ross, A. C., and Zolfaghari, R. (2004). Regulation of hepatic retinol metabolism: Perspectives from studies on vitamin A status. *J. Nutr.* **134,** 269S–275S.

Sauberlich, H. E., Hodges, R. E., Wallace, D. L., Kolder, H., Canham, J. E., Hood, J., Raica, N., Jr., and Lowry, L. K. (1974). Vitamin A metabolism and requirements in the human studied with the use of labeled retinol. *Vitam. Horm.* **32,** 251–275.

Sewell, H. B., Mitchell, G. E., Jr., Little, C. O., and Hayes, B. W. (1967). Mobilization of liver vitamin A in rats. *Int. Z. Vitaminforsch.* **37,** 301–306.

Soprano, D. R., and Soprano, K. J. (2003). Role of RARs and RXRs in mediating the molecular mechanism of action of vitamin A. *In* "Molecular Nutrition" (J. Zemplenic and H. Daniel, Eds.), pp. 135–149. CABI Publishing, Cambridge.

Soprano, D. R., Soprano, K. J., and Goodman, D. S. (1986). Retinol-binding protein messenger RNA levels in the liver and in extrahepatic tissues of the rat. *J. Lipid Res.* **27,** 166–171.

Stefanovski, D., Moate, P. J., and Boston, R. C. (2003). WinSAAM: A Windows-based compartmental modeling system. *Metabolism* **52,** 1153–1166.

Sundaresan, P. R. (1977). Rate of metabolism of retinol in retinoic acid-maintained rats after a single dose of radioactive retinol. *J. Nutr.* **107,** 70–78.

Tang, G., Qin, J., Dolnikowski, G. G., and Russell, R. M. (2003). Short-term (intestinal) and long-term (postintestinal) conversion of β-carotene to retinol in adults as assessed by a stable-isotope reference method. *Am. J. Clin. Nutr.* **78,** 259–266.

von Reinersdorff, D., Green, M. H., and Green, J. B. (1998). Development of a compartmental model describing the dynamics of vitamin A metabolism in men. *Adv. Exp. Biol. Med.* **445,** 207–223.

Wald, G. (1968). Molecular basis of visual excitation. *Science* **162,** 230–239.

Wastney, M. E., Angelus, P., Barnes, R. M., and Subramanian, K. N. (1996). Zinc kinetics in preterm infants: A compartmental model based on stable isotope data. *Am. J. Physiol.* **271,** R1452–R1459.

Wastney, M. E., Pattterson, B. H., Linares, O. A., Greif, P. C., and Boston, R. C. (1999). "Investigating Biological Systems Using Modeling: Strategies and Software." Academic Press, San Diego.

WinSAAM: The Simulation, Analysis and Modeling Software. http://www.WinSAAM.com

Wolbach, S. B., and Howe, P. R. (1925). Tissue changes following deprivation of fat soluble A vitamin. *J. Exp. Med.* **42,** 453–457.

Zile, M., Bunge, E. C., and DeLuca, H. F. (1977). Effect of vitamin A deficiency on intestinal cell proliferation. *J. Nutr.* **107,** 552–560.

Zile, M. H. (1992). Vitamin A homeostasis endangered by environmental pollutants. *Proc. Soc. Exp. Biol. Med.* **201,** 141–153.

8

Vitamin A Supplementation and Retinoic Acid Treatment in the Regulation of Antibody Responses *In Vivo*

A. Catharine Ross

Department of Nutritional Sciences and Huck Institute for Life Sciences
Pennsylvania State University, University Park, Pennsylvania 16802

I. Introduction
II. Rationale for Interest in VA Supplementation and Antibody Production
III. VA and the Response to Immunization in Children
IV. Experimental Studies of VA or RA Supplementation and Antibody Production *In Vivo*
   A. Experimental Models
   B. RA Treatment and Antibody Production in a VA-Deficient Model
   C. RA Treatment and Antibody Production in VA-Adequate Models
   D. RA Supplementation in a Neonatal Model
   E. Cytokine Production and Th1:Th2 Antibody Isotype Balance
V. Innate Immune Cells and Factors Regulated by VA and RA That May Affect Immunization Outcome

VI. Discussion and Perspectives
References

Vitamin A (VA, retinol) is essential for normal immune system maturation, but the effect of VA[1] on antibody production, the hallmark of successful vaccination, is still not well understood. In countries where VA deficiency is a public health problem, many children worldwide are now receiving VA along with immunizations against poliovirus, measles, diphtheria, pertussis, and tetanus. The primary goal has been to provide enough VA to protect against the development of VA deficiency for a period of 4–6 months. However, it is also possible that VA might promote the vaccine antibody response. Several community studies, generally of small size, have been conducted in children supplemented with VA at the time of immunization, as promoted by the World Health Organization/UNICEF. However, only a few studies have reported differences in antibody titers or seroconversion rates due to VA. However, VA status was not directly assessed, and in some communities children were often breast fed, another strategy for preventing VA deficiency. Some of the vaccines used induced a high rate of seroconversion, even without VA. In children likely to have been VA deficient, oral polio vaccine seroconversion rate was increased by VA. In animal models, where VA status was controlled and VA deficiency confirmed, the antibody response to T-cell-dependent (TD) and polysaccharide antigens was significantly reduced, congruent with other defects in innate and adaptive immunity. Moreover, the active metabolite of VA, retinoic acid (RA) can potentiate antibody production to TD antigens in normal adult and neonatal animals. We speculate that numerous animal studies have correctly identified VA deficiency as a risk factor for low antibody production. A lack of effect of VA in human studies could be due to a low rate of VA deficiency in the populations studied or low sample numbers. The ability to detect differences in antibody response may also depend on the vaccine–adjuvant combination used. Future studies of VA supplementation and immunization should include assessment of VA status and a sufficiently large sample size. It would also be worthwhile to test the effect of neonatal VA supplementation on the response to immunization given after 6 months to 1 year of age, as VA supplementation, by preventing the onset of VA deficiency, may improve the response to immunizations given later on. © 2007 Elsevier Inc.

---

[1]Abbreviations: CI, confidence interval; DPT, diphtheria–pertussis–tetanus; EPI, Expanded Program on Immunization; IFN, interferon; IU, international unit; LPS, lipopolysaccharide; NK, natural killer; NKT, natural kill T-(Cell); OPV, oral polio vaccine; PBMC, peripheral blood mononuclear cell(s); PIC, polyriboinosinic acid:polyribocytidylic acid; RA, retinoic acid; TD, T-cell dependent; TI, T-cell independent; TNF, tumor necrosis factor; VA, vitamin A; WHO, World Health Organization.

I. INTRODUCTION

Vitamin A (VA, retinol) has long been considered important for the maintenance of the immune system, but its role in antibody production is still being uncovered. Antibody production, the hallmark of a successful response to vaccination is, indeed, the only proven mechanism which vaccines protect against infectious disease (Beverley, 2002; Del Giudice, 2003). This chapter focuses on studies, in the past decade, on the effects of providing VA or its active metabolite, retinoic acid (RA), during the inductive phase of the antibody response *in vivo*. After discussing the rationale for the topics selected, the chapter then considers: (1) the effect of VA supplementation on the response to immunization in children, (2) experimental studies addressing mechanisms by which VA and/or RA may affect antibody production *in vivo*, (3) innate immune cells and factors regulated by VA and RA that may affect immunization outcome, and is followed by (4) a discussion of factors that may account for differences observed in human and animal studies of VA supplementation and the response to immunization. Other reviews have addressed VA deficiency and infection, and morbidity and mortality outcomes in VA supplementation studies (Semba, 2000; Stephensen, 2001; Villamor and Fawzi, 2005).

II. RATIONALE FOR INTEREST IN VA SUPPLEMENTATION AND ANTIBODY PRODUCTION

VA deficiency in young children is associated with increased morbidity and mortality, especially from measles and diarrheal diseases (Beaton *et al.*, 1994; Sommer and West, 1996). It has been estimated that improving VA status in children at risk of deficiency will reduce mortality by 23% (Beaton *et al.*, 1994), and avert >24,000 deaths per year (Ching *et al.*, 2000). The reduction in morbidity and mortality by VA is widely attributed to a decreased severity of infectious diseases (Beaton *et al.*, 1994; Semba, 1999; Villamor and Fawzi, 2005). In 1994, WHO established a policy of integrating VA supplementation as a part of the Expanded Program on Immunization (EPI) in countries where VA deficiency is still prevalent (World Health Organization, 1994). For 6- to 12-month-old infants, it is recommended that 100,000 IU of VA (equivalent to 30 mg of retinol) be given along with measles immunization, and for infants under 6 months of age, 25,000 IU of VA along with diphtheria–pertussis–tetanus (DPT) vaccines. Therefore, many children worldwide are now receiving VA along with immunizations against poliovirus, measles, diphtheria, pertussis, and tetanus. The primary goal has been to provide enough VA to protect against VA deficiency for a period of 4–6 months (Sommer and West, 1997; Underwood, 1995).

However, it is also possible that VA might promote the vaccine antibody response. To date, this has been tested relatively sporadically and only in trials much smaller than the mortality trials that led up to the policy to provide VA with immunization. Because cotreatment with VA at the time of immunization is widely practiced today, it is important to understand the effect that VA supplementation may have on the response to immunization. As discussed in the section below, only a few of the studies that have measured the antibody response to vaccination have shown evidence of a benefit of VA, but neither have they shown reduced titers or other evidence of unintended effects.

Animal experiments have demonstrated that an adequate level of VA is necessary to mount an efficient antibody response to many antigens; however, previous studies have also revealed that VA-deficient animals are capable of producing a strong antibody response to some antigens (Ross, 1996a,b). Thus, the requirement for VA may be better considered as *conditional*, as it apparently differs with and depends on the type of immune stimulus employed. Antigens that require T-cell help [T-cell-dependent (TD) antigens such as tetanus toxoid and cellular antigens] in the initiation phase of the antibody response, or antigens that, while considered T-cell independent are nonetheless regulated by T-cells (TI-2 antigens such as polysaccharides), were poorly immunogenic in VA-deficient animals, while TI-1 antigens [lipopolysaccharides (LPS)] were strongly immunogenic in VA-deficient animals as well as controls (Arora and Ross, 1994; Ross, 1996a,b). Indeed, in response to some types of antigenic challenge, VA-deficient animals have produced higher titers of certain antibodies. For example, anti-influenza IgG was elevated in VA-deficient mice infected with influenza virus (Stephensen *et al.*, 1996). These data provide evidence that the antibody response is *dysregulated* by VA deficiency rather than simply impaired. It is encouraging that VA-deficient animals remain to some extent immunocompetent, as this suggests that when they are provided with VA, the "machinery" for a competent immune response is present, and although dysregulated, it may potentially be reregulated to produce normal adaptive immune responses without significant delay.

RA, a natural bioactive metabolite of VA (Ross, 2006), is well known as a hormone capable of promoting the differentiation of a wide variety of cell types and as a regulator of many physiological processes. Animal experiments have demonstrated that supplementation with VA or RA *in vivo* can stimulate the antibody response to vaccination, even in normal, non-VA-deficient animals (Cui *et al.*, 2000; DeCicco *et al.*, 2001; Ma *et al.*, 2005). Therefore, VA and/or RA could potentially be useful as immunologic stimuli, but their effects need to be better understood at the cellular and whole-body levels. Little is known about the underlying mechanisms, especially related to lymphocyte activation and differentiation, through which supplemental VA or RA may stimulate antibody production. A successful antibody response requires a finely timed collaboration among several types of cells: antigen-presenting cells (dendritic cells, macrophages, and B-cells), T-helper (Th) cells; cytokine-producing cells [natural killer (NK) cells, and NK-T-cells], and naïve B-cells capable of developing into plasmacytes and then antibody-secreting plasma cells, or memory B-cells required for long-lived immunity and protection against a potential future encounter with the infectious agent for which the vaccine has been designed. An area of significant interest regarding VA and RA is whether, alone and when combined with other immune stimuli, they are effective in augmenting the humoral response to vaccination.

The past decade in immunology research has been remarkable for the greatly increased understanding of the role of innate immunity in shaping the outcome of adaptive immune responses. It had long been recognized that bacterial cell wall components—LPS, lipoteichoic acid, and other components—as well as viral nucleic acids provide "danger signals" to mammalian cells, and often are potent immune stimuli (Janeway and Medzhitov, 2002; Matzinger, 2002). These bacterial and viral components are typically composed of repeating subunit structures, which Janeway and coworkers (Janeway and Medzhitov, 2002) termed "pathogen-associated molecular patterns," or PAMPs, and these agents have now been shown to be ligands for Toll-like receptors [TLRs family molecules (Janeway and Medzhitov, 2002; Takeda and Akira, 2005)], or other "pattern recognition receptors" (PRRs) on the surface of mammalian cells. The interaction of PAMPs with PPRs triggers potent innate immune responses, including production of cytokines and oxidants, which are critical in the first line of defense against many microbial pathogens. It is now recognized that cells and factors considered part of the innate immune system, which typically act very early and are relatively nonspecific in the course of infection or immune stimulation, have a strong impact on cells of the adaptive immune system—B-cells and T-cells—that are activated more slowly but sustained over time to produce humoral immunity and the memory response that is critical to successful vaccination.

VA and RA have been shown to affect some of the functions of macrophages, NK cells, and other cells of the innate immune system in earlier studies (reviewed in Ross, 1996b), and to stimulate the functions of dendritic cells in culture (Hengesbach and Hoag, 2004). The exposure of cells of the innate immune system to VA and RA at the time of antigen exposure (priming *in vivo*) may stimulate the initial phase of antibody production and, if memory to antigen is formed, affect the response to vaccination over a much longer term. It, therefore, is also important to understand how retinoids affect the innate immune system to better understand the regulation of adaptive immunity.

III. VA AND THE RESPONSE TO IMMUNIZATION IN CHILDREN

Several studies in the last decade have added to the literature on VA supplementation and immunity in young children. The effect of VA given with measles immunization on serum antibody titers and seroconversion percentage was studied in a randomized controlled trial in 395 infants, 9- to 12-month old, in India (Cherian et al., 2003). Previous studies in 6- to 9-month-old infants (see references in Cherian et al., 2003) had shown no enhanced response to measles immunization in one study, while in two others the antibody response was higher but seroconversion rates were similar, and in another study antibody titer and seroconversion were increased in VA-supplemented infants. Supplemented with 100,000 IU of VA (30-mg retinol) did not affect the rate of seroconversion, but rates were 99% in both the supplemented and unsupplemented groups, so any additional effect could not have been detected. Antibody titers at 1 and 6 months postvaccination did not differ, nor did the proportion of infants with titers considered protective for measles at 6 months postvaccination. The authors commented that most infants in this study were breast fed, which would be expected to provide some protection against VA deficiency. However, VA did not enhance the response of low weight-for-age infants in the study, a group that may be considered more vulnerable to VA deficiency.

A small prospective study of 89 healthy breast-fed infants, randomized into 4 groups, examined VA supplementation, alone and combined with vitamin E, on the response to DPT immunization (Kutukculer et al., 2000). Infants were given 30,000 IU of VA (9-mg retinol) for 3 days just after each immunization, at 2–4 months, and they received a booster immunization (without supplementation) at 16–18 months. VA levels were determined and pre- and postimmunization and antibody titers were measured. No significant differences were found for serum VA, vitamin E, or anti-tetanus antibody titers at 2, 5, and 16 months. However, visual inspection of the data indicates that titers at 5 months were, on average, about 25% higher in the two groups that received VA, with and without vitamin E. Overall, this small study did not find a benefit of VA on the anti-tetanus antibody response, consistent with two earlier small studies. However, an earlier larger study had reported lower levels of anti-tetanus IgG in VA-deficient children. Together, the studies support either no observable benefit on immunization against tetanus or possibly a small benefit in some studies.

Bahl et al. (2002) reported on the effect of VA supplementation, given at EPI contacts, on the antibody response to oral polio vaccine (OPV) in Indian infants in a community with a significant rate of stunting and a high prevalence of subclinical VA deficiency (37% of children 1–5 years old with serum retinol <0.7 μmol/L). Thus it was believed that VA deficiency may exist prior to 6 months of age in this community, and therefore providing

VA before 6 months of age might prove beneficial. Three hundred ninety-nine infants received OPV at 6, 10, and 14 weeks of age, and half ($n = 194$) of the infants received 25,000 IU of VA (7.5-mg retinol) with each immunization. The mothers also received 60 mg of retinol 18–38 days postpartum. OPV titers were determined to poliovirus types 1–3 12 weeks after the third immunization. VA-supplemented infants had a significantly higher geometric mean antibody titer against OPV type 1 (relative risk ratio 1.55; 95% CI, 1.03–2.31), while the number of infants with protective titers against poliovirus was also increased. No differences, however, were observed for responses to type 2 or type 3 poliovirus immunization. The authors concluded that VA did not interfere with the response to any of the three types of poliovirus, while the response to type 1 poliovirus was enhanced. Although it is unknown why the beneficial response was limited to type 1 poliovirus, the authors noted that the proportion of infants in the placebo group with protective antibody titers was lower (71%) for type 1 poliovirus than for either type 2 or type 3 poliovirus (93% and 80%, respectively), while VA increased the percent of infants with protective titers against type 1 poliovirus to 82%. Thus, it may be that the effect of VA supplementation is limited to those antigens which tend to be less effective, and for which the response rate of the untreated group is low.

The effect of VA given simultaneously with measles immunization was examined in a study of 462 children in Guinea-Bissau who were randomized to receive either a two-dose schedule of measles vaccine at the ages of 6 and 9 months ($n = 150$ infants) or one dose of measles vaccine at age 9 months ($n = 312$ infants), the more common age for immunization in developing countries (Benn et al., 1997). Children were followed up to the age of 18 months when serum measles titers were determined. The rate of seroconversion was 98% among children who received two doses of vaccine. Neither the percentage of seroconversion nor geometric mean titer of anti-measles antibodies differed in children receiving VA compared with children receiving no supplement. Among children receiving only one dose of measles vaccine at age 9 months, seroconversion was also high, 95%. In this group, antibody titer was significantly higher in the children, and especially in boys, who received VA (relative risk ratio 1.52; CI, 1.22–1.88). Benn et al. (1997) interpreted their study as providing no indication that simultaneous administration of measles vaccine and VA supplements has a negative effect on measles immunity. Among the children who had received two doses of measles vaccine at the ages of 6 and 9 months, VA had no significant effect, while among children receiving only one dose of measles vaccine at age 9 months of age, 100,000 IU of VA increased antibody concentrations, especially for boys.

These children were then followed up and reexamined when they reached 6–8 years of age (Benn et al., 2002). At that time, fewer of the previously VA-supplemented children had nonprotective antibody concentrations ($P = 0.0095$), and among children with protective antibody levels,

VA-supplemented children tended to have higher antibody titers ($P = 0.09$). This result suggests the effect of VA on measles vaccine given at age 9 months may result in longer protection, although titer levels were not higher.

Wieringa et al. (2004) examined the effect of VA, or other micronutrients, administered *in vivo*, on whole blood cytokine production measured *ex vivo*, in a small study of 59 Indonesian infants without evidence of infection. Whole blood was treated with phytohemagglutinin and LPS, incubated at 37 °C for 10 h, and cytokine concentrations were measured. Serum retinol, zinc, and iron were measured as indicators of micronutrient status. The VA-deficient and nondeficient groups had serum retinol concentrations of 0.49 and 0.88 μmol/L, respectively. Interferon (IFN)γ production was lower ($P < 0.05$) in the VA-deficient group, while interleukin (IL)-12 production showed a similar mean difference, but was not significant, and neither IL-10 nor IL-6 showed any differences. VA did not affect leukocyte counts or differential blood cell counts. The lower production of IFNγ would be consistent with a reduced type 1 cytokine response; however, in mice, VA-deficient $CD4^+$ T-cells were reported to secrete more IFNγ per cell (Carman and Hayes, 1991), which was reduced RA. Since LPS is a ligand of TLR4 (Janeway and Medzhitov, 2002; Takeda and Akira, 2005), the IFNγ could have been produced by non-T-cells such as innate immune cells bearing TLR4 on their surface. Although this study was small, it is interesting for its exploration of a possible mechanism whereby VA might affect the balance of Th1:Th2 cytokine production.

IV. EXPERIMENTAL STUDIES OF VA OR RA SUPPLEMENTATION AND ANTIBODY PRODUCTION *IN VIVO*

A. EXPERIMENTAL MODELS

Experiments to examine the effects of VA or RA on antibody production *in vivo* have been conducted in several animal models. Animal models of VA deficiency provide a means to assess the effects of nutritional repletion with VA, or the effects due specifically to RA in the absence of significant levels of retinol when RA is given as a treatment. It is well appreciated that most, if not all, of the biological effects of VA outside of vision are attributed to RA, and RA, while not stored, is able to reverse the growth impairment and normalize epithelial functions in retinol-depleted animals (Dowling and Wald, 1960), so long as it is supplied continuously (Lamb et al., 1974). VA-adequate animal models provide a means to assess the potential immunostimulatory or immunoinhibitory activity of supplemental VA, or of RA used to simulate a therapeutic treatment, on antibody production and/or regulatory cytokines. Additionally, combinations of VA or RA together with other immune stimuli are of interest. For example, our laboratory has investigated the effects of VA or RA combined with LPS, now known as a ligand for TLR4, and of VA combined with tumor necrosis factor (TNF)-α, a ligand for several receptors of the TNF-α family that are critical for immune regulation. Each of these combinations increased the antibody response to tetanus toxoid in VA-deficient rats, and also elevated the level of anti-tetanus antibodies above the normal level in VA-adequate rats (Ross, 2000). We have also combined VA and RA with polyriboinosinic acid:polyribocytidylic acid (PIC), a double-stranded RNA that is a mimetic of double-stranded RNA viruses and a ligand of TLR3 (Janeway and Medzhitov, 2002; Takeda and Akira, 2005). PIC has been studied by others and ourselves for its ability to elicit type I IFNs and activate NK cells, but it is now known that PIC-stimulated dendritic cells and peripheral blood mononuclear cells produce an array of cytokines, including TNF-α and chemokines (Re and Strominger, 2004). As described below, PIC and RA are strong and sometimes synergistic regulators of antibody production. Factors like LPS, TNF-α, and PIC are most often considered "pro-inflammatory," yet LPS (from normal commensal microflora), and TNF and related molecules are necessary for a normal immune response. The concept of employing TLR ligands such as PIC, CpG, or other PAMPs as vaccine adjuvants, or of incorporating cytokine-expressing vectors such as IL-12 with vaccines, is now receiving serious attention (Del Giudice, 2003; Marciani, 2003; Schijns and Tangeras, 2005).

It is important to bear in mind that in the nutritional model of VA deficiency, deficiency develops gradually, generally over a course of weeks or longer, and thus changes in bone marrow hematopoiesis, and lymphoid organ cellularity may already be apparent, along with other physiological–biochemical changes. Retinoids are known to be important for the maintenance of stromal cell interactions with neighboring cells, and a dysfunction of cell–cell interactions may develop insidiously, as VA deficiency becomes more severe. Conversely, the response to treatment with VA and RA can be very rapid, as both forms of the vitamin are rapidly absorbed from the intestine, even in the VA-deficient state (Ross and Zolfaghari, 2004), and changes in the expression of some retinoid-responsive genes can be detected within hours after administration of RA to VA-deficient animals (Wang et al., 2001; Zolfaghari and Ross, 2002; Zolfaghari et al., 2002). Thus, the context of supplementation studies, especially the extent of deficiency prior to treatment, is likely to be an important factor in the outcomes observed.

B. RA TREATMENT AND ANTIBODY PRODUCTION IN A VA-DEFICIENT MODEL

Antibody responses to tetanus toxoid, a TD antigen, had previously been shown to be reduced during VA deficiency and restored by supplementation with VA (reviewed in Ross, 1996b, 2000). In a study to test whether RA alone, and RA combined with PIC, can increase antigen-specific antibody production, VA-deficient rats were immunized with tetanus toxoid and treated orally with all-*trans* RA, PIC, or the combination at the time of primary immunization (DeCicco *et al.*, 2000). After the primary response has subsided, all rats were reimmunized with tetanus toxoid; however, no additional RA or PIC was administered. VA-deficient rats produced low primary anti-tetanus IgG response (~20% of the VA-adequate control group) and a very low secondary anti-tetanus IgG response (<10% of the VA-adequate control group). The primary response was increased by RA alone and by PIC alone. The primary response was increased much more strongly, in a synergistic manner, by RA + PIC ($P < 0.0001$). Interestingly, the secondary response was equally augmented by RA + PIC, even though these treatments were given only at the time of first immunization (priming). PIC alone, however, did not promote the secondary anti-tetanus response, suggesting that despite an effect of PIC (or cytokines elicited by the binding of PIC to TLR3), on the primary immune response, the development of memory cells was not increased by PIC. However, when PIC was combined with RA, the anti-tetanus memory response was increased strongly and synergistically. These results suggest that RA is required for the differentiation and/or maintenance of memory T and memory B-cells, while PIC alone is not sufficient to induce or maintain these populations. In combination, however, the augmentation of antibody production is both stronger than for either agent alone and long-lasting.

Several cytokine mRNAs were measured in the spleen of VA-deficient rats to explore possible mechanisms for the differences in antibody production. In VA-deficient rats (DeCicco *et al.*, 2000), mRNA levels were low for IL-2 receptor-β, interferon regulatory factor-1, a downstream factor in IFN-regulated gene transcription, and signal transducer and activator of transcription (STAT)-1, a mediator of type I and type II IFN signal transduction and transcriptional activation. In contrast, RA together with PIC increased the expression of each of these factors ($P < 0.0001$ versus controls). Conversely, in VA deficiency, IL-12 and IL-10 mRNAs were both elevated, but reduced toward normal levels by RA. As noted later, an elevation in Th1/type 1 cytokines, either absolutely or in relationship to Th2/type 2 cytokines, has been consistently observed in the VA-deficient state, and the IL-12 results, but not the IL-10 results in this study, support this conclusion. RA normalized this cytokine mRNA imbalance by downregulating IL-12 mRNA to the level in VA-adequate controls. Because retinol levels were very low in the VA-deficient group throughout this experiment, the downregulation of type 1 responses in rats supplemented with RA is apparently directly due to this retinoid. Overall, this study provided support for the notion that RA together with PIC could be a promising combination for stimulating specific immunity *in vivo*.

C. RA TREATMENT AND ANTIBODY PRODUCTION IN VA-ADEQUATE MODELS

A study of similar design was conducted in nonimmunocompromised rats of normal VA status to assess the effects of RA, given at a dose resembling a chemotherapeutic regimen, on TD antibody production (DeCicco et al., 2001). Rats were treated with 100-μg RA $\pm$ 20-μg PIC on day 1 with continued administration of 100 μg of RA daily for 11 days, after which antibody production changes in lymphocyte populations, and cell proliferation were evaluated. In another study conducted for just 21 h, early changes in lymphocyte populations and gene expression were measured. Similar to previous results in VA-deficient rats, the combination of RA + PIC significantly potentiated anti-tetanus IgG levels in VA-adequate rats. This combination also increased the numbers of B-cells and MHC class II$^+$ cells in spleen and lymph nodes, determined by flow cytometry, and the number of NK cells in spleen and blood (see Section V). RA + PIC significantly increased the levels of IL-10, IL-12, and STAT-1 mRNA, and STAT-1 protein, suggestive of heightened immune stimulation. Because tetanus toxoid is a TD antigen, the proliferative response of T-cells *ex vivo* was further studied after short-term treatments administered *in vivo*, as a model for the early stages of T-cell activation *in vivo*. RA combined with PIC significantly increased T-cell proliferation stimulated by anti-CD3/phorbol myristyl acetate + IFNα *ex vivo*. These changes in antibody production, cell distribution, cytokine gene expression, and T-cell proliferation suggest that the combination of RA + PIC stimulates humoral and cell-mediated immunity. It was interesting, however, that the strong synergy between RA and PIC on anti-tetanus antibody production was not apparent in the VA-sufficient rat model.

To further investigate the potential for RA and PIC to promote immunity in the VA-adequate state, studies were conducted in adult mice immunized with tetanus toxoid and treated with RA and/or PIC at priming (Ma et al., 2005). Three independent studies of short and long duration were conducted to evaluate early responses to treatment and long-term outcomes on antibody titers and memory formation. Anti-tetanus IgG isotypes were measured to further assess the effect of treatment on Th1/type 1 immunity, associated with higher IgG2a responses, and Th2/type 2 immunity associated with higher IgG1 production. Whereas RA and PIC differentially regulated both primary and secondary anti-TT IgG isotypes, the combination of RA + PIC stimulated the highest level of total anti-TT IgG (Fig. 1A and B). Concomitantly, the ratio of IgG1 to IgG2a was similar to that of the control group, indicating that the combination of RA + PIC promoted a higher but normally balanced response.

Antibody production was strongly associated with type 1/type 2 cytokine gene expression, assessed as IFNγ and IL-12 mRNA as indicators of type 1 response, and IL-4 and IL-12 mRNA as indicators for type 2 response.

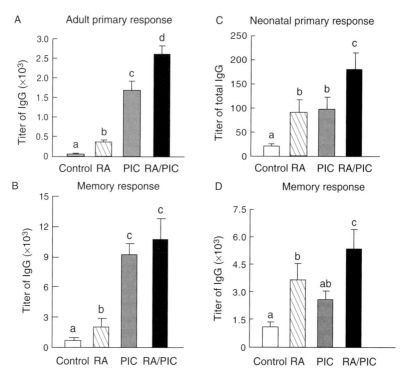

FIGURE 1. Primary and secondary (memory) anti-tetanus antibody responses of adult (A and B) and neonatal (C and D) mice treated at the time of initial dose (priming) with retinoic acid (RA, given orally), polyriboinosinic acid:polyribocytidylic acid (PIC, i.p.), or both RA + PIC. For determination of the memory response, animals were reimmunized with tetanus toxoid without additional treatment with RA or PIC. Data (A, B) from Ma et al. [Copyright (2005) The American Association of Immunologists, Inc., with permission] and (C, D) from Ma and Ross [Copyright (2005) National Academy of Sciences, U.S.A., with permission].

Whereas RA reduced type 1 cytokines (IFNγ and IL-12), PIC enhanced *both* type 1 and type 2 cytokines and cytokine-related transcription factors (T-bet, GATA-4). Despite the presence of PIC, the IL-4:IFNγ ratio was significantly elevated by RA, indicative of skewing toward Th2/type 2 immunity in RA-treated mice. In addition, RA and/or PIC modulated the level of expression of costimulatory molecules involved in B-cell activation, CD80/CD86, an effect that was evident 3 days after antigen priming. Overall RA, PIC, and RA + PIC rapidly and differentially increased the anti-tetanus IgG response. However, the greatest and yet well-balanced response was achieved with RA + PIC, which resulted in a robust, durable, and proportionate increase in all anti-TT IgG isotypes. Because no further treatment with RA or PIC was given after the primary immunization, the heightened secondary antibody response in RA + PIC-treated mice (Fig. 1B) must be attributable to reactivation of immune memory, which was augmented by RA, and RA + PIC, during the primary response.

D. RA SUPPLEMENTATION IN A NEONATAL MODEL

Neonates are highly susceptible to infectious diseases (Kovarik and Siegrist, 1998). In clinical trials in newborn babies, VA supplementation on days 1 and 2 after birth reduced neonatal mortality in the first 2 months of life, especially in infants of low birth weight (Rahmathullah et al., 2003). The VA status, in terms of liver VA reserves and plasma retinol of newborns, even full-term infants of well-nourished mothers, is low as compared to that of older children and adults. Thus, neonates might well be considered physiologically VA deficient, or as of marginal VA status (Ross, 2005). Neonates are known to respond poorly to conventional vaccines due to immaturity of the immune system (Siegrist, 2001), and this has stimulated a search for better adjuvants for neonatal vaccines. As indicated in Table I, the pattern of immune deficiency reported for neonates (Marshall-Clarke et al., 2000) resembles the pattern observed in VA-deficient adult rats and mice (Ross, 1996a), with low responses to TD and polysaccharide (TI-2) antigens, but a relatively normal response to TI-1 (LPS-type) antigens.

Because RA and PIC successfully promoted the antibody response of both VA-deficient rats (DeCicco et al., 2000) and non-VA-deficient adult rats (DeCicco et al., 2001) and mice (Ma et al., 2005), we hypothesized that RA, PIC, and both RA + PIC in combination would promote a stronger antitetanus response in neonatal mice. No previous studies of RA on the immune system in this age group had been reported. We modeled the treatments in our neonatal study (Ma and Ross, 2005) on our previous study of RA + PIC given to adult mice (Ma et al., 2005), scaling the doses of RA and PIC for neonatal mice based on body weight. Early-life treatments with RA and/or PIC were well tolerated, as indicated by equal growth rates in all groups. As was observed in adult mice, RA, PIC, and RA + PIC stimulated the primary anti-tetanus IgG response in neonatal mice (control < RA < PIC < RA + PIC; Fig. 1C). Neonates were maintained on a normal diet after weaning

TABLE I. Characteristics of Antibody Responses in Murine Adult and Neonatal Models, and in Adult VA-Deficient Rats

Antigen type	Adults[a]	Neonates[a]	VA-deficient adult rats[b]
T-cell-independent type 1 (TI-1)	++	++	+++
T-cell-independent type 2 (TI-2)	++	+	±
T-cell dependent (TD), total Ig			Weak
Isotype switching	+++	Weak	Weak
Affinity maturation	+++	Poor	ND
Heterogeneity	+++	Restricted	ND

[a]From Marshall-Clarke et al. (2000).
[b]From DeCicco et al. (2000), Ross (1996a,b).
ND—not determined.

and then, on day 40, they were reimmunized as young adult rats, without any further treatment with RA, PIC, or RA + PIC, to assess the formation of the anti-tetanus memory response. Treatment with RA at priming resulted in a durable increase in anti-tetanus IgG, which was about four times than that of the control group (Fig. 1D). PIC, a potent adjuvant in adult mice, also elevated the neonatal primary anti-tetanus IgG response and induced tetanus-specific IFNγ however, PIC alone failed to benefit the memory response (Fig. 1D). The combination of RA + PIC was more potent than either agent alone in elevating both primary and secondary anti-tetanus IgG responses, as well as all IgG isotypes. Nevertheless, the titer of the memory anti-tetanus response of mice that were treated at neonatal age with RA, PIC, and RA + PIC, during the priming phase of the immune response, was about half that of the memory response of mice primed as adults and treated comparably. Therefore, although anti-tetanus immunity was stimulated by RA + PIC in neonatal mice, the formation of B- and T-cell memory was lower compared to that of adult mice.

E. CYTOKINE PRODUCTION AND TH1:TH2 ANTIBODY ISOTYPE BALANCE

VA deficiency was shown previously to be associated with a predominance of type 1 cytokines (Carman and Hayes, 1991; Carman et al., 1989; Wiedermann et al., 1993). In recent studies of VA supplementation and RA treatment, several investigators measured cytokines, IgG isotypes, or both, that are considered signatures of type 1 and type 2 responses. A consistent finding across studies has been a relative increase in type 2 immunity compared to type 1 immunity after treatment with VA or RA, which has been observed either as an increase in the level of Th2/type 2 cytokines or as a decrease in Th1/type 1 cytokines. In either case, the ratio of Th2 to Th1 cytokines, and/or IgG isotypes, was increased by retinoid treatment. This pattern is robust, as it has been observed in various models of immunization and infection. A predominance of type 2 response was reported in a study of mice immunized intramuscularly with a DNA vector expressing human chorionic gonadotropin in which IgG production to the expressed antigen was monitored over the course of several weeks, during which mice were treated with or without RA continuously (Yu et al., 2005) so that exposure to RA was much higher in this study. The ratio of IgG1/IgG2a plasma antibody levels was somewhat increased in the RA-treated mice, suggesting skewing toward a Th2 response occurred in this model of immunization. In a study of respiratory infection, rather than immunization, in a mouse model of influenza, two groups of mice were fed high-VA diet either before and continuing after influenza infection or beginning at the time of infection to simulate the adjuvant therapy previously used in clinical trials (Cui et al., 2000). The production of IFNγ, a Th1 cytokine, was lower in the group fed the high level of VA continuously compared with the control group, whereas the production of IL-10, a Th2 cytokine, was higher. No differences in disease symptoms were found among the three groups. The authors noted that high-dose VA supplements may enhance Th2-mediated responses, which are beneficial in the case of extracellular bacterial and parasitic infections, and IgA-mediated responses to mucosal infections (Cui *et al.*, 2000).

In the immunization study discussed above in neonatal mice, tetanus-specific lymphocyte proliferation and type 1/type 2 cytokine production, measured as IL-5 and IFNγ in cell culture supernatants after restimulation with tetanus toxoid, were significantly augmented by the prior *in vivo* treatment with RA, PIC, or the combination of RA + PIC. RA alone selectively increased anti-TT IgG1 and IL-5, resulting in skewing toward a type 2 response. Additionally, in a 3-day study, RA and PIC significantly modulated the maturation and/or differentiation of neonatal B-cells, NK/natural killer T-cells (NKT cells) (see Section V), and antigen-presenting cells. These results imply that RA rapidly affects cell populations while effectively promoting predominantly type 2 responses in neonatal mice, an age group in which type 2 immunity is inherently predominant (Marshall-Clarke *et al.*, 2000). However, when RA was combined with PIC as a nutritional–immunologic intervention, the production of anti-tetanus IgG isotypes was well balanced, robust, and durable into adulthood, and both type 1 and type 2 cytokine responses were increased.

In summary, several experimental models are consistent in showing that adaptive immune responses are elevated by VA or RA but skewed in the type 2 direction. Retinoids combined with immune stimuli like PIC, however, stimulated a quantitatively higher level of immune response, as shown in both adult and neonatal mice. Qualitatively, the balance of Th1:Th2 cytokines and the production of IgG isotypes were similar to that of VA-adequate animals.

V. INNATE IMMUNE CELLS AND FACTORS REGULATED BY VA AND RA THAT MAY AFFECT IMMUNIZATION OUTCOME

NK cells are important effector cells of the innate immune system and also important regulators of adaptive immunity (Lanier, 2005; Papamichail *et al.*, 2004). NK cells are produced in bone marrow and enter blood as relatively immature cells, which then can rapidly mature under the influence of various cytokines, especially type I interferons (α/β), released early after viral infection, and IL-2, IL-12, and IL-18 which act synergistically to increase the activation state of NK cells and increase their cytotoxic activity against tumor cells (Baxevanis *et al.*, 2003). Unlike cytotoxic T-cells, NK cells do not require prior sensitization by exposure to the target cell to become cytotoxic.

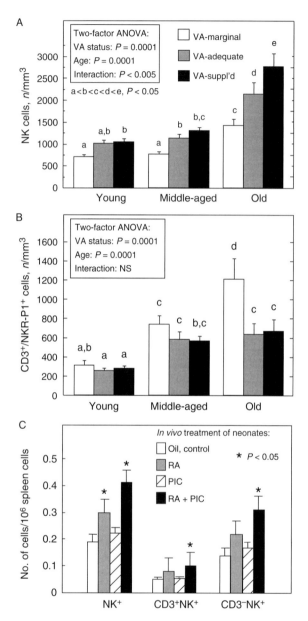

FIGURE 2. Innate immune cells in rats and mice are regulated by dietary VA and acute treatment with RA. In a study of long-term VA status (A and B), rats were fed either a VA-marginal, VA-adequate, or VA-supplemented diet from weaning until the age indicated (young, 2–3 months old; middle-aged, 8–10 months old; old, 18–22 months old). (A) Natural killer (NK) cells in peripheral blood were decreased in rats fed VA-marginal diet and increased in rats fed VA-supplemented diet, while age was also a factor for NK cell number. Data from Dawson et al. (1999). (B) Natural killer T-cells (NKT cells) in peripheral blood were regulated by diet in

Activated NK cells produce various cytokines and thus play an important immunomodulatory role in the production of antibodies. The marker for NK cells, NK1.1 in mice and NKR-P1 on human and rat cells (Lanier et al., 1994), is also expressed on a subset of T-lymphocytes, the NKT cells (Bendelac et al., 1997; Kronenberg, 2005). NK cells are known as an early source of IFNγ, whereas NKT cells have been shown to rapidly secrete IL-4 and IL-10 as well as IFNγ. Understanding the development, functionality, and activation potential of NK cells and NKT cells is currently of great interest both in relationship to antibody production and tumor immunity.

NK and NKT cells in spleen and peripheral blood were investigated in a study of aging rats fed diets that differed only in VA contents (marginal, adequate, and supplemented) from the time of weaning, throughout their life time (Dawson and Ross, 1999; Dawson et al., 1999). The study was designed to cover a wide range of VA consumption but to exclude states of clinically evident VA deficiency or toxicity. Rats fed these diets showed distinct differences in VA status, ranging from a state of marginal VA deficiency (depleted liver VA stores, reduced serum retinol, but normal growth, which became progressively lower as the rats aged), to normal VA status (normal plasma retinol and tissue VA reserves that gradually accumulated with age), to a state of excessive VA accumulation in the VA-supplemented group (elevated plasma retinol and tissues stores, which increased with age, but no overt toxicity) (Dawson and Ross, 1999; Dawson et al., 1999). To quantify NK and NKT cells, peripheral blood mononuclear cell (PBMC) and splenocytes were costained with antibodies against NKR-P1 and CD3 and analyzed by flow cytometry. Marginal VA status was associated with a reduction in the number of NK cells in peripheral blood and a lower percentage of NK cells compared total PBMCs (Dawson et al., 1999). The reduction in NK cells in VA-marginal rats is consistent with previous reports of low NK cells and reduced cytotoxicity in VA-deficient rats (Zhao et al., 1994). Conversely, VA supplementation and aging increased the percentage and number of NK cells above the values in VA-adequate rats (Dawson et al., 1999). Overall, the percentage of NK cells was significantly reduced in VA-marginal rats and increased in VA-supplemented rats, and the effect of diet was greater in old-aged rats (Fig. 2A). In contrast to NK cells, the percentage and number of NKT cells were both increased in peripheral

---

a reciprocal manner compared to NK cells; age was also a factor for NKT cells. Data from Dawson and Ross (1999). The percentage of NKT cells was inversely correlated with the ratio of CD4:CD8 T-cells in peripheral blood (not shown, Dawson and Ross, 1999). (C) Neonatal mice were treated in a short-term (3 days after priming with tetanus toxoid) study with RA (days −1, 0, 1, and 2 before cells were analyzed on day 3), PIC (day 0 only), or both RA + PIC. Total NK1.1$^+$ cells, CD3$^+$ NK1.1$^+$ (NKT), and CD3$^-$ NK1.1$^+$ (NK cells) were analyzed by flow cytometry after double staining with fluorescently labeled anti-NK1.1 and anti-CD3 antibodies. Data from Ma and Ross (2005).

blood of VA-marginal rats, but NKT cells did not differ between VA-adequate and VA-supplemented rats (Dawson and Ross, 1999) (Fig. 2B). The proportion of CD3+ cells (total T-cells) expressing NKR-P1 increased significantly with age, consistent with the higher proportion of these cells among total T-cells in adult humans compared to infants [15–40% in adults versus <5% in infants (Lanier et al., 1994)]. Overall in this long-term study in rats, marginal VA deficiency significantly increased the absolute number and the proportion of NKT cells relative to NK cells (Dawson and Ross, 1999; Dawson et al., 1999), especially as rats aged, and the marginal VA status became more tenuous (with serum declining from 0.97, to 0.63, to 0.38 in VA-marginal young, middle-aged, and old rats, respectively) (Dawson et al., 1999). Given the cytokine-producing function of NKT cells, it is tempting to speculate that an increase in NKT cells could be a compensatory mechanism that helps to counteract a reduced overall capacity for cytokine production in VA deficiency. However, it was reported earlier that IFNγ production is elevated in splenocytes of VA-deficient mice (Carman and Hayes, 1991). An increase in the production of IFNγ by activated NKT cells might explain, in part, the dysregulated antibody responses of VA-deficient mice and rats. However, IFNγ production was not observed to differ in this study of VA-marginal, adequate, and VA-supplemented aging rats, which may suggest that retinol must be nearly completely depleted before IFNγ production becomes dysregulated.

NK cell cytolytic activity in spleen and blood, measured as lysis of Yac-1 target cells, was proportional to the number of NK cells, and was therefore lower in VA-marginal rats, and in older rats. NK cell lytic efficiency (activity per NK cell) also fell with age but it was not affected by VA status. All groups showed a similar increase in NK cell lytic function when peripheral blood cells were incubated with IFNα (Dawson et al., 1999), a cytokine known to be released early in the response to viruses and to be a potent activator of NK cells (Asselin-Paturel and Trinchieri, 2005). This result suggests that the cell surface receptor activated by IFNα and the signaling pathways involved in NK cell proliferation and increased cytotoxicity are intact and functionally equivalent in rats of different ages and VA status. Although IL-2 production by peripheral blood mononuclear cells did not differ significantly with VA status, IL-2 production by splenocytes was lower in VA-marginal rats in each age group (Dawson et al., 1999). Overall, this study identified VA status as a factor in maintaining the NK to NKT cell ratio and suggested that even marginal VA deficiency, especially with advancing age, is a risk factor for reduced NK cell function. Unfortunately, it was unknown at the time this study was conducted that a substantial proportion of the lymphocytes residing in the liver are NKT cells (Wick et al., 2002), and thus this population was not analyzed. It would be of interest to characterize liver NKT cells in relationship to VA status because NKT cells are implicated in the rapid production of cytokines, especially

IFNγ and IL-4, which are both important in regulating adaptive immune responses, and as NKT cells in the liver may be important in the surveillance of virally infected or abnormal cells passing through the liver sinusoids (Kawamura et al., 1999; Wick et al., 2002).

Short-term studies were conducted of VA-adequate rats treated with RA, coincident with immunization with tetanus toxoid (DeCicco et al., 2001). RA alone, in an 11-day study, did not significantly increase the NK cell population in spleen or blood, but did increase the number of NK-cells induced by PIC, known to stimulate NK cell proliferation. In a more comprehensive study of the lymphocyte populations in VA-adequate adult mice (Ma et al., 2005), RA was not a significant factor for the number of NK1.1$^+$ NK cells, but, as in adult VA-adequate rats, RA further increased the positive effect of PIC on the NK cell population. RA also was a positive regulator of the CD3$^+$ NK1.1$^+$ (NKT) population, and it increased the NKT to NK cell ratio. Furthermore, the NKT to NK cell ratio in adult mouse spleen was positively correlated with the ratio of IL-4 to IFNγ mRNAs in the same tissue, which was higher in RA-treated mice, suggesting that an elevation of NKT cells could be a factor in the increase in Th2/type 2 antibody production (ratio of IgG1 to IgG2a) in the same animals (Ma et al., 2005), which is likely to be driven in part by IL-4. As discussed above, the ability of RA to promote type 2 immune responses has been a consistent finding in a number of studies. In neonatal mice treated at the time of antigen priming with RA and/or PIC (Ma and Ross, 2005), the population of NK1.1$^+$ cells in the spleen was less than half that in adult spleen (NK cells, 2.08% versus 4.33%; NKT cells, 0.7% versus 1.36%, respectively). Yet even with these small numbers, it was observed that RA combined with PIC rapidly increased the percentage and number of splenic NK and NKT cells (Fig. 2C). The changes in cell populations observed in normal rats (DeCicco et al., 2001) and mice (Ma and Ross, 2005; Ma et al., 2005) after short-term treatment with RA are likely to be due to a rapid release of cells already near maturation, or to changes in cell-surface molecules that could result in changes in the number of cells in certain compartments. It is interesting that RA has been shown to influence the homing of mouse T-cells to the gut (Iwata et al., 2004), through expression of integrins and other factors.

VI. DISCUSSION AND PERSPECTIVES

As the review above indicates, currently there is only scattered evidence for a positive effect on VA on antibody production in children, whereas, in animals, the evidence for a positive effect of VA in VA-deficient animals, and of RA, in both VA-deficient and VA-sufficient animals, is quite consistent. Several factors could possibly account for these differences and each should be considered including: species differences; the VA status of the host at the time of immunization; the timing of the dose; and the nature of the vaccine or antigen administered.

Although species differences are possible, it seems unlikely that humans and rodents differ in a fundamental way in their response to VA, or to vaccination, because these species are similar in their transport and metabolism of VA, they have mostly similar lymphocyte populations, and many of their genes are homologous.

The VA status of the host at the time of immunization could be a factor, as most animal studies of VA deficiency have been conducted after the animals have reached a state of nearly complete depletion of retinol, with low serum retinol (typically <0.2 μmol/L) and exhaustion of liver VA reserves. In the human studies, VA deficiency has been defined at a higher level of serum retinol (as high as <0.7 μmol/L), and mostly healthy, and often breast fed, children have been studied. It is therefore possible that an effect of VA supplementation on antibody production in children was not apparent in most of the studies because the VA status of the children enrolled was not low enough for differences between the VA-supplemented and placebo groups to have been discerned. This interpretation is supported by the results of an earlier animal study in VA-adequate neonatal rats in which VA supplementation neither increased nor decreased the anti-tetanus antibody response, although it was evident that the neonates responded to the immunization and produced a memory response (Gardner and Ross, 1995). It is interesting that the study of Bahl *et al.* (2002), in which 6-month-old children from a community with a high prevalence of low serum VA were immunized with OPV, showed a significant effect of VA for serum titers against OPV type 1. The VA status of these children may have been more tenuous than that of children in other studies where VA supplementation was given with immunization and the antibody response was measured.

The timing of the dose may also be a significant factor. In most of the human studies of VA and immunization, and in the WHO/EPI (World Health Organization, 1994) strategy for using immunization contacts to deliver VA and eliminate VA deficiency, VA has been, or is, administered at high dosage on a periodic basis. In animal studies, VA has been incorporated at a higher than usual level into the diet (Cui *et al.*, 2000), or if provided as an oral supplement, usually given more than once. These differences may be significant because, although a single large dose of VA can quickly restore plasma retinol to a normal level and replenish liver reserves, it does not provide VA continuously for absorption from the intestines. In comparison, a diet enriched in VA, or oral VA supplements given in smaller divided doses, would not only restore plasma retinol to a normal level and replenish liver reserves but would also provide retinol substrate for formation of retinyl esters in the intestine, which are released bound to chylomicrons. VA is also, to some extent, oxidized to RA in the intestines. While most chylomicron-associated retinyl ester is taken up by the liver, a proportion of the newly absorbed chylomicron retinyl ester is taken up into extrahepatic tissues, especially tissues that express lipoprotein lipase and are active in the metabolism of chylomicron triglycerides. Additionally, some of the newly absorbed VA is oxidized in the intestine and absorbed as RA into the portal system (Ross, 2006). Newly absorbed VA and intestinally formed metabolites may have a metabolic fate different from that of retinol bound to retinol-binding protein. It was shown in studies of chylomicron metabolism that chylomicron-associated VA is taken up, in an apparently transient manner, by bone marrow (Hussain et al., 1989a,b). However, the implications of this uptake process for hematopoiesis and immune function have not been studied. It is also interesting that, among several large-scale community trials on the effect of VA on child mortality, the study showing the largest reduction in all-cause mortality, 54% (Rahmathullah et al., 1990), delivered VA as a weekly dose at a level near the recommended dietary allowance. This reduction in mortality was more than twice the average reduction of 23% for eight studies combined (Beaton et al., 1994), most of which delivered VA to children as periodic large-dose supplements. The delivery of VA as periodic supplements using vaccination contacts is convenient, but it may be that smaller, more frequent doses, or dietary improvement (Underwood and Smitasiri, 1999), provide benefits that are missed with larger infrequent doses.

Another difference between the human and animal studies that could be important is the form of the immunizing dose. The goal of experiments in animals is to demonstrate potential effects and mechanisms, and thus studies are often designed to optimize the researcher's ability to discriminate differences. In most of the animal studies of VA and immunity, the immunizing dose has been provided without additional adjuvants. In contrast, vaccines for humans have undergone optimization to safely produce strong antibody responses, and most contain proprietary or known adjuvants (Beverley, 2002; Del Giudice, 2003). It thus may be that the vaccines used in human studies already contain enough extra "help," due to the adjuvants they contain, to promote a strong antibody response. In animals, the addition of bacterial LPS, TNF-α (Arora and Ross, 1994), or PIC (DeCicco et al., 2001, 2000; Ma and Ross, 2005; Ma et al., 2005) significantly increased antibody production in both VA-deficient and VA-adequate animal models, and also reduced the difference in antibody response due to differences in VA status. In the study of antibody production in children immunized with OPV, it is interesting that percentage seroconversion was high for OPV types 2 and 3 (Bahl et al., 2002), even in children who did not receive VA, while VA increased the response to OPV type 1, for which the rate of seroconversion was the lowest. Similarly, the seroconversion response to measles immunization was relatively low, and in this study VA was effective in at least some subgroups of children (Benn et al., 1997), and may have promoted protection over a longer period of time (Benn et al., 2002). If some vaccines promote a strong response regardless of VA, this may explain why an effect of VA was not consistently evident in all studies. Therefore, it may be that the experimental animal models correctly predict the positive impact of VA on antibody production, but that this impact is only evident in children if the response to the vaccine is not already strong.

Treating animals with RA provides another way to explore the impact of VA on the immune system. The model is also relevant to the human condition because RA and other retinoids, due to their ability to induce cell differentiation, are used therapeutically in the treatment of leukemias, other cancers, and dermatological diseases (Altucci and Gronemeyer, 2001). In animals given RA, rather than VA itself, the physiological controls that otherwise regulate and limit the conversion of VA to RA are bypassed. The results of several animal studies have demonstrated the potential of RA, at a well-tolerated therapeutic level, to augment antibody production. Therefore, these results suggest that RA could be useful in the treatment of some forms of immunodeficiency. The tendency of VA and RA to promote Th2/type 2 responses may be beneficial in the response to certain types of pathogens and infectious diseases, but not to others. However, when RA is combined with other agents, such as PIC, that promote a Th1/type 1 response, a higher and well-balanced antibody response can be achieved. These results suggest that combination therapies in which RA is coadministered with other immune stimuli could offer a range of possibilities for modulating the magnitude and the type of antibody response elicited by various vaccines.

ACKNOWLEDGMENTS

I thank all of the researchers whose projects over the years have made important contributions to the ideas discussed in this chapter. Supported by NIH grant DK-41479, and the Dorothy Foehr Huck Chair.

REFERENCES

Altucci, L., and Gronemeyer, H. (2001). The promise of retinoids to fight against cancer. *Nat. Rev.* **1**, 181–193.

Arora, D., and Ross, A. C. (1994). Antibody response against tetanus toxoid is enhanced by lipopolysaccharide or tumor necrosis factor-alpha in vitamin A-sufficient and -deficient rats. *Am. J. Clin. Nutr.* **59**, 922–928.

Asselin-Paturel, C., and Trinchieri, G. (2005). Production of type I interferons: Plasmacytoid dendritic cells and beyond. *J. Exp. Med.* **202**, 461–465.

Bahl, R., Bhandari, N., Wahed, M. A., Kumar, G. T., Bhan, M. K., and Vitami, W. C. I. L. (2002). Vitamin A supplementation of women postpartum and of their infants at immunization alters breast milk retinol and infant vitamin A status. *J. Nutr.* **132**, 3243–3248.

Baxevanis, C. N., Gritzapis, A. D., and Papamichail, M. (2003). *In vivo* antitumor activity of NKT cells activated by the combination of IL-12 and IL-18. *J. Immunol.* **171**, 2953–2959.
Beaton, G. H., Martorell, R., Aronson, K. A., Edmonston, B., McCabe, G., Ross, A. C., and Harvey, B. (1994). Vitamin A supplementation and child morbidity and mortality in developing countries. *Food Nutr. Bull.* **15**, 282–289.
Bendelac, A., Rivera, M. N., Park, S. H., and Roark, J. H. (1997). Mouse CD1-specific NK1 T cells: Development, specificity, and function. *Annu. Rev. Immunol.* **15**, 535–562.
Benn, C. S., Aaby, P., Balé, C., Olsen, J., Michaelsen, K. F., George, E., and Whittle, H. (1997). Randomised trial of effect of vitamin A supplementation on antibody response to measles vaccine in Guinea-Bissau, west Africa. *Lancet* **350**, 101–106.
Benn, C. S., Balde, A., George, E., Kidd, M., Whittle, H., Lisse, I. M., and Aaby, P. (2002). Effect of vitamin A supplementation on measles-specific antibody levels in Guinea-Bissau. *Lancet* **359**, 1313–1314.
Beverley, P. C. L. (2002). Immunology of vaccination. *Br. Med. Bull.* **62**, 15–28.
Carman, J. A., and Hayes, C. E. (1991). Abnormal regulation of IFN-γ secretion in vitamin A deficiency. *J. Immunol.* **147**, 1247–1252.
Carman, J. A., Smith, S. M., and Hayes, C. E. (1989). Characterization of a helper T-lymphocyte defect in vitamin A deficient mice. *J. Immunol.* **142**, 388–393.
Cherian, T., Varkki, S., Raghupathy, P., Ratnam, S., and Chandra, R. K. (2003). Effect of vitamin A supplementation on the immune response to measles vaccination. *Vaccine* **21**, 2418–2420.
Ching, P., Birmingham, M., Goodman, T., Sutter, R., and Loevinsohn, B. (2000). Childhood mortality impact and costs of integrating vitamin A supplementation into immunization campaigns. *Am. J. Public Health* **90**, 1526–1529.
Cui, D. M., Moldoveanu, Z., and Stephensen, C. B. (2000). High-level dietary vitamin A enhances T-helper type 2 cytokine production and secretory immunoglobulin A response to influenza A virus infection in BALB/c mice. *J. Nutr.* **130**, 1132–1139.
Dawson, H. D., and Ross, A. C. (1999). Chronic marginal vitamin A status affects the distribution and function of T cells and natural T calls in aging Lewis rats. *J. Nutr.* **129**, 1782–1790.
Dawson, H. D., Li, N. Q., DeCicco, K. L., Nibert, J. A., and Ross, A. C. (1999). Chronic marginal vitamin A status reduces natural killer cell number and function in aging Lewis rats. *J. Nutr.* **129**, 1510–1517.
DeCicco, K. L., Zolfaghari, R., Li, N.-Q., and Ross, A. C. (2000). Retinoic acid and polyriboinosinic acid:polyribocytidylic act synergistically to enhance the antibody response to tetanus toxoid during vitamin A deficiency: Possible involvement of interleukin-2 receptor beta, signal transducer and activator of transcription-1, and interferon regulatory factor-1. *J. Infect. Dis.* **182**, S29–S36.
DeCicco, K. L., Youngdahl, J. D., and Ross, A. C. (2001). All-*trans*-retinoic acid and polyriboinosinic:polyribocytidylic acid in combination potentiate specific antibody production and cell-mediated immunity. *Immunology* **104**, 341–348.
Del Giudice, G. (2003). Vaccination strategies. *An overview. Vaccine* **21**, S2/83–S82/88.
Dowling, J. E., and Wald, G. (1960). The biological function of vitamin A acid. *Proc. Natl. Acad. Sci. USA* **46**, 587–608.
Gardner, E. M., and Ross, A. C. (1995). Immunologic memory is established in nursling rats immunized with tetanus toxoid, but is not affected by concurrent supplementation with vitamin A. *Am. J. Clin. Nutr.* **62**, 1007–1012.
Hengesbach, L. M., and Hoag, K. A. (2004). Physiological concentrations of retinoic acid favor myeloid dendritic cell development over granulocyte development in cultures of bone marrow cells from mice. *J. Nutr.* **134**, 2653–2659.

Hussain, M. M., Mahley, R. W., Boyles, J. K., Fainaru, M., Brecht, W. J., and Lindquist, P. A. (1989a). Chylomicron-chylomicron remnant clearance by liver and bone marrow in rabbits. *J. Biol. Chem.* **264,** 9571–9582.

Hussain, M. M., Mahley, R. W., Boyles, J. K., Lindquist, P. A., Brecht, W. J., and Innerarity, T. L. (1989b). Chylomicron metabolism. Chylomicron uptake by bone marrow in different animal species. *J. Biol. Chem.* **264,** 17931–17938.

Iwata, M., Hirakiyama, A., Eshima, Y., Kagechika, H., Kato, C., and Song, S. Y. (2004). Retinoic acid imprints gut-homing specificity on T cells. *Immunity* **21,** 527–538.

Janeway, C. A., Jr., and Medzhitov, R. (2002). Innate immune recognition. *Annu. Rev. Immunol.* **20,** 197–216.

Kawamura, T., Seki, S., Takeda, K., Narita, J., Ebe, Y., Naito, M., Hiraide, H., and Abo, T. (1999). Protective effect of NK1.1$^+$ T cells as well as NK cells against intraperitoneal tumors in mice. *Cell. Immunol.* **193,** 219–225.

Kovarik, J., and Siegrist, C. A. (1998). Immunity in early life. *Immunol. Today* **19,** 150–152.

Kronenberg, M. (2005). Toward an understanding of NKT cell biology: Progress and paradoxes. *Annu. Rev. Immunol.* **23,** 877–900.

Kutukculer, N., Akil, T., Egemen, A., Kurugöl, Z., Aksit, S., Özmen, D., Turgan, N., Bayindir, O., and Çaglayan, S. (2000). Adequate immune response to tetanus toxoid and failure of vitamin A and E supplementation to enhance antibody response in healthy children. *Vaccine* **18,** 2979–2984.

Lamb, A. J., Apiwatanaporn, P., and Olson, J. A. (1974). Induction of rapid, synchronous vitamin A deficiency in the rat. *J. Nutr.* **104,** 1140–1148.

Lanier, L. L. (2005). NK cell recognition. *Annu. Rev. Immunol.* **23,** 225–274.

Lanier, L. L., Chang, C., and Phillips, J. H. (1994). Human NKR-P1A. A disulfide-linked homodimer of the C-type lectin superfamily expressed by a subset of NK and T lymphocytes. *J. Immunol.* **153,** 2417–2428.

Ma, Y., and Ross, A. C. (2005). The anti-tetanus immune response of neonatal mice is augmented by retinoic acid combined with polyriboinosinic:polyribocytidylic acid. *Proc. Natl. Acad. Sci. USA* **102,** 13556–13561.

Ma, Y. F., Chen, Q. Y., and Ross, A. C. (2005). Retinoic acid and polyriboinosinic: polyribocytidylic acid stimulate robust anti-tetanus antibody production while differentially regulating type 1/type 2 cytokines and lymphocyte populations. *J. Immunol.* **174,** 7961–7969.

Marciani, D. J. (2003). Vaccine adjuvants: Role and mechanisms of action in vaccine immunogenicity. *Drug Discov. Today* **8,** 934–943.

Marshall-Clarke, S., Reen, D., Tasker, L., and Hassan, J. (2000). Neonatal immunity: How well has it grown up? *Immunol. Today* **21,** 35–41.

Matzinger, P. (2002). An innate sense of danger. *Ann. NY Acad. Sci.* **961,** 341–342.

Papamichail, M., Perez, S. A., Gritzapis, A. D., and Baxevanis, C. N. (2004). Natural killer lymphocytes: Biology, development, and function. *Cancer Immunol. Immunother.* **53,** 176–186.

Rahmathullah, L., Underwood, B. A., Thulasiraj, R. D., Milton, R. C., Ramaswamy, K., Rahmathullah, R., and Babu, G. (1990). Reduced mortality among children in southern India receiving a small weekly dose of vitamin A. *N. Engl. J. Med.* **323,** 929–935.

Rahmathullah, L., Tielsch, J. M., Thulasiraj, R. D., Katz, J., Coles, C., Devi, S., John, R., Prakash, K., Sadanand, A. V., Edwin, N., and Kamaraj, C. (2003). Impact of supplementing newborn infants with vitamin A on early infant mortality: Community based randomised trial in southern India. *Br. Med. J.* **327,** 254–257.

Re, F., and Strominger, J. L. (2004). Heterogeneity of TLR-induced responses in dendritic cells: From innate to adaptive immunity. *Immunobiology* **209,** 191–198.

Ross, A. C. (1996a). The relationship between immunocompetence and vitamin A status. *In* "Vitamin A Deficiency: Health, Survival, and Vision" (A. Sommer and K. P. West, Jr., Eds.), pp. 251–273. Oxford University Press, Inc, New York.

Ross, A. C. (1996b). Vitamin A deficiency and retinoid repletion regulate the antibody response to bacterial antigens and the maintenance of natural killer cells. *Clin. Immunol. Immunopathol.* **80,** S36–S72.

Ross, A. C. (2000). Vitamin A, retinoids and immune responses. In "Vitamin A and Retinoids: An Update of Biological Aspects and Clinical Applications" (M. A. Livrea, Ed.), pp. 83–95. Birkhèuser Verlag, Basel.

Ross, A. C. (2005). Introduction to vitamin A: A nutritional and life cycle perspective. In "Carotenoids and Retinoids. Molecular Aspects and Health Issues" (L. Packer, U. Obermüller-Jevic, K. Kraemer, and H. Sies, Eds.), pp. 23–41. AOCS Press, Champaign, IL.

Ross, A. C. (2006). Vitamin A and carotenoids. In "Modern Nutrition in Health and Disease" (M. E. Shils, M. Shike, A. C. Ross, B. Caballero, and R. J. Cousins, Eds.), pp. 319–431. William & Wilkins, Baltimore.

Ross, A. C., and Zolfaghari, R. (2004). Regulation of hepatic retinol metabolism: Perspectives from studies on vitamin A status. *J. Nutr.* **134,** 269S–275S.

Schijns, V. E., and Tangeras, A. (2005). Vaccine adjuvant technology: From theoretical mechanisms to practical approaches. *Dev. Biol. (Basel)* **121,** 127–134.

Semba, R. D. (1999). Vitamin A as "anti-infective" therapy, 1920–1940. *J. Nutr.* **129,** 783–791.

Semba, R. D. (2000). Vitamin A and infectious diseases. In "Vitamin A and Retinoids: An Update of Biological Aspects and Clinical Applications" (M. A. Livrea, Ed.), pp. 97–108. Birkhèuser Verlag, Basel.

Siegrist, C. A. (2001). Neonatal and early life vaccinology. *Vaccine* **19,** 3331–3346.

Sommer, A., and West, K. P., Jr. (1996). "Vitamin A Deficiency: Health, Survival, and Vision." Oxford University Press, Inc, New York.

Sommer, A., and West, K. P., Jr. (1997). The duration of the effect of vitamin A supplementation. *Am. J. Public Health* **87,** 467–469.

Stephensen, C. B. (2001). Vitamin A, infection, and immune function. *Annu. Rev. Nutr.* **21,** 167–192.

Stephensen, C. B., Moldoveanu, Z., and Gangopadhyay, N. N. (1996). Vitamin A deficiency diminishes the salivary immunoglobulin A response and enhances the serum immunoglobulin G response to influenza A virus infection in BALB/c mice. *J. Nutr.* **126,** 94–102.

Takeda, K., and Akira, S. (2005). Toll-like receptors in innate immunity. *Int. Immunol.* **17,** 1–14.

Underwood, B. A. (1995). Editorial: The timing of high-dose vitamin A supplementation to children. *Am. J. Public Health* **85,** 1200–1201.

Underwood, B. A., and Smitasiri, S. (1999). Micronutrient malnutrition: Policies and programs for control and their implications. *Annu. Rev. Nutr.* **19,** 303–324.

Villamor, E., and Fawzi, W. W. (2005). Effects of vitamin A supplementation on immune responses and correlation with clinical outcomes. *Clin. Microbiol. Rev.* **18,** 446–464.

Wang, Y., Zolfaghari, R., and Ross, A. C. (2001). Cloning of rat cytochrome P450RAI (CYP26) cDNA and regulation of hepatic CYP26 gene expression by retinoic acid *in vivo*. *FASEB J.* **15,** A602.

Wick, M. J., Leithauser, F., and Reimann, J. (2002). The hepatic immune system. *Crit. Rev. Immunol.* **22,** 47–103.

Wiedermann, U., Hanson, L. Å., Kahu, H., and Dahlgren, U. I. (1993). Aberrant T-cell function *in vitro* and impaired T-cell dependent antibody response *in vivo* in vitamin A-deficient rats. *Immunology* **80,** 581–586.

Wieringa, F. T., Dijkhuizen, M. A., West, C. E., van der Ven-Jongekrijg, J., and van der Meer, J. W. (2004). Reduced production of immunoregulatory cytokines in vitamin A- and zinc-deficient Indonesian infants. *Eur. J. Clin. Nutr.* **58,** 1498–1504.

World Health Organization (1994). "Using Immunization Contacts as the Gateway to Eliminating Vitamin A Deficiency—a Policy Document," WHO EPI/GEN/94.9, WHO, Geneva.

Yu, S., Xia, M., Xu, W., Chu, Y., Wang, Y., and Xiong, S. (2005). All-trans retinoic acid biases immune response induced by DNA vaccine in a Th2 direction. *Vaccine* **23,** 5160–5167.

Zhao, Z., Murasko, D. M., and Ross, A. C. (1994). The role of vitamin A in natural killer cell cytotoxicity, number and activation in the rat. *Nat. Immunol.* **13,** 29–41.

Zolfaghari, R., and Ross, A. C. (2002). Lecithin:retinol acyltransferase expression is regulated by dietary vitamin A and exogenous retinoic acid in the lung of adult rats. *J. Nutr.* **132,** 1160–1164.

Zolfaghari, R., Wang, Y., Sancher, A., Chen, Q., and Ross, A. C. (2002). Cloning and molecular expression analysis of large and small lecithin:retinol acyltransferase mRNAs in the liver and other tissues of adult rats. *Biochem. J.* **368,** 621–631.

9

Physiological Role of Retinyl Palmitate in the Skin

Peter P. Fu,* Qingsu Xia,* Mary D. Boudreau,*
Paul C. Howard,* William H. Tolleson,*
and Wayne G. Wamer[†]

*National Center for Toxicological Research, Food and Drug Administration
Jefferson, Arkansas 72079
[†]Center for Food Safety and Applied Nutrition, Food and Drug Administration
College Park, Maryland 20740

I. Introduction
II. Structure and Physiological Functions of the Skin
III. Cutaneous Absorption and Deposition of Dietary and Topically Applied Retinol and Retinyl Esters
   A. Absorption and Deposition of Dietary Vitamin A by the Skin
   B. Absorption and Deposition of Topically Applied Retinol and Retinyl Palmitate
   C. Physiological and Environmental Factors Affecting Cutaneous Levels of Retinol and Retinyl Esters
IV. Mobilization and Metabolism of Retinol and Retinyl Esters in the Skin
V. Effects on Selected Biological Responses of the Skin
   A. Immune Response
   B. Wound Healing
   C. Aging
   D. Response to UV Light

VI. Summary
References

The skin is similar to other organs in how it absorbs, stores, and metabolizes vitamin A. However, because of the anatomical location of skin and the specialized physiological roles it plays, there are ways in which the skin is rather unique. The stratified structure of the epidermis results from the orchestration of retinoid-influenced cellular division and differentiation. Similarly, many of the physiological responses of the skin, such as dermal aging, immune defense, and wound healing, are significantly affected by retinoids. While much is known about the molecular events through which retinoids affect the skin's responses, more remains to be learned. Interest in the effects of retinol, retinyl palmitate, and other retinoids on the skin, fueled in part by the promise of improved dermatologic and cosmetic products, will undoubtedly make the effects of retinoids on skin a subject for continued intense investigation. © 2007 Elsevier Inc.

I. INTRODUCTION

For nearly a century, it has been recognized that vitamin A (i.e., retinol and its esters; Fig. 1) plays a critical role in the health of epithelial tissue. Clinical observations made by Bloch (1921) and experimental studies by Mori (1922) and Wolbach and Howe (1925) first established a link between a diet deficient in vitamin A and abnormal keratinization of epithelia. Since these early beginnings, studies on epithelial tissues and cells derived from them have provided an enormous amount of information on the roles of vitamin A in processes such as cellular division, differentiation, and transformation, as well as intracellular and intercellular signaling. The skin has been an epithelial tissue of particular interest as a target for the effects of vitamin A and other retinoids. This is due to several reasons. The skin is an easily accessible tissue for *in vivo* studies, and the structure and function of the skin is dependent on retinoid-influenced orchestration of cellular division, differentiation, and keratinization. In addition, interest in the skin is undoubtedly driven by the potential for development of dermatological products and cosmetics containing retinoids. Today, a wide range of dermatologic disorders are treated with a number of geometric isomers of retinoic acid and structurally similar synthetic retinoids (Sekula-Gibbs *et al.*, 2004). Retinoid-containing cosmetic products, particularly those marketed to reduce the appearance of skin aging and photoaging, continue to

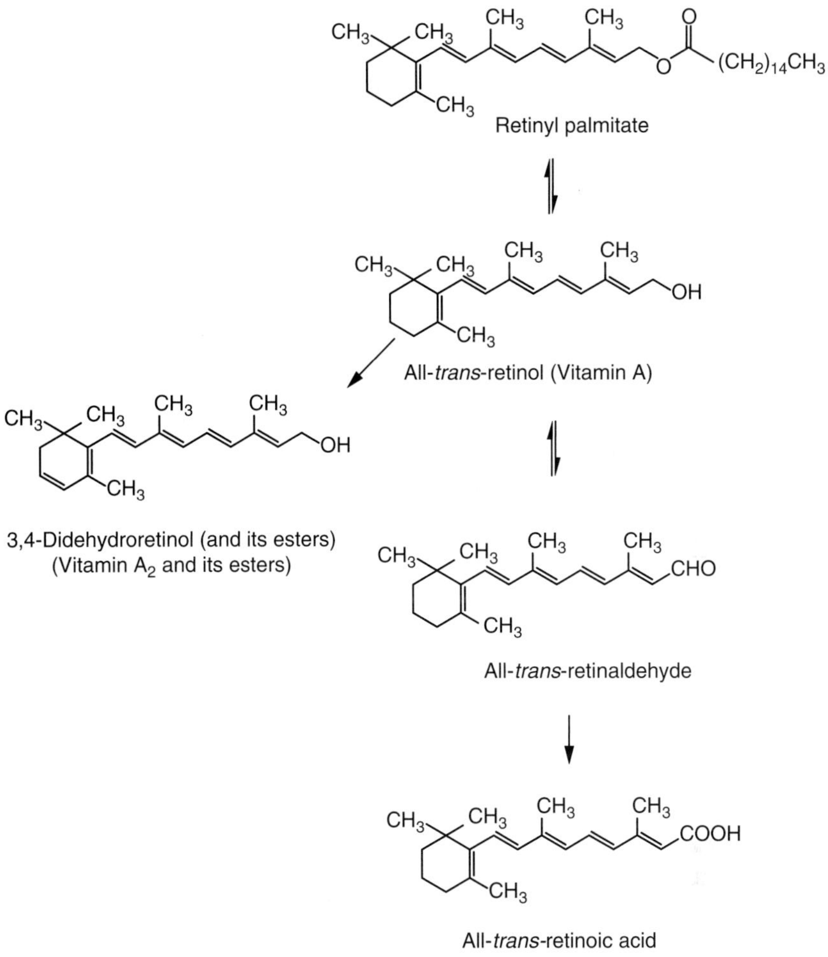

FIGURE 1. The structures of selected retinoids found in skin.

grow in popularity. Data available from FDA's Voluntary Cosmetics Registration Program on cosmetic products in the US market indicate that 102 cosmetic formulations in 1981, 355 cosmetic formulations in 1992, 667 formulations in 2000, and more than 700 products in 2004 contained retinyl palmitate (FDA, 2004). It may, therefore, be anticipated that the skin will remain a focus for studies on the biological activity of vitamin A and other retinoids.

The biological effects of vitamin A on the skin have been reviewed by a number of authors. Reviews emphasizing the use of retinol and retinyl palmitate in cosmetic products have appeared (Cosmetic Ingredient Review, 1987;

Ries and Hess, 1999). Clinical management of specific dermatological disorders by retinoids is the subject of a vast number of reviews. The specific molecular mechanisms underlying the biological effects of vitamin A and other retinoids on the skin appear to be less frequently reviewed. Reviews of the general role played by retinoids in the skin and of retinoid-mediated molecular events in the skin include those by Fisher and Voorhees (1996), Roos et al. (1998), Randolph and Siegenthaler (1999), and Kang et al. (2000). The present chapter is intended to update information regarding the physiological role of vitamin A and its biologically active metabolites in the skin. Throughout the discussion, retinyl palmitate will play a central role since retinyl palmitate is the primary retinyl ester found in the diet, stored in the body, and used in topically applied consumer products.

II. STRUCTURE AND PHYSIOLOGICAL FUNCTIONS OF THE SKIN

The skin is one of the human body's largest organs, with a surface area ~ 2 m^2 and a thickness that commonly varies between 0.5 and 4 mm (Kerr, 1999). Many of the specialized physiological functions of the skin are made possible by its unique, stratified structure (Fig. 2). The two major components of the skin are the epidermis and dermis. The primary architecture of the epidermis is formed through the division of keratinocytes, the predominant epidermal cell type, in the basal (bottom) layer of the epidermis followed by a process of differentiation and upward migration to the skin's surface forming the upper layers of the viable epidermis. The most superficial layer of the skin, the stratum corneum, is composed of lipids and corneocytes which are dead, enucleated cells in the final stage of differentiation (Madison, 2003). Differentiation of keratinocytes gives rise to the following histologically distinct strata in the epidermis: stratum basale (single row of mitotically active basal cells), stratum spinosum (about five rows of cells having spines, or desmosomes), stratum granulosum (about five rows of cells having cytoplasmic keratohyalin granules), stratum lucidum (present in thick skin and containing dead cells with abundant keratin proteins), and stratum corneum (lipids and a dead, anucleated, keratin-containing cell layer ranging in thickness from 10 to a few hundred cells). Although estimates can vary substantially, it is thought that the time interval between division of a keratinocyte in the stratum basale and the appearance of this newly formed cell as a corneocyte in the stratum corneum (i.e., the minimal transit time) is about 14 days and that complete replacement of the epidermis occurs in 52–75 days (Hoath and Leahy, 2003). In addition to keratinocytes, other types of cells are present in the epidermis and perform critical physiological functions. Melanocytes, which are typically located between every 5 and 10 basal keratinocytes and synthesize melanin, play a critical role in the protective function of the skin against sunlight.

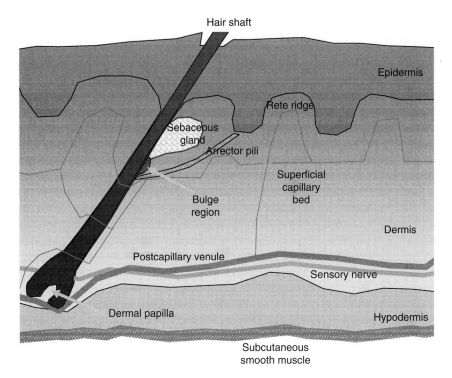

FIGURE 2. Major structural components of the skin. (See Color Insert.)

Langerhans cells, found in the lower epidermis, are an integral part of the skin's immunological defenses serving as antigen-presenting cells. Merkel cells, which are specialized sensory cells, are found adjacent to basal cells, particularly in the epidermis of the fingers, lips, and around hair follicles (Kerr, 1999).

The dermis is the lower layer of the skin and in large part serves to sustain and support the epidermis. The dermis interfaces with the epidermis through a layer of upward protrusions of dermal papillae (Fig. 2), also called rete pegs, which provide a firm anchor to physically connect the dermis with the epidermis. In addition, the papillary dermis contains a network of capillaries. Since the epidermis contains no blood vessels, all of the metabolic needs of the epidermis are met through diffusion of nutrients and waste products between the epidermis and capillaries in the dermis. The lower portion of the dermis, known as the reticular dermis, is less profusely vascularized. Dermal fibroblasts, the primary cell type in the dermis, are a heterogeneous population of cells whose subpopulations are unique to the dermal layer (Sorrell and Caplan, 2004). An important role of dermal fibroblasts is synthesis of the various components that constitute the extracellular matrix of the dermis. These proteins include collagens (mainly type I), elastin, and glycosaminoglycans. Dermal fibroblasts also produce matrix metalloproteinases (MMPs) that provide for a dynamic turnover of extracellular matrix proteins and also are important in wound healing (Pilcher *et al.*, 1999). These MMPs include collagenase-1 (MMP-1), stromelysin-1 (MMP-3), and gelatinase B (MMP-9). A balance between synthesis and enzymatic degradation of extracellular matrix proteins is required for homeostasis in the dermis. In aging skin, the synthesis of matrix proteins slows, while expression of MMPs is increased (West, 1994). These biochemical changes in the dermis can result in the sagging and wrinkles commonly observed in aged skin. Chronic exposure to sunlight can further exacerbate these biochemical changes in the dermis (Wlaschek *et al.*, 2001).

The unique structure of the skin allows it to perform a number of specialized functions. One fundamental function of the skin, in particular the stratum corneum, is to provide an essential barrier to penetration of environmental chemicals by limiting their diffusion into the skin (Madison, 2003; Monteiro-Riviere, 2004). In addition, cells in the viable epidermis are able to metabolize compounds that penetrate the stratum corneum and thereby moderate toxicity (Bronaugh *et al.*, 1994; Steinstrasser and Merkle, 1995). The skin also presents a barrier to penetration of sunlight (Kornhauser *et al.*, 2004). About 60% of incident sunlight in the damaging UVB (290–320 nm) region of the spectrum is reflected at the skin's surface or absorbed in the stratum corneum. Light that enters the viable epidermis is further attenuated by scattering and through absorption by epidermal chromophores such as melanin, proteins, and urocanic acid. Because of the constant exposure of skin to a diverse set of antigenic pathogens and environmental chemicals, the skin also provides a unique immune defense. Streilein (1983) proposed that skin-associated lymphoid tissues (SALT) provide immune surveillance in the skin. Components of SALT include epidermal Langerhans cells, with the capacity for antigen presentation, keratinocytes, which release cytokines and other mediators, infiltrating immunocompetent lymphocytes and strategically placed lymph nodes that accept signals derived from the skin. In addition to protection from environmental insults, the skin performs other important physiological functions such as thermoregulation and synthesis of vitamin D (Anderson and Parrish, 1981; Holick *et al.*, 1982).

III. CUTANEOUS ABSORPTION AND DEPOSITION OF DIETARY AND TOPICALLY APPLIED RETINOL AND RETINYL ESTERS

While vitamin A is required for development and maintenance of healthy skin, humans, as well as other animals, are incapable of *de novo* synthesis of compounds with vitamin A activity. These compounds must, therefore, be obtained exogenously. For the skin, compounds with vitamin A activity can be obtained from the diet or from topically applied products. Reviews by Randolph and Siegenthaler (1999) and Saurat et al. (1999) have described cutaneous uptake of vitamin A through dietary sources and topical application.

A. ABSORPTION AND DEPOSITION OF DIETARY VITAMIN A BY THE SKIN

Following ingestion of a meal rich in vitamin A, the skin and other organs in the body encounter a changing profile of blood-borne vitamin A. In humans, retinyl esters (mostly retinyl palmitate) in chylomicrons may predominate in the blood for 3–6 h following ingestion of a retinol-rich meal. Smaller amounts of retinyl esters may be found in VLDL (Lemieux et al., 1998). The clearance of chylomicrons from the bloodstream is preceded by chylomicron catabolism in the blood, which involves removal of triglycerides from chylomicrons catalyzed by blood-borne lipoprotein lipases producing smaller lipoprotein particles called chylomicron remnants (Goodman and Blaner, 1984). Most retinyl esters in chylomicron remnants are efficiently removed from circulation, primarily by the liver. However, up to 30% of chylomicron retinyl esters are cleared from the circulation by extrahepatic tissues (Vogel et al., 1999). It is unclear whether the skin plays any significant role in the clearance of chylomicrons. Postprandial clearance of chylomicrons containing retinyl esters results in the appearance holoRBP4, a complex formed between retinol and plasma retinol-binding protein (apoRBP4), as the predominant circulating form of retinol (Arnhold et al., 1996; Harrison, 2005). To avoid loss due to filtration by the kidney, holoRBP4 is complexed in the bloodstream with transthyretin (Zanotti and Berni, 2004). Plasma levels of retinol are homeostatically controlled and have been reported by Hartmann et al. (2001) to average 714 ng/mL (2.5 μM) and the plasma levels of retinyl palmitate, the most abundant retinyl ester in plasma, average 24.2 ng/mL (46.2 nM). Plasma levels of retinol metabolites, including retinoic acid, are much lower and have been reported to be between 0.26 and 7.72 ng/mL (Wiegand et al., 1998).

Although the mechanism(s) for uptake of blood-borne retinol by target cells in the skin has been intensely investigated, much remains to be learned. Since RBP has a low molecular weight (21 kDa), RBP is able to enter the interstitial fluid compartment of most tissues, including skin. The presence of RBP, ostensibly holoRBP4, in human skin has been demonstrated. Forsum et al. (1977) used immunofluorescence to show that RBP is present in the stratum corneum and in the viable layers of human epidermis. Törmä and Vahlquist (1983) have also shown that RBP is present in the epidermis by analysis of friction blister fluid derived from the intracellular space of the epidermis. The role of cutaneous retinol-binding protein and the mechanism(s) for cellular uptake of retinol bound in holoRBP4 have been investigated in skin.

Two major pathways for cellular uptake have been examined. First, investigators have provided evidence that cell surface receptors in keratinocytes play a role in the uptake of retinol from holoRBP4. Evidence for the presence of cellular receptors for RBP in the skin has been presented by several investigators. *In vitro* studies by Båvik *et al.* (1995) found a specific RBP-binding activity in the membranes of keratinocytes. In addition, these authors demonstrated that retinol, when complexed with RBP, was much more efficient in inhibiting the terminal differentiation of keratinocytes than retinol added directly to culture medium. Smeland *et al.* (1995) found that membranes prepared from undifferentiated rabbit skin keratinocytes had a high specific binding capacity for RBP while the binding capacity of membranes from differentiated skin keratinocytes was 44 times lower. In addition, membranes from skin fibroblasts showed no binding activity. Hinterhuber *et al.* (2004) showed that RPE65, a protein found in retinal pigment epithelium and a putative receptor for holoRBP4, is present on the cell surface of cultured human epidermal keratinocytes. Subsequent studies demonstrated that the gene for RPE65 is expressed in histological sections of normal human epidermis and that expression is downregulated in squamous cell carcinomas (Hinterhuber *et al.*, 2005). While these studies suggest that specific receptors for holoRBP4 play a role in the cellular uptake of retinol in skin, other investigators have provided evidence that receptor-mediated delivery of retinol to the skin is not physiologically significant. It has been demonstrated that epidermal keratinocytes, maintained in culture medium containing a physiological concentration of retinol, have an intracellular level of retinol similar to that in intact epidermis (Randolph and Simon, 1993). Consistent with this observation, *in vitro* studies by Hodam and Creek (1998) have shown that uptake and subsequent esterification of retinol by human keratinocytes was not facilitated by complexing retinol with RBP. These *in vitro* studies suggest that cellular uptake of retinol could be independent of a specific receptor and could involve free (i.e., unbound) retinol in equilibrium with holoRBP4. In general, the delivery of retinol to tissues and cellular uptake of retinol may involve multiple overlapping, redundant pathways that provide compensatory mechanisms under different physiological conditions (Paik *et al.*, 2004). It has been suggested that changes in physiological conditions, such as levels of vitamin A intake and storage, may influence the relative importance of receptor-mediated and receptor-independent cellular uptake of retinol (Paik *et al.*, 2004). Additional studies with relevant *in vivo* models may define more clearly the relative importance of these mechanisms for cellular uptake of retinol in the skin. Further investigations are also needed to establish whether RPE65 plays additional physiological roles in the skin. RPE65 has been shown to act as a retinyl ester-binding protein (Gollapalli *et al.*, 2003) that binds all-*trans* retinyl palmitate with a K_d of 20 pM, compared to weaker binding of 11-*cis* retinyl palmitate, 11-*cis* retinol, or all-*trans* retinol (K_d 14, 3.8, and 10.8 nM, respectively). Moiseyev et al. (2006) have characterized RPE65 as the isomerohydrolase that generates 11-*cis* retinol essential for the retinoid visual cycle. Naturally occurring RPE65 gene inactivating mutations are known for dogs and humans (Thompson et al., 2000; Veske et al., 1999) and are associated with Leber congenital amaurosis type 2. Congenital amaurosis represents a heritable retinal dystrophy with occasional extraocular manifestations, sometimes involving the skin (Fazzi et al., 2005; Yano et al., 1998). The potential physiological implications of isomerohydrolase activity, 11-*cis* retinol, and retinyl ester binding by RPE65 within the skin require further elucidation.

Under usual physiological conditions, retinol, taken up from the serum by cells in the skin, is found bound to cellular retinol-binding protein type 1 (CRBP-1). CRBP-1 solubilizes, protects, and, as described below, influences the fate of retinol in the cell. Using immunohistochemical methods, Busch et al. (1992) showed that levels of CRBP-1 are lowest in epidermal basal cells and increase in suprabasal layers of the skin. The highest levels of CRBP-1 were found in the stratum granulosum. There was no evidence for CRBP-1 in the stratum corneum. Consistent with an earlier report (Siegenthaler, 1986), CRBP-1 was also expressed in several dermal structures such as hair follicle epithelium, sebaceous gland epithelium, and vascular endothelium. Retinol is tightly bound in holoCRBP-1 (K_d 0.1 nM) (Napoli, 1999). Because of this tight binding, it may be expected that most retinol in the skin would be associated with CRBP-1. However, Randolph and Siegenthaler (1999) have estimated that the epidermal concentrations of CRBP-1 ($\sim$80 nM) and RBP4 ($\sim$50 nM) are lower than that needed to bind the amount of retinol estimated to be in the epidermis ($\sim$600 nM). Therefore, a substantial amount of free, that is, unbound, retinol ($\sim$470 nM) may be present in human epidermis (Randolph and Siegenthaler, 1999). In addition, there is some experimental evidence, involving the sensitivity of epidermal retinol to ultraviolet (UV) light, which supports the notion that a portion of the retinol in skin resides in a separate intracellular or extracellular pool, unprotected by binding to CRBP-1 or RBP4 (Sorg et al., 1999).

Three immediate fates are available for intracellular retinol: (1) direct utilization of retinol or maintenance of an intracellular pool of retinol (unbound or as holoCRBP-1); (2) acylation to form retinyl esters; or (3) metabolic transformation, including oxidation to form retinaldehyde and retinoic acid (Figs. 1 and 3). Chemical analysis of retinoids in human skin provides some insight into the disposition of retinol. Retinyl esters have been reported to comprise $\sim$70% of total vitamin A in human skin (Randolph and Siegenthaler, 1999; Törmä and Vahlquist, 1990; Vahlquist, 1982), and are widely considered to be the skin's storage form of vitamin A. Using skin and skin keratinocytes grown in culture, investigators have shown that, unlike retinyl esters in liver which are predominately retinyl palmitate, cutaneous retinyl esters are a mixture of primarily retinyl linoleate, retinyl myristate,

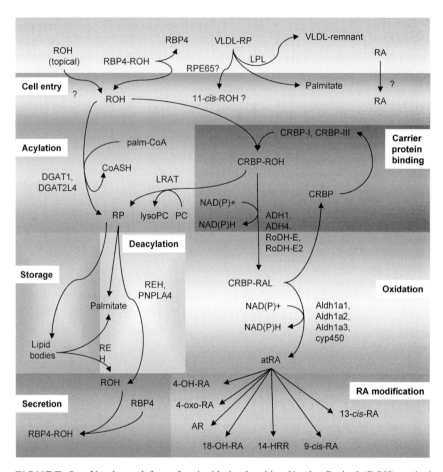

FIGURE 3. Uptake and fate of retinoids in the skin. *Uptake*: Retinol (ROH), retinyl palmitate (RP), or retinoic acid (RA) can be taken up by the skin in a receptor-dependent or independent process. *Storage*: ROH, bound to CRBP, may be acylated and stored in lipid bodies, or *Metabolism*: ROH undergoes a two-step oxidation to retinaldehyde to all-*trans* retinoic acid (atRA). *Retinoic acid modification*: Levels of intracellular atRA are modulated by oxidation to less bioactive metabolites. *Mobilization*: Retinyl esters can be hydrolyzed to form ROH for use in the cell or return to serum. (See Color Insert.)

retinyl oleate, retinyl palmitate, and retinyl stearate (Kang *et al.*, 1995; Kurlandsky *et al.*, 1996; Simmons *et al.*, 2002; Törmä and Vahlquist, 1990). The deposition of retinyl esters within the skin is not uniform. Because the capillary bed supplying blood-borne components to the skin lies entirely in the dermis, it is expected that the dermis and the lower epidermis have the most direct access to components in the blood including holoRBP4 (Fig. 2). Therefore, dermal fibroblasts and adipocytes have closer access to holoRBP4, followed by the well-perfused portions of mature cutaneous hair follicles and, finally, by the cells of the interfollicular epidermis. The cutaneous cells having the lowest access to plasma-derived holoRBP4 are expected to be found in the outermost layers of the epidermis. However, in spite of its anatomic disadvantage, the stratum corneum exhibits higher levels of retinol and retinyl esters than the stratum basale (Törmä and Vahlquist, 1990).

Esterification of cutaneous retinol appears to involve mainly two enzymes, lecithin:retinol acyltransferase (LRAT) and acyl-CoA:retinol acyltransferase (ARAT). LRAT is a microsomal enzyme that catalyzes the reversible transfer of the *sn-1* fatty acid from membrane-associated phosphatidyl choline to retinol bound to CRBP-1 (Ruiz et al., 1999). LRAT plays an important role in regulating retinol storage and diverting retinol from metabolism to more biologically active retinoids such as retinoic acid (Kurlandsky et al., 1996). Another microsomal enzyme, ARAT, catalyzes the reversible transfer of the fatty acid from acyl-CoA to free retinol, that is, retinol not bound to CRBP-1. While ARAT activity in the skin and other organs has been described, molecular identification and characterization of ARAT has remained elusive (O'Byrne et al., 2005). Some evidence suggests that the ARAT activity responsible for esterifying retinol may be attributed to acyl-CoA:diacylglycerol acyltransferase (DGAT) (Orland et al., 2005; Yen et al., 2005). High levels of expression for genes in the DGAT family have been demonstrated in interfollicular human skin and in sebaceous glands (Turkish et al., 2005; Yen et al., 2005). Clearly, much remains to be learned about the molecular identity of ARAT in the skin. It has been suggested that LRAT and ARAT are complementary, that is, that the relative importance of LRAT and ARAT varies in different compartments of the skin. Kurlandsky et al. (1996) studied retinol esterification in keratinocytes derived from different layers of human skin. They determined that keratinocytes from basal layer of the epidermis esterified retinol four times faster per cell than keratinocytes derived from suprabasal layers of the skin. Since holoCRBP-1 was required for retinol esterification by cells derived from the basal layer, this activity was attributed to LRAT. Additional experiments by Kurlandsky et al. (1996) supported the view that ARAT was primarily responsible for retinol esterification by keratinocytes in the suprabasal layers of the skin. It has been observed that the pH in the upper layers of the epidermis may additionally favor retinol esterification through a pathway involving ARAT, since it has been shown that CRBP-independent ARAT activity has a pH optimum between 5.5 and 6.0 (Törmä and Vahlquist, 1990) which is close to the pH at the skin's surface (Rothman, 1954). Kurlandsky et al. (1996) propose a model for retinol esterification within the skin similar to that depicted in Fig. 4.

Chemical analysis also reveals that, in addition to esterification, metabolism is another fate of retinol in skin. A major metabolic pathway in skin is

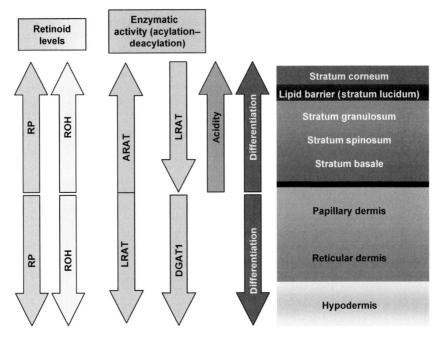

FIGURE 4. Acylation of retinol (ROH). ROH may be acylated by LRAT, ARAT which may be indistinguishable from DGAT. The relative importance of each acyltransferase may vary with pH and location in the skin. (See Color Insert.)

transformation of retinol to 3,4-didehydroretinol, also called vitamin A_2 (Fig. 1). Chemical analysis of retinoids indicates that 20–25% of the total retinoid content in normal human skin is 3,4-didehydroretinol or its esters (Vahlquist *et al.*, 1982). Rollman *et al.* (1993) determined that cultured human keratinocytes, but not cultured dermal fibroblasts or epidermal melanocytes, transform retinol to 3,4-didehydroretinol. In addition, it was shown that differentiation of keratinocytes results in increased formation of 3,4-didehydroretinol. Although 3,4-didehydroretinol is a major component of cutaneous retinoids, little is known about the dehydrogenases involved in its formation or the physiological role of 3,4-didehydroretinol in skin. Retinol may also be oxidized to yield retinaldehyde and retinoic acid (Fig. 1). Levels of retinaldehyde and retinoic acid in normal untreated human skin are very low and are usually below analytical detection. However, conversion of retinol to retinoic acid has been demonstrated using human keratinocytes grown in culture (Kurlandsky *et al.*, 1994) and skin derived from mice topically treated with retinol (Connor and Smit, 1987). Randolph and Siegenthaler (1999) have estimated the levels of epidermal retinoic acid to be less than 20 nM.

B. ABSORPTION AND DEPOSITION OF TOPICALLY APPLIED RETINOL AND RETINYL PALMITATE

Pioneering work by Montagna (1954) and Sobel et al. (1959) provided some of the earliest evidence that topically applied vitamin A penetrates the skin and can cause both local and systemic biological effects. Subsequently, many investigators have shown that topical application is an effective means for loading the skin with vitamin A and related retinoids. Studies of skin penetration and percutaneous absorption have been performed with a wide range of both naturally occurring and synthetic retinoids (Saurat et al., 1999; Schaefer, 1993). The discussion here will focus on the penetration and percutaneous absorption of topically applied retinol and retinyl palmitate.

The penetration and percutaneous absorption of retinol and retinyl palmitate has been studied by a number of investigators using both *in vitro* and *in vivo* approaches. *In vitro* methods have been used to assess both the penetration and metabolism of topically applied retinol and retinyl palmitate. Boehnlein et al. (1994) have investigated the *in vitro* percutaneous absorption of retinyl palmitate through excised human skin. After mounting excised skin in a flow-through diffusion cell, retinyl palmitate was applied (20 $\mu g/cm^2$) in acetone. The amount of applied material, or its metabolites, that had penetrated into the skin and the amount that had passed through the skin were assessed at different time points. It was determined that 24 h after application, 17.8% of the topically applied material had penetrated into, and was found in, the skin. Approximately 44% of the material found in the skin was found to be retinol, indicating hydrolysis of retinyl palmitate in the skin. A much smaller amount of the applied material (0.2%) had penetrated completely through the skin, and had completely been hydrolyzed to retinol. Penetration and percutaneous absorption was also examined using skin from hairless guinea pigs, which was found to be more permeable to retinyl palmitate than skin from humans. The work of Bailly et al. (1998) is unique in that they report the formation of measurable levels of retinoic acid following application of retinol to excised human skin. Skin biopsies were placed in a petri dish containing incubation media, and retinol was applied to the skin's surface. Following incubation for 24 h, the retinoid content of whole skin, epidermis and dermis, was determined. Up to 70% of the applied retinol was found to be absorbed by the skin. Approximately 75% of the absorbed material was found in the epidermis and was composed of ~60% retinol, ~18.5% retinyl esters, ~1.6% retinaldehyde, and ~3% retinoic acid. Approximately 20% of the absorbed material was found in the dermis. The relative abundance of retinol and its metabolites in the epidermis and dermis was similar; however, slightly more oxidized metabolites of retinol were found in the dermis. Antille et al. (2004) examined the penetration and subsequent metabolism of a number of retinoids after topical application to human skin explants mounted in Franz cells. Twenty-four hours following application (2.5 mg/cm^2) of a cream containing 0.02% retinol, an ~100-fold increase in retinol and 5-fold increase in retinyl esters (predominately retinyl palmitate) were observed in the skin. Application of retinyl palmitate under similar conditions resulted in a nearly 3-fold increase in the skin's level of retinol and 33-fold increase in the skin's level of retinyl palmitate. No increase in the levels of retinaldehyde or retinoic acid was observed after either topical application. These authors also examined *in vitro* and *in vivo* penetration into the skin of hairless mice, and found both retinol and retinyl palmitate penetrated mouse skin dramatically better than human skin. Abdulmajed and Heard (2004) have characterized the penetration and metabolism of topically applied retinyl palmitate in different layers of excised human skin. Approximately 60% of the applied retinyl palmitate was found in the skin or had penetrated through the skin. In addition, retinol was found in the epidermis and dermis. Taken together, these *in vitro* results indicate that topically applied retinol and retinyl palmitate readily penetrate into the skin and that esterification of retinol and hydrolysis of retinyl palmitate are major routes of metabolism after topical application in these *in vitro* systems for assessing skin penetration.

Consistent with the results of *in vitro* studies, clinical studies and *in vivo* experimental studies demonstrate that both retinol and retinyl palmitate penetrate the skin. Kang *et al.* (1995) reported that topically applied retinol not only penetrates subjects' skin, but is metabolized and elicits biochemical changes in the skin. One day after a single application (100 μL/18 cm^2) of a cream containing 1.6% retinol, the levels of cutaneous retinol increased 70-fold, levels of 13-*cis* retinol increased 280-fold, and levels of retinyl esters (predominately retinyl linoleate) increased 260-fold. No significant increase in retinoic acid or its metabolites was found. In addition, a single topical treatment with retinol at the levels described, and occluded for 4 days, resulted in significant increases in epidermal thickness with increased mitotic figures in the epidermis. Biochemical changes were also observed including an approximately threefold increase in level of both CRBP and cellular retinoic acid-binding protein (CRABP-2). Duell *et al.* (1996) examined the metabolism of retinol (0.3%) for periods up to 4 days after topical application under occlusion. Isomerization of the topically applied all-*trans* retinol to 13-*cis* retinol, presumably occurring at the skin surface prior to penetration, was the most prominent structural change seen. In addition, increases in cutaneous levels of retinol (3.7-fold), retinyl esters (150-fold), and 3,4-didehydroretinol (15-fold) were observed. Also, 14-hydroxy-4,14-*retro*-retinol was found in significant amounts. A later study by Duell *et al.* (1997) examined the effects of topically applied retinyl palmitate (0.6%) on cutaneous retinoid levels at 48 and 72 h after application. A time-dependent increase in levels of 14-hydroxy-4,14-*retro*-retinol, 13-*cis* retinol, all-*trans* retinol, retinyl palmitate, and retinyl linoleate was observed. The most dramatic increases were observed in levels of 13-*cis* retinol (~15-fold), all-*trans* retinol (~5-fold), and retinyl linoleate (~35-fold). The authors interpret the large increase in retinyl linoleate levels as an indication that topically applied retinyl palmitate is first hydrolyzed in the skin to retinol with subsequent re-esterification to retinyl linoleate. Duell *et al.* (1997) also showed that topical application of retinyl palmitate results in increased levels of retinoic acid 4-hydrolase, an enzyme essential for maintaining appropriate levels of retinoic acid in tissues. Although no increases in cutaneous levels of retinoic acid were directly measured, induction of retinoic acid 4-hydrolase is evidence that topical application of retinyl palmitate increases retinoic acid levels in skin.

The effects of topical application of retinol and retinyl palmitate have also been investigated using animal models. Connor and Smit (1987) demonstrated that application of 100 nmol of retinol to the dorsal surface of SKH-1 (hairless) mice resulted in dramatic increases in levels of retinol in the epidermis (95-fold increase) and in the dermis (16-fold increase). In contrast to human skin after topical application of retinol, elevated levels of retinoic acid were found in the epidermis and dermis of mice at up to 8 h (maximum level at 2 h) after application of retinol. In contrast, Antille *et al.* (2003) found no elevation of retinoic acid in the skin of hairless mice after topical application of retinol. However, these investigators assessed levels of retinoic acid 24 h after topical application of retinol, and the results of Connor and Smit (1987) suggest retinoic acid levels may have decreased below detectable limits at 24 h post application. A study provides additional information regarding the spatial distribution of retinoids in skin following topical application of retinyl palmitate. Yan *et al.* (2006a) have studied the profiles of retinol and retinyl palmitate in strata of the skin of SKH-1 mice treated daily for 4 consecutive days with 0.5% (w/w) retinyl palmitate in an oil-in-water cream. The levels of retinyl palmitate and retinol in the stratum corneum, viable epidermis, and dermis, as well as whole skin, were determined at time points from 24 h to 18 days following the final topical application. A rapid and sustained diffusion of retinyl palmitate into all the three skin layers was observed. As shown in Fig. 5, on the basis of unit weight, the epidermis at each measured time point contained the highest level of retinyl palmitate and retinol (Yan *et al.*, 2006a). Although the dermis contains an amount of retinyl palmitate and retinol per unit weight much lower than in epidermis, the dermal layer represented the largest tissue by mass. As a result, the total amounts of retinyl palmitate and retinol in dermis are higher than those in the epidermis (Fig. 6). Levels of retinyl palmitate and retinol were shown to decrease with time following the final application. Levels of retinyl palmitate in whole skin, 1 day after the final application of the 0.5% retinyl palmitate cream, were ~15-fold higher than the levels at 18 days after the last application. Compared with untreated control mice, the retinyl palmitate levels were significantly higher in the whole skin of mice sacrificed 18 days after treatment with the 0.5% retinyl palmitate cream. In contrast, the levels of retinol returned to control values within 11 days after

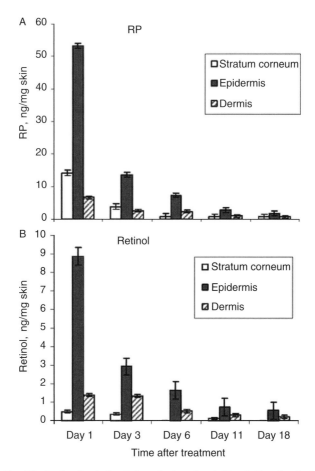

FIGURE 5. The levels of retinyl palmitate (ng/mg tissue) (Panel A) and retinol (Panel B) in stratum corneum, epidermis, and dermis of female SKH-1 mice topically treated with 0.5% retinyl palmitate cream for 4 consecutive days and sacrificed 1, 3, 6, 11, and 18 days after last retinyl palmitate cream application.

the last application of retinyl palmitate (Yan *et al.*, 2006a). Since the level of retinol in the skin of the mice treated with retinyl palmitate is significantly higher than that in the control animals, these results clearly indicate that retinyl palmitate was metabolized into retinol in the skin. These results demonstrate that elevated levels of both retinyl palmitate and retinol persist in skin after repeated topical treatment with retinyl palmitate. The biological consequence of persistently elevated levels of these retinoids is unclear at this time.

The results of these clinical and experimental studies uniformly demonstrate that topical application is an effective strategy for loading the skin

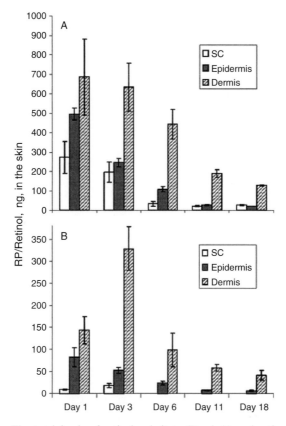

FIGURE 6. The total levels of retinyl palmitate (Panel A) and retinol (Panel B) in the whole skin of female SKH-1 mice topically treated with 0.5% retinyl palmitate cream for 4 consecutive days and sacrificed 1, 3, 6, 11, and 18 days after last retinyl palmitate cream application.

with substantial levels of vitamin A. In addition, these studies indicate that topically applied retinol and retinyl palmitate trigger biochemical changes in the skin that might be expected from perturbation of previously established retinoid homeostasis. These biochemical changes include increased expression of retinol and retinoic acid-binding proteins as well as increased levels of enzymes that metabolize retinoic acid.

Concerns have been raised about potential systemic toxicity resulting from the increased use of retinol and retinyl esters in topically applied products. Clinical studies have been reported that address this issue. Ries and Hess (1999) have reported that topical application of retinol at concentrations expected in some cosmetics (0.25%) causes no increase in plasma levels of retinol or its metabolites; however, details of topical exposure and methods of analysis were not provided. Nohynek *et al.* (2005) have published results from a clinical study involving 2 groups of 14 female volunteers. Prior to initiating topical applications, baseline plasma levels of retinol, retinyl esters, and acids were determined. Thereafter, one group of subjects received daily application ($\sim$1 mg/cm^2, over 3000 cm^2 of the back, hips, and legs) of a cream containing 0.34% retinol, while the other group received daily application of a cream containing 0.55% retinyl palmitate. Topical applications continued for 3 weeks. Plasma levels of retinoid were determined on the first and the last day of topical treatments. On each day designated for sampling plasma, multiple samples were drawn over a 24-h period to compensate for any diurnal variations in plasma retinoid levels. The authors report that neither the topical treatment with the cream containing retinol nor with the cream containing retinyl palmitate caused significant increases in plasma levels of retinol, retinyl esters, or acids.

C. PHYSIOLOGICAL AND ENVIRONMENTAL FACTORS AFFECTING CUTANEOUS LEVELS OF RETINOL AND RETINYL ESTERS

Although the cutaneous levels of retinol, retinyl esters, and metabolites of retinol are tightly controlled under normal conditions, several physiological and environmental factors have been found to affect these levels. While data are not available for humans, studies using mice suggests that age influences the levels of vitamin A in the skin. Törmä et al. (1987) determined that at ages between 1 day and 6 weeks, levels of both retinol and retinyl palmitate increased in the epidermis of hairless mice. These levels remained relatively constant thereafter. Yan et al. (2006b) determined that levels of retinyl palmitate increased in the stratum corneum, epidermis, and dermis of female hairless mice between 10 and 20 weeks of age. Smaller changes were observed in the levels of retinol. Older mice (60 and 68 weeks old) had lower levels of retinol and retinyl palmitate in all strata of the skin. It is unknown, at this time, whether analogous age-related effects on cutaneous levels of vitamin A are seen in humans, and, if so, whether these effects have biological significance.

Diseases of the skin, particularly those characterized by abnormal differentiation and hyperproliferation, can result in alterations in the concentrations of vitamin A and its metabolites. In Darier's disease (keratosis follicularis), a genetically determined disorder of keratinization, the levels of retinol and of 3,4-didehydroretinol in the skin are several fold higher than normal (Vahlquist et al., 1982). Levels of retinol are reported to be low, while levels of 3,4-didehydroretinol are determined to be high in hyperproliferative forms of ichthyosis. Furthermore, skin affected by this disorder has a ratio of retinol to 3,4-didehydroretinol more than 10 times higher normal, possibly suggesting a defect in epidermal vitamin A metabolism (Rollman and Vahlquist, 1985).

Exposure to UV light can also alter the levels of vitamin A in the skin. A characteristic feature of retinoids is their sensitivity to UV light. UVA light (320–400 nm) and UVB light (290–320 nm) have been shown to reduce the vitamin A content in human skin (Antille *et al.*, 2003). Pathways for photochemical decomposition of retinoids include photoisomerization, photodimerization, and photooxidation (Dillon *et al.*, 1996; Mousseron-Canet, 1971). *In vitro* and *in vivo* studies have demonstrated that retinol is significantly more photochemically labile than retinyl palmitate (Ihara *et al.*, 1999; Sorg *et al.*, 1999; Tang *et al.*, 1994). Retinol in rabbit skin *in vivo* and human skin *in vitro* (Berne *et al.*, 1984) was found to be unstable under UV irradiation in a dose-dependent manner. The maximum effect was noted with monochromatic irradiation at 334 nm, as compared to irradiation at 313, 365, and 405 nm. The retinol level was lowered immediately after a single irradiation treatment and the reduction was similar in frozen and fresh skin, suggesting independence from a cellular metabolism (Berne *et al.*, 1984). Sunlight-induced photodegradation of retinyl esters proceeds much faster than that of retinol, and it has been suggested that CRBPs-1 protect retinol from photodegradation (Tang *et al.*, 1994). However, studies using hairless mice treated topically with retinol before and after UVB exposure showed that retinol was depleted to a similar extent after UVB exposure of the pretreated animals as of untreated animals in spite of an induction of CRBP-1. The results in this animal model are inconsistent with protection of retinol from UVB-induced depletion by CRBP-1 (Tran *et al.*, 2001). Sorg *et al.* (2002) have suggested that UV-induced vitamin A depletion and lipid peroxidation in mouse epidermis are unrelated processes and that UV light destroys epidermal vitamin A through a photochemical reaction rather than via oxidative stress. Additional studies are needed to determine the mechanisms underlying the depletion of cutaneous retinoids by UV light.

IV. MOBILIZATION AND METABOLISM OF RETINOL AND RETINYL ESTERS IN THE SKIN

Vitamin A homeostasis in the skin requires that the levels of retinol, retinyl esters, and metabolites of retinol be tightly controlled. The biochemical mechanisms for establishing physiologically appropriate cutaneous levels of retinoids have been previously reviewed (Randolph and Siegenthaler, 1999; Roos *et al.*, 1998). An important step in maintaining adequate retinol levels in the skin is retrieval of retinyl esters from their storage sites in the skin followed by hydrolysis to form retinol. It should be noted that the intracellular site(s) for storage of retinyl esters in the skin is not known. In some tissues, such as retinal pigment epithelium, where retinoids serve a unique physiological function, retinyl esters are stored in specialized lipid bodies known as retinosomes or retinyl ester-storage particles (Imanishi *et al.*, 2004). Like other lipid bodies, retinosomes appear to originate in the endoplasmic reticulum where LRAT is localized and catalyzes esterification of retinol (Imanishi *et al.*, 2004). Subsequently, retinosomes bud off of the endoplasmic reticulum and are ultimately found near the plasma membrane. It is also possible that retinyl esters are stored in cytoplasmic neutral lipid droplets rather than in specialized structures. Cultured fibroblasts, incubated with physiological levels of retinol, have been shown to accumulate retinyl palmitate in lipid droplets (Takagawa and Hirosawa, 1989). In addition, the presence of neutral lipid bodies in keratinocytes has been described (Corsini *et al.*, 2003; Grubauer *et al.*, 1989).

Intracellular retinyl esters can be mobilized from their storage sites by retinyl ester hydrolases. It is noteworthy that intracellular retinyl esters may be the preferred source of retinol used for subsequent formation of powerfully bioactive metabolites such as retinoic acid. Randolph and Simon (1993, 1996) demonstrated that mobilization of retinyl esters to form retinol, rather than use of preexisting retinol, is preferred by keratinocytes for formation of retinoic acid. Despite their clear importance, the enzymes responsible for hydrolyzing intracellular retinyl esters have not been identified definitively. Gao and Simon (2005) described a keratinocyte retinyl ester hydrolase that also catalyzed palmitoyl CoA-dependent and -independent retinol esterification. Hydrolysis was found to be greater at neutral pH and esterification greater at acidic pH. Gao and Simon (2005) observed that these properties are consistent with the increased retinyl ester content that accompanies epidermal maturation. In addition, the pH dependence for the acyl-CoA-dependent retinol esterification (ARAT) activity is similar to that described by Törmä and Vahlquist (1990). Additional studies are needed to determine the precise role that the hydrolase described by Gao and Simon (2005) plays in accessing cutaneous depots of retinyl esters to form retinol.

The oxidative metabolism of retinol in skin cells follows the general, two-step pathway presented in Fig. 1 and produces retinoic acid, the prominent active form of vitamin A (Fisher and Voorhees, 1996). The first step in the formation of retinoic acid is the reversible oxidation of retinol to retinaldehyde. In the skin, two distinct enzymes may be involved in the formation of retinaldehyde. Haselbeck *et al.* (1997) have described a cytoplasmic alcohol/retinol dehydrogenase that is abundantly expressed in the basal layer of the epidermis. In addition, a microsomal, short-chain retinol dehydrogenase has been identified in the epidermis (Markova *et al.*, 2003). This microsomal dehydrogenase is detected predominately in the basal and the most differentiated living layers of the epidermis. Both unbound retinol and holoCRBP can serve as substrates for this enzyme. While the relative importance of the cytosolic and microsomal retinol dehydrogenases is unclear, some evidence suggests that the microsomal enzyme is the more important source of retinaldehyde in the epidermis (Markova *et al.*, 2003). Retinol oxidation to retinaldehyde is considered the rate-limiting step in the two-step conversion of retinol to retinoic acid. Retinaldehyde formed from retinol does not typically accumulate in skin because the second step, oxidation of retinaldehyde to retinoic acid, occurs at rapid rate.

The final step in the formation of retinoic acid from retinol is the irreversible oxidation of retinaldehyde to retinoic acid. Much remains to be learned about the mechanism for the formation of retinoic acid; however, cytochrome P450 is thought to play a significant role (Roos et al., 1998; Zhang et al., 2000). Five human cytochrome P450 families including CYP1A1, CYP1A2, CYP1B1, CYP3A4, and CYP3A5 are found to be expressed in the skin (Zhang et al., 2000).

Tight regulation of retinoic acid activity in the skin is imperative for maintaining epithelial homeostasis. Randolph and Siegenthaler (1999) have outlined four major modes for controlling cellular levels of retinoic acid: (1) control of the cellular levels of retinol through regulation of retinol uptake, esterification, and storage as esters; (2) control of retinoic acid synthesis by regulation of relevant enzyme activities; (3) metabolic transformation retinoic acid to biologically inactive forms (e.g., formation of 4-oxo-retinoic acid through oxidation of retinoic acid catalyzed by CYP26); and (4) control of access of retinoic acid to the nucleus through the regulatory role of CRABP-I and CRABP-II.

V. EFFECTS ON SELECTED BIOLOGICAL RESPONSES OF THE SKIN

Most of the effects of vitamin A on the skin are thought to involve its most bioactive metabolite, retinoic acid (Roos et al., 1998; Shroot et al., 1999). Retinoic acid triggers cutaneous biological effects by binding to two families of nuclear receptors. These nuclear receptors are known to be ligand-activated transcription factors. The family of retinoid acid receptors (RAR), including RARα, RARβ, and RARγ, is activated specifically by all-*trans* retinoic acid, while the family of retinoid X receptors (RXRα, RARβ, and RARγ) is activated by all-*trans* retinoic acid as well as 9- and 13-*cis* retinoic acids. These families of nuclear receptors are readily detected in human skin and epidermal keratinocytes. Human epidermal keratinocytes express three isoforms of RAR: RARα, RARγ, and RXRα, abundantly but all RAR and RXR isoforms can be detected immunohistochemically in normal human skin (Reichrath et al., 1997). Billoni et al. (1997) detected consistent RARβ expression and variable expression of RARα and RARγ in the dermal papilla of the hair follicle with moderate expression of RARα, RARβ, and RARγ in the dermal sheath fibroblasts. Rosdahl et al. (1997) reported expression of CRBP-1, CRABP-1, CRABP-2, RARα, RARβ, RARγ, and RXRα in cultured human epidermal melanocytes, along with metabolism of ROH to 9-*cis*, 13-*cis*, and all-*trans* retinoic acid. Activation of these nuclear receptors by retinoid ligands is known to influence many of the biological responses of the skin. Three of these functions (the immune response, wound healing, and aging) will be briefly discussed below. In addition to receptor mediated effects, evidence is emerging that vitamin A can elicit nonligand effects. The role of vitamin A in mediating some effects of UV light on the skin is discussed below as an example of a nonligand effect of vitamin A.

A. IMMUNE RESPONSE

The salutary effect of vitamin A on the immune system has been long recognized (Ross, 2000; Semba, 1999). Indeed, one of the earliest biological functions of vitamin A, the ability to reduce infections, was noted by Green and Mellanby (1928) who dubbed vitamin A an "anti-infective agent." Subsequent studies have provided a vast amount of evidence indicating that vitamin A, through receptor-mediated signaling pathways, can broadly modulate the immune response (Ross, 2000). Retinoids have been shown to have effects on antigenic activation of T-lymphocytes, activation of monocytes, and B-cell differentiation (Ross, 2000).

Vitamin A and its metabolites have been shown to modulate aspects of the skin's immune response. Several investigators have shown that contact hypersensitivity to topically applied chemicals is enhanced by vitamin A. Maisey and Miller (1986) found that mice fed a diet supplemented with vitamin A showed an enhanced response to topically applied sensitizers. Enhancement of contact hypersensitivity by dietary vitamin A has also been reported by Thorne *et al.* (1991). Sailstad *et al.* (2000) showed that dietary vitamin A can restore the contact hypersensitivity response in mice whose responses have been suppressed by exposure to UV light. These vitamin A-induced effects on contact hypersensitivity indicate that vitamin A influences cutaneous immune responses mediated by T-lymphocytes. The mechanism underlying these effects is not understood. Contact hypersensitivity is a two-phase process. The induction phase is initiated by application of a low-molecular-weight antigen. This antigen is processed by Langerhans cells, dendritic accessory cells in the epidermis, for presentation to T-lymphocytes (Sailstad *et al.*, 2000). Keratinocytes secrete cytokines that facilitate this process. Presentation of processed antigen to T-lymphocytes in local lymph nodes results in antigen-specific expansion of T-lymphocytes. The second phase in the contact hypersensitivity response is elicitation. During the elicitation phase, reexposure of skin previously exposed to the antigen triggers T-lymphocyte-dependent cytokine release and ultimately an inflammatory response. Some evidence suggests that dietary vitamin A can increase numbers of accessory cells needed to process antigens (Katz *et al.*, 1987). In addition, vitamin A may affect release of stimulatory cytokines by keratinocytes (Shroot *et al.*, 1999). The production of these cytokines by keratinocytes is known to be influenced by transcription factors, such as AP-1, whose levels are regulated by retinoid-responsive genes (Schule et al., 1991). Several proinflammatory cytokines are known to be retinoid-responsive through a receptor mechanism (Ross, 2000). Presently, much is unknown about the mechanisms underlying the effects of vitamin A on the skin's responses to environmental pathogens and chemicals.

B. WOUND HEALING

Wound healing is a complex process that can be divided into three phases: (1) the inflammatory phase, (2) the proliferation phase, and (3) the remodeling phase (Yamaguchi and Yoshikawa, 2001). Progression through these phases requires coordinated involvement of blood vessels (exchange of platelets, macrophages, and neutrophils with tissue), epidermis (stimulation of keratinocytes, melanocytes, and Langerhans cells), dermis (activation of fibroblasts and myofibroblasts), nerves, and subcutaneous fatty layers (adipocytes). The effects of vitamin A on wound healing have long been recognized. Clinical and experimental studies have shown that vitamin A deficiency can impair wound healing (Hunt, 1986). In addition, treatment with vitamin A and other retinoids, through dietary supplementation or topical application, has been shown to enhance wound healing (Ehrlich and Hunt, 1968; Gerber and Erdman, 1982; Tom et al., 2005). While much about the biochemical and molecular mechanisms through which vitamin A enhances wound healing remains unknown, it is clear that events in all three phases of the wound healing process are affected by vitamin A and retinoids. Effects on the inflammatory and proliferation phases of wound healing have been observed and ostensibly involve regulation in the release of cytokines and growth factors in the skin (Muehlberger et al., 2005; Yuen and Stratford, 2004). Studies have shown that vitamin A and retinoic acid influence production of cytokines and facilitators of keratinocyte mobility, such plasminogen activator, as well as growth factors, such as transforming growth factor-$\beta 1$ and insulin-like growth factor-1 (Braungart et al., 2001; Wicke et al., 2000; Yuen and Stratford, 2004). Effects of vitamin A and retinoids on events associated with the remodeling phase of wound healing are well documented. Many of the events in remodeling at the site of a wound involve dermal fibroblasts and the synthesis of components in the extracellular matrix of the dermis (Pilcher et al., 1999; Yamaguchi and Yoshikawa, 2001). Retinoic acid has been shown to enhance production of extracellular matrix components such as collagen in cultured fibroblasts and during wound healing (Varani et al., 1993; Wicke et al., 2000).

C. AGING

The effects of vitamin A on aging skin are an area of great interest. Aging is associated with changes throughout the skin; however, many of the most dramatic effects due to aging occur in the dermis. An age-related reduction in the number of interstitial dermal fibroblasts has been observed in human skin (Lavker, 1979; Lovell et al., 1987). Reduced synthesis of collagen is observed in aged skin, especially after the seventh decade (Lovell et al., 1987; Oikarinen, 1994). In addition, enhanced degradation of dermal collagen and elastin is observed in aging skin (West, 1994). These changes in the dermis are thought to be largely responsible for the thin, fragile, and finely wrinkled properties of naturally aged skin (Varani et al., 2000). Varani et al. (2000) have shown that topical application of vitamin A can moderate many of these biochemical changes in aged skin. Four age groups: 18–29, 30–59, 60–79, and 80+ years were examined. Cellular and histological markers for age-related changes were measured, and included: collagen synthesis/collagen gene expression, dermal MMPs levels, and fibroblast growth. A progressive decrease in numbers of dermal fibroblasts was noted with increasing age as well as thinning and increased disorganization of collagen bundles in the dermis. In addition, increased levels of the matrix metalloproteinases, MMP-1, MMP-2, and MMP-9, were observed. Treatment with 1% retinol for 7 days resulted in a reduction in levels of MMP-1 and MMP-9, an increase in fibroblast growth, and increased expression of dermal collagen. Retinoids have also been shown to have similar effects on photoaged skin (Fisher et al., 1996). While the molecular events underlying these effects have yet to be fully understood, it appears that retinoid receptor-mediated repression of transcription factors such as AP-1 and inhibition of metalloproteases play a role in effects retinoid effects on both aged and photoaged skin (Fisher et al., 1999).

D. RESPONSE TO UV LIGHT

Discussions of vitamin A in photosensitive biological systems usually focus on the retina and the photochemical transformations involved in vision. However, interactions between light and vitamin A in the skin may also have important biological consequences. The skin contains several important chromophores for visible and UV light. Cutaneous chromophores include DNA, proteins, melanin, porphyrin, and urocanic acid (Kornhauser et al., 2004). Emerging evidence suggests that retinol and its esters, which have an absorbance maximum at ~325 nm, may also be important chromophores in the epidermis. Hoyos et al. (2002) have shown that retinol can influence the activity of proteins known to participate in the "mammalian UV response" (Devary et al., 1992; Karin and Gallagher, 2005). These proteins, which include protein kinases and cellular transcription factors, initiate an altered pattern of gene expression observed following exposure to UV light. Hoyos et al. (2002) showed that the signaling serine/threonine protein kinase, CRAF, has a site in its C1 zinc finger domain that binds retinol, 14-hydroxy-4,14-retro-retinol, retinoic acid, and anhydroretinol with equivalent affinities (K_d 22.0, 30.7, 19.8, and 36.4 nM, respectively).

Activation of CRAF by UV light was observed when retinol or 14-hydroxy-4,14-retro-retinol was bound to CRAF, while UV light caused inactivation of complexes between CRAF and retinoic acid or anhydroretinol. Activation of CRAF by UV light required the presence of RAS, another protein involved in the mammalian UV response. Therefore, binding of retinoids to CRAF may modulate the response of this important protein kinase. In addition, Hoyos *et al.* (2005) have demonstrated that retinol binding to CRAF may be involved in the redox-sensing function of this protein. Additional studies are needed to establish the role of retinol as a chromophore involved in initiating early signaling events in response to UV light, and the role of retinol in responding to oxidative stress in the skin.

The role of vitamin A as a sunscreen has also been suggested. Antille *et al.* (2003) have reported that human skin, treated once daily for 2 days with 2% retinyl palmitate, was protected against UV light-induced erythema. A sun protection factor (SPF) of ~ 20 was associated with application of 2% retinyl palmitate. In addition, application of retinyl palmitate protected against UV light-induced DNA damage. Subsequent studies showed that absorption of UV light, rather than effects on retinoid-responsive genes, gave rise to the observed protective effects (Sorg *et al.*, 2005).

The authors observed that while epidermal levels of retinyl esters resulting from topical treatment with 2% retinyl palmitate ($\sim 100~\mu M$) provide protection against sunburn, endogenous levels of retinyl esters in untreated skin ($\sim 1~\mu M$) may be too low to be protective. However, as noted earlier, the distribution of retinyl esters in the epidermis is not uniform, raising the possibility that locally high concentrations of retinyl esters in the outer epidermis may have some biologically significant sunscreen effects. Additional studies are needed to more fully investigate the biological significance of sun protection associated with endogenous and topically applied retinyl esters.

Recent *in vitro* studies indicate that retinyl palmitate can photosensitize damage to cells and biomolecules. Mei *et al.* (2005) studied the photomutagenicity of retinyl palmitate and UVA light (320–400 nm) in L5178Y/$Tk^{+/-}$ mouse lymphoma cells. Treatment of the cells with 1- to 25-μg/ml retinyl palmitate and UVA light (82.8 mJ/cm^2/min for 30 min) produced a dose-dependent induction of mutations. Examination of heterozygosity on chromosome 11, on which the *Tk* gene is located, provided evidence that retinyl palmitate was acting as a photoclastogen (Mei *et al.*, 2005). These authors interpret the photomutagenicity of retinyl palmitate as a consequence of oxidative damage to chromosomes photosensitized by retinyl palmitate. Consistent with this observation, Yan *et al.* (2005) have shown that treatment with retinyl palmitate, or with either of two photodecomposition products of retinyl palmitate, anhydroretinol or 5,6-epoxy-retinyl palmitate, made Jurkat T-cells more sensitive to toxicity elicited by UVA or visible light. In addition, treatment sensitized cellular DNA to damage measured by the Comet assay. Additional studies suggested photoexcitation of these retinoids caused DNA damage through a free radical mechanism (Cherng et al., 2005; Xia et al., 2006; Yan et al., 2005). Generation of free radicals and reactive oxygen species accompanying irradiation of retinol, retinyl palmitate, and their photodecomposition products has also been demonstrated by several authors (Cherng et al., 2005; Klamt et al., 2003). Figure 7 outlines some of the events that have been observed *in vitro* following photoexcitation of retinoids. It should be noted that the skin possesses substantial constitutive and inducible defenses against free radicals and oxidative damage (Vessey, 1993). In addition, as described earlier in this chapter, retinoids found in tissues are often separated from the cellular milieu through binding to specialized proteins. Therefore, the biological significance of free radicals and reactive oxygen species formed after photoexcitation of retinoids remains to be established. It is noteworthy that vitamin A is often viewed as an antioxidant vitamin (Livrea and Packer, 1994). However, the above *in vitro* studies suggest that under some circumstances, vitamin A may be a prooxidant.

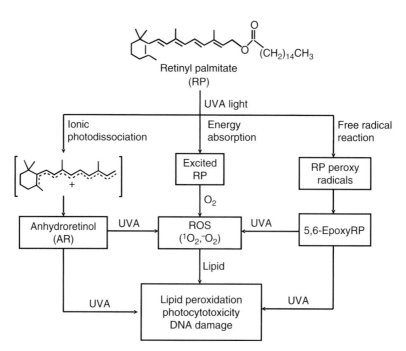

FIGURE 7. Events observed following exposure retinyl palmitate or its photodecomposition products to UVA light.

VI. SUMMARY

The skin has played a central role in the discovery of the biological activities of vitamin A. Though much is known about vitamin A homeostasis in the skin, and the effects of retinoids on the physiology of the skin, much remains to be learned. Most vitamin A-induced responses in the skin are thought to involve retinoic acid functioning as a ligand for cutaneous retinoid receptors. The network of cutaneous genes, whose expression is effected by retinoids, is only partially understood. Undoubtedly, our knowledge of the human genome and high throughput genomic and proteomic methods will accelerate the exploration of retinoid-influenced physiological pathways in the skin. In addition, there is emerging evidence that mechanisms independent from receptor binding may contribute to the activity of vitamin A in the skin. Because of these significant gaps in our knowledge of how vitamin A affects the skin, and because of the potential for development of dermatological and cosmetic products containing retinoids, it may be anticipated that the skin will remain a focus for future studies of vitamin A.

REFERENCES

Abdulmajed, K., and Heard, C. M. (2004). Topical delivery of retinyl ascorbate co-drug 1. Synthesis, penetration into and permeation across human skin. *Int. J. Pharm.* **280**, 113–124.

Anderson, R. R., and Parrish, J. A. (1981). The optics of skin. *J. Invest. Dermatol.* **77**, 13–19.

Antille, C., Tran, C., Sorg, O., Carraux, P., Didierjean, L., and Saurat, J. H. (2003). Vitamin A exerts a photoprotective action in skin by absorbing ultraviolet B radiation. *J. Invest. Dermatol.* **121**, 1163–1167.

Antille, C., Tran, C., Sorg, O., and Saurat, J.-H. (2004). Penetration and metabolism of topical retinoids in *ex vivo* organ-cultured full-thickness human skin explants. *Skin Pharmacol. Physiol.* **17**, 124–128.

Arnhold, T., Tzimas, G., Wittfoht, W., Plonait, S., and Nau, H. (1996). Identification of 9-*cis*-retinoic acid and 9,13-di-*cis*-retinoic acid, and 14-hydroxy-4,14-retro-retinol in human plasma after liver consumption. *Life Sci.* **59**, PL169–PL177.

Bailly, J., Crettaz, M., Schifflers, M. H., and Marty, J. P. (1998). *In vitro* metabolism by human skin and fibroblasts of retinol, retinaldehyde and retinoic acid. *Exp. Dermatol.* **7**, 27–34.

Båvik, C. O., Peterson, P. A., and Ericksson, U. (1995). Retinol-binding protein mediates uptake of retinol to cultured human keratinocytes. *Exp. Cell Res.* **216**, 358–362.

Berne, B., Nilsson, M., and Vahlquist, A. (1984). UV irradiation and cutaneous vitamin A: An experimental study in rabbit and human skin. *J. Invest. Dermatol.* **83**, 401–404.

Billoni, N., Gautier, B., Mahe, Y. F., and Bernard, B. A. (1997). Expression of retinoid nuclear receptor superfamily members in human hair follicles and its implication in hair growth. *Acta Derm. Venereol.* **77**, 350–355.

Bloch, C. E. (1921). Clinical investigation of xeropthalmia and dystrophy in infants and young children *(xerophthalmia et dystrophia alipogenetica)*. *J. Hyg.* **19**, 283–301.

Boehnlein, J., Sakr, A., Lichtin, J. L., and Bronaugh, R. L. (1994). Characterization of esterase and alcohol dehydrogenase activity in skin. Metabolism of retinyl palmitate to retinol (Vitamin A) during percutaneous absorption. *Pharm. Res.* **11**, 1155–1159.

Braungart, E., Magdolen, V., and Degritz, K. (2001). Retinoic acid upregulates the plasminogen activator system in human epidermal keratinocytes. *J. Invest. Dermatol.* **116**, 778–784.

Bronaugh, R. L., Collier, S. W., Macpherson, S. E., and Kraeling, M. E. (1994). Influence of metabolism in skin on dosimetry after topical application. *Environ. Health Perspect.* **102**(Suppl.), 71–74.

Busch, C., Siegenthaler, G., Vahlquist, A., Nordlinder, H., Sundelin, J., Saksena, P., and Eriksson, U. (1992). Expression of cellular retinoid-binding proteins during normal and abnormal epidermal differentiation. *J. Invest. Dermatol.* **99**, 795–802.

Cherng, S. H., Xia, Q., Blankenship, L. R., Freeman, J. P., Wamer, W. G., Howard, P. C., and Fu, P. P. (2005). Photodecomposition of retinyl palmitate in ethanol by UVA light-formation of photodecomposition products, reactive oxygen species, and lipid peroxides. *Chem. Res. Toxicol.* **18**, 129–138.

Connor, M. J., and Smit, M. H. (1987). The formation of all-*trans*-retinoic acid from all-*trans*-retinol in hairless mouse skin. *Biochem. Pharmacol.* **36**, 919–924.

Cosmetic Ingredient Review (1987). Final report on the safety assessment of retinyl palmitate and retinol. *J. Am. Coll. Toxicol.* **6**, 279–320.

Corsini, E., Viviani, B., Zancanella, O., Lucchi, L., Visioli, F., Serrero, G., Bartesaghi, S., Galli, C. L., and Marinovich, M. (2003). Induction of adipose differentiation related protein and neutral lipid droplet accumulation in keratinocytes by skin irritants. *J. Invest. Dermatol.* **121**, 337–344.

Devary, Y., Gottlieb, R. A., Smeal, T., and Karin, M. (1992). The mammalian ultraviolet response is triggered by activation of src tyrosine kinases. *Cell* **71**, 1081–1091.

Dillon, J., Gaillard, E. R., Bilski, P., Chignell, C. F., and Reszka, K. J. (1996). The photochemistry of retinoids as studied by steady-state and pulsed methods. *Photochem. Photobiol.* **63**, 680–685.

Duell, E. A., Derguini, F., Kang, S., Elder, J. T., and Voorhees, J. J. (1996). Extraction of human epidermis treated with retinol yields *retro*-retinoids in addition to free retinol and retinyl esters. *J. Invest. Dermatol.* **107**, 178–182.

Duell, E. A., Kang, S., and Voorhees, J. J. (1997). Unoccluded retinol penetrates human skin *in vivo* more effectively than unoccluded retinyl palmitate or retinoic acid. *J. Invest. Dermatol.* **109**, 301–305.

Ehrlich, H. P., and Hunt, T. K. (1968). Effects of cortisone and vitamin A on wound healing. *Ann. Surg.* **167**, 324–328.

Fazzi, E., Signorini, S. G., Uggetti, C., Bianchi, P. E., Lanners, J., and Lanzi, G. (2005). Towards improved clinical characterization of Leber congenital amaurosis: Neurological and systemic findings. *Am. J. Med. Genet. A* **132**, 13–19.

FDA (2004). The U.S. Food and Drug Administration's Voluntary Cosmetics Registration Program. www.fda.gov

Fisher, G. J., and Voorhees, J. J. (1996). Molecular mechanisms of retinoid actions in skin. *FASEB J.* **10**, 1002–1013.

Fisher, G. J., Datta, S. C., Talwar, H. S., Wang, Z.-Q., Varani, J., Kang, S., and Voorhees, J. J. (1996). Molecular basis of sun-induced premature skin ageing and retinoid antagonism. *Nature* **379**, 335–379.

Fisher, G. J., Talwar, H. S., Lin, J., and Voorhees, J. J. (1999). Molecular mechanisms of photoaging in human *in vivo* and their prevention by all-*trans* retinoic acid. *Photochem. Photobiol.* **69**, 154–157.

Forsum, U., Rask, L., Tjernlund, U. M., and Peterson, P. A. (1977). Detection of the vitamin A carrier plasma proteins in human epidermis. *Arch. Dermatol. Res.* **258**, 85–88.

Gao, J., and Simon, M. (2005). Identification of a novel keratinocyte retinyl ester hydrolase as a transacylase and lipase. *J. Invest. Dermatol.* **124**, 1259–1266.

Gerber, L. E., and Erdman, J. W., Jr. (1982). Effect of dietary retinyl acetate, beta-carotene and retinoic acid on wound healing in rats. *J. Nutr.* **112,** 1555–1564.

Gollapalli, D. R., Maiti, P., and Rando, R. R. (2003). RPE65 operates in the vertebrate visual cycle by stereospecifically binding all-trans-retinyl esters. *Biochemistry* **42,** 11824–11830.

Goodman, D. S., and Blaner, W. S. (1984). Biosynthesis, absorption, and hepatic metabolism of retinol. *In* "The Retinoids" (M. B. Sporn, A. B. Roberts, and D. S. Goodman, Eds.), Vol. 2, pp. 1–39. Academic Press, Orlando, FL.

Green, H. N., and Mellanby, E. (1928). Vitamin A as an anti-infective agent. *Br. Med.* **2,** 691–696.

Grubauer, G., Feingold, K. R., Harris, R. M., and Elias, P. M. (1989). Lipid content and lipid type as determinants of the epidermal permeability barrier. *J. Lipid Res.* **30,** 89–96.

Hartmann, S., Froescheis, O., Ringenbach, F., Wyss, R., Bucheli, F., Bischof, S., Bausch, J., and Wiegand, U. W. (2001). Determination of retinol and retinyl esters in human plasma by high-performance liquid chromatography with automated column switching and ultraviolet detection. *J. Chromatogr. B: Biomed. Sci. Appl.* **751,** 265–275.

Harrison, E. H. (2005). Mechanisms of digestion and absorption of dietary vitamin A. *In* "Annual Review of Nutrition" (R. J. Cousins, D. M. Bier, Assoc., and B. A. Bowman, Assoc., Eds.), Vol. 25, pp. 87–103. Annual Reviews, Palo Alto, CA.

Haselbeck, R. J., Ang, H. L., and Duester, G. (1997). Class IV alcohol/retinol dehydrogenase localization in epidermal basal layer: Potential site of retinoic acid synthesis during skin development. *Dev. Dyn.* **208,** 447–453.

Hinterhuber, G., Cauza, K., Brugger, K., Dingelmaier-Hovorka, R., Horvat, R., Wolff, K., and Foedinger, D. (2004). RPE65 of retinal pigment epithelium, a putative receptor molecule for plasma retinol-binding protein, is expressed in human keratinocytes. *J. Invest. Dermatol.* **122,** 406–413.

Hinterhuber, G., Cauza, K., Dingelmaier-Hovorka, R., Diem, E., Horvat, R., Wolff, K., and Foedinger, D. (2005). Expression of RPE65, a putative receptor for plasma retinol-binding protein, in nonmelanocytic skin tumours. *Br. J. Dermatol.* **153,** 785–789.

Hoath, S. B., and Leahy, D. G. (2003). The organization of human epidermis: Functional epidermal units and phi proportionality. *J. Invest. Dermatol.* **121,** 1440–1446.

Hodam, J. R., and Creek, K. E. (1998). Comparison of the metabolism of retinol delivered to human keratinocytes either bound to serum retinol-binding protein or added directly to the culture medium. *Exp. Cell Res.* **238,** 257–264.

Holick, M. F., MacLaughlin, J. A., Parrish, J. A., and Anderson, R. R. (1982). The photochemistry and photobiology of vitamin D_3. *In* "The Science of Photomedicine" (J. D. Regan and J. A. Parrish, Eds.), pp. 195–218. Plenum Press, New York.

Hoyos, B., Imam, A., Korichneva, I., Levi, E., Chua, R., and Hammerling, U. (2002). Activation of c-Raf kinase by ultraviolet light. Regulation by retinoids. *J. Biol. Chem.* **277,** 23949–23957.

Hoyos, B., Jiang, S., and Hammerling, U. (2005). Location and functional significance of retinol-binding sites on the serine/threonine kinase, c-Raf. *J. Biol. Chem.* **280,** 6872–6878.

Hunt, T. K. (1986). Vitamin A and wound healing. *J. Am. Acad. Dermatol.* **15,** 815–821.

Ihara, H., Hashizume, N., Hirase, N., and Suzue, R. (1999). Esterification makes retinol more labile to photolysis. *J. Nutr. Sci. Vitaminol. (Tokyo)* **45,** 353–358.

Imanishi, Y., Gerke, V., and Palczewski, K. (2004). Retinosomes: New insights into intracellular managing of hydrophobic substances in lipid bodies. *J. Cell Biol.* **166,** 447–453.

Kang, S., Duell, E. A., Fisher, G. J., Datta, S. C., Wang, Z. Q., Reddy, A. P., Tavakkol, A., Yi, J. Y., Griffiths, C. E. T., Elder, J. T., and Voorhees, J. J. (1995). Application of retinol to human skin *in vivo* induces epidermal hyperplasia and cellular retinoid binding proteins characteristic of retinoic acid but without measurable retinoic acid levels or irritation. *J. Invest. Dermatol.* **105,** 549–556.

Kang, S., Fisher, G. J., and Voorhees, J. J. (2000). Pharmacology and molecular mechanisms of retinoid action in skin. *In* "Vitamin A and Retinoids: An Update of Biological Aspects and Clinical Applications" (M. A. Livrea, Ed.), pp. 151–159. Birkhäuser Verlag, Basel.

Katz, D. R., Mukherjee, S., Maisey, J., and Miller, K. (1987). Vitamin A acetate as a regulator of accessory cell function in delayed-type hypersensitivity responses. *Int. Arch. Allergy Appl. Immunol.* **82,** 53–56.

Karin, M., and Gallagher, E. (2005). From JNK to pay dirt: Jun kinases, their biochemistry, physiology and clinical importance. *IUBMB Life* **57,** 283–295.

Kerr, J. B. (1999). "Atlas of Functional Histology." Mosby International, London.

Klamt, F., Dal-Pizzol, F., Bernard, E. A., and Moreira, J. C. F. (2003). Enhanced UV-mediated free radical generation; DNA and mitochondrial damage caused by retinol supplementation. *Photochem. Photobiol. Sci.* **2,** 856–860.

Kornhauser, A., Wamer, W. G., and Lambert, L. A. (2004). Light-induced dermal toxicity: Effects on the cellular and molecular levels. *In* "Dermatotoxicology" (H. Zhai and H. I. Maibach, Eds.), 6th ed., pp. 1105–1178. CRC Press, Boca Raton, FL.

Kurlandsky, S. B., Xiao, J.-H., Duell, E. A., Voorhees, J. J., and Fisher, G. J. (1994). Biological activity of all-trans retinol requires metabolic conversion to all-trans retinoic acid and is mediated through activation of nuclear retinoid receptors in human keratinocytes. *J. Biol. Chem.* **269,** 32821–32827.

Kurlandsky, S. B., Duell, E. A., Kang, S., Voorhees, J. J., and Fisher, G. J. (1996). Autoregulation of retinoic acid biosynthesis through regulation of retinol esterification in human keratinocytes. *J. Biol. Chem.* **271,** 15346–15352.

Lavker, R. M. (1979). Structural alterations in exposed and unexposed aged skin. *J. Invest. Dermatol.* **73,** 559–566.

Lemieux, S., Tontani, R., Uffelman, K. D., Lewis, G. F., and Steiner, G. (1998). Apolipoprotein B-48 and retinyl palmitate are not equivalent markers of postprandial intestinal lipoproteins. *J. Lipid Res.* **39,** 1964–1971.

Livrea, M. A., and Packer, L. (1994). Vitamin A as an antioxidant *in vitro* and *in vivo*. *In* "Retinoids: From Basic Science to Clinical Applications" (M. A. Livrea and L. Packer, Eds.), pp. 293–303. Birkhäuser Verlag, Basel.

Lovell, C. R., Smolenski, K. A., Duance, V. C., Light, N. D., Young, S., and Dysons, M. (1987). Type I and III collagen content in normal human skin during aging. *Br. J. Dermatol.* **117,** 419–428.

Madison, K. C. (2003). Barrier function of the skin: "La Raison d'Être" of the epidermis. *J. Invest. Dermatol.* **121,** 231–241.

Maisey, J., and Miller, K. (1986). Assessment of the ability of mice fed vitamin A supplemented diet to respond to a variety of potential contact sensitizers. *Contact Dermatitis* **15,** 17–23.

Markova, N. G., Pinkas-Sarafova, A., Karaman-Jurukovska, N., Jurukovski, V., and Simon, M. (2003). Expression pattern and biochemical characteristics of a major epidermal retinol dehydrogenase. *Mol. Genet. Metab.* **78,** 119–135.

Mei, N., Xia, Q., Chen, L., Moore, M. M., Fu, P. P., and Chen, T. (2005). Photomutagenicity of retinyl palmitate by ultraviolet A irradiation in mouse lymphoma cells. *Toxicol. Sci.* **88,** 142–149.

Moiseyev, G., Takahashi, Y., Chen, Y., Gentleman, S., Redmond, T. M., Crouch, R. K., and Ma, J. X. (2006). RPE65 is an iron(II)-dependent isomerohydrolase in the retinoid visual cycle. *J. Biol. Chem.* **281,** 2835–2840.

Montagna, W. (1954). Penetration and local effect of vitamin A on the skin of the guinea pig. *Proc. Soc. Exp. Biol. Med.* **86,** 668–672.

Monteiro-Riviere, N. A. (2004). Anatomical factors affecting barrier function. *In* "Dermatotoxicology" (H. Zhai and H. I. Maibach, Eds.), 6th ed., pp. 43–70. CRC Press, Boca Raton, FL.

Mori, S. (1922). The changes in para-ocular glands which follow the administration of diets low in fat-soluble A; with notes of effect of the same diets on the salivary glands and the mucosa of the larynx and trachea. *Bull. Johns Hopkins Hosp.* **33**, 357–358.

Mousseron-Canet, M. (1971). Photochemical transformation of vitamin A. *In* "Methods of Enzymology" (D. B. McCormick and L. D. Wright, Eds.), Vol. 19 (Part C), pp. 591–615. Academic Press, NY.

Muehlberger, T., Moresi, J. M., Schwarze, H., Hristopoulos, G., Laenger, F., and Wong, L. (2005). The effect of topical tretinoin on tissue strength and skin components in a murine incisional wound model. *J. Am. Acad. Dermatol.* **52**, 583–588.

Napoli, J. L. (1999). Interactions of retinoid binding proteins and enzymes in retinoid metabolism. *Biochim. Biophys. Acta* **1440**, 139–162.

Nohynek, G. J., Meuling, W. J. A., Vaes, W. H. J., Lawrence, R. S., Shapiro, S., Schulte, S., Steilung, W., Bausch, J., Gerber, E., Sasa, H., and Nau, H. (2005). Repeated topical treatment, in contrast to single oral doses, with Vitamin A-containing preparations does not affect plasma concentrations of retinol, retinyl esters or retinoic acids in female subjects of child-bearing age. *Toxicol. Lett.* **163**, 65–76.

O'Byrne, S. M., Wongsiriroj, N., Libien, J., Vogel, S., Goldberg, I. J., Baehr, W., Palczewski, K., and Blaner, W. S. (2005). Retinoid absorption and storage is impaired in mice lacking lecithin: Retinol acyltransferase (LRAT). *J. Biol. Chem.* **280**, 35647–35657.

Oikarinen, A. (1994). Aging of the skin connective tissue: How to measure the biochemical and mechanical properties of aging dermis. *Photodermatol. Photoimmunol. Photomed.* **10**, 47–52.

Orland, M. D., Anwar, K., Cromley, D., Chu, C.-H., Chen, L., Billheimer, J. T., Hussain, M. M., and Cheng, D. (2005). Acyl coenzyme A dependent retinol esterification by acyl coenzyme A:diacylglycerol acyltransferase 1. *Biochim. Biophys. Acta* **1737**, 76–82.

Paik, J., Vogel, S., Quadro, L., Piantedosi, R., Gottesman, M., Lai, K., Hamberger, L., de Morais Vieria, M., and Blaner, W. S. (2004). Vitamin A: Overlapping delivery pathways to tissues from the circulation. *J. Nutr.* **134**, 276S–280S.

Pilcher, B. K., Wang, M., Qin, X. J., Parks, W. C., Senior, R. M., and Welgus, H. G. (1999). Role of matrix metalloproteinases and their inhibition in cutaneous wound healing and allergic contact hypersensitivity. *Ann. NY Acad. Sci.* **878**, 12–24.

Randolph, R. K., and Siegenthaler, G. (1999). Vitamin A homeostasis in human epidermis: Native retinoid composition and metabolism. *In* "Retinoids. The Biochemical and Molecular Basis of Vitamin A and Retinoid Action" (H. Nau and W. S. Blaner, Eds.), pp. 491–520. Springer-Verlag, Berlin.

Randolph, R. K., and Simon, M. (1993). Characterization of retinol metabolism in cultured human epidermal keratinocytes. *J. Biol. Chem.* **268**, 198–205.

Randolph, R. K., and Simon, M. (1996). All-*trans*-retinoic acid regulates retinol and 3,4-didehydroretinol metabolism in cultured human epidermal keratinocytes. *J. Invest. Dermatol.* **106**, 168–175.

Reichrath, J., Mittmann, M., Kamradt, J., and Muller, S. M. (1997). Expression of retinoid-X receptors (-alpha, -beta, -gamma) and retinoic acid receptors (-alpha, -beta, -gamma) in normal human skin: An immunohistological evaluation. *Histochem. J.* **29**, 127–133.

Ries, G., and Hess, R. (1999). Retinol: Safety considerations for its use in cosmetic products. *J. Toxicol. Cutan. Ocular Toxicol.* **18**, 169–185.

Rollman, O., and Vahlquist, A. (1985). Vitamin A in skin and serum-studies of acne vulgaris, atopic dermatitis, ichthyosis vulgaris and lichen planus. *Br. J. Dermatol.* **113**, 405–413.

Rollman, O., Wood, E. J., Olsson, M. J., and Cunliffe, W. J. (1993). Biosynthesis of 3,4-didehydroretinol from retinol by human skin keratinocytes in culture. *Biochem. J.* **293**, 675–682.

Roos, T. C., Jugert, F. K., Merk, H. F., and Bickers, D. R. (1998). Retinoid metabolism in the skin. *Pharmacol. Rev.* **50**, 315–333.

Rosdahl, I., Andersson, E., Kagedal, B., and Törmä, H. (1997). Vitamin A metabolism and mRNA expression of retinoid-binding protein and receptor genes in human epidermal melanocytes and melanoma cells. *Melanoma Res.* **7,** 267–274.

Ross, A. C. (2000). Vitamin A, retinoids and immune responses. In "Vitamin A and Retinoids: An Update of Biological Aspects and Clinical Applications" (M. A. Livrea, Ed.), pp. 83–95. Birkhäuser Verlag, Basel.

Rothman, S. (1954). pH of sweat and skin surface. In "Physiology and Biochemistry of the Skin" (S. Rothman, Ed.), pp. 221–232. University of Chicago Press.

Ruiz, A., Winston, A., Lim, Y. H., Gilbert, B. A., Rando, R. R., and Bok, D. (1999). Molecular and biochemical characterization of lecithin retinol acyltransferase. *J. Biol. Chem.* **274,** 3834–3841.

Sailstad, D. M., Boykin, E. H., Slade, R., Doerfler, D. L., and Selgrade, M. K. (2000). The effect of a vitamin A acetate diet on ultraviolet radiation-induced immune suppression as measured by contact hypersensitivity in mice. *Photochem. Photobiol.* **72,** 766–771.

Saurat, J.-H., Sorg, O., and Didierjean, L. (1999). New concepts for delivery of topical retinoid activity to human skin. In "Retinoids. The Biochemical and Molecular Basis of Vitamin A and Retinoid Action" (H. Nau and W. S. Blaner, Eds.), pp. 521–538. Springer-Verlag, Berlin.

Schaefer, H. (1993). Penetration and percutaneous absorption of topical retinoids. A review. *Skin Pharmacol.* **6**(Suppl. 1), 17–23.

Schule, R., Rangarajan, P., Yang, N., Kliewer, S., Ransone, L. J., Bolando Verma, I. M., and Evans, R. M. (1991). Retinoic acid is a negative regulator of AP-1 responsive genes. *Proc. Natl. Acad. Sci. USA* **88,** 6092–6096.

Sekula-Gibbs, S., Uptmore, D., and Otillar, L. (2004). Retinoids. *J. Am. Acad. Dermatol.* **50,** 405–415.

Semba, R. D. (1999). Vitamin A as "anti-infective" therapy, 1920–1940. *J. Nutr.* **129,** 783–791.

Shroot, B., Gibson, D. F. C., and Lu, X.-P. (1999). Retinoid receptor-selective agonists and their action in skin. In "Retinoids. The Biochemical and Molecular Basis of Vitamin A and Retinoid Action" (H. Nau and W. S. Blaner, Eds.), pp. 539–559. Springer-Verlag, Berlin.

Siegenthaler, G. (1986). Cellular retinol-binding protein (CRBP) is measurable in human dermis. *J. Invest. Dermatol.* **87,** 295–296.

Simmons, D. P., Andreola, F., and De Luca, L. M. (2002). Human melanomas of fibroblast and epithelial morphology differ widely in their ability to synthesize retinyl esters. *Carcinogenesis* **23,** 1821–1830.

Smeland, S., Bjerknes, T., Malaba, L., Eskild, W., Norum, K. R., and Blomhoff, R. (1995). Tissue distribution of the receptor for plasma retinol-binding protein. *Biochem. J.* **305,** 419–424.

Sobel, A. E., Sherman, B. S., Bradley, D. K., and Parnell, J. P. (1959). The influence of concentration on the percutaneous and oral absorption of vitamin A. *J. Invest. Dermatol.* **32,** 569–576.

Sorg, O., Tran, C., Carraux, P., Didierjean, L., and Saurat, J.-H. (1999). Retinol and retinyl ester epidermal pools are not identically sensitive to UVB irradiation and anti-oxidant protective effect. *Clin. Lab. Invest.* **199,** 302–307.

Sorg, O., Tran, C., Carrau, P., Didierjean, L., Falson, F., and Saurat, J.-H. (2002). Oxidative stress-independent depletion of epidermal vitamin A by UVA. *J. Invest. Dermatol.* **118,** 513–518.

Sorg, O., Tran, C., Carraux, P., Grand, D., Hugin, A., Didierjean, L., and Saurat, J. H. (2005). Spectral properties of topical retinoids prevent DNA damage and apoptosis after acute UV-B exposure in hairless mice. *Photochem. Photobiol.* **81,** 830–836.

Sorrell, J. M., and Caplan, A. I. (2004). Fibroblast heterogeneity: More than skin deep. *J. Cell Sci.* **117,** 667–675.

Steinstrasser, I., and Merkle, H. P. (1995). Dermal metabolism of topically applied drugs: Pathways and models reconsidered. *Pharm. Acta Helv.* **70,** 2–24.
Streilein, J. W. (1983). Skin-associated lymphoid tissues (SALT): Origins and functions. *J. Invest. Dermatol.* **80**(Suppl.), 12s–16s.
Takagawa, K., and Hirosawa, K. (1989). Uptake of retinol by cultured fibroblasts. *Cell Struct. Funct.* **14,** 353–362.
Tang, G., Webb, A. R., Russell, R. M., and Holick, M. F. (1994). Epidermis and serum protect retinol but not retinyl esters from sunlight-induced photodegradation. *Photodermatol. Photoimmunol. Photomed.* **10,** 1–7.
Tom, W. L., Peng, D. H., Allaei, A., Hsu, D., and Hata, T. R. (2005). The effect of short-contact topical tretinoin therapy for foot ulcers in patients with diabetes. *Arch. Dermatol.* **141,** 1373–1377.
Thompson, D. A., Gyurus, P., Fleischer, L. L., Bingham, E. L., McHenry, C. L., Apfelstedt-Sylla, E., Zrenner, E., Lorenz, B., Richards, J. E., Jacobson, S. G., Sieving, P. A., and Gal, A. (2000). Genetics and phenotypes of RPE65 mutations in inherited retinal degeneration. *Invest. Ophthalmol. Vis. Sci.* **41,** 4293–4299.
Thorne, P. S., Hawk, C., Kaliszewski, S. D., and Guiney, P. D. (1991). The noninvasive mouse ear swelling assay. I. Refinement for detecting weak contact sensitizer. *Fundam. Appl. Toxicol.* **17,** 790–806.
Törmä, H., and Vahlquist, A. (1983). Vitamin A transporting proteins in human epidermis and blister fluids. *Arch. Dermatol. Res.* **275,** 324–328.
Törmä, H., and Vahlquist, A. (1990). Vitamin A esterification in human epidermis: A relationship to keratinocyte differentiation. *J. Invest. Dermatol.* **94,** 132–138.
Törmä, H., Brunnberg, L., and Vahlquist, A. (1987). Age-related variations in acyl-CoA:retinol acyltransferase activity and vitamin A concentrations in the liver and epidermis of hairless mice. *Biochim. Biophys. Acta* **921,** 254–258.
Tran, C., Sorg, O., Carraux, P., Didierjean, L., and Saurat, J. H. (2001). Topical delivery of retinoids counteracts the UVB-induced epidermal vitamin A depletion in hairless mouse. *Photochem. Photobiol.* **73,** 425–431.
Turkish, A. R., Henneberry, A. L., Cromley, D., Padamsee, M., Oelkers, P., Bazzi, H., Christiano, A. M., Billheimer, J. T., and Sturley, S. L. (2005). Identification of two novel human acyl-CoA wax alcohol acyltransferases: Members of the diacylglycerol acyltransferase 2 (DGAT2) gene superfamily. *J. Biol. Chem.* **280,** 14755–14764.
Vahlquist, A. (1982). Vitamin A in human skin: Detection and identification of retinoids in normal epidermis. *J. Invest. Dermatol.* **79,** 89–93.
Vahlquist, A., Lee, J. B., Michaëlsson, G., and Rollman, O. (1982). Vitamin A in human skin: II Concentrations of carotene, retinol and dehydroretinol in various components of normal skin. *J. Invest. Dermatol.* **79,** 94–97.
Varani, J., Larson, B. K., Perone, P., Inman, D. R., Fligiel, S. E. G., and Voorhees, J. J. (1993). All-*trans*-retinoic acid and extracellular Ca^{2+} differentially influence extracellular matrix production by human skin in organ culture. *Am. J. Pathol.* **142,** 1813–1822.
Varani, J., Warner, R. L., Gharaee-Kermani, M., Phan, S. H., Kang, S., Chung, J.-H., Wang, Z.-Q., Datta, S. C., Fisher, G. J., and Voorhees, J. J. (2000). Vitamin A antagonizes decreased cell growth and elevated collagen-degrading matrix metalloproteinases and stimulates collagen accumulation in naturally aged human skin. *J. Invest. Dermatol.* **114**(3), 480–486.
Veske, A., Nilsson, S. E., Narfstrom, K., and Gal, A. (1999). Retinal dystrophy of Swedish briard/briard-beagle dogs is due to a 4-bp deletion in RPE65. *Genomics* **57,** 57–61.
Vessey, D. A. (1993). The cutaneous antioxidant system. *In* "Oxidative Stress in Dermatology" (J. Fuchs and L. Packer, Eds.), pp. 81–103. Marcel Dekker, Inc., NY.
Vogel, S., Gamble, M. V., and Blaner, W. S. (1999). Retinoid uptake, metabolism and transport. *In* "The Handbook of Experimental Pharmacology, The Retinoids" (H. Nau and W. S. Blaner, Eds.), pp. 31–96. Springer Verlag, Heidelberg.

West, M. D. (1994). The cellular and molecular biology of skin aging. *Arch. Dermatol.* **130,** 87–95.

Wicke, C., Halliday, B., Allen, D., Roche, N. S., Scheuenstuhl, H., Spencer, M. M., Roberts, A. B., and Hunt, T. K. (2000). Effects of steroids and retinoids on wound healing. *Arch. Surg.* **135,** 1265–1270.

Wiegand, U. W., Hartmann, S., and Hummler, H. (1998). Safety of vitamin A: Recent results. *Int. J. Vitam. Nutr. Res.* **68,** 411–416.

Wlaschek, M., Tantcheva-Poór, I., Naderi, L., Ma, W., Schneider, L. A., Razi-Wolf, Z., Schüller, J., and Scharffetter-Kochanek, K. (2001). Solar UV irradiation and dermal photoaging. *J. Photochem. Photobiol. B Biol.* **63,** 41–51.

Wolbach, S. B., and Howe, P. R. (1925). Tissue changes following deprivation of fat-soluble A vitamin. *J. Exp. Med.* **43,** 753–777.

Xia, Q., Yin, J. J., Cherng, S. H., Wamer, W. G., Boudreau, M., Howard, P. C., and Fu, P. P. (2006). UVA photoirradiation of retinyl palmitate—formation of singlet oxygen and superoxide, and their role in induction of lipid peroxidation. *Toxicol. Lett.* **163,** 30–43.

Yamaguchi, Y., and Yoshikawa, K. (2001). Cutaneous wound healing: An update. *J. Dermatol.* **28,** 521–534.

Yan, J., Xia, Q., Cherng, S. H., Wamer, W. G., Howard, P. C., Yu, H., and Fu, P. P. (2005). Photo-induced DNA damage and photocytotoxicity of retinyl palmitate and its photodecomposition products. *Toxicol. Ind. Health* **21,** 167–175.

Yan, J., Wamer, W. G., Howard, P. C., Boudreau, M., and Fu, P. P. (2006a). Levels of retinyl palmitate and retinol in stratum corneum, epidermis, and dermis of female SKH-1 mice topically treated with retinyl palmitate. *Toxicol. Ind. Health* **22,** 181–191.

Yan, J., Xia, Q., Webb, P., Warbritton, A. R., Wamer, W. G., Howard, P. C., Boudreau, M., and Fu, P. P. (2006b). Levels of retinyl palmitate and retinol in stratum corneum, epidermis, and dermis of SKH-1 mice. *Toxicol. Ind. Health* **22,** 103–112.

Yano, S., Oda, K., Watanabe, Y., Watanabe, S., Matsuishi, T., Kojima, K., Abe, T., and Kato, H. (1998). Two sib cases of Leber congenital amaurosis with cerebellar vermis hypoplasia and multiple systemic anomalies. *Am. J. Med. Genet.* **78,** 429–432.

Yen, C. L., Monetti, M., Burri, B. J., and Farese, R. V., Jr. (2005). The triacylglycerol synthesis enzyme DGAT1 also catalyzes the synthesis of diacylglycerols, waxes, and retinyl esters. *J. Lipid. Res.* **46,** 1502–1511.

Yuen, D. E., and Stratford, A. F. (2004). Vitamin A activation of transforming growth factor-β_1 enhances porcine ileum wound healing *in vitro*. *Pediatr. Res.* **55,** 935–939.

Zhang, Q.-Y., Dunbar, D., and Kaminsky, L. (2000). Human cytochrome P-450 metabolism of retinals to retinoic acid. *Drug Metab. Dispos.* **28,** 292–297.

Zanotti, G., and Berni, R. (2004). Plasma retinol-binding protein: Structure and interactions with retinol, retinoids, and transthyretin. *Vitam. Horm.* **69,** 271–295.

10

Retinoic Acid and the Heart

Jing Pan and Kenneth M. Baker

Division of Molecular Cardiology, The Texas A&M University System Health Science Center, Cardiovascular Research Institute, College of Medicine Central Texas Veterans Health Care System, Temple, Texas 76504

I. Introduction
II. Role of RA in Heart Development and Congenital Heart Defects
   A. Normal Heart Development
   B. RA Signaling
   C. RA and Heart Development
III. Postnatal Development Effects of RA in the Heart
   A. Antihypertrophic Effects of RA in Neonatal Cardiomyocytes
   B. Effects of RA Signaling in Cardiac Remodeling
   C. Role of RA Signaling in the Regulation of the Renin–Angiotensin System
IV. Conclusions
   References

Retinoic acid (RA), the active derivative of vitamin A, by acting through retinoid receptors, is involved in signal transduction pathways regulating embryonic development, tissue homeostasis, and cellular differentiation and proliferation. RA is important for the development of the heart.

The requirement of RA during early cardiovascular morphogenesis has been studied in targeted gene deletion of retinoic acid receptors and in the vitamin A-deficient avian embryo. The teratogenic effects of high doses of RA on cardiovascular morphogenesis have also been demonstrated in different animal models. Specific cardiovascular targets of retinoid action include effects on the specification of cardiovascular tissues during early development, anteroposterior patterning of the early heart, left/right decisions and cardiac situs, endocardial cushion formation, and in particular, the neural crest. In the postdevelopment period, RA has antigrowth activity in fully differentiated neonatal cardiomyocytes and cardiac fibroblasts. Recent studies have shown that RA has an important role in the cardiac remodeling process in rats with hypertension and following myocardial infarction. This chapter will focus on the role of RA in regulating cardiomyocyte growth and differentiation during embryonic and the postdevelopment period. © 2007 Elsevier Inc.

I. INTRODUCTION

Retinoic acid (RA) is the bioactive metabolite of vitamin A, and there is considerable evidence implicating RA as an important signaling molecule during embryonic and fetal cardiovascular development (Lammer *et al.*, 1985; Moore, 1965; Wolf, 1984). RA signaling is important for both cardiac development and for differentiation into adult cardiac muscle cells (Kastner *et al.*, 1997b; Subbarayan *et al.*, 2000). Changes in RA homeostasis result in severe malformations during cardiogenesis. Both, a lack or an excess of RA during embryonic development results in congenital cardiovascular malformations. The retinoid dependence of cardiogenesis was first demonstrated in vitamin A-deficient (VAD) rats (Wilson and Warkany, 1950a,b). Similar to the cardiovascular abnormalities caused by vitamin A insufficiency, retinoic acid receptor (RARs) and retinoic X receptor (RXRs) knockouts in mice lead to abnormal cardiogenesis (Chen *et al.*, 1998; Ghyselinck *et al.*, 1998; Gruber *et al.*, 1996; Kastner *et al.*, 1994, 1997b; Mendelsohn *et al.*, 1994a; Sucov *et al.*, 1994). These observations demonstrate that RA is required in normal embryonic heart development. An excess of RA has also been found to cause teratogenic effects during early heart development (Colbert *et al.*, 1997; Dickman and Smith, 1996; Drysdale *et al.*, 1997; Osmond *et al.*, 1991), indicating that a precisely regulated supply of RA is essential for normal cardiogenesis. Studies have shown that RA-dependent signal transduction appears to preserve the normal differentiated phenotype of cardiomyocytes, by antagonizing the effect of various hypertrophic stimuli in the postdevelopment period (Palm-Leis *et al.*, 2004; Wang *et al.*, 2002; Wu *et al.*, 1996; Zhou *et al.*, 1995). RA signaling also has an important role in cardiac remodeling during the development of hypertension and following myocardial infarction (de Paiva *et al.*, 2003; Lu *et al.*, 2003; Paiva *et al.*, 2005). Thus, gaining an expanded understanding of the signaling pathways mediated by RA may lead to the development of novel agents for the prevention and treatment of cardiovascular diseases.

II. ROLE OF RA IN HEART DEVELOPMENT AND CONGENITAL HEART DEFECTS

A. NORMAL HEART DEVELOPMENT

The vertebrate heart is the first organ to form and function during embryonic development. Heart morphogenesis can generally be broken down into the following steps: cardiomyocyte determination and specification; formation of the heart tube; looping, chamber development, and growth; endocardial cushion, valve, and septal formation (Fig. 1). Heart precursor cells or cardiac progenitors are among the first mesodermal cells to gastrulate through the

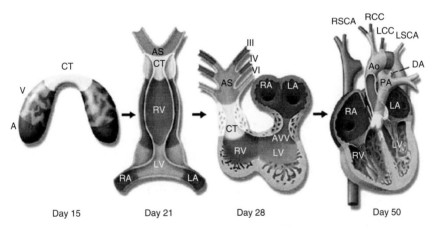

FIGURE 1. Steps of heart morphogenesis. The illustrations depict cardiac development with color coding of morphologically related regions, seen from a ventral view. Cardiogenic precursors form a crescent (left-most panel) that forms specific segments of the linear heart tube, which is patterned along the anterior–posterior axis to form the various regions and chambers of the looped and mature heart. Each cardiac chamber balloons out from the outer curvature of the looped heart tube in a segmental fashion. Neural crest cells populate the bilaterally symmetrical aortic arch arteries (III, IV, and VI) and aortic sac (AS) that together contribute to specific segments of the mature aortic arch. Mesenchymal cells form the cardiac valves from the conotruncal (CT) and atrioventricular valve (AVV) segments. Corresponding days of human embryonic development are indicated. A, atrium; Ao, aorta; DA, ductus arteriosus; LA, left atrium; LCC, left common carotid; LSCA, left subclavian artery; LV, left ventricle; PA, pulmonary artery; RA, right atrium; RCC, right common carotid; RSCA, right subclavian artery; RV, right ventricle; and V, ventricle. (See Color Insert.)

primitive streak (Garcia-Martinez and Schoenwolf, 1993; Icardo, 1996; Schoenwolf and Garcia-Martinez, 1995). Beginning soon after gastrulation, anterolateral endoderm and anterocentral mesoderm induce cells of the posterior primitive streak to differentiate into cardiac myocytes, indicating that anterior endoderm contains signaling molecules that can induce cardiac myocyte specification of early primitive streak cells (Schultheiss et al., 1995). These cells migrate to an anterolateral position known as the cardiogenic area or heart field. Heart progenitor cells are recognized in these areas based on expression of several cardiac-specific genes, including Nkx-2.5, GATA4, MEF-2, and Tbx5 (Fishman and Chien, 1997; Kostetskii et al., 1999; Lyons, 1996; Schultheiss et al., 1995; Searcy et al., 1998). Anterior segments of the heart-forming region will become the ventricles, and posterior segments will become the atria of the four-chambered heart (Olson and Srivastava, 1996; Yutzey and Bader, 1995). Soon after specification, cardiac muscle cells converge along the ventral midline of the embryo to fuse into a beating linear heart tube, patterned along the anteroposterior axis to make up the conotruncus (outflow tract), right (pulmonary) and left (systemic) ventricles, and atria. As fusion continues, the heart tube bulges out and begins to loop to the right (Fig. 1). At the time of looping, the single heart tube is composed of distinct myocardial and endocardial layers separated by an extracellular matrix known as the cardiac jelly. The endocardium is critical to the generation of the heart valves. The cardiac cushions are the primordia of the valves and membranous septa (Eisenberg and Markwald, 1995). As looping occurs, cushion tissues that have formed in the atrioventricular canal and the outflow track regions are brought together at the inner curvature of the looped heart tube. This repositioning facilitates septation by bringing atrioventricular and outflow track cushion tissues together for formation of the valves, atrioventricular septum, and the outflow track septum. The atrioventricular septum divides the inflow tract, known as the common atrioventricular canal, into a right and left atrioventricular canal. As the atrioventricular septum shifts to the right to lie above the ventricular septum, the atrioventricular canals follow and become aligned over the respective ventricles. Cardiac septation and heart chamber formation take place at three levels: the atrium, the ventricle, and the arterial pole. Correct looping and the extracardiac contribution of neural crest cells have an essential role in normal septation. Cardiac chamber formation can be separated into two phases: the initial specification of separate preatrial and preventricular lineages and the subsequent differentiation events that produce distinct atrial and ventricular chambers (Fig. 1).

B. RA SIGNALING

RA is the most biologically active member of the family of retinoids, all of which are derived from vitamin A. Studies have demonstrated that the pleiotropic effects of vitamin A (except for its role in vision) are mediated by RA and possibly some of its metabolites via specific nuclear retinoid receptors. Cells take up RA from the blood, where it circulates as retinol bound to retinol-binding protein (Fig. 2). Inside the cell, the sequestered retinol is enzymatically converted to retinal by the retinol or alcohol dehydrogenases, and retinal is further converted to RA by the retinaldehyde dehydrogenases (RALDHs) (Duester, 2000). RA is further metabolized to inactive products such as 4-oxo-RA, 4-OH-RA, 18-OH-RA, and 5,8-epoxy-RA by two cytochrome P450 enzymes, CYP26A1 and CYP26B1 (Fujii *et al.*, 1997; White *et al.*, 1996) (Fig. 2). Retinoid receptors comprise two subfamilies composed of three RA receptors (RARα, RARβ, and RARγ) and three retinoid X receptors (RXRα, RXRβ, and RXRγ) (Chambon, 1996;

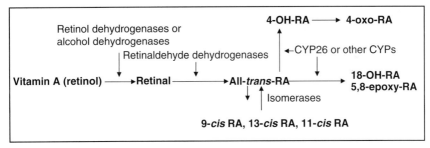

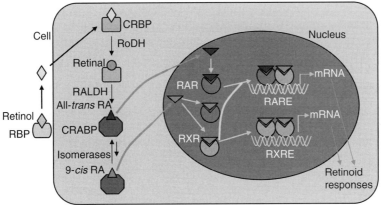

FIGURE 2. The metabolic pathway and the cellular mechanism of RA action. The metabolic pathway of RA. Retinol is taken up from the blood, where it binds to retinol-binding proteins (RBPs) and rebinds intracellularly to cellular retinol-binding proteins (CRBPs). Intracellularly, retinol is converted to retinal by the retinol or alcohol dehydrogenases (RoDH), and retinal is further metabolized to RA by RALDHs. RA is bound in the cytoplasm by cellular RA-binding protein (CRABP). Alternatively, RA and its 9-*cis* isomer enter the nucleus and bind to RARs or RXRs, respectively. On dimerization of these receptors, RAR/RXR heterodimer or RXR/RXR homodimer, the activated receptors bind to a sequence of DNA known as the RA-response element (RARE). This latter action either activates or represses transcription of the target gene. (See Color Insert.)

Mangelsdorf et al., 1995; Petkovich, 1992). Retinoid receptors were cloned in the late 1980s and early 1990s (Benbrook et al., 1988; Brand et al., 1988; Hamada et al., 1989; Leid et al., 1992; Mangelsdorf et al., 1992; Ragsdale et al., 1989). RARs are activated by all-*trans* RA and its 9-*cis* isomer, while RXRs are only activated by 9-*cis* RA. These receptors act as ligand-dependent transcription factors. The RARs and RXRs function as heterodimers (Kastner et al., 1997b; Mangelsdorf and Evans, 1995). RAR–RXR heterodimers bind to specific genomic DNA sequences designated as RAREs, which are characterized by two half sites with the consensus sequence AGGTCA. These are generally arranged as direct repeats (DR) separated by two or five nucleotides. These heterodimers have two distinct functions: first, they modulate the frequency of transcription initiation of target genes after binding to RAREs in the promoters; and second, they affect the efficiency of other signaling pathways (cross talk) by unknown mechanisms. This cross talk suggests that retinoid receptors are also targets of other pathways (Chen et al., 1995; Zechel et al., 1994). Retinoids do not act solely through the two subunits of the RAR–RXR heterodimer. RXR is a promiscuous heterodimerization partner for various nuclear receptors such as the thyroid hormone receptors, vitamin D receptors, peroxisomal proliferator-activated receptor, and several other orphan receptors (Kliewer et al., 1992a,b; Schrader et al., 1993). Therefore, RXR ligands have the potential to affect the signaling of many pathways. The functions of RARs and RXRs are not limited to a direct transactivation process, as they also include transrepressive activity. Transcriptional interference may result when the receptor, bound to ligand, prevents other transcription factors, such as AP-1, from interacting with the transcription initiation complex (Yang-Yen et al., 1991; Zhou et al., 1999). This transrepressive function is likely responsible for a large part of the biological effects of retinoids. The nongenomic effects of RA have been demonstrated in recent studies. RA acts by promoting the activation of cytoplasmic-signaling cascades that control the activity of specific genes. It has been shown that in acute promyelocytic leukemia cells, the RA-induced differentiation process is associated with a rapid increase in the level of intracellular cAMP as well as protein kinase A activity. In contrast, no such change was observed in RA-resistant cells (Zhao et al., 2004). In SH-SY5Y neuroblastoma cells and NIH3T3 cells, RA treatment induces increased phosphoinositide 3-kinase (PI3K) activity and a rapid increase in phosphorylation of AKT in Ser-473 and extracellular-regulated kinase (ERK). RA-induced differentiation was impaired by inhibition of PI3K, indicating that RA, by activating the PI3K/AKT-signaling pathway, has an important role in the regulation of neuronal cell survival (Antonyak et al., 2002; Lopez-Carballo et al., 2002). In this respect, RA behaved much like estrogens, which have been reported to induce rapid activation of signaling pathways, such as ERK and AKT (Christ et al., 1999; Gerdes et al., 2000). With many interactions and numerous effects in target cells, these findings suggest that RA is a critical regulator of embryonic development and cellular and tissue homeostasis.

C. RA AND HEART DEVELOPMENT

The significance of RA for embryonic and fetal heart development has been established in numerous studies (Clagett-Dame and DeLuca, 2002; Wilson and Warkany, 1950a; Wilson *et al.*, 1953; Zile, 2004). Embryonic exposure to either an excess or a deficiency of vitamin A leads to abnormal development, suggesting that the embryo requires a precisely regulated amount of RA. Evidence gathered from the study of retinoid-induced embryopathy and of RAR gene knockouts and mutations suggests multiple roles for RA in embryonic heart development.

1. Retinoic Acid-Mediated Signaling Is Required for Normal Heart Development

RA is critical in the development of the heart. It was recognized in the 1930s that maternal insufficiency of vitamin A, during pregnancy, results in fetal death and severe congenital malformations in the offspring, including aberrant heart development (Hale, 1937; Mason, 1935). The retinoid dependence of cardiogenesis was first shown in VAD rats, which displayed specific aortic arch and ventricular septal deficits (Wilson and Warkany, 1950a,b; Wilson *et al.*, 1953). RA function during embryonic and fetal heart development at the molecular level has been extensively studied, including the use of transgenic mice with mutations in retinoid receptor genes, or retinoid-ligand knockout models (Ghyselinck *et al.*, 1998; Kastner *et al.*, 1994; Luo *et al.*, 1996; Mic *et al.*, 2002; Niederreither *et al.*, 2001; Smith *et al.*, 1998). Using cultured embryonic stem cells or cardiomyocytes, it has been demonstrated that RA accelerates expression of cardiac-specific genes, enhances the development of ventricular cardiomyocytes, and promotes cardiomyocyte differentiation, indicating that RA-mediated signaling has an important role in embryonic cardiomyocyte proliferation and differentiation (Aranega *et al.*, 1999; Hidaka *et al.*, 2003; Honda *et al.*, 2005; Wobus *et al.*, 1997). Specific functions of the different retinoid receptors during heart embryogenesis have been identified over the past decade. Although the different receptor isoforms have unique distributions during development, knocking out one specific isoform of the RAR family does not cause developmental defects analogous to those observed in the fetal vitamin A deficiency syndrome, demonstrating a redundancy between members of each receptor subtype. Mice deficient for individual RARα1, RARβ2, and RARγ2 isoforms or all RARβ isoforms appear normal (Li *et al.*, 1993; Lohnes *et al.*, 1993; Lufkin *et al.*, 1993; Luo *et al.*, 1995; Mendelsohn *et al.*, 1994b). Although RARα and RARγ gene null mutants display early postnatal lethality or growth deficiency, the heart development is normal (Lohnes *et al.*, 1993; Lufkin *et al.*, 1993).

In contrast with RAR single mutants, compound null mutations of RAR genes lead to significant heart malformations (Lohnes et al., 1994; Mendelsohn et al., 1994a). A marked defect observed in the heart and outflow tract of double mutant fetuses was persistent truncus arteriosus. The truncus arteriosus is a transient structure, which by 14.5 dpc should be completely divided into the ascending aorta and pulmonary trunk, by the spiral-shaped aorticopulmonary septum (Fananapazir and Kaufman, 1988; Vuillemin and Pexieder, 1989). Failure of this division to take place or to become complete results in a persistent truncus arteriosus, receiving blood from both ventricles. In addition, ventricular septal defects, myocardial deficiency, persistent atrioventricular canal, and an abnormal aortic arch pattern were found in RAR double mutants (Lee et al., 1997; Luo et al., 1996; Mendelsohn et al., 1994a). All of these abnormalities have been described in the offspring from VAD rats (Wilson and Warkany, 1950a; Wilson et al., 1953) (Table I). These results demonstrate that RAR-mediated retinoid signaling is essential for proper myocardial growth, aorticopulmonary and ventricular septation, and patterning of the aortic arches.

RXRs exert multiple functions in several signaling systems (Mangelsdorf and Evans, 1995), and RAR/RXR heterodimers bind more efficiently to RAREs than homodimers of either RAR or RXR. Furthermore, RXRs enhance the binding of other nuclear receptors (Kastner et al., 1994; Leid et al., 1992). Thus, RXRs act as partners of multiple nuclear receptors and are pleiotropic in cellular effects. Disruption of the RXRα gene results in prominent cardiac defects, including hypoplasia of the ventricular compact zone and muscular ventricular septal defects. Embryonic heart failure was also displayed in the RXR$\alpha^{-/-}$ embryo (Dyson et al., 1995; Kastner et al., 1994; Sucov et al., 1994). Atrial and ventricular chamber-specific myosin light chain 2 (MLC-2) genes have been used as molecular markers for the process of chamber maturation and specification (Kubalak et al., 1994; O'Brien et al., 1993). The expression of the atrial isoform MLC-2a appears to be uniform throughout the linear heart tube and is selectively down-regulated in the ventricular chamber during the process of expansion of the compact zone and onset of trabeculation. An aberrant, persistent expression of MLC-2a in the thin-walled ventricular chambers was identified in the RXR$\alpha^{-/-}$ mutant, indicating that the RXRα-mediated pathway is important for differentiation of ventricular cardiomyocytes and is required in the progression of development of the ventricular region of the heart from its early atrial-like form to the thick-walled adult ventricle. The conduction system disturbances found in RXR$\alpha^{-/-}$ embryos indicate that an RXRα-mediated signaling pathway is required in the developing conduction system. Interestingly, RXRβ and RXRγ gene knockout mice develop normal heart formation. Moreover, double mutation of RXR$\beta^{-/-}$/RXR$\gamma^{-/-}$ or triple mutation of RXR$\alpha^{+/-}$/RXR$\beta^{-/-}$/RXR$\gamma^{-/-}$ phenotypes are viable, displaying no obvious congenital heart abnormalities except a marked growth

TABLE I. Cardiovascular Defects in Vitamin A Deficiency or Excess and Retinoid Receptor Knockout Embryos

Heart defects	Persistant truncus arteriosus	Tubular heart	Ventricular septal defect	Atrial septal defect	Abnormal aortic arch pattern	Atrio-ventricular canal defect	Ventricular chamber hypoplasia
VAD	+	+	+	−	+	+	+
Excess RA	+	+	+	+	+	+	−
RARα$^{-/-}$	−	−	−	−	−	−	−
RARβ$^{-/-}$	−	−	−	−	−	−	−
RARγ$^{-/-}$	−	−	−	−	−	−	−
RXRα$^{-/-}$	+	+	+	−	+	+	+
RXRβ$^{-/-}$	−	−	−	−	−	−	−
RXRγ$^{-/-}$	−	−	−	−	−	−	−
RARα$^{-/-}$ RARβ$^{-/-}$	+	+	+	−	+	+	−
RARα$^{-/-}$ RARγ$^{-/-}$	+	+	+	−	+	+	+
RARβ$^{-/-}$ RARγ$^{-/-}$	−	−	−	−	−	−	−
RARα$^{-/-}$ RXRα$^{-/-}$	+	+	+	−	+	+	+
RARβ$^{-/-}$ RXRα$^{-/-}$	+	+	+	−	+	+	+
RARγ$^{-/-}$ RXRα$^{-/-}$	+	+	+	−	+	+	+
RXRα$^{-/-}$ RXRγ$^{-/-}$	+	+	+	−	+	+	+
RXRβ$^{-/-}$ RXRγ$^{-/-}$	−	−	−	−	−	−	−
RXRα$^{+/-}$ RXRβ$^{-/-}$ RXRγ$^{-/-}$	−	−	−	−	−	−	−

deficiency and male sterility due to loss of function of RXRβ (Krezel et al., 1996). In contrast, RXR$\alpha^{-/-}$/RXR$\gamma^{-/-}$ mutants exhibited the same cardiac and ocular defects found in RXRα null mutants, with no additional abnormalities observed, suggesting that one copy of RXRα is sufficient for most of the functions of the RXRs. Compound null mutations of RXRα with RARs display a marked synergistic effect, as a large number of developmental defects, found mainly in RAR single and compound mutants, and are recapitulated in specific RXRα/RAR compound mutants. Several malformations are observed only in one type of RXRα/RAR mutant combination, whereas others are observed in several types of RXRα/RAR double knockout mutants. But no such synergy is observed when RXRβ or RXRγ mutations are combined with any of the RAR mutations (Kastner et al., 1994, 1997a) (Table I). These data suggest that RXRα/RAR heterodimers are essential for most of the events during embryogenesis and fetal development and that RXRα is the main RXR implicated in the developmental functions of RARs.

The requirement of embryonic synthesis of RA for early heart development has also been demonstrated by targeted disruption of the retinaldehyde dehydrogenase 2 (Raldh2) gene (Niederreither et al., 1999). Raldh2 deficiency results in a block in early embryonic RA production, and induces a complete failure of embryo survival and early morphogenesis. The heart of mutant embryos consists of a single, dilated ventricle-like cavity, lacking heart looping and chamber morphogenesis. Maternal RA administration rescued the early morphogenesis. These data demonstrate that local embryonic RA synthesis by Raldh2 is essential for early postimplantation mouse development.

Together the molecular and genetic identification of the RA-signaling pathway has provided valuable insights into the pleiotropic effects of RA, demonstrating a critical role of retinoid receptors in embryonic heart malformation. Interpretation of the data has been hampered by receptor redundancy, and thus the receptor knockouts alone cannot provide definitive answers to RA function.

The significance of RA for early embryogenesis and heart development is most clearly demonstrated in the VAD avian embryo (Dersch and Zile, 1993; Dong and Zile, 1995; Heine et al., 1985; Kostetskii et al., 1998; Thompson et al., 1969). With the VAD avian model, which is devoid of any form of vitamin A from the beginning of fertilization, it is possible to examine morphological, anatomical, and molecular biological aspects that are attributable to RA. The ability to rescue the VAD embryo at a precise time during development makes this model a powerful tool for the elucidation of the physiological functions of RA during early heart development. The VAD quail embryo is grossly abnormal in many developmental aspects, with the misshapen heart positioned to the left side. The heart does not loop, is enlarged and ballooned, has no chambers, is closed at the site of the inflow tract, and the formation of the extraembryonal circulatory system was blocked (Zile, 2004; Zile et al., 2000). An essential aspect of heart development is the establishment of proper heart sidedness. The asymmetry is set up early in embryogenesis and is regulated by many stage-specific genes. The cardiogenic cells do not function appropriately in the absence of vitamin A and migration to either left or right of the midline is random. In the VAD embryo the orientation of the heart is often abnormal, as 72% of embryos had reversed cardiac situs, with only 28% of the embryos demonstrating normal heart positioning. These studies demonstrate a requirement of retinoids for avian heart development and in establishing cardiac left/right asymmetry. Other studies have shown that RA-induced activation of retinoid receptors is required during early avian heart development (Heine et al., 1985; Romeih et al., 2003; Zile et al., 2000). A specific role for RARα2 in cardiac inflow tract morphogenesis and for RARγ in cardiac left/right orientation and looping morphogenesis has been demonstrated (Romeih et al., 2003). Blocking the function of RARα2, RARγ, and RXRα recapitulates the VAD phenotype. These studies provide strong evidence that critical RA-requiring developmental events in the early avian embryo are regulated by distinct retinoid receptor-signaling pathways.

2. Retinoic Acid-Induced Heart Malformations

Although RA is required for normal embryonic and fetal development and cardiogenesis, embryonic exposure to an excess of RA leads to abnormal development. The teratogenic effects of a high dose of vitamin A were first demonstrated in pregnant rats (Cohlan, 1953, 1954). RA was shown to be a more potent teratogen than retinol in several animal models (Creech Kraft et al., 1989; Morriss and Steele, 1977; Shenefelt, 1972a). RA has previously been shown to have teratogenic effects on heart development in mammalian embryos (Fantel et al., 1977; Kalter and Warkany, 1961; Robens, 1970; Shenefelt, 1972b; Taylor et al., 1980). Dependent on the species, stage, and mode of administration, excess RA can result in different types of heart malformations. During the early stage of development, excess RA exposure restricted the cardiac progenitor pool and cardiac specification in the zebrafish embryo (Keegan et al., 2005). In the chick embryo, excess RA inhibits normal precardiac mesoderm migration and the formation of the normal heart tube. Similarly, local application of RA to the heart-forming area disrupts the formation of the cardiogenic crescent and subsequent development of a single midline heart tube (Osmond et al., 1991). Late primitive streak-stage chick embryos exposed to excess RA result in cardiac bifida and clustered heart tissue formation (Dickman and Smith, 1996). Exposure of *Xenopus* embryos to continuous low levels of RA (1 μM), starting at the time of neural fold closure, blocks expression of myocardial differentiation markers and the heart tube fails to loop during subsequent development, never developing into beating tissue (Drysdale et al., 1997). All vertebrates develop with left/right asymmetry with formation of the left/right body axis being a critical early step in embryogenesis. The heart loop is one of the first clearly recognizable morphological asymmetries. RA-induced asymmetry defects in mammalian (hamster, rat, and mouse) embryos have been reported (Fujinaga, 1997).

Excess RA administration can cause situs inversus in avian embryos, resulting in randomization of heart looping and defects in anteroposterior patterning in the mouse embryo (Chazaud et al., 1999; Smith et al., 1997; Wasiak and Lohnes, 1999). These results suggest that alterations of RA signaling affect the left/right situs, as well as heart morphogenesis in embryonic heart development. Excess RA exposure after heart specification results in congenital heart malformations, including ventricular septal defects, double outlet right ventricle, and persistent truncus arteriosus. Transgenic mice that overexpress a constitutively active RARα in fetal ventricles developed a dilated cardiomyopathy. Lesions included biventricular chamber dilation and left atrial thrombosis, the incidence and severity increasing with copy number. Hypertrophic markers (α-skeletal actin and atrial natriuretic factor) were also upregulated. In contrast, animals that overexpressed a constitutively active RARα in developing atria and/or in postnatal ventricles developed no signs of malformations (Colbert et al., 1997). The overlap of the teratological symptoms of vitamin A deficiency and excess suggests common targets and an important role for RA in the development of many organs, including the cardiovascular system.

The above observations suggest that RA-mediated signaling pathways are required at early stages of cardiac development to prevent differentiation, support cell proliferation, and control the shape of ventricular myocytes. Both RXRs and RARs participate in the mediation of these functions. In the postdevelopment period, RA-dependent signal transduction appears to preserve the normal differentiated phenotype of cardiomyocytes by antagonizing the effect of various hypertrophic stimuli (Wang et al., 2002; Wu et al., 1996; Zhou et al., 1995).

III. POSTNATAL DEVELOPMENT EFFECTS OF RA IN THE HEART

The role of retinoids in promoting the ventricular phenotype in the embryonic heart suggests that retinoids may also be important in maintaining normal ventricular phenotype in the postnatal state. We and others have demonstrated that RA inhibits mechanical stretch and G-protein–coupled receptor (GPCR)-mediated hypertrophy in neonatal cardiomyocytes (Palm-Leis et al., 2004; Wang et al., 2002; Wu et al., 1996; Zhou et al., 1995). It has also been shown that RA inhibits left ventricular fibrosis and has a role in ventricular remodeling during the development of hypertension in spontaneously hypertensive rats (SHR) and in the myocardial infarction rat model (de Paiva et al., 2003; Lu et al., 2003; Paiva et al., 2005). RA signaling could potentially serve as an alternative target in the prevention and treatment of pathological hypertrophy and heart failure.

A. ANTIHYPERTROPHIC EFFECTS OF RA IN NEONATAL CARDIOMYOCYTES

Cardiac hypertrophy occurs as an adaptive response to many forms of cardiac disease, including high blood pressure (hypertension), myocardial infarction, cardiac arrhythmias, genetic defects in cardiac contractile proteins, and endocrine disorders. Cardiac hypertrophy is characterized by an increase in myocyte cell size (in the absence of cell division) and by a number of qualitative and quantitative changes in gene expression. Many pathological stimuli induce the heart to undergo adaptive hypertrophic growth. Although the initial hypertrophic response may be beneficial, sustained hypertrophy often results in a transition to heart failure, which is a leading cause of mortality and morbidity worldwide, and is characterized by a progressive deterioration in cardiac function. Cardiac hypertrophy is generally associated with the expression of atrial natriuretic peptide (ANP) that is restricted to the atria shortly after birth, and is reexpressed in the ventricles following hemodynamic overload (Chien *et al.*, 1991; Izumo *et al.*, 1988). Several genes switch to the expression of fetal isoforms, such as transition from cardiac α-actin to skeletal α-actin and from the α-myosin heavy chain (α-MHC) to β-MHC in rodents (Schwartz *et al.*, 1986). These "atrialization" or fetal type changes of hypertrophic ventricles raise the question as to whether retinoid-dependent pathways function in the adult myocardium to actively maintain the ventricular phenotype. Using an *in vitro* cultured cardiomyocyte model, an antihypertrophic effect of RA has been demonstrated (Zhou *et al.*, 1995). Studies have shown that RA-mediated pathways suppress the acquisition of specific features of the hypertrophic phenotype, following exposure to an α-adrenergic receptor agonist. At physiological concentrations, RA suppresses the increase in cell size and induction of a genetic marker for hypertrophy, the ANF gene. We and others have also demonstrated that RA signaling suppresses cardiac hypertrophic features (including increased total protein content, protein synthesis, cell size, and myofibrillar reorganization) in response to mechanical stretch, angiotensin II (Ang II), and endothelin-1 (Palm-Leis *et al.*, 2004; Wang *et al.*, 2002; Wu *et al.*, 1996). These observations suggest that RA-mediated signaling pathways have an important role in regulating hypertrophic stimuli-induced cardiomyocyte growth. Using isolated, spontaneously beating neonatal rat cardiac myocytes, it has been shown that RA has a protective effect against cardiac arrhythmias induced by isoproterenol, lysophosphatidylcholine or ischemia, and reperfusion, indicating a novel role of RA as a potential antiarrhythmic agent (Kang and Leaf, 1995).

Although the antihypertrophic effect of RA in neonatal cardiomyocytes has been demonstrated, the molecular mechanisms involved in RA-dependent growth inhibition are not well understood. Cardiac hypertrophy is associated with the activation of numerous signal transduction factors, including GPCRs, receptor tyrosine kinases, the PI3K/AKT pathway, protein kinase C, calcineurin, and members of the mitogen-activated protein kinase (MAPK)-signaling cascade (Dorn and Brown, 1999; Hefti et al., 1997; Hunter and Chien, 1999; Olson and Molkentin, 1999; Rapacciuolo et al., 2001; Uozumi et al., 2001). As a negative regulator of cardiac hypertrophy, the dual-specificity mitogen-activated protein kinase phosphatase 1 (MKP-1) has been shown to limit the cardiac hypertrophic response *in vitro* and *in vivo* via dephosphorylation and inactivation of MAP kinases (Bueno et al., 2001; Fuller et al., 1997; Thorburn et al., 1995). These latter results suggest that enhancement of inhibitory regulators may provide clinical benefit in the prevention and treatment of pathological hypertrophy and heart failure. Studies have revealed that neonatal rat ventricular myocardial cells express functional RA receptors of both the RAR and RXR subtypes. Using synthetic agonists or antagonists of RA, which selectively bind to RXR or RAR, we and others have demonstrated that RAR/RXR heterodimers have an important role in mediating suppression of cardiac hypertrophy (Palm-Leis et al., 2004; Zhou et al., 1995). Studies have shown that RXRα represses ANF promoter activity in cardiomyocytes, by interacting directly with the transcription factor GATA-4 and the corepressor FOG-2 (Clabby et al., 2003). These results identify a mechanism whereby retinoid signaling modulates gene expression in the heart and reveals a new permutation in the retinoid-signaling pathway. Our group has demonstrated that cyclic stretch or Ang II-induced activation of extracellular signal-regulated kinase 1/2 (ERK1/2), c-Jun N-terminal kinase (JNK), and p38 mitogen-activated protein kinase (MAP kinase) were dose- and time-dependently inhibited by RA. This inhibitory effect was not mediated at the level of mitogen-activated protein kinase kinases (MKKs), since RA had no effect on stretch- or Ang II-induced phosphorylation of MEK1/2, MKK4, and MKK3/6. We further demonstrated that the upregulated expression of MKP-1 has an important role in mediating the MAP kinase inhibition and antihypertrophic effect of RA in cyclic stretch and Ang II-stimulated neonatal cardiomyocytes (Palm-Leis et al., 2004). These data indicate that upregulation of MKPs and inhibition of MAP kinase-signaling pathways may serve as one of the signaling mechanisms involved in mediating the inhibitory effects of RA in cardiac hypertrophy. The regulating mechanisms of RA on the expression of MKPs remain unclear. There is no RARE reported in the promoter/enhancer regions of human, mouse, and rat MKP genes (Kwak et al., 1994; Ryser et al., 2001). One possible explanation is that RA-bound RAR/RXR may induce MKPs expression via binding to a RARE-like sequence that has not been identified. Another possibility is that the induction of MKPs by RA is mediated by RARE-independent mechanisms or by binding to other transcription factors which contain RARE in the promoter region (indirect regulation). These questions remain important mechanistic area to be addressed.

B. EFFECTS OF RA SIGNALING IN CARDIAC REMODELING

Left ventricular (LV) remodeling has a major role in the progression to heart failure. This process is associated with alterations in ventricular mass, chamber size, shape, and function that result from myocardial injury, pressure, or volume overload. At the cellular level, LV remodeling is characterized by myocyte hypertrophy, fibroblast hyperplasia accompanied by an increase in collagen deposition within the interstitial matrix (fibrosis), and cell death. Although remodeling is initially an adaptive response to maintain normal cardiac function, it often becomes maladaptive and leads to progressive decompensation. Elucidating the mechanisms responsible for preventing and/or reversing the process of LV remodeling may lead to identifying novel therapeutic targets for the treatment of heart failure. In normal rats, RA treatment improved the systolic and diastolic function of isolated papillary muscle (de Paiva *et al.*, 2003). Although RA treatment produced an increase in myocyte cross-sectional area, the myocardial collagen volume fraction was similar to controls. These results demonstrate that small, physiological doses of RA induce ventricular remodeling resembling compensated volume-overload hypertrophy in rats. *In vivo* studies have shown that chronic RA treatment prevented hypertrophy of intramyocardial and intrarenal arteries and ventricular fibrosis during the development of hypertension in the SHR (Lu *et al.*, 2003). However, RA treatment did not lower blood pressure or left ventricular weight and left ventricular weight-to-body weight ratio and had no influence on LV ANP levels, cardiac geometry, and function in the SHR. The failure to observe an inhibitory effect of RA on LV hypertrophy and blood pressure may be due to a suboptimal dose, since in another study, RA treatment significantly lowered the increased blood pressure and attenuated myocardial damage of the LV (including myocardial mitochondria swelling, crest disruption, and myofilaments derangement) in the SHR (Zhong *et al.*, 2005). The dosage of RA (5–10 mg/kg/day) used by Lu *et al.* is lower than that (10–20 mg/kg/day) used by Zhong *et al.* The effect of RA on the impaired heart function would likely be observed at 12 months in the SHR; but not at the age of 4 months (which Lu *et al.* studied), which may be too early to see an impairment of ventricular function. Our preliminary data have shown that RA treatment (30 mg/kg/day) improved both systolic and diastolic heart function in chronic pressure overload-induced hypertrophy. In rats with myocardial infarction, RA significantly reduced the cross-sectional area of the myocyte and interstitial collagen fraction, and increased the maximum rate of rise of LV pressure ($+dp/dt$), indicating that RA treatment attenuates cardiac remodeling in this experimental model (Paiva *et al.*, 2005). Cardiac fibrosis has an important role in the process of LV remodeling. It has been shown that Ang II-induced cardiac fibroblast proliferation and collagen synthesis are inhibited by RA (He *et al.*, 2006), indicating that RA-mediated signaling is involved in regulating cardiac fibrosis. In a pacing-induced heart failure model in the dog, activity of two key enzymes of free fatty acid (FFA) oxidation: carnitine palmitoyl transferase-I and medium-chain acyl-coenzyme A dehydrogenase (MCAD) were significantly reduced in left ventricular tissue from failing hearts, a finding which correlated to reduced expression of RXRα, suggesting that RXRα downregulation could be responsible for the impairment of the FFA oxidative pathway in the failing heart (Osorio et al., 2002). It has been demonstrated that KLF5, a Krüppel-like zinc-finger transcription factor, is an essential transcription factor in cardiovascular remodeling (Nagai et al., 2005). Ang II-induced cardiac hypertrophy and fibrosis are attenuated in heterozygotes of KLF5 knockout mice. Interestingly, an RARα agonist suppresses KLF5 and cardiovascular remodeling, whereas an RARα antagonist activates KLF5 and induces angiogenesis, suggesting that RAR ligands exert protective effects against cardiovascular remodeling via transrepression of KLF5 (Shindo et al., 2002). The above results suggest that RA has a role in preventing cardiac remodeling and the transition from adaptive cardiac hypertrophy to maladaptive heart failure; however, the signaling mechanisms remain to be elucidated.

C. ROLE OF RA SIGNALING IN THE REGULATION OF THE RENIN–ANGIOTENSIN SYSTEM

1. Renin–Angiotensin System

The renin–angiotensin system (RAS) is a regulatory cascade that is critical for the maintenance of blood pressure, electrolyte, and volume homeostasis. Inappropriate stimulation of the RAS has been associated with hypertension, cardiac hypertrophy, myocardial infarction, and stroke. Ang II is the main effector molecule of the RAS (Bernstein and Berk, 1993). Renin, which is synthesized by the kidney and secreted into the blood, hydrolyses the decapeptide Ang I from the N-terminus of angiotensinogen. Ang I is converted to the octapeptide Ang II by the dipeptidyl carboxypeptidase, angiotensin-converting enzyme (ACE). Findings have shown that Ang II directly mediates cell growth, regulates gene expression, and activates multiple intracellular-signaling pathways in cardiovascular and renal cells (Baker and Aceto, 1990; Booz et al., 2002; Dostal et al., 1997; Schunkert et al., 1995). The importance of the RAS in cardiac remodeling is well documented. Numerous studies have shown that the RAS is activated in response to hemodynamic overload and that activation of the (local) RAS contributes to myocardial hypertrophy, fibrosis, and dysfunction (Cohn et al., 2000; Pfeffer et al., 1995; Schnee and Hsueh, 2000). In addition, animal studies (Kim et al., 2001; Nakamura et al., 2003) and clinical trials in humans (Cohn and Tognoni, 2001; Flather et al., 2000; Pfeffer et al., 2003) have shown that inhibition of Ang II by ACE inhibitors or AT_1 receptor antagonists prevents or reverses ventricular remodeling and improves survival in patients with heart failure. ACE2 is a recently discovered homologue of ACE with tissue-restricted expression, including heart and kidney endothelium (Tipnis et al., 2000). ACE2 cleaves Ang I to generate Ang 1–9 and converts Ang II to Ang 1–7, a vasodilator (Donoghue et al., 2000; Ferrario et al., 1997). ACE2 has been implicated in heart function, hypertension, renal disease, and diabetes, with effects being mediated in part, through the ability to convert Ang II into Ang (1–7). Targeted disruption of ACE2 in the mouse results in a severe cardiac contractility defect, increased Ang II levels, and upregulation of hypoxia-induced genes in the heart, indicating that ACE2 is an essential regulator of heart function (Crackower et al., 2002). Studies have shown that Ang (1–7) acts as an endogenous inhibitor of Ang II, providing a negative feedback mechanism for the regulation of the actions of Ang II.

2. RA and the Renin–Angiotensin System

There is evidence that RA regulates the gene expression of RAS components, including renin, ACE, ACE2, and AT_1 receptor. RA influences the renal RAS components in rats with experimental nephritis (Dechow et al., 2001). In the renal cortex of nephritic rats, pretreatment with RA significantly reduced mRNAs of all the examined renal RAS components (angiotensinogen, renin, ACE, and AT_1 receptor), but in glomeruli it increased ACE gene and protein expression. In vascular smooth muscle cells (VSMCs), RA dose-dependency inhibits Ang II-induced cell proliferation as well as DNA and protein synthesis. Ang II-induced gene expression of c-Fos and transforming growth factor-β_1 mRNA is abrogated by RA treatment. Downregulation of AT_1 receptor mRNA and repressed Ang II-stimulated AT_1 receptor promoter activity are observed in RA-treated VSMCs (Haxsen et al., 2001; Takeda et al., 2000). These findings demonstrate that retinoids are potent inhibitors of the actions of Ang II on VSMCs. It has been shown that RA downregulates the expression level of AT_1, and upregulates the expression of ACE2 in the heart of SHR; but not in normal WKY rats (Zhong et al., 2004, 2005). Thus, RA-mediated signaling is involved in regulating RAS components during the development of hypertension. Further studies are necessary to elucidate the molecular mechanisms of the effects of RA on AT_1 and ACE2 expression and the reason for a different role in hypertensive and normotensive rats.

IV. CONCLUSIONS

RA, as the active form of vitamin A, has a complex role in embryonic and postnatal heart development and remodeling. Using the vitamin A deficiency animal model and retinoid receptor knockout transgenic model, multiple studies have implicated RA as essential for normal cardiovascular development. The requirement for RA in the embryo begins at the time of formation of the primordial heart tube. This time of development in the avian embryo correlates with the first 2–3 weeks of human pregnancy, indicating the importance of adequate vitamin A nutrition during the initial stages of pregnancy. Lack of RA during this time, will result in gross abnormalities in heart development and early embryo lethality. Embryos exposed to excess RA also exhibit abnormal heart development, analogous to those seen in VAD animals. Thus, vitamin A supplements should be used cautiously during early pregnancy to avoid teratogenic risks. A precisely regulated supply of RA appears to be necessary for normal embryonic development. Retinoid receptor knockout studies have shown that RXRα is the major partner for the RARs in RA signaling affecting developmental function. Congenital heart disease is the most prevalent human birth defect, with about 3% of all children born in the United States having major heart malformations at birth. The etiology of the vast majority of congenital heart diseases (70%) remains unknown (Srivastava, 2000). Determining the molecular mechanisms and function of RA during embryonic development will be important in enhancing our understanding of congenital heart diseases.

The effects of RA on postnatal and adult cardiac cells suggest potential novel targets in the prevention and treatment of cardiac hypertrophy and heart failure. *In vitro* studies have demonstrated that RA has an inhibitory effect on cell growth in response to various hypertrophic stimuli in fully differentiated cardiomyocytes and cardiac fibroblasts. RA-mediated signaling prevents or improves cardiac remodeling, during the development of hypertension and following myocardial infarction, in rat models. The inhibitory effect of RA on neonatal cardiac cell growth is not observed in normal or hypertensive rats. Additional studies appear warranted to determine the effect of RA on regulation of cardiac cell growth and the signaling mechanisms under pathological conditions. Studies have shown that RA signaling is involved in the regulation of the expression and/or activation of the RAS; however, the effects of RA on the circulating and/or the local cardiac RAS, during the process of cardiac remodeling, have not been elucidated. The signaling mechanisms by which RA regulates the expression of the RAS and the correlation with cardiac remodeling also warrant investigation. Although the involved signaling mechanisms of RA in the cardiovascular system are not well elucidated, the results presented above suggest the potential utility of RA in the prevention and treatment of cardiovascular diseases.

REFERENCES

Antonyak, M. A., Boehm, J. E., and Cerione, R. A. (2002). Phosphoinositide 3-kinase activity is required for retinoic acid-induced expression and activation of the tissue transglutaminase. *J. Biol. Chem.* **277,** 14712–14716.

Aranega, A. E., Velez, C., Prados, J., Melguizo, C., Marchal, J. A., Arena, N., Alvarez, L., and Aranega, A. (1999). Modulation of alpha-actin and alpha-actinin proteins in cardiomyocytes by retinoic acid during development. *Cells Tissues Organs* **164,** 82–89.

Baker, K. M., and Aceto, J. F. (1990). Angiotensin II stimulation of protein synthesis and cell growth in chick heart cells. *Am. J. Physiol.* **259,** H610–H618.

Benbrook, D., Lernhardt, E., and Pfahl, M. (1988). A new retinoic acid receptor identified from a hepatocellular carcinoma. *Nature* **333,** 669–672.

Bernstein, K. E., and Berk, B. C. (1993). The biology of angiotensin II receptors. *Am. J. Kidney Dis.* **22,** 745–754.

Booz, G. W., Day, J. N., and Baker, K. M. (2002). Interplay between the cardiac renin angiotensin system and JAK-STAT signaling: Role in cardiac hypertrophy, ischemia/reperfusion dysfunction, and heart failure. *J. Mol. Cell. Cardiol.* **34,** 1443–1453.

Brand, N., Petkovich, M., Krust, A., Chambon, P., de The, H., Marchio, A., Tiollais, P., and Dejean, A. (1988). Identification of a second human retinoic acid receptor. *Nature* **332,** 850–853.

Bueno, O. F., De Windt, L. J., Lim, H. W., Tymitz, K. M., Witt, S. A., Kimball, T. R., and Molkentin, J. D. (2001). The dual-specificity phosphatase MKP-1 limits the cardiac hypertrophic response *in vitro* and *in vivo*. *Circ. Res.* **88,** 88–96.

Chambon, P. (1996). A decade of molecular biology of retinoic acid receptors. *FASEB J.* **10,** 940–954.

Chazaud, C., Chambon, P., and Dolle, P. (1999). Retinoic acid is required in the mouse embryo for left-right asymmetry determination and heart morphogenesis. *Development* **126,** 2589–2596.

Chen, J., Kubalak, S. W., and Chien, K. R. (1998). Ventricular muscle-restricted targeting of the RXRalpha gene reveals a non-cell-autonomous requirement in cardiac chamber morphogenesis. *Development* **125,** 1943–1949.

Chen, J. Y., Penco, S., Ostrowski, J., Balaguer, P., Pons, M., Starrett, J. E., Reczek, P., Chambon, P., and Gronemeyer, H. (1995). RAR-specific agonist/antagonists which dissociate transactivation and AP1 transrepression inhibit anchorage-independent cell proliferation. *EMBO J.* **14,** 1187–1197.

Chien, K. R., Knowlton, K. U., Zhu, H., and Chien, S. (1991). Regulation of cardiac gene expression during myocardial growth and hypertrophy: Molecular studies of an adaptive physiologic response. *FASEB J.* **5,** 3037–3046.

Christ, M., Haseroth, K., Falkenstein, E., and Wehling, M. (1999). Nongenomic steroid actions: Fact or fantasy? *Vitam. Horm.* **57,** 325–373.

Clabby, M. L., Robison, T. A., Quigley, H. F., Wilson, D. B., and Kelly, D. P. (2003). Retinoid X receptor alpha represses GATA-4-mediated transcription via a retinoid-dependent interaction with the cardiac-enriched repressor FOG-2. *J. Biol. Chem.* **278,** 5760–5767.

Clagett-Dame, M., and DeLuca, H. F. (2002). The role of vitamin A in mammalian reproduction and embryonic development. *Annu. Rev. Nutr.* **22,** 347–381.

Cohlan, S. Q. (1953). Excessive intake of vitamin A as a cause of congenital anomalies in the rat. *Science* **117,** 535–536.

Cohlan, S. Q. (1954). Congenital anomalies in the rat produced by excessive intake of vitamin A during pregnancy. *Pediatrics* **13,** 556–567.

Cohn, J. N., and Tognoni, G. (2001). A randomized trial of the angiotensin-receptor blocker valsartan in chronic heart failure. *N. Engl. J. Med.* **345,** 1667–1675.

Cohn, J. N., Ferrari, R., and Sharpe, N. (2000). Cardiac remodeling—concepts and clinical implications: A consensus paper from an international forum on cardiac remodeling. Behalf of an International Forum on Cardiac Remodeling. *J. Am. Coll. Cardiol.* **35,** 569–582.

Colbert, M. C., Hall, D. G., Kimball, T. R., Witt, S. A., Lorenz, J. N., Kirby, M. L., Hewett, T. E., Klevitsky, R., and Robbins, J. (1997). Cardiac compartment-specific overexpression of a modified retinoic acid receptor produces dilated cardiomyopathy and congestive heart failure in transgenic mice. *J. Clin. Invest.* **100,** 1958–1968.

Crackower, M. A., Sarao, R., Oudit, G. Y., Yagil, C., Kozieradzki, I., Scanga, S. E., Oliveira-dos-Santos, A. J., da Costa, J., Zhang, L., Pei, Y., Scholey, J., Ferrario, C. M., *et al.* (2002). Angiotensin-converting enzyme 2 is an essential regulator of heart function. *Nature* **417,** 822–828.

Creech Kraft, J., Lofberg, B., Chahoud, I., Bochert, G., and Nau, H. (1989). Teratogenicity and placental transfer of all-trans-, 13-cis-, 4-oxo-all-trans-, and 4-oxo-13-cis-retinoic acid after administration of a low oral dose during organogenesis in mice. *Toxicol. Appl. Pharmacol.* **100,** 162–176.

Dechow, C., Morath, C., Peters, J., Lehrke, I., Waldherr, R., Haxsen, V., Ritz, E., and Wagner, J. (2001). Effects of all-trans retinoic acid on renin-angiotensin system in rats with experimental nephritis. *Am. J. Physiol. Renal Physiol.* **281,** F909–F919.

de Paiva, S. A., Zornoff, L. A., Okoshi, M. P., Okoshi, K., Matsubara, L. S., Matsubara, B. B., Cicogna, A. C., and Campana, A. O. (2003). Ventricular remodeling induced by retinoic acid supplementation in adult rats. *Am. J. Physiol. Heart Circ. Physiol.* **284,** H2242–H2246.

Dersch, H., and Zile, M. H. (1993). Induction of normal cardiovascular development in the vitamin A-deprived quail embryo by natural retinoids. *Dev. Biol.* **160,** 424–433.

Dickman, E. D., and Smith, S. M. (1996). Selective regulation of cardiomyocyte gene expression and cardiac morphogenesis by retinoic acid. *Dev. Dyn.* **206,** 39–48.

Dong, D., and Zile, M. H. (1995). Endogenous retinoids in the early avian embryo. *Biochem. Biophys. Res. Commun.* **217,** 1026–1031.

Donoghue, M., Hsieh, F., Baronas, E., Godbout, K., Gosselin, M., Stagliano, N., Donovan, M., Woolf, B., Robison, K., Jeyaseelan, R., Breitbart, R. E., and Acton, S. (2000). A novel angiotensin-converting enzyme-related carboxypeptidase (ACE2) converts angiotensin I to angiotensin 1–9. *Circ. Res.* **87,** E1–E9.

Dorn, G. W., 2nd, and Brown, J. H. (1999). Gq signaling in cardiac adaptation and maladaptation. *Trends Cardiovasc. Med.* **9,** 26–34.

Dostal, D. E., Hunt, R. A., Kule, C. E., Bhat, G. J., Karoor, V., McWhinney, C. D., and Baker, K. M. (1997). Molecular mechanisms of angiotensin II in modulating cardiac function: Intracardiac effects and signal transduction pathways. *J. Mol. Cell. Cardiol.* **29,** 2893–2902.

Drysdale, T. A., Patterson, K. D., Saha, M., and Krieg, P. A. (1997). Retinoic acid can block differentiation of the myocardium after heart specification. *Dev. Biol.* **188,** 205–215.

Duester, G. (2000). Families of retinoid dehydrogenases regulating vitamin A function: Production of visual pigment and retinoic acid. *Eur. J. Biochem.* **267,** 4315–4324.

Dyson, E., Sucov, H. M., Kubalak, S. W., Schmid-Schonbein, G. W., DeLano, F. A., Evans, R. M., Ross, J., Jr., and Chien, K. R. (1995). Atrial-like phenotype is associated with embryonic ventricular failure in retinoid X receptor $\alpha^{-/-}$ mice. *Proc. Natl. Acad. Sci. USA* **92,** 7386–7390.

Eisenberg, L. M., and Markwald, R. R. (1995). Molecular regulation of atrioventricular valvuloseptal morphogenesis. *Circ. Res.* **77,** 1–6.

Fananapazir, K., and Kaufman, M. H. (1988). Observations on the development of the aorticopulmonary spiral septum in the mouse. *J. Anat.* **158,** 157–172.

Fantel, A. G., Shepard, T. H., Newell-Morris, L. L., and Moffett, B. C. (1977). Teratogenic effects of retinoic acid in pigtail monkeys (Macaca nemestrina). I. General features. *Teratology* **15,** 65–71.

Ferrario, C. M., Chappell, M. C., Tallant, E. A., Brosnihan, K. B., and Diz, D. I. (1997). Counterregulatory actions of angiotensin-(1–7). *Hypertension* **30,** 535–541.

Fishman, M. C., and Chien, K. R. (1997). Fashioning the vertebrate heart: Earliest embryonic decisions. *Development* **124,** 2099–2117.

Flather, M. D., Yusuf, S., Kober, L., Pfeffer, M., Hall, A., Murray, G., Torp-Pedersen, C., Ball, S., Pogue, J., Moye, L., and Braunwald, E. (2000). Long-term ACE-inhibitor therapy in patients with heart failure or left-ventricular dysfunction: A systematic overview of data from individual patients. ACE-Inhibitor Myocardial Infarction Collaborative Group. *Lancet* **355**, 1575–1581.

Fujii, H., Sato, T., Kaneko, S., Gotoh, O., Fujii-Kuriyama, Y., Osawa, K., Kato, S., and Hamada, H. (1997). Metabolic inactivation of retinoic acid by a novel P450 differentially expressed in developing mouse embryos. *EMBO J.* **16**, 4163–4173.

Fujinaga, M. (1997). Development of sidedness of asymmetric body structures in vertebrates. *Int. J. Dev. Biol.* **41**, 153–186.

Fuller, S. J., Davies, E. L., Gillespie-Brown, J., Sun, H., and Tonks, N. K. (1997). Mitogen-activated protein kinase phosphatase 1 inhibits the stimulation of gene expression by hypertrophic agonists in cardiac myocytes. *Biochem. J.* **323**(Pt. 2), 313–319.

Garcia-Martinez, V., and Schoenwolf, G. C. (1993). Primitive-streak origin of the cardiovascular system in avian embryos. *Dev. Biol.* **159**, 706–719.

Gerdes, D., Christ, M., Haseroth, K., Notzon, A., Falkenstein, E., and Wehling, M. (2000). Nongenomic actions of steroids—from the laboratory to clinical implications. *J. Pediatr. Endocrinol. Metab.* **13**, 853–878.

Ghyselinck, N. B., Wendling, O., Messaddeq, N., Dierich, A., Lampron, C., Decimo, D., Viville, S., Chambon, P., and Mark, M. (1998). Contribution of retinoic acid receptor beta isoforms to the formation of the conotruncal septum of the embryonic heart. *Dev. Biol.* **198**, 303–318.

Gruber, P. J., Kubalak, S. W., Pexieder, T., Sucov, H. M., Evans, R. M., and Chien, K. R. (1996). RXR alpha deficiency confers genetic susceptibility for aortic sac, conotruncal, atrioventricular cushion, and ventricular muscle defects in mice. *J. Clin. Invest.* **98**, 1332–1343.

Hale, F. (1937). Relation of maternal vitamin A deficiency to microphthalmia in pigs. *Texas State J. Med.* **33**, 228–232.

Hamada, K., Gleason, S. L., Levi, B. Z., Hirschfeld, S., Appella, E., and Ozato, K. (1989). H-2RIIBP, a member of the nuclear hormone receptor superfamily that binds to both the regulatory element of major histocompatibility class I genes and the estrogen response element. *Proc. Natl. Acad. Sci. USA* **86**, 8289–8293.

Haxsen, V., Adam-Stitah, S., Ritz, E., and Wagner, J. (2001). Retinoids inhibit the actions of angiotensin II on vascular smooth muscle cells. *Circ. Res.* **88**, 637–644.

He, Y., Huang, Y., Zhou, L., Lu, L. M., Zhu, Y. C., and Yao, T. (2006). All-trans retinoic acid inhibited angiotensin II-induced increase in cell growth and collagen secretion of neonatal cardiac fibroblasts. *Acta Pharmacol. Sin.* **27**, 423–429.

Hefti, M. A., Harder, B. A., Eppenberger, H. M., and Schaub, M. C. (1997). Signaling pathways in cardiac myocyte hypertrophy. *J. Mol. Cell. Cardiol.* **29**, 2873–2892.

Heine, U. I., Roberts, A. B., Munoz, E. F., Roche, N. S., and Sporn, M. B. (1985). Effects of retinoid deficiency on the development of the heart and vascular system of the quail embryo. *Virchows Arch. B Cell Pathol. Mol. Pathol.* **50**, 135–152.

Hidaka, K., Lee, J. K., Kim, H. S., Ihm, C. H., Iio, A., Ogawa, M., Nishikawa, S., Kodama, I., and Morisaki, T. (2003). Chamber-specific differentiation of Nkx2.5-positive cardiac precursor cells from murine embryonic stem cells. *FASEB J.* **17**, 740–742.

Honda, M., Hamazaki, T. S., Komazaki, S., Kagechika, H., Shudo, K., and Asashima, M. (2005). RXR agonist enhances the differentiation of cardiomyocytes derived from embryonic stem cells in serum-free conditions. *Biochem. Biophys. Res. Commun.* **333**, 1334–1340.

Hunter, J. J., and Chien, K. R. (1999). Signaling pathways for cardiac hypertrophy and failure. *N. Engl. J. Med.* **341**, 1276–1283.

Icardo, J. M. (1996). Developmental biology of the vertebrate heart. *J. Exp. Zool.* **275**, 144–161.

Izumo, S., Nadal-Ginard, B., and Mahdavi, V. (1988). Protooncogene induction and reprogramming of cardiac gene expression produced by pressure overload. *Proc. Natl. Acad. Sci. USA* **85**, 339–343.

Kalter, H., and Warkany, J. (1961). Experimental production of congenital malformations in strains of inbred mice by maternal treatment with hypervitaminosis A. *Am. J. Pathol.* **38,** 1–21.

Kang, J. X., and Leaf, A. (1995). Protective effects of all-trans-retinoic acid against cardiac arrhythmias induced by isoproterenol, lysophosphatidylcholine or ischemia and reperfusion. *J. Cardiovasc. Pharmacol.* **26,** 943–948.

Kastner, P., Grondona, J. M., Mark, M., Gansmuller, A., LeMeur, M., Decimo, D., Vonesch, J. L., Dolle, P., and Chambon, P. (1994). Genetic analysis of RXR alpha developmental function: Convergence of RXR and RAR signaling pathways in heart and eye morphogenesis. *Cell* **78,** 987–1003.

Kastner, P., Mark, M., Ghyselinck, N., Krezel, W., Dupe, V., Grondona, J. M., and Chambon, P. (1997a). Genetic evidence that the retinoid signal is transduced by heterodimeric RXR/RAR functional units during mouse development. *Development* **124,** 313–326.

Kastner, P., Messaddeq, N., Mark, M., Wendling, O., Grondona, J. M., Ward, S., Ghyselinck, N., and Chambon, P. (1997b). Vitamin A deficiency and mutations of RXRalpha, RXRbeta and RARalpha lead to early differentiation of embryonic ventricular cardiomyocytes. *Development* **124,** 4749–4758.

Keegan, B. R., Feldman, J. L., Begemann, G., Ingham, P. W., and Yelon, D. (2005). Retinoic acid signaling restricts the cardiac progenitor pool. *Science* **307,** 247–249.

Kim, S., Yoshiyama, M., Izumi, Y., Kawano, H., Kimoto, M., Zhan, Y., and Iwao, H. (2001). Effects of combination of ACE inhibitor and angiotensin receptor blocker on cardiac remodeling, cardiac function, and survival in rat heart failure. *Circulation* **103,** 148–154.

Kliewer, S. A., Umesono, K., Mangelsdorf, D. J., and Evans, R. M. (1992a). Retinoid X receptor interacts with nuclear receptors in retinoic acid, thyroid hormone and vitamin D3 signalling. *Nature* **355,** 446–449.

Kliewer, S. A., Umesono, K., Noonan, D. J., Heyman, R. A., and Evans, R. M. (1992b). Convergence of 9-cis retinoic acid and peroxisome proliferator signalling pathways through heterodimer formation of their receptors. *Nature* **358,** 771–774.

Kostetskii, I., Yuan, S. Y., Kostetskaia, E., Linask, K. K., Blanchet, S., Seleiro, E., Michaille, J. J., Brickell, P., and Zile, M. (1998). Initial retinoid requirement for early avian development coincides with retinoid receptor coexpression in the precardiac fields and induction of normal cardiovascular development. *Dev. Dyn.* **213,** 188–198.

Kostetskii, I., Jiang, Y., Kostetskaia, E., Yuan, S., Evans, T., and Zile, M. (1999). Retinoid signaling required for normal heart development regulates GATA-4 in a pathway distinct from cardiomyocyte differentiation. *Dev. Biol.* **206,** 206–218.

Krezel, W., Dupe, V., Mark, M., Dierich, A., Kastner, P., and Chambon, P. (1996). RXR gamma null mice are apparently normal and compound RXR alpha +/−/RXR beta −/−/RXR gamma −/− mutant mice are viable. *Proc. Natl. Acad. Sci. USA* **93,** 9010–9014.

Kubalak, S. W., Miller-Hance, W. C., O'Brien, T. X., Dyson, E., and Chien, K. R. (1994). Chamber specification of atrial myosin light chain-2 expression precedes septation during murine cardiogenesis. *J. Biol. Chem.* **269,** 16961–16970.

Kwak, S. P., Hakes, D. J., Martell, K. J., and Dixon, J. E. (1994). Isolation and characterization of a human dual specificity protein-tyrosine phosphatase gene. *J. Biol. Chem.* **269,** 3596–3604.

Lammer, E. J., Chen, D. T., Hoar, R. M., Agnish, N. D., Benke, P. J., Braun, J. T., Curry, C. J., Fernhoff, P. M., Grix, A. W., Jr., Lott, I. T., Richard, J. M., and Shyan, C. S. (1985). Retinoic acid embryopathy. *N. Engl. J. Med.* **313,** 837–841.

Lee, R. Y., Luo, J., Evans, R. M., Giguere, V., and Sucov, H. M. (1997). Compartment-selective sensitivity of cardiovascular morphogenesis to combinations of retinoic acid receptor gene mutations. *Circ. Res.* **80,** 757–764.

Leid, M., Kastner, P., Lyons, R., Nakshatri, H., Saunders, M., Zacharewski, T., Chen, J. Y., Staub, A., Garnier, J. M., Mader, S., and Chambon, P. (1992). Purification, cloning, and

RXR identity of the HeLa cell factor with which RAR or TR heterodimerizes to bind target sequences efficiently. *Cell* **68**, 377–395.

Li, E., Sucov, H. M., Lee, K. F., Evans, R. M., and Jaenisch, R. (1993). Normal development and growth of mice carrying a targeted disruption of the alpha 1 retinoic acid receptor gene. *Proc. Natl. Acad. Sci. USA* **90**, 1590–1594.

Lohnes, D., Kastner, P., Dierich, A., Mark, M., LeMeur, M., and Chambon, P. (1993). Function of retinoic acid receptor γ in the mouse. *Cell* **73**, 643–658.

Lohnes, D., Mark, M., Mendelsohn, C., Dolle, P., Dierich, A., Gorry, P., Gansmuller, A., and Chambon, P. (1994). Function of the retinoic acid receptors (RARs) during development (I). Craniofacial and skeletal abnormalities in RAR double mutants. *Development* **120**, 2723–2748.

Lopez-Carballo, G., Moreno, L., Masia, S., Perez, P., and Barettino, D. (2002). Activation of the phosphatidylinositol 3-kinase/Akt signaling pathway by retinoic acid is required for neural differentiation of SH-SY5Y human neuroblastoma cells. *J. Biol. Chem.* **277**, 25297–25304.

Lu, L., Yao, T., Zhu, Y. Z., Huang, G. Y., Cao, Y. X., and Zhu, Y. C. (2003). Chronic all-trans retinoic acid treatment prevents medial thickening of intramyocardial and intrarenal arteries in spontaneously hypertensive rats. *Am. J. Physiol. Heart. Circ. Physiol.* **285**, H1370–H1377.

Lufkin, T., Lohnes, D., Mark, M., Dierich, A., Gorry, P., Gaub, M. P., LeMeur, M., and Chambon, P. (1993). High postnatal lethality and testis degeneration in retinoic acid receptor alpha mutant mice. *Proc. Natl. Acad. Sci. USA* **90**, 7225–7229.

Luo, J., Pasceri, P., Conlon, R. A., Rossant, J., and Giguere, V. (1995). Mice lacking all isoforms of retinoic acid receptor beta develop normally and are susceptible to the teratogenic effects of retinoic acid. *Mech. Dev.* **53**, 61–71.

Luo, J., Sucov, H. M., Bader, J. A., Evans, R. M., and Giguere, V. (1996). Compound mutants for retinoic acid receptor (RAR) beta and RAR alpha 1 reveal developmental functions for multiple RAR beta isoforms. *Mech. Dev.* **55**, 33–44.

Lyons, G. E. (1996). Vertebrate heart development. *Curr. Opin. Genet. Dev.* **6**, 454–460.

Mangelsdorf, D. J., and Evans, R. M. (1995). The RXR heterodimers and orphan receptors. *Cell* **83**, 841–850.

Mangelsdorf, D. J., Borgmeyer, U., Heyman, R. A., Zhou, J. Y., Ong, E. S., Oro, A. E., Kakizuka, A., and Evans, R. M. (1992). Characterization of three RXR genes that mediate the action of 9-cis retinoic acid. *Genes Dev.* **6**, 329–344.

Mangelsdorf, D. J., Thummel, C., Beato, M., Herrlich, P., Schutz, G., Umesono, K., Blumberg, B., Kastner, P., Mark, M., Chambon, P., and Evans, R. M. (1995). The nuclear receptor superfamily: The second decade. *Cell* **83**, 835–839.

Mason, K. (1935). Fetal death, prolonged gestation, and difficult parturition in the rat as a result of vitamin A deficiency. *Am. J. Anat.* **57**, 303–349.

Mendelsohn, C., Lohnes, D., Decimo, D., Lufkin, T., LeMeur, M., Chambon, P., and Mark, M. (1994a). Function of the retinoic acid receptors (RARs) during development (II). Multiple abnormalities at various stages of organogenesis in RAR double mutants. *Development* **120**, 2749–2771.

Mendelsohn, C., Mark, M., Dolle, P., Dierich, A., Gaub, M. P., Krust, A., Lampron, C., and Chambon, P. (1994b). Retinoic acid receptor beta 2 (RAR beta 2) null mutant mice appear normal. *Dev. Biol.* **166**, 246–258.

Mic, F. A., Haselbeck, R. J., Cuenca, A. E., and Duester, G. (2002). Novel retinoic acid generating activities in the neural tube and heart identified by conditional rescue of Raldh2 null mutant mice. *Development* **129**, 2271–2282.

Moore, T. (1965). Vitamin A deficiency and excess. *Proc. Nutr. Soc.* **24**, 129–135.

Morriss, G. M., and Steele, C. E. (1977). Comparison of the effects of retinol and retinoic acid on postimplantation rat embryos *in vitro*. *Teratology* **15**, 109–119.

Nagai, R., Suzuki, T., Aizawa, K., Shindo, T., and Manabe, I. (2005). Significance of the transcription factor KLF5 in cardiovascular remodeling. *J. Thromb. Haemost.* **3,** 1569–1576.

Nakamura, Y., Yoshiyama, M., Omura, T., Yoshida, K., Izumi, Y., Takeuchi, K., Kim, S., Iwao, H., and Yoshikawa, J. (2003). Beneficial effects of combination of ACE inhibitor and angiotensin II type 1 receptor blocker on cardiac remodeling in rat myocardial infarction. *Cardiovasc. Res.* **57,** 48–54.

Niederreither, K., Subbarayan, V., Dolle, P., and Chambon, P. (1999). Embryonic retinoic acid synthesis is essential for early mouse post-implantation development. *Nat. Genet.* **21,** 444–448.

Niederreither, K., Vermot, J., Messaddeq, N., Schuhbaur, B., Chambon, P., and Dolle, P. (2001). Embryonic retinoic acid synthesis is essential for heart morphogenesis in the mouse. *Development* **128,** 1019–1031.

O'Brien, T. X., Lee, K. J., and Chien, K. R. (1993). Positional specification of ventricular myosin light chain 2 expression in the primitive murine heart tube. *Proc. Natl. Acad. Sci. USA 90* **515,** 5157–5161.

Olson, E. N., and Molkentin, J. D. (1999). Prevention of cardiac hypertrophy by calcineurin inhibition: Hope or hype? *Circ. Res.* **84,** 623–632.

Olson, E. N., and Srivastava, D. (1996). Molecular pathways controlling heart development. *Science* **272,** 671–676.

Osmond, M. K., Butler, A. J., Voon, F. C., and Bellairs, R. (1991). The effects of retinoic acid on heart formation in the early chick embryo. *Development* **113,** 1405–1417.

Osorio, J. C., Stanley, W. C., Linke, A., Castellari, M., Diep, Q. N., Panchal, A. R., Hintze, T. H., Lopaschuk, G. D., and Recchia, F. A. (2002). Impaired myocardial fatty acid oxidation and reduced protein expression of retinoid X receptor-alpha in pacing-induced heart failure. *Circulation* **106,** 606–612.

Paiva, S. A., Matsubara, L. S., Matsubara, B. B., Minicucci, M. F., Azevedo, P. S., Campana, A. O., and Zornoff, L. A. (2005). Retinoic acid supplementation attenuates ventricular remodeling after myocardial infarction in rats. *J. Nutr.* **135,** 2326–2328.

Palm-Leis, A., Singh, U. S., Herbelin, B. S., Olsovsky, G. D., Baker, K. M., and Pan, J. (2004). Mitogen-activated protein kinases and mitogen-activated protein kinase phosphatases mediate the inhibitory effects of all-trans retinoic acid on the hypertrophic growth of cardiomyocytes. *J. Biol. Chem.* **279,** 54905–54917.

Petkovich, M. (1992). Regulation of gene expression by vitamin A: The role of nuclear retinoic acid receptors. *Annu. Rev. Nutr.* **12,** 443–471.

Pfeffer, J. M., Fischer, T. A., and Pfeffer, M. A. (1995). Angiotensin-converting enzyme inhibition and ventricular remodeling after myocardial infarction. *Annu. Rev. Physiol.* **57,** 805–826.

Pfeffer, M. A., McMurray, J. J., Velazquez, E. J., Rouleau, J. L., Kober, L., Maggioni, A. P., Solomon, S. D., Swedberg, K., Van de Werf, F., White, H., Leimberger, J. D., Henis, M., *et al.* (2003). Valsartan, captopril, or both in myocardial infarction complicated by heart failure, left ventricular dysfunction, or both. *N. Engl. J. Med.* **349,** 1893–1906.

Ragsdale, C. W., Jr., Petkovich, M., Gates, P. B., Chambon, P., and Brockes, J. P. (1989). Identification of a novel retinoic acid receptor in regenerative tissues of the newt. *Nature* **341,** 654–657.

Rapacciuolo, A., Esposito, G., Prasad, S. V., and Rockman, H. A. (2001). G protein-coupled receptor signalling in in vivo cardiac overload. *Acta Physiol. Scand.* **173,** 51–57.

Robens, J. F. (1970). Teratogenic effects of hypervitaminosis A in the hamster and the guinea pig. *Toxicol. Appl. Pharmacol.* **16,** 88–99.

Romeih, M., Cui, J., Michaille, J. J., Jiang, W., and Zile, M. H. (2003). Function of RARgamma and RARalpha2 at the initiation of retinoid signaling is essential for avian embryo survival and for distinct events in cardiac morphogenesis. *Dev. Dyn.* **228,** 697–708.

Ryser, S., Tortola, S., van Haasteren, G., Muda, M., Li, S., and Schlegel, W. (2001). MAP kinase phosphatase-1 gene transcription in rat neuroendocrine cells is modulated by a calcium-sensitive block to elongation in the first exon. *J. Biol. Chem.* **276,** 33319–33327.

Schnee, J. M., and Hsueh, W. A. (2000). Angiotensin II, adhesion, and cardiac fibrosis. *Cardiovasc. Res.* **46,** 264–268.

Schoenwolf, G. C., and Garcia-Martinez, V. (1995). Primitive-streak origin and state of commitment of cells of the cardiovascular system in avian and mammalian embryos. *Cell. Mol. Biol. Res.* **41,** 233–240.

Schrader, M., Bendik, I., Becker-Andre, M., and Carlberg, C. (1993). Interaction between retinoic acid and vitamin D signaling pathways. *J. Biol. Chem.* **268,** 17830–17836.

Schultheiss, T. M., Xydas, S., and Lassar, A. B. (1995). Induction of avian cardiac myogenesis by anterior endoderm. *Development* **121,** 4203–4214.

Schunkert, H., Sadoshima, J., Cornelius, T., Kagaya, Y., Weinberg, E. O., Izumo, S., Riegger, G., and Lorell, B. H. (1995). Angiotensin II-induced growth responses in isolated adult rat hearts. Evidence for load-independent induction of cardiac protein synthesis by angiotensin II. *Circ. Res.* **76,** 489–497.

Schwartz, K., de la Bastie, D., Bouveret, P., Oliviero, P., Alonso, S., and Buckingham, M. (1986). Alpha-skeletal muscle actin mRNA's accumulate in hypertrophied adult rat hearts. *Circ. Res.* **59,** 551–555.

Searcy, R. D., Vincent, E. B., Liberatore, C. M., and Yutzey, K. E. (1998). A GATA-dependent nkx-2.5 regulatory element activates early cardiac gene expression in transgenic mice. *Development* **125,** 4461–4470.

Shenefelt, R. E. (1972a). Gross congenital malformations. Animal model: Treatment of various species with a large dose of vitamin A at known stages in pregnancy. *Am. J. Pathol.* **66,** 589–592.

Shenefelt, R. E. (1972b). Morphogenesis of malformations in hamsters caused by retinoic acid: Relation to dose and stage at treatment. *Teratology* **5,** 103–118.

Shindo, T., Manabe, I., Fukushima, Y., Tobe, K., Aizawa, K., Miyamoto, S., Kawai-Kowase, K., Moriyama, N., Imai, Y., Kawakami, H., Nishimatsu, H., Ishikawa, T., et al. (2002). Kruppel-like zinc-finger transcription factor KLF5/BTEB2 is a target for angiotensin II signaling and an essential regulator of cardiovascular remodeling. *Nat. Med.* **8,** 856–863.

Smith, S. M., Dickman, E. D., Thompson, R. P., Sinning, A. R., Wunsch, A. M., and Markwald, R. R. (1997). Retinoic acid directs cardiac laterality and the expression of early markers of precardiac asymmetry. *Dev. Biol.* **182,** 162–171.

Smith, S. M., Dickman, E. D., Power, S. C., and Lancman, J. (1998). Retinoids and their receptors in vertebrate embryogenesis. *J. Nutr.* **128,** 467S–470S.

Srivastava, D. (2000). Congenital heart defects: Trapping the genetic culprits. *Circ. Res.* **86,** 917–918.

Subbarayan, V., Mark, M., Messadeq, N., Rustin, P., Chambon, P., and Kastner, P. (2000). RXRalpha overexpression in cardiomyocytes causes dilated cardiomyopathy but fails to rescue myocardial hypoplasia in RXRalpha-null fetuses. *J. Clin. Invest.* **105,** 387–394.

Sucov, H. M., Dyson, E., Gumeringer, C. L., Price, J., Chien, K. R., and Evans, R. M. (1994). RXR alpha mutant mice establish a genetic basis for vitamin A signaling in heart morphogenesis. *Genes Dev.* **8,** 1007–1018.

Takeda, K., Ichiki, T., Funakoshi, Y., Ito, K., and Takeshita, A. (2000). Downregulation of angiotensin II type 1 receptor by all-trans retinoic acid in vascular smooth muscle cells. *Hypertension* **35,** 297–302.

Taylor, I. M., Wiley, M. J., and Agur, A. (1980). Retinoic acid-induced heart malformations in the hamster. *Teratology* **21,** 193–197.

Thompson, J. N., Howell, J. M., Pitt, G. A., and McLaughlin, C. I. (1969). The biological activity of retinoic acid in the domestic fowl and the effects of vitamin A deficiency on the chick embryo. *Br. J. Nutr.* **23,** 471–490.

Thorburn, J., Carlson, M., Mansour, S. J., Chien, K. R., Ahn, N. G., and Thorburn, A. (1995). Inhibition of a signaling pathway in cardiac muscle cells by active mitogen-activated protein kinase kinase. *Mol. Biol. Cell* **6,** 1479–1490.

Tipnis, S. R., Hooper, N. M., Hyde, R., Karran, E., Christie, G., and Turner, A. J. (2000). A human homolog of angiotensin-converting enzyme. Cloning and functional expression as a captopril-insensitive carboxypeptidase. *J. Biol. Chem.* **275,** 33238–33243.

Uozumi, H., Hiroi, Y., Zou, Y., Takimoto, E., Toko, H., Niu, P., Shimoyama, M., Yazaki, Y., Nagai, R., and Komuro, I. (2001). gp130 plays a critical role in pressure overload-induced cardiac hypertrophy. *J. Biol. Chem.* **276,** 23115–23119.

Vuillemin, M., and Pexieder, T. (1989). Normal stages of cardiac organogenesis in the mouse. I. Development of the external shape of the heart. *Am. J. Anat.* **184,** 101–113.

Wang, H. J., Zhu, Y. C., and Yao, T. (2002). Effects of all-trans retinoic acid on angiotensin II-induced myocyte hypertrophy. *J. Appl. Physiol.* **92,** 2162–2168.

Wasiak, S., and Lohnes, D. (1999). Retinoic acid affects left-right patterning. *Dev. Biol.* **215,** 332–342.

White, J. A., Guo, Y. D., Baetz, K., Beckett-Jones, B., Bonasoro, J., Hsu, K. E., Dilworth, F. J., Jones, G., and Petkovich, M. (1996). Identification of the retinoic acid-inducible all-trans-retinoic acid 4-hydroxylase. *J. Biol. Chem.* **271,** 29922–29927.

Wilson, J. G., and Warkany, J. (1950a). Cardiac and aortic arch anomalies in the offspring of vitamin A deficient rats correlated with similar human anomalies. *Pediatrics* **5,** 708–725.

Wilson, J. G., and Warkany, J. (1950b). Congenital anomalies of heart and great vessels in offspring of vitamin A-deficient rats. *Am. J. Dis. Child.* **79,** 963.

Wilson, J. G., Roth, C. B., and Warkany, J. (1953). An analysis of the syndrome of malformations induced by maternal vitamin A deficiency. Effects of restoration of vitamin A at various times during gestation. *Am. J. Anat.* **92,** 189–217.

Wobus, A. M., Kaomei, G., Shan, J., Wellner, M. C., Rohwedel, J., Ji, G., Fleischmann, B., Katus, H. A., Hescheler, J., and Franz, W. M. (1997). Retinoic acid accelerates embryonic stem cell-derived cardiac differentiation and enhances development of ventricular cardiomyocytes. *J. Mol. Cell. Cardiol.* **29,** 1525–1539.

Wolf, G. (1984). Multiple functions of vitamin A. *Physiol. Rev.* **64,** 873–937.

Wu, J., Garami, M., Cheng, T., and Gardner, D. G. (1996). 1,25(OH)2 vitamin D3, and retinoic acid antagonize endothelin-stimulated hypertrophy of neonatal rat cardiac myocytes. *J. Clin. Invest.* **97,** 1577–1588.

Yang-Yen, H. F., Zhang, X. K., Graupner, G., Tzukerman, M., Sakamoto, B., Karin, M., and Pfahl, M. (1991). Antagonism between retinoic acid receptors and AP-1: Implications for tumor promotion and inflammation. *New Biol.* **3,** 1206–1219.

Yutzey, K. E., and Bader, D. (1995). Diversification of cardiomyogenic cell lineages during early heart development. *Circ. Res.* **77,** 216–219.

Zechel, C., Shen, X. Q., Chen, J. Y., Chen, Z. P., Chambon, P., and Gronemeyer, H. (1994). The dimerization interfaces formed between the DNA binding domains of RXR, RAR and TR determine the binding specificity and polarity of the full-length receptors to direct repeats. *EMBO J.* **13,** 1425–1433.

Zhao, Q., Tao, J., Zhu, Q., Jia, P. M., Dou, A. X., Li, X., Cheng, F., Waxman, S., Chen, G. Q., Chen, S. J., Lanotte, M., Chen, Z., et al. (2004). Rapid induction of cAMP/PKA pathway during retinoic acid-induced acute promyelocytic leukemia cell differentiation. *Leukemia* **18,** 285–292.

Zhong, J. C., Huang, D. Y., Yang, Y. M., Li, Y. F., Liu, G. F., Song, X. H., and Du, K. (2004). Upregulation of angiotensin-converting enzyme 2 by all-trans retinoic acid in spontaneously hypertensive rats. *Hypertension* **44,** 907–912.

Zhong, J. C., Huang, D. Y., Liu, G. F., Jin, H. Y., Yang, Y. M., Li, Y. F., Song, X. H., and Du, K. (2005). Effects of all-trans retinoic acid on orphan receptor APJ signaling in spontaneously hypertensive rats. *Cardiovasc. Res.* **65,** 743–750.

Zhou, M. D., Sucov, H. M., Evans, R. M., and Chien, K. R. (1995). Retinoid-dependent pathways suppress myocardial cell hypertrophy. *Proc. Natl. Acad. Sci. USA* **92,** 7391–7395.

Zhou, X. F., Shen, X. Q., and Shemshedini, L. (1999). Ligand-activated retinoic acid receptor inhibits AP-1 transactivation by disrupting c-Jun/c-Fos dimerization. *Mol. Endocrinol.* **13,** 276–285.

Zile, M. H. (2004). Vitamin a requirement for early cardiovascular morphogenesis specification in the vertebrate embryo: Insights from the avian embryo. *Exp. Biol. Med. (Maywood)* **229,** 598–606.

Zile, M. H., Kostetskii, I., Yuan, S., Kostetskaia, E., St. Amand, T. R., Chen, Y., and Jiang, W. (2000). Retinoid signaling is required to complete the vertebrate cardiac left/right asymmetry pathway. *Dev. Biol.* **223,** 323–338.

11

Tocotrienols in Cardioprotection

Samarjit Das,* Kalanithi Nesaretnam,[†] and Dipak K. Das*

*Cardiovascular Research Center, University of Connecticut School of Medicine
Farmington, Connecticut 06030
[†]Malaysian Palm Oil Board, Kuala Lumpur, Malaysia

I. Introduction
II. A Brief History of Vitamin
   A. Vitamin E, Now and Then
   B. Tocotrienols Versus Tocopherols
   C. Sources of Tocotrienols
III. Tocotrienols and Cardioprotection
IV. Atherosclerosis
V. Tocotrienols in Free Radical Scavenging and Antioxidant Activity
VI. Tocotrienols in Ischemic Heart Disease
VII. Conclusions
   References

I. INTRODUCTION

Tocotrienols, a group of vitamin E stereoisomers, offer many health benefits including their ability to lower cholesterol levels and provide anticancer and tumor-suppressive activities. A diet rich in tocotrienols, especially dietary tocotrienols from a tocotrienol-rich fraction (TRF) of palm oil, reduced the concentration of plasma cholesterol and apolipoprotein B, platelet factor 4, and thromboxane B_2, indicating its ability to protect against platelet aggregation and endothelial dysfunction (Qureshi *et al.*, 1991a). Red palm oil is one of the richest sources of carotenoids; together with vitamin E, tocotrienols, and ascorbic acid present in this oil, it represents a powerful network of antioxidants, which can protect tissues and cells from oxidative damage (Edem, 2002; Hendrich *et al.*, 1994; Krinsky, 1992; Packer, 1992). For rat hearts, the isomers of α-tocotrienol were more proficient in the protection against oxidative stress induced by ischemia-reperfusion than α-tocopherol (Serbinova *et al.*, 1992). Tocotrienols are found to be more effective in central nervous system protection compared to α-tocopherol itself (Sen *et al.*, 2004). In another study, TRF is found to inhibit the glutamate-induced pp60c-src kinase activation in HT4 neuronal cells (Sen *et al.*, 2000). A recent study has indicated that TRF was able to reduce myocardial infarct size and improve postischemic ventricular dysfunction, and reduce the incidence of ventricular arrhythmias (Das *et al.*, 2005b). TRF was also shown to stabilize 20S and 26S proteasome activities and reduce the ischemia-reperfusion-induced increase in c-Src phosphorylation (Das *et al.*, 2005b).

The growing interest in tocotrienols among all other vitamin E isoforms is the purpose of this chapter.

II. A BRIEF HISTORY OF VITAMIN

Hippocrates (460–377 B.C.), the father of Medicine said, "Let food be thy medicine and medicine be thy food." In the eighteenth century, it was found that the intake of citrus fruits can reduce the development of scurvy. In 1905, a British clinician, William Fletcher, who was working with the disease beriberi, discovered that taking unpolished rice prevented beriberi and taking polished rice did not. On the basis of this finding he concluded that if some special factors were removed from the foods, there are high chances to have diseases. The very next year, Dr. Fletcher's hypothesis became stronger when another British biochemist, Sir Frederick Gowland Hopkins, found that foods contained necessary "accessory factors" in addition to proteins, carbohydrates, fats, minerals, and water.

In 1911, Polish chemist Casimir Funk discovered that the anti-beriberi substance in unpolished rice was an amine, so Dr. Funk named the special amine as "vitamine" for "vita amine" after "vita" means life and "amine" which he found in the unpolished rice, a nitrogen containing substance. It was later discovered that many vitamins do not contain nitrogen, and, therefore, not all vitamins are amine. Because of its widespread use, Funk's term continued to be applied, but the final letter "e" was dropped.

In 1912, Hopkins and Funk further advanced the vitamin hypothesis of deficiency, a theory that postulates that the absence of sufficient amounts of a particular vitamin in a system may lead to certain diseases. During the early 1900s, through experiments in which animals were deprived of certain types of foods, scientists succeeded in isolating and identifying the various vitamins recognized today.

A. VITAMIN E, NOW AND THEN

In 1922 at Berkeley University in California, a physician scientist Dr. Herbert M. Evans and his assistant Katherine S. Bishop discovered a fat-soluble alcohol that functioned as an antioxidant, which they named "Factor X" (Papas, 1999). Evans and Bishop were feeding rats a semi-purified diet when they noticed that the female rats were unable to produce offspring because the pups died in the womb. They then fed the female rats lettuce and wheat germ, and observed that healthy offspring were produced. During their research, Evans and Bishop discovered that "Factor X" was contained in the lipid extract of the lettuce and concluded that this "Factor X" was fat soluble (Papas, 1999). In 1924, Dr. Bennett Sure renamed "Factor X" as vitamin E. The first component identified was α-tocopherol. It was named as such from the Greek *tokos* (*offspring*) and *pheros* (*to bear*) and the *ol* ending was added to indicate the alcoholic properties of the molecules. For over more than 30 years, it was well believed that vitamin E existed in only one form, α-tocopherol. As a result vitamin E named as tocopherol. It is the most abundant form of vitamin E found in blood and body tissue. But in 1956, scientist J. Green discovered the eight isoforms of vitamin E, four tocopherol isomers (α-, β-, γ-, and δ-) and four tocotrienol isomers (α-, β-, γ-, and δ-), split into two different categories: tocopherols and tocotrienols which are corresponding stereoisomers. Tocopherols and tocotrienols are very similar, except for the fact that tocopherols have a saturated phytyl tail, and tocotrienols have an isoprenoid tail with three unsaturated points. In addition, on the chromanol nucleus, the various isoforms differ in their methyl substitutions (Fig. 1). Tocotrienols are initially named as ζ-, ε-, or η-tocopherols. The δ-form has one methyl group, the γ- and β-forms have two methyl groups, and α-form contains three methyl groups on its chromanol head. Tocopherols and tocotrienols share a common chromanol head and a side chain at the C-2 position (Theriault *et al.*, 1999). Very recently two new isomers of tocotrienols have been found and are present in TRF of rice bran oil, desmethyl (D-P_{21}-T3) and didesmethyl (D-P_{25}-T3) tocotrienols (Qureshi *et al.*, 2001b).

The therapeutic application of vitamin E was first shown by Kamimura (1977). The inhibitory effect of the unsaturated fatty acid by α-tocopherol was well established by Tappel (1953, 1954, 1955). This observation was

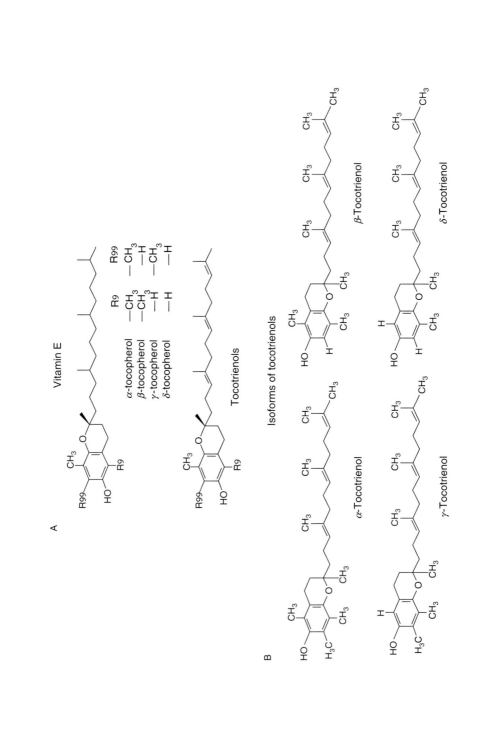

repeated in humans with the same result by Horwitt *et al.* (1956), in the very next year when Tappel *et al.* identified the fact that the deficiency of α-tocopherol may lead to the high levels of oxidative lipid damage. Antioxidative effect of vitamin E can be due to the equal contribution of phenolic head as well as the phytyl tails was explained in both *in vitro* and *in vivo* studies (Burton and Ingold, 1989). The important discoveries of various aspects of α-tocopherols are listed in Table I. Since α-tocopherol is the most abundant vitamin E in the body, its activity as an antioxidant and its role in protection from oxidative stress have been studied more extensively than other forms of vitamin E. Studies show that α-tocopherols are protective against atherosclerosis. A study (Devaraj and Jialal, 2005) of α-tocopherol's effect on important pro-inflammatory cytokine, tumor necrosis factor (TNFα) release from human monocytes, found that α-tocopherols inhibited the release of TNF via inhibition of 5-lipoxygenase. Inhibition of 5-lipoxygenase also significantly reduced TNF mRNA and NF-κB-binding activity. Other studies (Meydani, 2004) show how α-tocopherol inhibits the activation of endothelial cells stimulated by high levels of low-density lipoprotein (LDL) cholesterol and pro-inflammatory cytokines. This inhibition is associated with the suppression of chemokines, the expression of cell surface adhesion molecules, and the adhesion of leukocytes to endothelial cells, all of which contribute to the development of lesions in the arterial wall. While the benefits of tocopherols have been studied for years, health benefits of the other seven forms of vitamin E are only recently being explored. Just like cholesterol, tocopherols also influence the biophysical membrane characteristics, like fluidity (Sen *et al.*, 2006).

But for the last few years, researchers have been focusing toward tocotrienols more as compared to tocopherols because of the fact that tocotrienols have a more potent antioxidative property than α-tocopherols (Serbinova and Packer, 1994; Serbinova *et al.*, 1991). Still there is not enough research going on tocotrienols as compared to the extensive work done on tocopherols.

B. TOCOTRIENOLS VERSUS TOCOPHEROLS

There are at least eight isoforms that are commonly found to have vitamin E's activity: α-, β-, γ-, and δ-tocopherol; α-, β-, γ-, and δ-tocotrienol (Fig. 1). Tocotrienols differ from tocopherols by having a farnesyl (isoprenoid) structure compared to saturated phytyl side chain. Yet, the focus on tocopherols is much higher than that of tocotrienols. Out of the studies done

---

FIGURE 1. (A) Chemical structures of two different isoforms of vitamin E, tocopherol and tocotrienols. (B) Four different isoforms of tocotrienols, α, β, γ, and δ, differ by their methyl group position in their respective ring structure.

TABLE I. Background Information of Vitamin E

Year	Event
1922	Food factor X discovered by H. M. Evans and L. S. Bishop as a substance essential for rat pregnancy
	Food factor X found in yeast and lettuce by H. A. Martill
1923	Food factor X found in alfalfa, wheat, oats, and butter by H. M. Evans et al.
1924	Food factor X named vitamin E by B. Sure
1936	α-Tocopherol extracted from wheat-germ oil by H. M. Evans et al.
1938	Chemical structure of vitamin E determined by E. Fenholz
	DL-α-tocopherol synthesized by P. Karrer
1950	Research on application of vitamin E in treating frostbite started by M. Kamimura
1956	Free radical theory of the aging process proposed by D. Harmann
	Eight homologues of vitamin E (tocopherols and tocotrienols) discovered by J. Green
1961	Vitamin E (DL-α-tocopherol) admitted to the Japanese Pharmacopoeia
1962	Antioxidant activity in the body suggested by A. L. Tappel
1968	Recommended dietary allowance (RDA) of vitamin E set at 30 IU (20-mg α-TE) in the United States
1972	Recommended dietary allowance (RDA) of vitamin E revised to 10 IU (7-mg α-TE) in the United States
1988	The approval standards for vitamin products revised and the daily intake of vitamin E as an OTC product set at 300 mg/day in Japan
1991	Vitamin E shown in MONICA Study to reduce risk of coronary disorders
	α-Tocopherol transport protein (α-TTP), which selectively transports α-tocopherol, isolated from the liver
1993	Familial vitamin E deficiency reported by C. Ben Hamida et al.
1994	Vitamin E intake reported to reduce mortality from coronary heart disorders by M. C. Bellizzi et al.—European PARADOX
1996	Vitamin E shown in CHAOS Study to reduce risk of myocardial infarction
1997	Vitamin E reported by M. Sano et al. to delay progression of Alzheimer's disease
	Vitamin E reported by S. N. Meydani et al. to activate immunologic competence in the elderly
	Vitamin E reported by A. Herday et al. to improve liver function in patients with hepatitis C
1999	Vitamins C and E reported by L. C. Chappell et al. to relieve preeclampsia
	Recommended dietary allowances of vitamin E set at 10 mg for males and 8 mg for females in Japan by the sixth revision of Nutrition Requirements
2000	Vitamins E reported by M. Boaz et al. to reduce risk of cardiovascular disease in hemodialysis patients
2004	Vitamin E reported by S. N. Meydani et al. to lower the incidence of common colds in elderly nursing home residents

Source: www.eisai.co.jp

on tocopherols only 1% have been done on tocotrienols (Sen et al., 2006). But for the last few years, there has been a growing interest among researchers on tocotrienols as compared to tocopherols.

The abundance of α-tocopherol in the living cells compares to other isoforms and of course the maximum half life period of the same isoform may be the major cause of its research importance among the various disciplines of clinical research. But it is well established that the antioxidative power of tocotrienols is 1600 times more than that of α-tocopherol (Serbinova and Packer, 1994). There is evidence that tocotrienols are more potent when compared with tocopherols. The reason of this increased efficacy is the unsaturation in the aliphatic tail which facilitates easier penetration into the tissue (Suzuki et al., 1993) and also because of unsaturation in the aliphatic tail tocotrienols are a more potent antioxidant than tocopherols. These important findings may be attracting many other researchers to consider tocotrienols as a better therapeutic agent than tocopherols. It has already been proved by various research groups that tocotrienols possess neuroprotective, anticancer, and also cholesterol lowering properties as compared to its other isoform, tocopherols.

It is only tocotrienols, which at nanomolar concentration protect the neuronal cells from glutamate-induced cell death (Khanna et al., 2003; Roy et al., 2002; Sen et al., 2000). In a very interesting study, Sen et al. (2000) showed that tocotrienol, but not tocopherol, inhibits the activation of pp60 (c-Src), which is a key regulator of glutamate-induced neuronal cell death. In another study, it was found that tocotrienols, and not tocopherols, protect the neurons from glutamate-induced 12-lipoxygenase (12-Lox) activation (Khanna et al., 2003). This 12-Lox takes very important part in signal transduction pathway to kill the neurons. The molecular level of target for the neuroprotective effect of tocotrienols, mainly α-tocotrienol, is cytosol, but not at the nucleus (Khanna et al., 2003; Sen et al., 2000). It is now a well-established fact that, it is tocotrienols, mainly α-tocotrienol, which possess a potent neuroprotection at very low concentration, but not any other tocopherol (Khanna et al., 2005).

The anti-carcinogenic property of tocotrienols has been established. Many studies have shown tocotrienols provide better protection against cancer than tocopherols do. In mice, tocotrienols were compared with α-tocopherols, and interestingly it was found that i.p. administration of α- and γ-tocotrienols, and not α-tocopherols, showed a slight life-prolonging effect in mice from transplanted tumors (Komiyama et al., 1989). Similar observation was found by Gould et al. (1991) in rats for chemoprevention of chemically induced mammary tumors. Also, in human study it was shown that tocotrienols significantly suppress growth of breast cancer cells in culture, whereas tocopherols fail to show similar action under identical conditions (Nesaretnam et al., 1995). In another study, it was shown that the anti-carcinogenic property of tocotrienols may be a better option than tamoxifen from breast cancer prevention (Guthrie et al., 1997). γ- and δ-tocotrienols are considered as the most effective isoform among all the eight isoforms of vitamin E, for there physiological role in modulating normal mammary gland growth, function, and remodeling (McIntyre et al., 2000). Relative to tocopherols, tocotrienols are a more potent suppressor of EGF-dependent normal mammary epithelial cell growth, the mechanism of which is by the deactivation of PKCα, compared to tocopherols (Sylvester et al., 2002).

The hypocholesterolemic effect of tocotrienols is also found to be more potent than that of tocopherols. Due to the presence of three double bonds in the isoprenoid chain, tocotrienols can lower cholesterol levels much more effectively compared to tocopherols (Qureshi et al., 1986). Tocotrienols significantly reduced the concentration of plasma cholesterol and apolipoprotein B, platelet factor 4, and thromboxane B_2, indicating its ability to protect against platelet aggregation and endothelial dysfunction (Qureshi et al., 1991a,b). It was found that tocotrienols and not tocopherols suppress the 3-hydroxy-3-methylglutaryl Coenzyme A (HMG-CoA) reductase, which directly inhibits the biosynthesis of cholesterol (Parker et al., 1993; Pearce et al., 1992, 1994). Later on, the significant hypocholesterolemic effect of tocotrienols was compared with tocopherols in humans (Qureshi et al., 1995, 2001a, 2002), chicken (Qureshi and Peterson, 2001), hypercholesterolemic rat (Iqbal et al., 2003), swine (Qureshi et al., 2001b), and hamster's plasma (Raederstorff et al., 2002). In conclusion, researchers have shown that for lowering of the cholesterol, tocotrienols are the better option than tocopherols.

C. SOURCES OF TOCOTRIENOLS

Tocotrienols are mainly found in the seed endosperm of almost all the monocots such as wheat, rice, barley oat, rye, and sour cherry. In some dicots, endosperm also contains tocotrienols as in some Apiaceae species and also in some Solanaceae species such as tobacco (Sen et al., 2006). But tocotrienols are not the only member of vitamin E present in the endosperm, but it is always present as a mixture of tocopherol–tocotrienols. That is why researchers normally use TRF, the ratio between tocotrienols to tocopherols. The TRF of rice bran oil, 90:10, is the maximum so far identified. In this particular oil, apart from the normal four isoforms of tocotrienols, there are two new isoforms also found as well, desmo- and didesmo-tocotrienols. Crude palm oil extract from the fruit of Elaeis guineensis also contain higher concentration of TRF, almost 80:20. Normally, the major components of palm-derived TRF extract contain mainly 36% γ-tocotrienol, 26–30% α-tocotrienol and 20–22% α-tocopherol and 12% δ-tocotrienol (Kamat et al., 1997).

III. TOCOTRIENOLS AND CARDIOPROTECTION

Since cardiovascular disease contributes in a major way to the morbidity and mortality, it is becoming a strain on the economy of many countries worldwide. Various factors have been identified as possible causes of different cardiac diseases such as heart failure and ischemic heart disease. As discussed earlier, tocotrienols are very poorly studied compared to tocopherols. Due to this reason there is very few evidence of cardioprotective effect of tocotrienols, whereas the cardioprotective effect of tocopherols is immense.

IV. ATHEROSCLEROSIS

Atherosclerosis is the process by which the deposition of cholesterol plaques takes place on the wall of blood vessels and makes those vessels narrow, ultimately getting blocked by those fatty deposits. Atherosclerosis finally leads ischemia to the heart muscle and can cause damage to the heart muscle. The complete blockage of the arteries leads to myocardial infarction (MI). According to the World Health Organization, the major cause of death in the world as a whole by the year 2020 will be acute coronary occlusion (Murray and Lopez, 1997).

As mentioned earlier, tocotrienols differ from tocopherols only in three double bonds in the isoprenoid chain which appear to be essential for the inhibition of cholesterogenesis by higher cell penetration and followed by better interaction with the deposited plaques (Qureshi et al., 1986). In some clinical trials with hypercholesterolemics patients, tocotrienols significantly reduced the serum cholesterols (Qureshi et al., 1991b). In another similar kind of clinical trial, tocotrienols lowered both the serum cholesterols, total cholesterol (TC), and more interestingly the LDL (Tan et al., 1991). In the late 1980s, it was found that one of the major cause of lipid oxidation was the oxidation of LDL. Therefore, Tan et al.'s (1991) observation draws many researchers attention toward tocotrienols as a better anti-lipid oxidative agent. Later on, a diet rich in tocotrienols, especially dietary tocotrienols from a TRF palm oil, reduced the concentration of plasma cholesterol and apolipoprotein B, platelet factor 4, and thromboxane B_2, indicating its ability to protect against platelet aggregation and endothelial dysfunction (Qureshi et al., 1991a,b). In mammalian cells, HMG-CoA reductase enzyme was found to regulate the cholesterol production. Tocotrienols, mainly γ-isoform or the tocotrienols mixture, significantly suppress the secretion of HMG-CoA reductase, ultimately lowering the production of cholesterols in the cells (Parker et al., 1993; Pearce et al., 1992, 1994). Another possible mechanism of protection from lipid peroxidation by tocotrienols was found by isoprenoid-mediated suppression of mevalonate synthesis that depletes tumor tissues of two intermediate products, farnesyl pyrophosphate and geranylgeranyl pyrophosphate, which are incorporated posttranslationally into growth control-associated proteins (Elson and Qureshi, 1995). From the above observations, researchers have also started to compare tocotrienols with any statin group of medicine. In one of the study, Qureshi et al. (2001b) showed that when tocotrienols were applied with lovastatin or when lovastatin was compared to tocotrienol action, there was no difference in terms of cholesterol lowering power in chicken. Very interestingly, it was found apart from these two mechanism of tocotrienols; tocotrienols are also protecting from hypercholesterolemic phase by activating the conversion of LDL to high-density lipoprotein (HDL) through the inter phase VLDL–VDL and finally HDL (Qureshi et al., 1995, 2001a). In hypercholesterolemic phase, it was also observed either γ-tocotrienol or the tocotrienols mixture increases the number of HDL, which then interact with LDL to reduce the concentration of LDL in the plasma (Qureshi et al., 1995), HDL may also go by phagocytosis to lower the LDL concentration. In another clinical trial, 100 mg/day of TRF derived from rice bran oil effectively lower the serum cholesterol in hypercholesterolemic patients (Qureshi et al., 2002). The same study showed that α-tocopherol induce the HMG-CoA reductase and that is why in higher doses of TRF, the opposite effect is observed to some extent compared to 100 mg/day of TRF (Qureshi et al., 2002). This may be due to the fact that tocotrienols were found to go on conversion into tocopherols *in vivo* (Qureshi et al., 2001a). This study clearly showed it is only tocotrienols which are responsible for the lowering of serum cholesterol, but not tocopherols. Tocopherols may increase the cholesterol level by inducing HMG-CoA reductase (Qureshi et al., 2002).

V. TOCOTRIENOLS IN FREE RADICAL SCAVENGING AND ANTIOXIDANT ACTIVITY

TRF has excellent free radical scavenging capacity (Kamat et al., 1997). Numerous studies (Ikeda et al., 2003; Kamat et al., 1997) show that it is a potent inhibitor of lipid peroxidation and protein peroxidation in rat microsomes and mitochondria. At low concentrations of 5 μM, TRF, mainly γ-tocotrienol and to a lesser extent α- and δ-tocotrienols, significantly inhibited oxidative damage to both lipids and proteins in rat brain mitochondria. Studies of the effect of γ-tocotrienols on endothelial nitric oxide synthase (eNOS) activity in spontaneously hypertensive rats have reported that on treatment with antioxidant γ-tocotrienol increased the nitric oxide (NO) activity and concomitantly reduced the blood pressure and enhanced total antioxidant status in plasma and blood vessels (Ikeda et al., 2003). In general, TRF has significantly higher antioxidant ability as compared to tocopherols. This can be explained by the structural difference between the saturated side chain of tocopherols and the unsaturated side chain of tocotrienols. The molecular mobility of polyenoic lipids in the membrane bilayer (composed mainly of unsaturated fatty acid) is much higher than that of saturated lipids, and hence tocotrienols are more mobile and less restricted in their interaction with lipid radicals in membranes than tocopherols.

This is further supported by the higher effectiveness of tocotrienols in processes that may involve oxidative stress such as in red blood cells where tocotrienols have more potency against oxidative hemolysis than α-tocopherols (Kamat et al., 1997). In an *in vitro* study, the potent free radical scavenging property of α-tocopherol was found 1600 times more as compared to free radical scavenging property of α-tocotrienols (Serbinova and Packer, 1994). In another study, it was found the potent antioxidative property of γ-tocotrienols significantly protect the spontaneously hypertensive rats (Newaz et al., 2003).

VI. TOCOTRIENOLS IN ISCHEMIC HEART DISEASE

Ischemia is a stage when there is no blood flow in a cell; as blood is the only carrier of air or oxygen, cells become subject to a lot of stress due to lack of oxygen. When this kind of situation arises in the heart, the disease is known as ischemia heart disease. Apart from atherosclerotic plaque deposition, oxidative stress is also considered as one of the major causes of ischemic heart disease.

The excellent free radical scavenging property of tocotrienols attenuates the oxidative stress better compared to tocopherols. That is why, recently researchers are considering tocotrienols as a better therapeutic option from ischemic heart disease compared to tocopherols.

γ-Tocotrienols are found to act as a myocardial preconditioning agent by activating the eNOS expression (Ikeda et al., 2003). eNOS is considered one of the major cause of intracellular NO generator. This NO then goes on vasodialation and protects the heart from ischemic phase. Due to the eNOS regulating property of γ-tocotrienol is now considered as an important pharmacological preconditioning agent. In a very recent study, it was shown for the first time that beneficial effects of tocotrienol derived from palm oil are due to its ability to reduce c-Src activation, which is linked with the stabilization of proteasomes, mainly 20S and 26S (Das et al., 2005b). Tocotrienols have extremely short half-lives; after oral ingestion, they are not recognized by α-tocotrienol transport protein, which also accounts for their low bioavailability. For this reason, TRF was used in an acute experiment to determine its immediate effects on the ischemic-reperfused myocardium. The results indicate that tocotrienol readily blocks the ischemia-reperfusion-mediated increase in Src kinase activation and proteasome inactivation, thereby providing cardioprotection (Das et al., 2005b). After this observation, in the continuing study by the same group, but this time with gavaging of the TRF derived from palm oil, for 15 days protection of the heart from ischemia-reperfusion injury was observed (Das et al., 2005a). In this chronic experiment, it was also observed that the key mechanism may be the inhibition of Src activation by TRF. Myocardial ischemia-reperfusion caused an induction of the expression of c-Src protein (Hattori *et al.*, 2001) inhibition of c-Src with PPI reduces the extent of cellular injury. The ability of TRF to block the increased phosphorylation of c-Src appears to play a crucial role in its ability to protect the heart from ischemia reperfusion injury.

VII. CONCLUSIONS

It should be clear from the above discussion that tocotrienols, as TRF, provide cardioprotection not only by its cholesterol-lowering property or by its reducing oxidative stress but also through their ability to performing redox signaling by potentiating an anti-death signal through the reduction of proapoptotic factors, at least c-Src was identified, thereby leading to the decrease in cardiomyocytes apoptosis. Out of a minimum of four different isoforms of tocotrienols, α- and γ-tocotrienols are considered as the effective isoforms, especially, which possess the cardioprotective abilities. Both α- and γ-isoforms are found to possess anti-atherosclerotic properties, not only by reducing the LDL cholesterols but also by increasing the number of HDL cholesterols and also simultaneous induction of HMG-CoA reductase activity. Apart from anti-atherosclerotic property, TRF was found to be protective both acutely and chronically, from ischemia-reperfusion-mediated cardiac dysfunction by inhibiting the phosphorylation of c-Src expression significantly with both 20S and 26S proteasome stabilization.

ACKNOWLEDGMENTS

This study was supported by NIH HL 34360, HL 22559, HL 33889, and HL 56803.

REFERENCES

Burton, G. W., and Ingold, K. U. (1989). Vitamin E as an *in vitro* and *in vivo* antioxidant. *Ann. NY Acad. Sci.* **570,** 7–22.

Das, S., Nesaretnam, K., and Das, D. K. (2005a). Cardioprotective abilities of palm oil derived tocotrienol rich factor. Proceedings of Malaysian Palm Oil Board.

Das, S., Powell, S. R., Wang, P., Divald, A., Nesaretnam, K., Tosaki, A., Cordis, G. A., Maulik, N., and Das, D. K. (2005b). Cardioprotection with palm tocotrienol: Andioxidant activity of tocotrienol is linked with its ability to stabilize proteasomes. *Am. J. Physiol. Heart Circ. Physiol.* **289,** H361–H367.

Devaraj, S., and Jialal, I. (2005). Alpha-tocopherol decreases tumor necrosis factor-alpha mRNA and protein from activated human monocytes by inhibition of 5-lipoxygenase. *Free Radic. Biol. Med.* **38,** 1212–1220.

Edem, D. O. (2002). Palm oil: Biochemical, physiological, nutritional, hematological, and toxicological aspects: A review. *Plant Foods Hum. Nutr.* **57,** 319–341.

Elson, C. E., and Qureshi, A. A. (1995). Coupling the cholesterol- and tumor-suppressive actions of palm oil to the impact of its minor constituents on 3-hydroxy-3-methylglutaryl coenzyme A reductase activity. *Prostaglandins Leukot. Essent. Fatty Acids* **52,** 205–207.

Gould, M. N., Haag, J. D., Kennan, W. S., Tanner, M. A., and Elson, C. E. (1991). A comparison of tocopherol and tocotrienol for the chemoprevention of chemically induced rat mammary tumors. *Am. J. Clin. Nutr.* **53**(4 Suppl.), 1068S–1070S.

Guthrie, N., Gapor, A., Chambers, A. F., and Carroll, K. K. (1997). Inhibition of proliferation of estrogen receptornegative MDA-MB-435 and -positive MCF-7 human breast cancer cells by palm oil tocotrienols and tamoxifen, alone and in combination. *J. Nutr.* **127,** 544S–548S.

Hattori, R., Otani, H., Uchiyama, T., Imamura, H., Cui, J., Maulik, N., Cordis, G. A., Zhu, L., and Das, D. K. (2001). Src tyrosine kinase is the trigger but not the mediator of ischemic preconditioning. *Am. J. Physiol. Heart Circ. Physiol.* **281,** H1066–H1074.

Hendrich, S., Lee, K., Xu, X., Wang, H., and Murphy, P. A. (1994). Defining food components as new nutrients. *J. Nutr.* **124,** 1789S–1792S.

Horwitt, M. K., Harvey, C. C., Duncan, G. D., and Wilson, W. C. (1956). Effects of limited tocopherol intake in man with relationships to erythrocyte hemolysis and lipid oxidations. *Am. J. Clin. Nutr.* **4,** 408–419.

Ikeda, S., Tohyama, T., Yoshimura, H., Hamamura, K., Abe, K., and Yamashita, K. (2003). Dietary alpha-tocopherol decreases alpha-tocotrienol but not gamma-tocotrienol concentration in rats. *J. Nutr.* **133,** 428–434.

Iqbal, J., Minhajuddin, M., and Beg, Z. H. (2003). Suppression of 7,12-dimethylbenz[alpha]-anthracene-induced carcinogenesis and hypercholesterolaemia in rats by tocotrienol-rich fraction isolated from rice bran oil. *Eur. J. Cancer Prev.* **12,** 447–453.

Kamat, J. P., Sarma, H. D., Devasagayam, T. P., Nesaretnam, K., and Basiron, Y. (1997). Tocotrienols from palm oil as effective inhibitors of protein oxidation and lipid peroxidation in rat liver microsomes. *Mol. Cell. Biochem.* **170,** 131–137.

Kamimura, M. (1977). Physiology and clinical use of vitamin E (author's translation). *Hokkaido Igaku Zasshi* **52,** 185–188.

Khanna, S., Roy, S., Ryu, H., Bahadduri, P., Swaan, P. W., Ratan, R. R., and Sen, C. K. (2003). Molecular basis of vitamin E action: Tocotrienol modulates 12-lipoxygenase, a key mediator of glutamate-induced neurodegeneration. *J. Biol. Chem.* **278,** 43508–43515.

Khanna, S., Roy, S., Slivka, A., Craft, T. K., Chaki, S., Rink, C., Notestine, M. A., DeVries, A. C., Parinandi, N. L., and Sen, C. K. (2005). Neuroprotective properties of the natural vitamin E alpha-tocotrienol. *Stroke* **36,** 2258–2264.

Komiyama, K., Iizuka, K., Yamaoka, M., Watanabe, H., Tsuchiya, N., and Umezawa, I. (1989). Studies on the biological activity of tocotrienols. *Chem. Pharm. Bull. (Tokyo)* **37,** 1369–1371.

Krinsky, N. I. (1992). Mechanism of action of biological antioxidants. *Proc. Soc. Exp. Biol. Med.* **200,** 248–254.

McIntyre, B. S., Briski, K. P., Tirmenstein, M. A., Fariss, M. W., Gapor, A., and Sylvester, P. W. (2000). Antiproliferative and apoptotic effects of tocopherols and tocotrienols on normal mouse mammary epithelial cells. *Lipids* **35,** 171–180.

Meydani, M. (2004). Vitamin E modulation of cardiovascular disease. *Ann. NY Acad. Sci.* **1031,** 271–279.

Murray, C. J., and Lopez, A. D. (1997). Alternate projections of mortality and disability by cause 1990–2020: Global burden of disease study. *Lancet* **349,** 1498–1504.

Nesaretnam, K., Guthrie, N., Chambers, A. F., and Carroll, K. K. (1995). Effect of tocotrienols on the growth of a human breast cancer cell line in culture. *Lipids* **30,** 1139–1143.

Newaz, M. A., Yousefipour, Z., Nawal, N., and Adeeb, N. (2003). Nitric oxide synthase activity in blood vessels of spontaneously hypertensive rats: Antioxidant protection by gamma-tocotrienol. *J. Physiol. Pharmacol.* **54,** 319–327.

Packer, L. (1992). Interactions among antioxidants in health and disease. Vitamin E and the redox cycle. *Proc. Soc. Exp. Biol. Med.* **200**, 271–276.

Papas, A. M. (1999). "The Vitamin E Factor." Harper Collins Publishers Inc., New York.

Parker, R. A., Pearce, B. C., Clark, R. W., Gordon, D. A., and Wright, J. J. (1993). Tocotrienols regulate cholesterol production in mammalian cells by post-transcriptional suppression of 3-hydroxy-3-methylglutarylcoenzyme A reductase. *J. Biol. Chem.* **268**, 11230–11238.

Pearce, B. C., Parker, R. A., Deason, M. E., Qureshi, A. A., and Wright, J. J. (1992). Hypocholesterolemic activity of synthetic and natural tocotrienols. *J. Med. Chem.* **35**, 3595–3606.

Pearce, B. C., Parker, R. A., Deason, M. E., Dischino, D. D., Gillespie, E., Qureshi, A. A., Volk, K., and Wright, J. J. (1994). Inhibitors of cholesterol biosynthesis. 2. Hypocholesterolemic and antioxidant activities of benzopyran and tetrahydronaphthalene analogues of the tocotrienols. *J. Med. Chem.* **37**, 526–541.

Qureshi, A. A., and Peterson, D. M. (2001). The combined effects of novel tocotrienols and lovastatin on lipid metabolism in chickens. *Atherosclerosis* **156**, 39–47.

Qureshi, A. A., Burger, W. C., Peterson, D. M., and Elson, C. E. (1986). The structure of an inhibitor of cholesterol biosynthesis isolated from barley. *J. Biol. Chem.* **261**, 10544–10550.

Qureshi, A. A., Qureshi, N., Hasler-Rapacz, J. O., Weber, F. E., Chaudhary, V., Crenshaww, T. D., Gapor, A., Ong, A. S., Chong, Y. H., Peterson, D., and Rapacz, J. (1991a). Dietary tocotrienols reduce concentrations of plasma cholesterol, apolipoprotein B, thromboxane B_2, and platelet factor 4 in pigs with inherited hyperlipidemias. *Am. J. Clin. Nutr.* **53**(Suppl. 4), 1042S–1046S.

Qureshi, A. A., Qureshi, N., Wright, J. J., Shen, Z., Kramer, G., Gapor, A., Chong, Y. H., DeWitt, G., Ong, A., and Peterson, D. M. (1991b). Lowering of serum cholesterol in hypercholesterolemic humans by tocotrienols (palmvitee). *Am. J. Clin. Nutr.* **53**, 1021S–1026S.

Qureshi, A. A., Bradlow, B. A., Brace, L., Manganello, J., Peterson, D. M., Pearce, B. C., Wright, J. J., Gapor, A., and Elson, C. E. (1995). Response of hypercholesterolemic subjects to administration of tocotrienols. *Lipids* **30**, 1171–1177.

Qureshi, A. A., Sami, S. A., Salser, W. A., and Khan, F. A. (2001a). Synergistic effect of tocotrienol-rich fraction (TRF(25)) of rice bran and lovastatin on lipid parameters in hypercholesterolemic humans. *J. Nutr. Biochem.* **12**, 318–329.

Qureshi, A. A., Peterson, D. M., Hasler-Rapacz, J. O., and Rapacz, J. (2001b). Novel tocotrienols of rice bran suppress cholesterogenesis in hereditary hypercholesterolemic swine. *J. Nutr.* **131**, 223–230.

Qureshi, A. A., Sami, S. A., Salser, W. A., and Khan, F. A. (2002). Dose-dependent suppression of serum cholesterol by tocotrienol-rich fraction (TRF(25)) of rice bran in hypercholesterolemic humans. *Atherosclerosis* **161**, 199–207.

Raederstorff, D., Elste, V., Aebischer, C., and Weber, P. (2002). Effect of either gamma-tocotrienol or a tocotrienol mixture on the plasma lipid profile in hamsters. *Ann. Nutr. Metab.* **46**, 17–23.

Roy, S., Lado, B. H., Khanna, S., and Sen, C. K. (2002). Vitamin E sensitive genes in the developing rat fetal brain: A high-density oligonucleotide microarray analysis. *FEBS Lett.* **530**, 17–23.

Sen, C. K., Khanna, S., Roy, S., and Packer, L. (2000). Molecular basis of vitamin E action. Tocotrienol potently inhibits glutamate-induced pp60(c-Src) kinase activation and death of HT4 neuronal cells. *J. Biol. Chem.* **275**, 13049–13055.

Sen, C. K., Khanna, S., and Roy, S. (2004). Tocotrienol: The natural vitamin E to defend the nervous system? *Ann. NY Acad. Sci.* **1031**, 127–142.

Sen, C. K., Khanna, S., and Roy, S. (2006). Tocotrienols: Vitamin E beyond tocopherols. *Life Sci.* **78,** 2088–2098.

Serbinova, E., Kagan, V., Han, D., and Packer, L. (1991). Free radical recycling and intramembrane mobility in the antioxidant properties of alpha-tocopherol and alpha-tocotrienol. *Free Radic. Biol. Med.* **10,** 263–275.

Serbinova, E., Khwaja, S., Catudioc, J., Ericson, J., Torres, Z., Gapor, A., Kagan, V., and Packer, L. (1992). Palm oil vitamin E protects against ischemia/reperfusion injury in the isolated perfused Langendorff heart. *Nutr. Res.* **12,** S203–S215.

Serbinova, E. A., and Packer, L. (1994). Antioxidant properties of alpha-tocopherol and alpha-tocotrienol. *Methods Enzymol.* **234,** 354–366.

Suzuki, Y. J., Tsuchiya, M., Wassall, S. R., Choo, Y. M., Govil, G., Kagan, V. E., and Packer, L. (1993). Structural and dynamic membrane properties of alpha-tocopherol and alpha-tocotrienol: Implication to the molecular mechanism of their antioxidant potency. *Biochemistry* **32,** 10692–10699.

Sylvester, P. W., Nachnani, A., Shah, S., and Briski, K. P. (2002). Role of GTP-binding proteins in reversing the antiproliferative effects of tocotrienols in preneoplastic mammary epithelial cells. *Asia Pac. J. Clin. Nutr.* **11**(Suppl. 7), S452–S459.

Tan, D. T., Khor, H. T., Low, W. H., Ali, A., and Gapor, A. (1991). Effect of a palm-oil-vitamin E concentrate on the serum and lipoprotein lipids in humans. *Am. J. Clin. Nutr.* **53**(Suppl. 4), 1027S–1030S.

Tappel, A. L. (1953). The inhibition of hematin-catalyzed oxidations by alpha-tocopherol. *Arch. Biochem. Biophys.* **47,** 223–225.

Tappel, A. L. (1954). Studies of the mechanism of vitamin E action. II. Inhibition of unsaturated fatty acid oxidation catalyzed by hematin compounds. *Arch. Biochem. Biophys.* **50,** 473–485.

Tappel, A. L. (1955). Studies of the mechanism of vitamin E action. III. *In vitro* copolymerization of oxidized fats with protein. *Arch. Biochem.* **54,** 266–280.

Theriault, A., Chao, J. T., Wang, Q., Gapor, A., and Adeli, K. (1999). Tocotrienol: A review of its therapeutic potential. *Clin. Biochem.* **32,** 309–319.

12

Cytodifferentiation by Retinoids, a Novel Therapeutic Option in Oncology: Rational Combinations with Other Therapeutic Agents

Enrico Garattini, Maurizio Gianni', and Mineko Terao

Laboratorio di Biologia Molecolare, Centro Catullo e Daniela Borgomainerio Istituto di Ricerche Farmacologiche "Mario Negri," via Eritrea 62 20157 Milano, Italy

I. Premise and Scope: Differentiation Therapy with Retinoids Is a Significant Goal in the Management of the Neoplastic Diseases
II. The Classical Nuclear RAR Pathway Is Complex and Has Led to the Development of Different Types of Synthetic Retinoids
III. Retinoids Promote Differentiation in Numerous Types of Neoplastic Cells
    A. Myeloid Leukemia
    B. Neuroblastoma
    C. Head and Neck Cancer
    D. Breast Carcinoma

E. Teratocarcinoma
  F. Melanoma
 IV. Retinoids Exert Pleiotropic Effects Interacting with
     Multiple Intracellular Pathways: An Opportunity for
     Combination Therapy
     A. Growth Factors and Cytokines: G-CSF, TGF-β,
        and Interferons
     B. THE cAMP Pathway
     C. The MAP Kinase Pathway
     D. THE PI3K/AKT Pathway
     E. Protein Kinase C
     F. Histone Acetylation and DNA Methylation
  V. Retinoid-Based Differentiation Therapy, General
     Observations, and Conclusion
     References

Retinoic acid (RA) and derivatives are promising antineoplastic agents endowed with both therapeutic and chemopreventive potential. Although the treatment of acute promyelocytic leukemia with all-*trans* retinoic acid is an outstanding example, the full potential of retinoids in oncology has not yet been explored and a more generalized use of these compounds is not yet a reality. One way to enhance the therapeutic and chemopreventive activity of RA and derivatives is to identify rational combinations between these compounds and other pharmacological agents. This is now possible given the information available on the biochemical and molecular mechanisms underlying the biological activity of retinoids. At the cellular level, the antileukemia and anticancer activity of retinoids is the result of three main actions, cytodifferentiation, growth inhibition, and apoptosis. Cytodifferentiation is a particularly attractive modality of treatment and differentiating agents promise to be less toxic and more specific than conventional chemotherapy. This is the result of the fact that cytotoxicity is not the primary aim of differentiation therapy. At the molecular level, retinoids act through the activation of nuclear retinoic acid receptor-dependent and -independent pathways. The cellular pathways and molecular networks relevant for retinoid activity are modulated by a panoply of other intracellular and extracellular pathways that may be targeted by known drugs and other experimental therapeutics. This chapter aims to summarize and critically discuss the available knowledge in the field. © 2007 Elsevier Inc.

I. PREMISE AND SCOPE: DIFFERENTIATION THERAPY WITH RETINOIDS IS A SIGNIFICANT GOAL IN THE MANAGEMENT OF THE NEOPLASTIC DISEASES

Natural retinoic acids (RAs) and their synthetic derivatives, collectively known as retinoids, are promising agents in the chemoprevention and treatment of the neoplastic disease. The natural isomer, 13-*cis* retinoic acid (13-*cis* RA), alone or in combination with interferon is used in the chemoprevention and treatment of head and neck cancer (Hong *et al.*, 1986; Lippman *et al.*, 1993). Fenretinide, a synthetic retinoid, has shown efficacy in the secondary chemoprevention of breast cancer (Veronesi *et al.*, 1999). Other prevention studies with retinoids have been performed with positive results in the following conditions: xeroderma pigmentosum, basal cell carcinoma, and squamous cell carcinoma of the skin (Gravis *et al.*, 1999; Kraemer *et al.*, 1988; Sankowski *et al.*, 1987). Etretinate, a synthetic analogue of RA decreases the recurrence of superficial bladder carcinoma (Studer *et al.*, 1984, 1995). Chemoprevention of precancerous lesions is another setting where different types of retinoids have shown efficacy. Treatment of oral leukoplakia with 13-*cis* RA underscores the potential of retinoids for this type of clinical application (Hong *et al.*, 1986).

As already mentioned, chemoprevention is not the only setting in which retinoids are useful. This class of agents finds clinical application or is proposed also in the treatment of various neoplastic diseases. In this setting, the most significant example of the use of retinoids is represented by acute promyelocytic leukemia (APL) (Castaigne *et al.*, 1990; Huang *et al.*, 1988; Warrell *et al.*, 1991). APL is a rare form of acute myeloid leukemia (AML), which accounts for no more than 10% of all the AMLs in the adult. A single course of all-*trans* retinoic acid (atRA) induces complete clinical remission in the vast majority of APL patients. After treatment with a single monotherapeutic course of atRA, remissions are generally short-lived, with median durations of 6–8 months. This is followed by relapses, which are often resistant to further treatment with the retinoid (Ding *et al.*, 1998; Muindi *et al.*, 1992). The problem prevents the use of atRA as a single agent in the treatment of APL. However, this has not undermined the therapeutic and clinical significance of atRA in the disease, as the compound has been adopted as an important component of the standard chemotherapeutic regimen used in APL (Leone *et al.*, 1998; Sanz *et al.*, 2004; Tallman *et al.*, 2002). Indeed, addition of atRA to the cocktail of cytotoxic drugs already in use has proven superior to chemotherapy alone in the treatment of this type of leukemia.

The original observation that atRA induces remission in APL through a mechanism of action that is distinct from cytotoxicity, is regarded as a milestone in the history of medicine (Huang *et al.*, 1988). The retinoid is the first and only example of clinically successful cytodifferentiating agent (Garattini and Terao, 2001) and the results obtained in APL represent proof of principle that cytodifferentiation is a viable option for the treatment of leukemia and possibly other types of cancer. Indeed, the neoplastic cell is characterized by multiple deficits not only in the control of proliferation and survival but also in the processes of cell maturation and differentiation. Thus, control of the leukemic and cancer cell can be achieved by reducing the proliferation rate, diminishing the survival advantage, primarily through activation of the apoptotic program, or favoring maturation along the physiological differentiation pathway. Cytodifferentiation is a particularly attractive modality of treatment and differentiating agents promise to be less toxic and more specific than conventional chemotherapy. This is the result of the fact that cytotoxicity is not the primary aim of differentiation therapy.

Currently, the promise of differentiation therapy is only partially met and a more general use of atRA and other retinoids as differentiating agents in oncology is hampered by a number of problems, including natural and acquired resistance as well as local and systemic toxicity. The problems can be overcome by increasing the efficacy and/or the therapeutic index of atRA and congeners. This can be achieved by potentiating retinoid differentiating activity using combination therapy approaches. In fact, the pleiotropic activity of retinoids and the multiplicity of targets these compounds act on give ample opportunities in terms of combination therapy. atRA and derivatives modulate the activity of numerous genes and intracellular pathways. On the other hand the activity of nuclear retinoic acid receptors (RARs) is controlled by various signals, including different types of kinase cascades. Often, the cross talk between atRA-dependent and other intracellular pathways modulates the cytodifferentiating activity of the retinoid. The knowledge on the molecular mechanisms underlying this cross talk has increased tremendously over the course of the last few years. This knowledge can be exploited to design rational combinations of atRA or retinoids and other biologically active compounds that may have significance in the context of differentiation therapy.

The aim of this chapter of the book is to give a brief overview of the potential of retinoid-based differentiation therapy, focusing on the possibility to increase the therapeutic efficacy of retinoids by rational combination approaches. With this in mind, we will summarize and discuss the current knowledge on the cross talk between the retinoid and other intracellular pathways. We will consider specific pathways involved in the process of cytodifferentiation, with particular reference to those with the potential to be targeted pharmacologically. We believe that the data provided will serve as a useful starting point for the design of combination strategies aimed at improving retinoid-based differentiation therapy.

II. THE CLASSICAL NUCLEAR RAR PATHWAY IS COMPLEX AND HAS LED TO THE DEVELOPMENT OF DIFFERENT TYPES OF SYNTHETIC RETINOIDS

atRA and its natural isomers, 13-*cis* RA and 9-*cis* retinoic acid (9-*cis* RA) bind and activate a small subset of receptors belonging to the superfamily of steroid nuclear receptors (Chambon, 1996; Kastner *et al.*, 1995; Lohnes *et al.*, 1992). As the topic is treated in great details in other sections of this book, we provide only the information strictly necessary to understand the remainder of the chapter. Nuclear RARs are ligand-dependent transactivation factors and control the transcriptional activity of numerous genes via binding to their DNA regulatory elements. Two types of nuclear RARs are known, RARs and RXRs. Three main isoforms of each type of receptors have been described, RARα, RARβ, and RARγ, as well as RXRα, RXRβ, and RXRγ. Each isotype is coded for by a separate gene and splicing variants of each receptor isoform differing at the N-terminus are known (Chambon, 1996). Two types of transcriptionally active complexes have been described, RXR/RAR heterodimers and RXR/RXR homodimers. atRA and 13-*cis* RA bind to and transactivate only RXR/RAR complexes, while 9-*cis* RA interacts with both RXR/RAR and RXR/RXRs. RXR/RAR dimers are believed to be nonpermissive complexes, in which RXRs act as silent partners. In other words when the cognate ligand is bound to the RAR moiety, the RXR counterpart loses the ability to bind its corresponding ligand (Mangelsdorf and Evans, 1995; Mangelsdorf *et al.*, 1993). Both RARs and RXRs are modular proteins consisting of a ligand-independent transactivating domain (AB region), a DNA-binding domain (D region), and an E region containing the ligand-binding as well as ligand-dependent transactivating functions. The structure of the two receptors is completed by two regions of uncharacterized function, the C and F domains.

In basal conditions, the RXR/RAR dimer is bound to the cognate DNA sequence (RARE, retinoic acid-responsive element) and interacts with a multiprotein complex known as the corepressor (Wei, 2004; Weston *et al.*, 2003; Xu *et al.*, 1999a). The corepressor contains proteins endowed with histone deacetylase (HDAC) and DNA-methylating activity that concur to keep the surrounding chromatin structure in a "closed" state, effectively suppressing the transcriptional activity of RNA polymerase II. On ligand binding, the corepressor is released from the RXR/RAR dimer and substituted by the "coactivator," which consists of a multiprotein complex with histone acetylase and demethylase activity (Edwards, 1999; Glass *et al.*, 1997; Gronemeyer and Miturski, 2001; Shibata *et al.*, 1997; Westin *et al.*, 2000; Xu *et al.*, 1999a). These two last enzymatic activities are involved in opening the chromatin structure and igniting transcription. The activity of the RXR/RAR dimer is controlled not only by its interaction with the ligand but also by a number of accessory signals in which phosphorylation events stand out (Gianni et al., 2002a,b, 2003, 2006; Parrella et al., 2004; Rochette-Egly, 2003). A further layer of control is represented by the rate of proteolytic degradation of the RXR/RAR dimer and the various components of the corepressor and coactivator complexes (Gianni et al., 2006). Degradation is predominantly mediated by the proteasome pathway and is controlled by some of the phosphorylation events mentioned above. Little is known about the RXR/RXR pathway, which is far less studied than the RXR/RAR counterpart.

The progress in the knowledge of the RA nuclear receptor pathway and the clinical results obtained with retinoids in the therapeutic and chemopreventive setting has resulted in the synthesis and development of a large number of synthetic retinoid molecules endowed with specific properties (Beehler et al., 2004; Benbrook et al., 1997; Brtko and Thalhamer, 2003; Chandraratna, 1997, 1998a,b; Crowe, 2002). The current armamentarium of synthetic retinoids consists of the following classes of molecules:

1. Pan-RAR agonists, that is, synthetic retinoids capable of interacting with similar affinity with RARα, RARβ, and RARγ. These molecules have the same spectrum of receptor selectivity as atRA and are sometimes more potent and often more toxic *in vivo* than the parent compound. The prototype of this class is TTNPB which is a more powerful ligand of RARs than atRA and is one of the first compounds developed.

2. RARα agonists and antagonists (Tamura et al., 1990). These compounds are characterized by a more rigid structure than atRA and discriminate RARα from RARβ and RARγ very well. AM580, an RARα agonist, is one of the earliest molecules to be synthesized. We showed that the molecule is a much more powerful cytodifferentiating agent than atRA in APL cells and activates the oncogenic fusion protein PML-RARα better than RARα itself (Gianni et al., 1996a).

3. RARβ agonists and antagonists. The crystal structure of the ligand-binding domain of RARβ has been used to identify selective agonists and antagonists (Germain et al., 2004). With this strategy, a ligand that shows RARβ selectivity with a 100-fold higher affinity to RARβ than to α or γ isotypes was identified. RARβ agonists and antagonists may be useful to address pharmacologically the tumor suppressor role of RARβ *in vitro* and in animal models.

4. RARγ agonists and antagonists. Numerous RARγ agonists are known. However, some of the molecules synthesized are not simply RARγ agonists and are classified as atypical retinoids (Garattini et al., 2004). The prototypes of atypical retinoids are CD437 and ST1926 (Garattini et al., 2004; Mologni et al., 1999; Ponzanelli et al., 2000). The two molecules are endowed with strong and selective apoptotic properties that are largely unrelated to their

ability of interacting with RARγ. These types of compounds are interesting anticancer agents on their own, but will not be considered in this article, as they are largely devoid of cytodifferentiating activity.

5. RXR agonists and antagonists (Alvarez *et al.*, 2004). These agents act on the silent partner of the RXR/RAR heterodimer or the RXR/RXR homodimer. RXR agonists are also known as rexinoids and have been used alone or in combinations with RAR-specific retinoids and other nuclear receptor ligands such as 1,25-dihydroxy vitamin D3 (Kang *et al.*, 1997; Li *et al.*, 1997, 1999). In the myeloid leukemia setting, activation of the RXR/RXR pathway by rexinoids induces an apoptotic but not a differentiating response (Boehm *et al.*, 1995; Gottardis *et al.*, 1996; Li *et al.*, 1999; Nagy *et al.*, 1995).

6. Dissociated or anti-AP-1 retinoids (Chen *et al.*, 1995). This type of retinoid binds to nuclear RARs of the RAR type but is unable to transactivate them. However, dissociated retinoids maintain the ability to transrepress the c-Fos/c-Jun-containing transcriptional complexes known as AP-1. Transrepression of the AP-1 complexes is a well-known phenomenon and is believed to play a role in the antiproliferative activity exerted by retinoids in certain cellular contexts. Indeed, activation of AP-1 serves as a potent mitogenic stimulus in various situations (Shaulian and Karin, 2002).

In spite of the availability of a remarkable number of synthetic retinoids possessing different properties, the quest for novel molecules is not over. Further synthetic molecules of potential clinical significance are expected as a consequence of the progress on the elucidation of the tridimensional structure of RARs and RXRs and the molecular mechanisms of action of retinoids.

III. RETINOIDS PROMOTE DIFFERENTIATION IN NUMEROUS TYPES OF NEOPLASTIC CELLS

As discussed above, induction of terminal differentiation of the neoplastic cell is an achievable and important goal in cancer therapeutics. atRA and other retinoids promote cellular maturation of numerous cell types and in various experimental conditions. Thus, this class of compounds is endowed with general cytodifferentiating properties. In cell cultures, the cytodifferentiating activity of retinoids is almost invariably accompanied by growth inhibition and the two processes are difficult to dissociate (Gianni *et al.*, 2000). On the other hand, retinoid-dependent cytodifferentiation is not necessarily associated with cell death or apoptosis. Indeed, classical retinoids are relatively weak apoptotic agents and, in some cases, they even exert a prosurvival action (Lomo *et al.*, 1998). These aspects of retinoid pharmacology need to be considered when discussing the use of such agents in oncology. An antiproliferative effect superimposed to cytodifferentiation is highly desirable, whereas an antiapoptotic action should be avoided.

A. MYELOID LEUKEMIA

Most of the available data on the cytodifferentiating properties of atRA and derivatives were obtained in the context of AML. Thus, the characteristics of this heterogeneous collection of diseases will be described in some detail. According to the French American British (FAB) classification, AMLs can be grouped into seven entities (M1–M7) on the basis of the phenotypic and morphological appearance of the leukemic blasts. AMLs classified as M1–M3 are committed along the granulocytic or monocytic pathways and differentiate in either direction according to the stimulus applied. The HL-60 myeloid cell line is representative of an M2 AML blast maturing along the granulocytic and monocytic pathway, if exposed to atRA and vitamin D3, respectively. APL is classified as M3, is exquisitely sensitive to the cytodifferentiating action of pharmacological concentrations of atRA, and undergoes terminal differentiation along the granulocytic pathway (Melnick and Licht, 1999; Sirulnik et al., 2003). Paradoxically, the APL blast is characterized by specific chromosomal translocations involving RARα, the major form of nuclear RARs expressed in the hematopoietic system (Melnick and Licht, 1999). The vast majority of APL patients carries a balanced translocation (t15:17) involving chromosomes 15 and 17, leading to the synthesis of the oncogenic fusion protein PML–RARα. In the fusion protein, the very N-terminal region of RARα is substituted by a portion of the oncosuppressor PML. This results in the synthesis of a retinoid receptor that retains ligand-dependent transactivating activity, but acquires novel properties relative to RARα. Very rare APL variants, in which the translocation partners of chromosome 17 are chromosomes 11 (PLZF–RARα fusion protein), and 5 (STAT5b–RARα fusion protein) (Dong and Tweardy, 2002) or (Npm–RARα fusion protein), have been described (Chen et al., 1994; Guidez et al., 1994). Interestingly, APL blasts expressing the PLZF–RARα fusion protein are completely resistant to the cytodifferentiating activity of atRA (Guidez et al., 1994; Melnick and Licht, 1999). This has led to the concept that PML–RARα is not only a classical oncogene involved in the first stages of the leukemic process but also mediates the cytodifferentiating response observed following challenge of the APL blast with pharmacological concentrations of atRA. Indeed, a widely accepted view is that expression of PML–RARα confers a growth and prosurvival advantage to the leukemic blast, interfering with the physiological signal transduction pathways regulated by RARα and/or PML. Interference with the physiological RARα pathway is shared by PML–RARα and the other rare fusion proteins observed in APL. This phenomenon is likely to be relevant for the maturation block at the level of the promyelocyte observed in APL, since physiological concentrations of atRA are known to regulate myelopoiesis (Collins, 2002; Gaines and Berliner, 2003). By contrast PML–RARα is the only fusion protein that maintains transcriptional activation by supraphysiological (pharmacological) concentrations of atRA, explaining sensitivity to the cytodifferentiating effect of the retinoid. Although, this may represent an oversimplified picture, it fits with the observation that atRA-resistant relapses of APL are often associated with mutations in the ligand-binding domain of PML–RARα (Cote *et al.*, 2000).

The maturation of M4 and M5 blasts is blocked at the level of committed monocyte precursor cells. Treatment of this type of blasts with atRA *in vitro* does not result in signs of morphological differentiation along the monocytic pathway. M6 and M7 blasts represent committed precursors of the erythrocyte and megakaryocyte pathways, respectively, and are similarly refractory to atRA-induced cytodifferentiation. Although there are reports showing that non-M3 AML blasts synthesize functional RARs (Grande *et al.*, 2001) and atRA-dependent monocytic differentiation of M2 blasts can be obtained (Manfredini *et al.*, 1999), an overwhelming amount of data indicate that the only type of AML which is reproducibly sensitive to the cytodifferentiating activity of retinoids is APL. At present, it is unclear whether sensitivity to atRA in this particular type of leukemia is the consequence of the expression of PML–RARα or is due to other as yet unidentified and cell-context-specific factors. The maturation response observed in the APL blast is recapitulated quite faithfully in the NB4 cellular model (Lanotte *et al.*, 1991). This is an immortalized cell line developed from an APL patient, maintaining expression of an active form of PML–RARα and showing sensitivity to atRA. NB4 cells are a unique tool to study the antileukemic activity of retinoids and have been extensively used to define the molecular mechanisms underlying the process of differentiation triggered by the retinoid in the myeloid context (Bastie *et al.*, 2005; Cao *et al.*, 2005; Gianni *et al.*, 1994, 1995a,b,c, 1997, 2001; Idres *et al.*, 2001; Parrella *et al.*, 2004).

B. NEUROBLASTOMA

As already alluded to, atRA exerts cytodifferentiating effects on a broad range of neoplastic cell types. Neuroblastoma is a pediatric tumor and represents the most frequent form of peripheral neuronal malignancy. Human neuroblastoma cell lines are sensitive to the differentiating action of retinoids. The earliest data in this context were obtained with 13-*cis* RA, one of the natural isomers of vitamin A (Reynolds *et al.*, 2003). In more recent studies, 13-*cis* RA has been substituted with the more powerful retinoid atRA, or RAR and RXR selective agonists. Although differences in the complement of expressed RARs may occur in different neuroblastoma cell lines (Carpentier *et al.*, 1997; Joshi *et al.*, 2006; Lovat *et al.*, 1997; Melino *et al.*, 1997; Nguyen *et al.*, 2003), it seems that this cell type synthesizes predominantly RARα, RARγ, and at least RXRα in a constitutive fashion. As observed in various other cell types, RARβ is induced by challenge with retinoids. The available data indicate that activation of RXR/RARα and

RXR/RARβ is sufficient to trigger the neuronal differentiation program (Carpentier *et al.*, 1997; Joshi *et al.*, 2006; Lovat *et al.*, 1997; Melino *et al.*, 1997; Nguyen *et al.*, 2003). Induction of RARβ and activation of the RXR/RARβ complex may be of particular therapeutic interest, as high expression of the receptor has been associated with good outcome in neuroblastoma (Cheung *et al.*, 1998). In spite of all this, it has been reported that costimulation of the RXR/RAR and RXR/RXR pathways by 9-*cis* RA or combinations of RAR and RXR agonists results in stronger cytodifferentiation (Chu *et al.*, 2003). Neuronal maturation of neuroblastoma cells is accompanied by growth arrest and a variable level of apoptosis. atRA-dependent growth inhibition is associated with downregulation of the specific oncogene MYCN regardless of its state of gene amplification (Castel and Canete, 2004). While growth arrest is always associated with neuronal differentiation, the apoptotic response can be dissociated. This last observation suggests that the molecular mechanisms underlying atRA-dependent cytodifferentiation and apoptosis in neuroblastoma cells are distinct and may involve a different set of nuclear RAR complexes. The cytodifferentiating effect of retinoids in neuroblastoma cell lines is likely to be of therapeutic significance, as 13-*cis* RA has proven efficacious in the treatment of neuroblastoma, when given after chemotherapy (Niles, 2000). It is interesting to notice that 13-*cis* RA rather than atRA is the preferred retinoid in the disease. This may be explained by the fact that, at clinically achievable drug levels, 13-*cis* RA has proven equal or superior to atRA in inducing morphological differentiation and growth arrest of neuroblastoma cell lines (Veal *et al.*, 2002).

C. HEAD AND NECK CANCER

A wealth of data on the cytodifferentiating, growth inhibitory, and apoptotic effects of retinoids is available in the case of head and neck cancer (Lotan, 1997). This is mainly the result of the fact that retinoids have been proposed for the chemoprevention of secondary tumors and premalignant lesions of the aerodigestive cavities. Head and neck squamous cell carcinomas (HNSCC) are characterized by an aberrant form of squamous differentiation, which is not observed in the normal epithelial counterpart. Retinoids suppress the growth and squamous differentiation of HNSCC both *in vitro* (Higuchi *et al.*, 2003) and *in vivo* (Satake *et al.*, 2003). Suppression of the squamous phenotype is believed to be the result of a cytodifferentiating effect, which is also observed in premalignant lesions of the oral cavity known as leukoplakia (Lee *et al.*, 2000; Shin *et al.*, 1997; Youssef *et al.*, 2004). Most of the HNSCC cell lines express RARα, RARγ, and RXRα in basal conditions (Klaassen *et al.*, 2001; Lotan, 1996; Sun *et al.*, 2000). Retinoids are capable of inducing the expression of RARβ in some of these cell lines. Suppression of the expression of RARβ has been described in premalignant oral lesions and HNSCC cell lines (Lotan, 1997), suggesting that the receptor plays a role in maintaining the normal phenotype of the epithelial cells lining the aerodigestive cavities. A survey of various RAR isotype-specific agonists indicates that activation of all the possible forms of RXR/RAR complexes is potentially capable of mediating growth inhibition by retinoids (Sun *et al.*, 2000). In contrast RXR/RXR agonists are generally devoid of antiproliferative effects (Sun *et al.*, 1999). Suppression of squamous differentiation may be the result of both RARγ and RARβ activation, as transfection of both types of receptors potentiates atRA-dependent inhibition of squamous molecular markers, such as cytokeratin 1 and transglutaminase (Xu *et al.*, 1999b).

D. BREAST CARCINOMA

Retinoids have been proposed in the adjuvant treatment of breast carcinoma for their ability to inhibit growth and induce morphological or phenotypic differentiation of breast carcinoma cell lines (Paik *et al.*, 2003; Yang *et al.*, 2002). As in many other cellular models, retinoid-dependent growth inhibition and cytodifferentiation are tightly linked processes. Most of the studies conducted with breast carcinoma cell lines have focused on the antiproliferative activity of retinoids (del Rincon *et al.*, 2003; Gottardis *et al.*, 1996). This is explained by the fact that differentiation of a breast cancer cell line is difficult to define given the general lack of accepted markers. Morphological criteria (Seewaldt *et al.*, 1999), cytokeratin (Jing *et al.*, 1996) and lactogenic markers (Di Lorenzo *et al.*, 1993), as well as components of the cell/cell and cell/matrix adhesion machinery (Pellegrini *et al.*, 1995; Petersen *et al.*, 1998) have been used to assess the cytodifferentiating properties of retinoids. It is generally accepted that breast cancer cells express RARα, RARγ, and RXRα. As in the case of head and neck cancer, suppression of RARβ expression is related to the etiology or progression of breast carcinoma (Widschwendter *et al.*, 1997; Yang *et al.*, 2001). RARα agonists are as efficient as pan-RAR agonists in terms of growth arrest (Schneider *et al.*, 2000), implicating this receptor as the main determinant of retinoid sensitivity in breast cancer cells (Schneider *et al.*, 2000). An important and peculiar characteristic of the breast carcinoma model is the dependence of retinoid sensitivity on the estrogen receptor (ER) status of the neoplastic cell line. Cell lines derived from ER^+ tumors, which are generally associated with favorable outcome and are sensitive to tamoxifen, respond to retinoids *in vitro*, while the ER^- counterpart is generally refractory (Rousseau *et al.*, 2004). This suggests that ER controls the activity of nuclear RARs in this cellular context. The molecular mechanisms underlying this phenomenon are completely unknown, although they may be related to the control of the expression levels of RARα in the neoplastic cell.

E. TERATOCARCINOMA

The remarkable amount of data available in the teratocarcinoma cell model is related to the widespread use of this system as an *in vitro* surrogate for the study of the developmental effects of retinoids. In fact, the majority of the data available have been obtained in two mouse teratocarcinoma cell lines, F9 and P19 (Alonso *et al.*, 1991; Bain *et al.*, 1994; Grabel *et al.*, 1998; Ingraham *et al.*, 1989; Lehtonen *et al.*, 1989). *In vitro* treatment of F9 cells with physiological or pharmacological concentrations of atRA and derivatives results in endodermal differentiation (Gianni *et al.*, 1991, 1993). This process can be followed both morphologically and through determination of the levels of molecular markers, such as laminin B and collagen type IV (Boylan *et al.*, 1995). The cell line can be reprogrammed along the visceral endodermal pathway on addition of cAMP-elevating agents to atRA or other retinoids (Rochette-Egly and Chambon, 2001). Differentiation to the primitive endoderm is strictly under the control of RARγ, as demonstrated with F9 sublines genetically deleted of the corresponding gene (Taneja *et al.*, 1997). In the F9 model, differentiation can be dissociated, albeit partially, from growth inhibition, which seems to be mainly under the control of RARβ (Faria *et al.*, 1999). In spite of all these data, recent studies conducted with F9 cells by our group both *in vitro* and *in vivo*, indicate that retinoid-dependent cytodifferentiation does not contribute significantly to the overall antitumor activity of active retinoids (E.G. and M.T., unpublished observations). Regardless of what observed in the F9 model, the most frequent type of cytodifferentiation program activated in teratocarcinoma leads to maturation along the neuronal pathway (Bain *et al.*, 1994; Ingraham *et al.*, 1989). Differentiation can be studied morphologically by counting the number and measuring the length of axons and dendrites induced by retinoid treatment (McBurney *et al.*, 1988; Parnas and Linial, 1997; Pyle *et al.*, 2001). Indeed, teratocarcinoma cells have a flat appearance and they show no evidence of axonal filaments when cultured in standard conditions. Addition of retinoids to the culture medium results in the appearance and growth of cell arborizations that are typical of neurons. Morphological differentiation is accompanied by the expression of neuronal molecular markers, such as GAP 43, which is often used to follow neuronal maturation of atRA-treated teratocarcinoma cells (Shen *et al.*, 2004). Given the embryonal stem (ES) cell characteristics of the teratocarcinoma cells, it does not come as a surprise that atRA induces neuronal maturation also of normal embryonal ES cells (Gajovic *et al.*, 1997; Renoncourt *et al.*, 1998). This last type of phenomenon may be of therapeutic relevance more in the context of the use of ES cells for the treatment of neurological disorders than in the realm of oncology.

F. MELANOMA

Metastatic melanoma is a devastating type of neoplasia for which no specific therapeutic options are available. Retinoids exert antiproliferative and cytodifferentiating effects in the mouse B16 melanoma cell line (Desai and Niles, 1997; Huang *et al.*, 2003; O'Connor and Fujita, 1995). Human melanoma cell lines are generally resistant to the effects of retinoids. In this cellular setting, cytodifferentiation is studied by determining the levels of enzymes or products involved in the synthesis of melanin, which is considered as a marker of phenotypic maturation (Niles, 2003). Once again, retinoid-dependent growth inhibition and differentiation of melanoma cell lines along the melanocytic pathway are tightly associated processes. The complement of retinoid nuclear receptors expressed in melanoma cells does not seem to be different from that of many other solid tumor counterparts, that is, RARα and RARγ constitutive expression as well as retinoid-dependent induction of RARβ. Activation of RARα and RARγ seems to be of particular relevance for the growth-inhibitory effects exerted by atRA in melanoma cells, as indicated by the results obtained with selective receptor agonists and antagonists in the human melanoma cell line, SK MEL 28 (Emionite *et al.*, 2003). A couple of human melanoma cell lines, S91 and A375, were the object of an interesting study aimed at defining the mechanisms underlying natural resistance to retinoids. S91 are sensitive to the growth inhibitory and cytodifferentiating effects of atRA, while A375 are totally unresponsive to the agent. The two cell lines express comparable levels of various components of the retinoid-signaling pathway. However, A375 cells have substantially higher levels of intracellular reactive oxygen species (ROS). ROS seems to exert inhibitory effects on the transactivation properties of RXR/RAR transcriptional complexes, explaining refractoriness to retinoids (Demary *et al.*, 2001). Interestingly, lowering ROS levels by culturing cells in hypoxic conditions or in the presence of antioxidants results in RXR/RAR reactivation. This represents a novel epigenetic control mechanism of retinoid receptor activity.

IV. RETINOIDS EXERT PLEIOTROPIC EFFECTS INTERACTING WITH MULTIPLE INTRACELLULAR PATHWAYS: AN OPPORTUNITY FOR COMBINATION THERAPY

The clinical results obtained with atRA in APL and the growing body of evidence indicating that retinoids are large-spectrum cytodifferentiating agents have raised enthusiasm over the use of these compounds in the management of different types of oncologic diseases. However, a generalized use of retinoids in oncology is hampered by a number of unresolved problems including natural and induced resistance as well as toxicity. These problems have a major impact in the context of both the therapeutic and chemopreventive use of retinoids. Toxicity issues are of particular evidence, given the fact that differentiation therapy with retinoids requires prolonged administration of the agents. This is particularly true in the case of the chemopreventing setting, where the effect of retinoids may last as long as the compounds are administered. Chronic exposure to retinoids is accompanied by serious effects at the level of the central nervous and hepatic systems (headache, pseudotumor cerebri, and hypertriglyceridemia), as well as the well-known teratogenic problems typically associated with the administration of this class of compounds. Clearly these problems call for strategies aimed at increasing the efficacy and the therapeutic index of retinoids. There are two possible ways to achieve these goals: (a) to develop novel, more powerful and less toxic synthetic retinoids and (b) to identify non-retinoid agents capable of potentiating the pharmacological activity of retinoids without affecting their toxicity. It is our opinion that the second approach is particularly promising and likely to result in significant progress. In fact, the nuclear retinoid receptor pathway interacts with numerous other intracellular pathways, some of which are of obvious significance from a therapeutic point of view. The knowledge acquired on the major biochemical pathways modulated by retinoids or having a regulatory action on nuclear RARs is reviewed in the next sections of this chapter.

A. GROWTH FACTORS AND CYTOKINES: G-CSF, TGF-β, AND INTERFERONS

Retinoids modulate the activity of various exocrine and endocrine factors with different types of activities on the neoplastic cell. Some of these factors, such as granulocyte colony-stimulating factor (G-CSF), granulocyte macrophage colony-stimulating factor (GM-CSF), or transforming growth factors (TGFs) are involved in the physiological differentiation and maturation of specific cell lineages. Targeting this type of molecules and the corresponding signal transduction pathways is very attractive from a therapeutic prospective. In fact, interfering with these pathways gives the opportunity to enhance the cytodifferentiating activity of retinoids in the neoplastic and the normal cells of origin rather specifically, resulting in a higher probability to increase the therapeutic index.

A good example of what mentioned above is the cross talk between retinoids and G-CSF. G-CSF is a growth factor controlling the maturation of myeloid precursors along the granulocytic pathway. Addition of G-CSF to atRA in the culture medium of freshly isolated APL cells or the derived NB4 cell line results in more rapid and more complete granulocytic maturation over what observed in the presence of the retinoid alone (Gianni *et al.*, 1994, 1995a; Higuchi *et al.*, 2004; Imaizumi *et al.*, 1994). These types of effects are observed in the presence of sub-optimal concentrations of atRA and with concentrations of G-CSF that are completely devoid of differentiating activity on their own. Enhanced differentiation is evident both at the morphological level and in terms of the expression of granulocytic maturation markers (Gianni *et al.*, 1994). For instance, leukocyte alkaline phosphatase, a specific marker of the terminally differentiated granulocyte, is turned on only in the presence of atRA and G-CSF. Enhanced differentiation is accompanied by a more pronounced growth arrest (Gianni et al., 1994). Similar effects are observed in blasts obtained from a few cases of chronic myeloid leukemia (CML) and in some cases of non-M3 AML, but not in atRA-resistant myeloid leukemia cell lines. From a therapeutic perspective, one potential drawback of a combined treatment of the leukemic blast with atRA + G-CSF is a decrease in the secondary apoptotic response to the retinoid. Indeed, G-CSF is a well-known survival factor for the differentiated granulocyte (Colotta et al., 1992). However, at least in vitro, the apoptotic effect of atRA seems to dominate and mask the antiapoptotic activity of G-CSF. The molecular mechanisms underlying the potentiating effect of G-CSF are likely to be straightforward and the result of retinoid-dependent effects on the G-CSF signal transduction pathway. In fact, atRA causes a marked increase in the surface expression of the G-CSF membrane receptor (G-CSFR) (Gianni et al., 1994; Tkatch et al., 1995). As a result of this, G-CSF binding to G-CSF-R results in a more pronounced activation of the downstream signaling pathway, with increased phosphorylation of signal transduction and activator of transcription (STAT3) (E.G., unpublished observations). So far there are no reports describing a modulation of the nuclear RAR pathway by G-CSF. The enhancing effect of the cytokine is relatively specific, as other molecules of the same family, like GM-CSF, TNFα, and TGF-β, do not potentiate the differentiating activity of atRA in APL cells. As a matter of fact, TNFα seems to exert an inhibitory action on the atRA-induced granulocytic maturation of myeloid leukemia cells orienting them along the monocytic pathway (Witcher et al., 2004). It remains to be proved that this combination is therapeutically significant in preclinical and clinical models of myeloid leukemia. However, it is interesting to notice that there is an anecdoctal report on a complete remission achieved in a relapsed APL patient, following combined treatment with atRA and G-CSF (Shimodaira et al., 1999).

The superfamily of TGFs-β includes a large number of structurally related polypetides with prominent roles in cell proliferation, differentiation, and death. Interactions between retinoids and TGF-β have been the object of numerous studies. In particular, TGF-β is induced in a variety of cell types and the effects of combinations between the growth factor and atRA on cellular differentiation have been studied in a number of models (Cao et al., 2003; Roberts and Sporn, 1992; Sporn and Roberts, 1985; Sporn et al., 1989; Wakefield et al., 1990). In skin epidermis, the actions of TGF-β and atRA on normal keratinization are synergistic, whereas those on abnormal differentiation associated with hyperproliferation are antagonistic (Choi and Fuchs, 1990). TGF-β enhances neuronal differentiation of human neuroblastoma SH-SY5Y cells treated with atRA, as determined by the increase in the length and density of neurite outgrowth as well as the measurement of a specific maturation marker like tyrosine hydroxylase (Gomez-Santos et al., 2002).

TGF-β has been shown to activate the monocytic differentiation of AML cell lines (De Benedetti et al., 1990; Turley et al., 1996; Walz et al., 1993). When HL-60 cells are treated simultaneously with atRA and TGF-β, a mixture of granulocytic and monocytic cells is observed. In this paradigm, atRA and the growth factor compete with each other and commit cells toward two different types of maturation processes. Antagonistic interactions may be the consequence of a retinoid-induced phosphatase mediating the dephosphorylation and inactivation of SMAD2, a downstream effector of TGF-β (Cao et al., 2003). The growth of retinoid sensitive MCF-7 breast carcinoma cells is inhibited synergistically by combinations of TGF-β and atRA (Turley et al., 1996; Valette and Botanch, 1990). However, to the best of our knowledge, interactions between the two compounds on the differentiation of this type of tumor cells have not been reported. From a mechanistic point of view, it is interesting to notice that TGF-β induces the expression of AIB1 (Lauritsen et al., 2002), a well-known coactivator of the RXR/RAR transcriptional complex, suggesting that the growth factor has the potential to modulate the retinoid-signaling pathway in a positive fashion.

Interferons (IFNs) are probably the cytokines or growth factors that have been studied more extensively in relation to the biological and pharmacological activity of retinoids (Bollag, 1994; Chelbi-Alix and Pelicano, 1999; Eisenhauer et al., 1994; Fossa et al., 2004; Smith et al., 1992). This is explained by the fact that interferons were among the first cytokines to be identified and characterized. In addition, type I IFNs in combination with 13-cis RA have been used in chemoprevention studies of head and neck cancer with encouraging results (Shin et al., 2001). IFN synthesis is activated as part of the cellular response to viral infection. Various types of IFNs have been described and classified as type I (IFNα and IFNβ) or type II (IFNγ). While IFNα and IFNβ are synthesized by the majority of cell types, IFNγ is produced only by cells of the immune system. IFNs act predominantly as autocrine or exocrine factors and bind to specific membrane receptors which are coupled to the JAK/STAT and mitogen-activated protein kinase (MAPK) pathway. In particular, binding and phosphorylation of IFN receptors lead to recruitment and activation of JAK kinases and subsequent phosphorylation of STAT1 or other types of STATs. Activated STAT1, in the form of a homodimer (IFNγ) or a heterotrimer (STAT1-IGF3-p48, IFNα, or IFNβ), migrates to the nucleus and acts as a transcription factor regulating the expression of numerous genes (Ramana et al., 2000, 2002; Stark et al., 1998). IFNs have also been shown to activate the MAPK pathway via p38MAPK (Kovarik et al., 1999). Type I IFNs exert predominantly growth inhibitory effects acting on various intracellular systems.

Numerous reports demonstrate cross talk between the retinoid and IFN pathway in various cell types, both in terms of cytodifferentiation and growth inhibition (Chelbi-Alix and Pelicano, 1999; Dimberg et al., 2000,

2003; Garattini et al., 1998; Gianni et al., 1996b, 1997; Lembo et al., 1992; Matikainen et al., 1996, 1997, 1998; Pelicano et al., 1997). The most interesting results on the cytodifferentiating effects of these types of combinations have been reported in the context of the APL or the AML blast. In NB4 cells, we demonstrated that atRA and synthetic RARα agonists are endowed with an IFN-mimetic action. Treatment of this cell type causes a rapid phosphorylation and activation of STAT1, which results in the transcription of IFN-responsive genes. STAT1 activation is followed by transcriptional stimulation of the corresponding gene and subsequent increases in the levels of the encoded protein (Gianni et al., 1997). A late event triggered by atRA is represented by stimulation of IFNα secretion, which suggests that retinoids may trigger an autocrine loop involving the cytokine (Gianni et al., 1997). Induction of STAT1 by retinoids is not cell specific, as it is observed also in breast carcinoma cells (Kambhampati et al., 2004), and may be mediated by rapid induction of the transcription factor interferon-responsive factor 1 (IRF1) (Percario et al., 1999) or activation of PKCδ (Kambhampati et al., 2003). In breast carcinoma cells, atRA-dependent STAT1 induction is associated with reversion of natural resistance to the growth inhibitory action of IFNs (Kolla et al., 1996, 1997; Lindner et al., 1997; Moore et al., 1994). Combinations of type I or type II IFNs with atRA in myeloid leukemia blasts causes enhanced cytodifferentiation, at least in terms of expression of some granulocytic maturation markers (Chelbi-Alix and Pelicano, 1999; Gianni et al., 1996b). STAT1 and IRF1 induction is relevant not only for the observed interactions between retinoids and IFNs but also for the cytodifferentiating effect of atRA in APL cells. Indeed, selective suppression of STAT1 and/or IRF1 results in inhibition of the myeloid differentiation program activated by atRA in this cell type (Dimberg et al., 2003). Further points of contact between retinoids and IFNs are represented by the fact that both types of stimuli activate the p38MAPK and PKC pathways (see Section IV.C). Although largely unexplored these phenomena may also be at the basis of the observed additive or synergistic intercations between the retinoid and IFN systems. For instance, PKCδ is required for the generation of the synergistic effects of IFNα and atRA on gene transcription. Such regulatory effects on transcription are mediated by the atRA-inducible, PKCδ-dependent upregulation of STAT1 protein expression.

As a concluding remark, it must be underscored that combinations of IFNs and retinoids have been the basis of numerous clinical trials in different types of neoplasia and have provided some encouraging results. In the CML context, where IFN is one of the treatment mainstay, some interesting data have been published (Egyed et al., 2003). The cytogenetic responses during the chronic phase of 11 patients with CML treated with atRA + IFN were compared with those of 9 other CML patients treated with IFN alone. The preliminary results suggest that the atRA + IFN combination may be superior in achieving cytogenetic remission in the first chronic phase of

CML (Egyed *et al.*, 2003). It remains to be established whether the positive clinical effects reported for the combinations of retinoids and IFNs in this as well as other clinical contexts is related to enhanced cytodifferentiation of the neoplastic cell.

B. THE cAMP PATHWAY

The intracellular second messenger, cAMP, controls many aspects of the cellular homeostasis, including proliferation, differentiation, and apoptosis. Adenyl-cyclases are coupled to many membrane receptors and control the synthesis of cAMP from ATP. The levels of intracellular cAMP are modulated in a negative fashion by phosphodiesterases, a large family of enzymes catalyzing the hydrolysis of cAMP. Elevation of intracellular cAMP results in the activation of the cAMP-dependent protein kinase (PKA), which phosphorylates and controls the state of activation of a number of protein substrates. PKA is a dimer consisting of a regulatory and a catalytic subunit. Binding of cAMP to the regulatory moiety causes release and activation of the catalytic subunit. A second and less studied effector molecule stimulated by cAMP is EPAC, aGTP-exchanging factor, mediating some of the biological effects of the cyclic nucleotide.

Most of the knowledge on the cross talk between the retinoid and the cAMP-signaling pathways have been acquired in two systems: AML and neuroblastoma. In AML, stimuli capable of increasing the intracellular levels of cAMP potentiate the cytodifferentiating and growth inhibitory effects of atRA (Breitman *et al.*, 1994; Imaizumi and Breitman, 1987; Parrella *et al.*, 2004; Yang *et al.*, 1998). Cotreatment with atRA and cell permeable cAMP analogues, like dibutyryl-cAMP or 8Cl-cAMP as well as specific phosphodiesterase IV inhibitors, like piclamilast, results in a more rapid and more efficient granulocytic maturation of HL-60 and NB4 cells relative to treatment with the retinoid alone (Altucci *et al.*, 2005; Garattini and Gianni, 1996; Gianni *et al.*, 1995c; Guillemin *et al.*, 2002; Kamashev *et al.*, 2004; Parrella *et al.*, 2004; Taimi *et al.*, 2001). This is accompanied by the expression of terminally differentiated granulocytic maturation markers, like leukocyte alkaline phosphatase, which are not induced by atRA alone (Garattini and Gianni, 1996; Gianni *et al.*, 1995c). Interestingly, treatment with cAMP-elevating agents in the absence of retinoids is associated only with a mild growth inhibitory effect in NB4 cells. In contrast, HL-60 cells undergo a certain level of granulocytic maturation on treatment with cell permeable analogues of cAMP alone (Breitman *et al.*, 1994; Imaizumi and Breitman, 1987). Combinations of dibutyryl-cAMP and atRA induce myeloid maturation in a subset of freshly isolated non-M3 AML and chronic myeloid leukemia cells (Gianni *et al.*, 1995c). These data suggest that cAMP-elevating agents not only potentiate the activity of atRA in retinoid sensitive cells but also sensitize otherwise refractory blasts to the cytodifferentiating action of retinoids.

The potentiating effect of cAMP on retinoid activity requires the presence of active RARs, as similar phenomena are not observed in HL-60 and NB4 sublines made resistant to atRA and showing inactivating mutations at the level of the ligand-binding sites of the nuclear receptors, RARα or PML–RARα.

The retinoid-sensitizing effect observed on elevation of intracellular cAMP augments neuronal differentiation of neuroblastoma, suggesting that the phenomenon is not limited to particular cell contexts and may be of more general therapeutic interest (Abemayor and Sidell, 1989; Holtzer et al., 1985). In F9 mouse teratocarcinoma cells, addition of cAMP-elevating agents to atRA reprograms cells along the visceral endoderm maturation pathway, suggesting that combinations of the two stimuli causes not only quantitative but also qualitative changes in the process of cellular differentiation (Rochette-Egly and Chambon, 2001).

The molecular mechanisms underlying the cross talk between the cAMP and the retinoid pathways are still incompletely defined. However, in myeloid leukemia cells, it is clear that potentiation is the result of a PKA-dependent and not of an EPAC-dependent event. In fact, only selective PKA inhibitors suppress induction of the granulocytic maturation markers triggered by the combination of cAMP-elevating agents and atRA. Whether this is related to the rapid and short-lived activation of PKA observed in HL-60 and NB4 cells treated with the retinoid alone is not yet known. A plausible mechanism at the basis of the cross talk is represented by the direct effects exerted by PKA on the nuclear RARs (Parrella et al., 2004; Rochette-Egly et al., 1995). RARα is phosphorylated by PKA and this phosphorylation seems to be necessary for the full activation of the receptor. In fact, cAMP-elevating agents, like piclamilast, cause a marked stimulation of the ligand-dependent activation of RARα. Similar effects are observed when RARα is substituted with the APL-specific fusion protein, PML–RARα. Relatively specific PKA inhibitors block the stimulating effects of cAMP analogues or phosphodiesterase IV inhibitors on the ligand-dependent activation of RARα or PML–RARα (Parrella et al., 2004). This is consistent with the presence of a PKA-sensitive and key serine residue (Ser/369) in the ligand-binding domain of the nuclear RAR (Rochette-Egly et al., 1995). Mutation of this serine to an alanine knocks down the potentiating effect afforded by piclamilast on the ligand-dependent activation of RARα (Parrella et al., 2004). If this mechanism of action is really operative, the prediction is that cAMP-elevating agents should have important effects on the expression of the majority if not the totality of RXR/RAR-dependent genes.

An interesting observation relates to the ability of cAMP to activate not only RXR/RAR heterodimers but also the RXR/RXR homodimers. Activation of RXR/RXR complexes by rexinoids is associated with induction of an apoptotic response in HL-60 cells (Nagy et al., 1995). This phenomenon is not accompanied by granulocytic maturation of the leukemic blast (Nagy et al., 1995). Surprisingly addition of cAMP cell permeable analogues to rexinoids induces cytodifferentiation of retinoid sensitive as well as retinoid-insensitive or resistant myeloid blasts (Altucci and Gronemeyer, 2002; Altucci et al., 2005; Benoit et al., 1999). The observation is of potential therapeutic significance, although the molecular mechanisms underlying RXR/RXR activation by cAMP are not well understood.

A key question is whether all these observations and findings can be translated into therapeutic effects in vivo. In preclinical models of APL, this seems to be the case. Indeed, we demonstrated that combinations of piclamilast and atRA are superior to the single components of the mixture in increasing the survival of immunodeficient SCID animals transplanted with NB4 cells. At present the contribution of cytodifferentiation to the overall antileukemic effect of combinations between piclamilast and atRA has not been determined. Nevertheless, administration of this association is well tolerated and does not produce significant toxicity. Although the effect is significant, the dosage and schedule of the combined treatment need to be optimized. Similar results were reported in a different model after prolonged administration of 8-Cl-cAMP and atRA using infusion pumps (Guillemin et al., 2002). The results obtained are promising and suggest that the approach leads to an increase in the therapeutic index of atRA. The approach needs to be tested in other leukemia and cancer models. This is particularly relevant as cotreatment with cell permeable analogues of cAMP and atRA sensitizes human neuroblastoma cells to the cytotoxic actions of chemotherapics like doxorubicin, melphalan, and BCNU (Carystinos et al., 2001). The observation suggests that cytodifferentiating therapy with cAMP-elevating agents and atRA could be combined effectively to classical chemotherapy of the neoplastic disease.

C. THE MAP KINASE PATHWAY

MAPKs are a group of enzymes that relay a variety of extracellular signals inside the cell (Chang and Karin, 2001). Three types of MAPKs are known, extracellular regulated kinases (ERKs), p38MAPK, and Jun N-terminal kinase (JNK). MAPKs are the terminal substrates of kinase cascades involving at least two other types of upstream kinases MAPKKKs and MAPKKs. The end result of this cascade is phosphorylation and activation of anyone of the three MAPKs mentioned above. ERKs are often associated with the transduction of mitotic signals generated by growth and survival factors (Shimada et al., 2006; Tibbles and Woodgett, 1999). JNK and p38MAPK mediate, among others, apoptotic signals generated by different types of cellular stresses (Beere, 2005; Bogoyevitch et al., 1996; Mehta and Miller, 1999; Ouyang et al., 2005; Roux and Blenis, 2004; Shen and Liu, 2006). In spite of these well-characterized effects, MAP kinases are involved in a panoply of cellular processes, including cytodifferentiation. The effects exerted by retinoids on MAPKs are multiple and an extensive discussion is beyond the scope of this chapter.

1. ERKs

The ERK pathway is known to be activated or inhibited by retinoids according to the specific cell context considered. In myeloid leukemia cells, such as HL-60, protracted activation of ERKs is believed to be necessary for the atRA-dependent granulocytic maturation and growth inhibition of the blast (Miranda *et al.*, 2002; Wang and Studzinski, 2001; Yen *et al.*, 1998, 1999). In fact, pharmacological inhibition of the ERK pathway suppresses the cytodifferentiating response of HL-60, NB4, and freshly isolated APL blasts to atRA (Miranda *et al.*, 2002; Parrella *et al.*, 2004; Wang and Studzinski, 2001; Yen *et al.*, 1998, 1999). Activation of ERKs is also instrumental in inducing the differentiation of the mouse F9 teratocarcinoma cell along the primitive endoderm. However, further differentiation into parietal endodermal cells is hampered by ERK activation (Verheijen *et al.*, 1999). Growth inhibition of breast carcinoma cells is accompanied and possibly mediated by atRA-dependent inhibition of the ERK pathway (Nakagawa *et al.*, 2003). In neuroblastoma cells, ERKs do not seem to play any role in the process of neuronal differentiation activated by atRA (Miloso *et al.*, 2004). At present, it is unclear whether the negative and positive actions of atRA on the ERK-signaling pathway require activation of the RXR/RAR complexes. This is possible in the case of the long-term activation of ERK in myeloid leukemia cells, while it is unlikely when atRA-triggered ERK phosphorylation and dephosphorylation events are rapid and require minutes to be completed. Though retinoids exert multiple effects on the ERK pathway, there is only one report demonstrating modulation of nuclear RARs' activity by this type of kinases. In T-cells, ERK induces the ligand-dependent transactivation of RXRs (Ishaq *et al.*, 2000).

2. JNKs

AP-1 are transcriptional complexes whose principal components are homo- or heterodimers of the c-Fos/c-Jun type (Eferl and Wagner, 2003). The activation of AP-1 complexes is controlled by phosphorylation events triggered by JNK. AP-1 is involved in cellular responses to stress, cytokine, and proliferative stimuli (Karin and Shaulian, 2001). As already mentioned, ligand-bound RXR/RAR complexes have long been known to exert anti-AP-1 activity through a mechanism known as transrepression (Allenby, 1995; Fisher *et al.*, 1998). More recently, the anti-AP-1 activity of retinoids has been associated with the ability of these compounds to inhibit JNK (Caelles *et al.*, 1997; Gonzalez *et al.*, 1999).

Given the importance of AP-1 complexes in the processes of cellular proliferation, it is not surprising that there is a relatively vast literature correlating the antiproliferative activity of retinoids with their inhibitory effects on JNK.

However, JNK inhibition by atRA has functional consequences not only for the proliferation but also for the differentiation of certain types of neoplastic cells. In AML, atRA reduces the basal level of JNK phosphorylation/activation and this effect may hinder the retinoid-dependent granulocytic maturation of the leukemia blast. The retinoid-potentiating agent ST1346 relieves the down-regulation of JNK afforded by atRA and stimulates retinoid-dependent granulocytic maturation. In addition, a specific JNK inhibitor blocks the enhancing effect of ST1346 on atRA-induced maturation of NB4 cells (Pisano et al., 2002).

JNK plays a central role in the process of cell differentiation activated by retinoids in embryocarcinoma, neuronal, and myeloid cells. In these cellular contexts, atRA stimulates rather than inhibit JNK phosphorylation and activation. Though the functional consequences are unknown, atRA has been reported to activate JNK also in head and neck squamous cell carcinoma (HNSCC). Interestingly, this effect is enhanced by cotreatment with 5-fluorouracil, a well-known chemotherapeutic agent, and correlated with the apoptotic responses induced by combinations of 5-fluorouracil and atRA (Masuda et al., 2002). Increased JNK activation was observed also in breast and prostate carcinoma cells. This phenomenon is associated with atRA-stimulated apoptotic responses to taxotere (Wang and Wieder, 2004).

JNK modulates the activity of nuclear RARs. In non-small cell lung carcinoma (NSCLC) cells, activation of JNK may be at the basis of the resistance to the action of retinoids (Lee et al., 1999). Activation of JNK contributes to RAR dysfunction by phosphorylating RARα and inducing degradation through the ubiquitin-proteasomal pathway. Interestingly, mice that develop lung cancer from activation of a latent K-ras oncogene have high intratumoral JNK activity, low RARα levels, and are resistant to treatment with RAR ligands. JNK inhibition in a human lung cancer cell line enhances RARα levels, ligand-induced activity of RXR-RAR dimers, and growth inhibition by atRA (Srinivas et al., 2005). In spite of these data, the major target of JNK activity seems to be the RXR moiety of the RAR/RXR complex. Overexpression and UV activation of JNK1 and JNK2 hyperphosphorylate mouse RXRα. This inducible hyperphosphorylation involves Ser61 and Ser75 as well as Thr87 in the B region and Ser265 in the ligand-binding domain (E region) (Adam-Stitah et al., 1999; Bour et al., 2005). Other serine residues in the A and C region of RXRα have been implicated in JNK-dependent hyperphosphorylation triggered by arsenic trioxide (Mann et al., 2005). The functional consequences of RXRα phosphorylation by JNK are controversial. In one report, hyperphosphorylation by JNKs has been shown to exert no significant effect on the transactivation properties of either RXRα homodimers or RXRα/RARα heterodimers (Adam-Stitah et al., 1999). In two other reports, phosphorylation has been associated with inhibition of RXRα-mediated transcription (Bruck et al., 2005; Mann et al., 2005). At present, Ser32 in the A region or Ser265 in the omega loop of the E region are the main candidate residues involved in JNK-dependent inhibition of RAR/RXRα-mediated transcription.

3. p38MAPK

p38MAPK is activated as a consequence of numerous extracellular signals. Growth factors, trophic factors, inflammatory cytokines, FAS ligand, UV, and γ radiations as well as heat shock activate the pathway through mechanisms that are not fully elucidated. The activation of p38MAPK leads to the phosphorylation and activation of numerous transcription factors such as c-myc, STAT1, ATF-2, and CHOP, which induce specific transcriptional responses. p38MAPK has been proposed to play a central role in apoptosis.

atRA and other retinoids modulate the p38MAPK pathway. The pathway is activated in an atRA-dependent manner in the NB4, APL cell line. This effect is not observed in atRA-resistant cell clones. Treatment with p38MAPK inhibitors enhances atRA-dependent induction of cell differentiation and atRA-regulated growth inhibitory responses (Alsayed et al., 2001). The phenomenon is reproduced in HL-60 cells, where cotreatment with atRA and the p38MAPK inhibitor results in synergistic upregulation of the two granulocytic maturation markers, CD11B and CD11C (M.G. and E.G., unpublished observations). These results indicate that activation of p38MAPK by atRA exerts a negative role on the granulocytic maturation of AML cells. Negative modulation of the process may result from specific actions of p38MAPK on atRA regulated genes involved in myeloid maturation. One such gene may be the transcription factor CHOP, otherwise known as GADD153. In NB4 and HL-60 cells, CHOP, a downstream phosphorylation target of p38MAPK (Wang and Ron, 1996) is induced by atRA. Interestingly, activated CHOP inhibits the transcriptional activity of two members of the cEBP family of transcription factors, cEBPα and cEBPε, which are known positive regulators of granulocytic maturation (Gery et al., 2004; Tenen, 2001; Verbeek et al., 1999).

Stimulation of the p38MAPK pathway by retinoids does not seem to be limited to AML, as similar phenomena are observed in breast carcinoma and adipocytic cell lines (Alsayed et al., 2001; Teruel et al., 2003). Furthermore, atRA phosphorylates the p38MAPK in normal human keratinocytes, mouse MC3T3-E1 fibroblasts and chondroblast (Dai et al., 2004). In the last cell type, atRA-dependent stimulation of p38MAPK leads to differentiation. Thus, dependent on the cellular context, activation of the kinase by the retinoid acts as a positive or negative determinant of cell maturation. At present, the molecular mechanisms underlying the effect of atRA on p38MAPK are unknown; however, they are clearly not the result of transcriptional effects induced via the RXR/RAR-dependent pathway. In spite of all these reports, activation of p38MAPK by atRA is not a general phenomenon, as the retinoid is capable of inhibiting the system in certain cellular contexts (Palm-Leis et al., 2004).

The most interesting aspect of the cross talk between retinoids and p38MAPK relates to the ability of the kinase to exert direct or indirect effects on at least two types of RAR isoforms (Gianni et al., 2006). atRA-stimulated p38MAPK controls the rate of degradation of SRC3, which represents a component of the coactivator complex associated with RARα. Direct phosphorylation of SRC3 by the MAPK is an important signal that directs the protein along the proteasome-dependent degradation pathway. Ligand binding of RARα recruits SRCR3 to the transcriptional complex. This is coupled to p38MAPK-dependent phosphorylation of SRC3, which is polyubiquitinylated and degraded by the proteasome. This process is likely to be of physiological importance putting a brake on the transcriptional activity of the RARα complex. If the activity of p38MAPK is suppressed in a specific fashion, the atRA-dependent RARα/SRC3 interaction is stabilized and prolonged, maintaining the RARα transcriptional complex in the active state. This has important therapeutic consequences, as the mechanism is likely to play a significant role in the potentiating effect of p38MAPK inhibitors on the cytodifferentiating and growth inhibitory activity of atRA in myeloid leukemia cells (Alsayed et al., 2001; Gianni et al., 2006). The action of p38MAPK on nuclear RARs is highly dependent on the RAR isotype considered, as what is observed in the case of RARα is not replicated in the case of RARγ. In fact, the turnover of RARγ is linked to transactivation (Gianni et al., 2002a, 2003). Unlike RARα, atRA-dependent RARγ phosphorylation by p38MAPK signals proteasomal degradation of the receptor. Unexpectedly and counterintuitively, specific inhibition of RARγ degradation decreases the ligand-dependent transcriptional activity of the receptor.

4. Modulation of MAPK: Therapeutic Prospectives

From a therapeutic prospective, the direct negative effects determined by p38MAPK and JNK in the processes of ligand-dependent transactivation of the RXR/RAR complexes are the most relevant. In the context of the treatment or chemoprevention of head and neck cancer, pharmacological inhibition of JNK may sensitize the neoplastic cell to the differentiating, growth inhibitory, and/or apoptotic action of retinoids. The anti-inflammatory potential of JNK inhibitors may also add value to these compounds given the importance that inflammation has during the tumor promotion phase of the carcinogenic process. The JNK pathway represents a "druggable" target and specific compounds like CEP-1347, an inhibitor of the MLK family of JNK pathway activators, or SP600125, a direct inhibitor of JNK activity, have been developed and used *in vivo*. These inhibitors have demonstrated efficacy *in vivo*, decreasing brain damage in animal models (CEP-1347) and ameliorating some of the symptoms of arthritis in other animal models (SP600125) (Bogoyevitch, 2005). It remains to be established whether JNK inhibitors can be added to retinoids without causing too much toxicity. Indeed, it is possible that the unrestrained activity of RA nuclear receptors may enhance the untoward actions of these compounds at the level of the central nervous system or liver. Furthermore, from a theoretical point of view, long-term and generalized inhibition of JNK may block some of the most important physiological responses to the external insults in normal cells. Long-term JNK inhibition may have adverse consequences in the setting of head and neck cancer where normal epithelial cells of the aerodigestive cells may become particularly susceptible to the DNA damage caused by environmental carcinogens. In spite of these concerns, cotreatment of SCID mice bearing NB4-derived myeloid leukemia with atRA and the experimental bisindol ST1346, which inhibits retinoid-induced JNK activation in APL blasts, results in increased life span relative to treatment with the single components of the combination. It is reassuring that this type of combination is not associated with signs of overt toxicity (Pisano *et al.*, 2002).

Several selective inhibitors of p38MAPK are available and they are proposed as anti-inflammatory agents. However, this type of inhibitors is currently studied also for the treatment of multiple myeloma (MM). MM is a type of cancer formed by the immune cells that normally produce infection-fighting antibodies. The protein p38MAPK appears to stimulate and promote an environment supportive of MM cell growth (Hideshima *et al.*, 2003). Inhibitors of p38MAPK are currently in the process of being investigated as potential therapeutic agents for MM. atRA has also been shown to inhibit the growth of MM blasts, possibly by acting at the level of the IL-6 autocrine loop, which is important for the proliferation of the neoplastic cell (Koskela *et al.*, 2004; Musto *et al.*, 1995; Sidell *et al.*, 1991). Thus, atRA and p38MAPK may represent a rational therapeutic combination to be tested in this setting.

D. THE PI3K/AKT PATHWAY

Protein kinase B or AKT is a serine/threonine kinase, which is believed to play an important role in cell differentiation and survival. The AKT pathway is complex and conveys signals from the extracellular compartment to various intracellular sites including the nucleus. A number of growth receptors are coupled to phosphatidylinositol-3-kinases (PI3K), which phosphorylate the phosphatidylinositol-diphosphate (PIP2), resulting in the synthesis of the second messenger phosphatidylinositol-triphosphate (PIP3). This phosphorylation event is limited by the activity of the PTEN phosphatase, which is a classical tumor suppressor and is mutated or inactivated in certain types of cancer cells. PIP3 activates the kinase, phosphatidyl-inositide-dependent kinase 1 (PDK1), which, in turn phosphorylates and activates AKT.

atRA modulates the activity of the PI3K/AKT pathway in various cellular systems. In turn, AKT regulates the activity of the classical nuclear RAR pathway. In neuroblastoma cells, atRA treatment activates the PI3K/AKT-signaling pathway, resulting in elevated PI3K activity and a rapid increase in the phosphorylation of AKT (Ser/473) (Lopez-Carballo et al., 2002). Inhibition of PI3K by LY294002 impairs atRA-induced differentiation. The human endometrial cell line, CAC-1 differentiates on treatment with pharmacological doses of both atRA and 13-cis RA, as evidenced by actin filament reorganization and cell enlargement (Carter and Madden, 2000). Pretreatment with the PI3K inhibitor wortmannin prevents retinoid-dependent actin reorganization and cell enlargement. In F9 mouse teratocarcinoma cells, atRA stimulates AKT at early time points (Bastien et al., 2006). Specific suppression of AKT induction with selective pharmacological inhibitors or siRNAs blocks the cytodifferentiation of F9 cells along the endodermal pathway. These data demonstrate that the PI3K/AKT pathway plays a fundamental role in the process of cytodifferentiation activated by retinoids in neoplastic cells of different nature and origin. Furthermore, they indicate that atRA and derivatives activate PI3K–AKT in certain cellular contexts. At present the molecular mechanisms underlying this activation are not completely defined. In F9 cells, stimulation is the consequence of an upregulation of the p85 regulatory subunit of PI3K, which is dependent on a transcriptional effect under the control of RARγ. However, this mechanism is not general, as a similar phenomenon is not observed in various breast carcinoma cell lines (Bastien et al., 2006).

Although retinoids activate the PI3K/AKT pathway in certain cellular context, the opposite effect has also been reported. Suppression of AKT is observed in HL-60 cells treated with atRA (Ishida et al., 1994). Interestingly, pharmacological inhibitors of AKT sensitize cells to the antiproliferative action of atRA in an HL-60 clone with a constitutively active PI3K/AKT pathway (Martelli et al., 2003; Neri et al., 2003). This suggests that the basal tone of AKT regulates the retinoid pathway in a negative fashion. Binding of atRA to cellular retinoic acid-binding protein 1 (CRBP1) suppresses the heterodimerization of the p85/p110 subunits of PI3K, inhibiting the kinase activity of the enzyme in breast carcinoma cells (Farias et al., 2005). The observation indicates a potential mechanism underlying retinoid-dependent down-modulation of the PI3K/AKT pathway. Another possible mechanism through which atRA interferes and downmodulates the PI3K/AKT pathway was discovered in F9 teratocarcinoma cells. In this system early activation of the PI3K/AKT pathway by atRA is followed by downregulation of the system after 3 days of treatment (Bastien et al., 2006). Downregulation is the consequence of inhibition of the PI3K p85a subunit and activation (phosphorylation) of PTEN. Similar effects on PTEN were observed in HL-60 cells. In this cell line, the retinoid increases PTEN levels, and increased expression of the phosphatase appears to parallel terminal differentiation of the myeloid cell line (Hisatake et al., 2001). atRA inhibits AKT indirectly in breast cancer cells, downregulating the expression of IRS1 and corresponding kinase activity. These two proteins are known to couple the insulin growth factor I (IGF-I) receptor to AKT (del Rincon et al., 2003).

As already mentioned, the cross talk between the retinoid and the PI3K/AKT pathway is a two-way process and does not involve simple perturbations in the levels and basal or induced state of activity of PI3K and/or AKT by retinoids. The ligand-dependent transactivating properties of RARγ2 are augmented by inhibition of AKT with dominant-negative constructs of the kinase or specific pharmacological inhibitors of PI3K (Gianni et al., 2002b). Transfection of constitutive active forms of AKT exerts the opposite effects. RARγ2 inhibition by AKT has been associated with decreased phosphorylation of the p38MAPK-dependent phosphorylation of the receptor (Gianni et al., 2002b). Similar effects on the transactivating properties of RARα have been reported (Srinivas et al., 2006). Srinavas et al. demonstrated that AKT has the potential to physically interact with RARα and phosphorylate the receptor directly on Ser96, a residue located in the DNA-binding domain. This phosphorylation reduces the transactivating potential of RARα, through an as yet unidentified mechanism. Overexpression or constitutive activation of AKT may underlie resistance to retinoids in leukemia, breast, and NSCLC cells (Srinivas et al., 2006). In breast cancer cells, this is linked to overexpression of epidermal growth factor receptor, Her2/Neu, which activates AKT and inhibits the binding of RARs to the cognate DNA regulatory sequences (Siwak et al., 2003; Tari et al., 2002). In CALU-1, a NSCLC cell line with high constitutive levels of the enzyme, transfection of a dominant-negative construct of AKT increases the sensitivity of the antiproliferative action of atRA (Srinivas et al., 2006). Another potential point of contact between the retinoid and the PI3K/AKT pathways is at the level of the forkhead receptor in rhabdomyosarcoma (FKHRL1). FKHRL1 is a transcriptional factor and a downstream target of AKT. AKT phosphorylates and inactivates the factor, inhibiting transcriptional upregulation of the FAS ligand, a protein endowed with proapoptotic properties. FKHRL1 acts as a coactivator of the RAR/RXR and stimulates ligand-dependent transactivation of responsive constructs (Zhao et al., 2001).

Besides PI3K, AKT, and PTEN, retinoids exert actions on other components of the pathway. atRA influences the activity of PP2A, a multisubstrate phosphatase, which dephosphorylates and inhibits AKT. PP2A is one of the proteins upregulated by the retinoid in acute myeloid leukemia cells (Harris et al., 2004). PP2A is downregulated during atRA-induced differentiation of HL-60 cells into granulocytes (Tawara et al., 1993). This effect may have relevance for the granulocytic maturation process induced by atRA in this cell type. In fact, okadaic acid, a relatively specific inhibitor of PP2A augments atRA-induced granulocytic differentiation of HL-60 cells (Morita et al., 1992). Contrary to what observed in the myeloid cellular context, atRA-dependent increases in PP2A activity were observed in ovarian carcinoma cells. The phenomenon was linked to retinoid-induced repression of the transcription complex AP-1, which is involved in cellular proliferation (Ramirez *et al.*, 2005).

The available data clearly demonstrate cross talk between retinoids and the PI3K/AKT pathway. This interaction is relevant in terms of growth inhibition, cytodifferentiation, and apoptosis. Thus, pharmacologic modulation of the PI3K/AKT pathway may translate into a potentiation of retinoids' antitumor and antileukemic activity. The inhibitory effects of retinoids on PI3K/AKT are of potential therapeutic significance, as they may lead to death of the neoplastic cell. Furthermore, inhibition of AKT mediates suppression of the telomerase activity present in HSC-1 human epidermal cells (Kunisada *et al.*, 2005). As telomerase activity regulates cell proliferation and senescence in a positive and negative manner, respectively, its inhibition by atRA may translate into an antineoplastic effect. In addition, while overexpression of CRBP-I in breast epithelial cells increases retinoic receptor activity, inhibits anoikis, promotes acinar differentiation, and inhibits tumorigenicity, the protein is often downregulated or absent in breast cancer cells.

E. PROTEIN KINASE C

The PKC family of serine/threonine kinases includes several members (Newton, 1997). The classification of distinct members of the PKC family in discrete isoform groups relies on the requirements that the different isoforms exhibit for activation of their kinase domains. One group includes PKC isoforms that require increases in intracellular calcium for their activation. The three known conventional PKC isoforms are PKCα, PKCβ, and PKCγ (Newton, 1997). The second group of PKC isozymes is the group of novel PKC isoforms, which do not require Ca^{2+} (Newton, 1997). PKCδ, PKCε, PKCθ, PKCη, and PKCμ are included in this group. Finally, a third group of atypical PKC isoforms exists, whose members are Ca^{2+}-independent and are insensitive to phorbol esters, a well-known family of PKC activators. PKCζ and PKCλ are the two known atypical PKC isoforms (Newton, 1997). Extensive studies have shown that these kinases play critical roles in the regulation of several important cellular responses such as differentiation, cell growth, and apoptosis (Newton, 1997). It is of interest that different PKC isoforms mediate different responses and, in some cases, appear to exhibit opposing effects on cell proliferation and apoptosis.

atRA and other retinoids modulate the activity of PKCs (Aggarwal *et al.*, 2006; Cho *et al.*, 1997; Kambhampati *et al.*, 2003; Khuri *et al.*, 1996; Yang *et al.*, 1994). This family of kinases is involved in the process of cytodifferentiation activated in neuroblastoma and myeloid leukemia cells. In neuroblastoma cells, the specific PKC inhibitor, GF 109203X, inhibits atRA-induced neuritogenesis and cell survival (Miloso *et al.*, 2004). In the myeloid leukemia context, pretreatment of HL-60 cells with staurosporine, a well-known PKC inhibitor, results in enhancement of the granulocytic maturation of the leukemic blast induced by atRA (Yung and Hui, 1995). More recent results indicate that activation of specific PKC isoforms may enhance the cytodifferentiating activity of atRA in the same cellular context (Kambhampati *et al.*, 2003). PKCδ mediates atRA-dependent induction of differentiation in NB4 cells. atRA activates PKCδ and specific inhibition of the kinase isoform results in suppression of the atRA-dependent differentiative response. Furthermore, forced expression of a constitutively active form of PKCδ enhances retinoid-dependent granulocytic maturation. These phenomena may be explained by the fact that PKCδ is involved in the ligand-dependent transactivation of the RXR/RAR transcriptional complexes. In fact, the kinase is present in RAR nuclear complexes bound to RAREs, as demonstrated by coimmunoprecipitation experiments and chromatin immunoprecipitation assays. Pharmacological inhibition of the kinase results in suppression of ligand-dependent transactivation of the complex.

In spite of all these results, the role of PKC activation in the regulation of the RXR/RAR pathway is far from being clear and is highly dependent on the type of PKC isoform considered. Relatively old studies had implicated a PKC isoform in RA-dependent gene transcription, as evidenced by the fact that depletion of the enzymatic activity by treatment with phorbol esters leads to loss of ligand-dependent transcription (Tahayato *et al.*, 1993). However, the identity of the PKC isoform involved was unknown at the time (Tahayato *et al.*, 1993). Other studies have demonstrated that PKCα- or PKCγ-dependent phosphorylation of RARα at Ser/157 correlates with a decreased ability of RARα to heterodimerize with RXRα, resulting in decreased transcriptional activity (Delmotte *et al.*, 1999). As other studies have established that different PKC isoforms have opposing effects in the induction of certain responses, it is possible that PKCδ acts as a positive modulator of retinoid-dependent gene transcription and opposes the effects of PKCα and/or PKCγ. A similar phenomenon appears to occur in the regulation of the RXRs in T-lymphocytes. In this cellular context, PKCθ synergizes with calcineurin to induce RXR-dependent activation, whereas such activation is antagonized by the PKCα isoform (Ishaq *et al.*, 2002). Another PKC isoform implicated in the process of retinoid-dependent granulocytic maturation of AML cells is PKCζ. atRA induces a parallel increase of ceramide and catalytically active PKC-zeta into the nuclear compartment of HL-60 cells. However, transient transfection of the PKC-zeta cDNA demonstrates that the overexpression of catalytically active PKCζ is not accompanied by the appearance of a differentiated morphology. Altogether these findings suggest that nuclear PKC-zeta is necessary but not sufficient to induce granulocytic differentiation of HL-60 myeloid malignant cells (Bertolaso *et al.*, 1998).

The process of retinoid-induced differentiation varies in different cell types and may involve activation of selected PKC isoforms in a cell-specific fashion.

PKCθ is involved in the differentiation process activated by atRA in LAN-5 neuroblastoma cells. In fact, PKCθ, which is expressed as a nuclear and perinuclear protein, is induced and redistributed inside the cell by atRA. More importantly, PKCθ antisense oligonucleotides reduce the expression level of the kinase and the cell response to atRA (Sparatore *et al.*, 2000). In F9 embryonal carcinoma cells, PKCα seems to play an important role in atRA-induced parietal endoderm differentiation. Undifferentiated stem cells express PKCβ but not PKCα, whereas differentiated parietal endoderm cells express PKCα but not PKCβ. Constitutive expression of PKCα or inhibition of PKCβ expression in F9 stem cells enhances atRA induced differentiation. In addition, expressing PKCβ in a parietal endoderm cell line causes these cells to retrodifferentiate into stem cells. On the basis of these results, it was proposed that PKCβ and PKCα are key targets for RA-regulated gene expression. In particular, PKCα plays an important, active role in inducing and maintaining the parietal endoderm phenotype, and PKCβ activity is incompatible with maintaining the differentiated state of these cells (Cho *et al.*, 1998). Similarly, PKCα mediates melanocytic differentiation in melanoma cells treated with atRA. Indeed, the retinoid induces expression of PKCα selectively. Furthermore, forced expression of PKCα decreases the ability of B16 cells to form colonies in soft agar and increases melanin production. This translates into a therapeutic effect, as atRA decreases tumorigenicity in mice transplanted subcutaneously with B16 melanoma cells (Niles, 2003). Clearly, the prodifferentiating effects of PKCα observed in F9 and B16 cells treated with atRA must be related to mechanisms other than modulation of the RXR/RAR complex, given the mentioned negative role exerted by the kinase on the transcriptional activity of the heterodimer (Delmotte *et al.*, 1999).

The active role of PKCα or other PKC isoforms in atRA-induced cytodifferentiation is likely to be cell-context-specific. In fact, there are at least two situations in which cytodifferentiation and growth inhibition are associated with down- rather than upregulation of the enzyme. CAC-1 is a poorly differentiated human endometrial adenocarcinoma cell line characterized by an epithelial appearance and the expression of epithelial molecular markers. CAC-1 cells undergo morphological differentiation and growth retardation in response to atRA. In concomitance with these effects, atRA delocalizes PKCα from the plasma membrane to the cytosol, effectively inactivating the enzyme and downregulating the system (Radominska-Pandya *et al.*, 2000). The growth of the estrogen receptor-negative MDA-MB-231 cell line is strongly inhibited by retinoids in combination with a PKC inhibitor. While neither RA nor the PKC inhibitor, GF109203X, has a significant growth inhibitory effect in these cells, the combination of the two compounds potently suppresses proliferation. This is accompanied by an apoptotic response. Expression of phosphorylated as well as total PKC is decreased by GF109203X and the phenomenon is potentiated by atRA (Pettersson *et al.*, 2004).

At this time, the precise upstream regulatory events that ultimately result in PKC activation and inactivation or up- and downregulation are not known. The phosphorylation/activation of PKCδ by atRA may reflect engagement of a signaling loop following the formation of atRA-RAR complexes or could be regulated by other early cellular events induced by the retinoid. PKCδ has been reported to be activated down-stream of PI3K via the kinase PDK1 (Balendran *et al.*, 2000; Le Good *et al.*, 1998). It is therefore possible that the atRA-dependent pathway involves sequential activation of a PI3K/PDK1/PKCδ cascade. On the other hand, direct and RXR/RAR-independent effects activated by retinoids cannot be ruled out. Amino acid alignments and crystal structure analysis of atRA-binding proteins reveals a putative atRA-binding motif in PKC. This is supported by photolabeling studies showing concentration- and UV-dependent photoincorporation of [(3)H]ATRA into PKCα, which is effectively protected by 4-OH-atRA, 9-*cis* RA, and atRA glucuronide, but not by retinol. Photoaffinity labeling demonstrates strong competition between atRA and phosphatidylserine for binding to PKCα (Radominska-Pandya *et al.*, 2000).

Although the available data indicate numerous points of contact between the retinoid and PKC intracellular pathways, it is difficult to predict how pharmacological modulation of the cross talk may affect the cytodifferentiating and therapeutic action of retinoids. From a pharmacological point of view, the most significant data are those demonstrating the negative effects of certain PKC-isoforms on RXR/RAR transcriptional activity and those reporting on the retinoid-dependent downregulation of the PKC system. These observations suggest that inhibition of the PKC system and of particular PKC isoforms may potentiate the cytodifferentiating and therapeutic activity of atRA and derivatives. However, future efforts in this direction require the availability of selective inhibitors for the various PKC isoforms, PKCα in particular. In spite of these caveats, combinations between selective PKC inhibitors and atRA may be rational in the context of chemoprevention, given the purported role of the PKC system in the promotion phase of the carcinogenic process.

F. HISTONE ACETYLATION AND DNA METHYLATION

From a basic standpoint, a major and recent breakthrough in the elucidation of retinoids' molecular mechanisms of action is the observation that chromatin structure controls the transcriptional activity of nuclear RARs via changes in the levels of local histone acetylation and DNA methylation. On the one hand, it is now clear that the state of DNA methylation and the level of acetylation of histones and other chromatin-associated factors in the local chromatin environment modulates the transcriptional activity of the DNA-bound retinoid receptors complexes (Di Croce et al., 2002; Fazi et al., 2005; He et al., 2001; Jones et al., 2004; Pandolfi, 2001a,b; Salomoni and Pandolfi, 2000; Warrell et al., 1998). On the other hand, ligand binding of the receptors changes the state of DNA methylation and chromatin acetylation via switching of the corepressor with the coactivator multiprotein complexes. These two concepts are at the basis of the current efforts aimed at using demethylating agents as well as deacetylase inhibitors to potentiate the therapeutic activity of retinoids in certain types of neoplastic disease.

Demethylating agents, such as 5-aza-2′-deoxycytidine (5-aza-CdR), or deacetylase inhibitors, such as phenyl butyrate and valproic acid, have been combined to retinoids with the aim of increasing the therapeutic effect obtained. A lack of RARβ2 mRNA expression has been observed in solid tumor cells, including lung carcinoma, squamous cell carcinoma, and breast cancer (Hanabata et al., 2004; Ivanova et al., 2002; Widschwendter et al., 2000). A growing body of evidence supports the hypotheses that the RARβ2 gene is a tumor suppressor gene (Ekmekci et al., 2004; Fackler et al., 2003; Segura-Pacheco et al., 2003) and that the chemopreventive effects of retinoids are due to induction of RARβ2 (Hsu et al., 2000). The demethylating agent 5-aza-CdR restores RARβ (Cote and Momparler, 1997; Youssef et al., 2004) inducibility by atRA in head and neck as well as colon cancer cell lines, which posses a methylated RARβ promoter. In the latter cell lines, this effect is associated with increased growth inhibition after combined treatment with 5-aza-CdR and atRA (Youssef et al., 2004). Potentiation of atRA-induced cytodifferentiation of N-18 neuroblastoma cells by 5-aza-CdR has been reported. In this cell line, atRA fails to induce choline-acetyltransferase, a marker of neuronal differentiation. Treatment with the DNA demethylating agent alone also does not affect choline-acetyltransferase activity. However, after pretreatment of the cells with the DNA demethylating agents, the activity of the enzyme is greatly increased by atRA (Okuse et al., 1993).

HDAC-dependent transcriptional repression of the RA-signaling pathway underlies the differentiation block of APL. Cotreatment of retinoid-refractory AML blasts with combinations of HDAC inhibitors and atRA is associated with increased myeloid maturation. One of the first deacetylase inhibitors to be used in combination with atRA is phenylbutyrate (PB). In the millimolar range, PB induces histone acetylation in all the cell lines tested. The impact of combining PB with atRA was studied in the ML-1 myeloid leukemia cell line. In this model, PB augments atRA-induced differentiation, cell cycle arrest, and apoptosis. PB increases atRA induction of the myelomonocytic marker CD11b at all doses tested. Compared to PB alone, the combination with atRA induces greater G0/G1 cell cycle arrest (Yu et al., 1999). Combinations of PB and atRA or Ro 41-5253 were tested in two human breast cancer cell lines: the retinoid-sensitive MCF-7 and the retinoid-resistant MDA-MB-231. In both cell lines, when the retinoids were combined with PB, synergistic effects on growth inhibition were observed (Emionite *et al.*, 2004). A similar combination, consisting of PB and 13-*cis* RA, has been considered for the treatment of prostate carcinoma. The combination of PB and 13-*cis* RA inhibits cell proliferation and increases apoptosis *in vitro* in an additive fashion as compared with single agents. Similar to what observed in the case of treatment with demethylating agents in breast carcinoma, prostate tumor cells treated with both PB and 13-*cis* RA reveal increased expression of RARβ, suggesting a molecular mechanism for the biological additive effect. The combination of PB and 13-*cis* RA also inhibits prostate tumor growth *in vivo* as compared with single agents. Histological examination of tumor xenografts reveals decreased *in vivo* tumor cell proliferation, an increased apoptosis rate, and a reduced microvessel density in the animals treated with combined drugs, suggesting an antiangiogenesis effect of this combination (Pili *et al.*, 2001). One of the main problems associated with this type of combinations is the need to achieve millimolar concentrations of PB to obtain effective deacetylase inhibition. In fact, millimolar concentrations of PB are difficult to maintain in the patient for the relatively long period of times required.

APL with the t(11;17) chromosomal translocation, expressing the PLZF–RARα gene fusion, is a rare variant of the disease. This variant has been associated with poor clinical response to atRA treatment. *In vitro* studies indicate that blast differentiation is potentiated by the addition of the HDAC inhibitor, tricostatin A, to atRA (Petti *et al.*, 2002). The combination is of potential therapeutic interest, as indicated by preclinical data obtained in relevant transgenic animals. Transgenic mice harboring the PLZF–RARα fusion gene develop forms of leukemia that faithfully recapitulate both the clinical features and the response to atRA observed in humans with the corresponding translocations. Combinations of HDAC inhibitors and atRA induce leukemia remission and prolonged survival, without apparent toxic side effects (He *et al.*, 2001). Transcriptional repression of atRA signaling seems to be a common mechanism in AMLs. HDAC inhibitors restore atRA-dependent transcriptional activation and trigger terminal differentiation of primary blasts from 23 AML patients. Accordingly, AML1/ETO, the commonest AML-associated fusion protein, is an HDAC-dependent repressor of atRA signaling. These findings relate alteration of the atRA pathway to myeloid leukemogenesis and underscore the potential of transcriptional/differentiation therapy in AML (Ferrara *et al.*, 2001). The combined use of HDAC inhibitors and atRA has been taken all the way to the clinic. In a recent clinical trial performed on AML patients who were too old and/or medically unfit to receive intensive chemotherapy, combinations between valproic acid and atRA were administered. Combined treatment with the two agents resulted in clinical benefit in $\sim$16% of the patients (Kuendgen *et al.*, 2004).

V. RETINOID-BASED DIFFERENTIATION THERAPY, GENERAL OBSERVATIONS, AND CONCLUSION

In conclusion, what reviewed in the article indicates that retinoids act on and regulate the activity of numerous intracellular pathways involved in the processes of differentiation of numerous cell types. Rational modulation of these pathways by pharmacological or other kinds of strategies is likely to result in increased therapeutic efficacy of atRA and derivatives in the chemoprevention or therapy of leukemia and solid tumors. However, a number of considerations regarding the general and unresolved problems associated with the differentiation approach in oncology. Obviously, these apply not only to retinoids but also to other classes of differentiating agents.

Differentiation therapy relies on the concept that the neoplastic cell, like the normal counterpart, can be induced to proceed along a relatively physiological maturation pathway by the application of appropriate exogenous stimuli. Maturation eventually results in the appearance of neoplastic cells which have acquired phenotypic markers typical of the terminally differentiated normal cell of origin. Operationally, a cell that expresses a number of specific differentiation-associated markers and has lost its ability to proliferate is defined as a terminally differentiated cell. Thus, terminal differentiation is accompanied or followed by growth arrest and loss of proliferative potential, which may result in subsequent cell death. Clearly, the goal of differentiation therapy is to eliminate the neoplastic cell by acting only indirectly on the processes of cellular proliferation and survival. As already discussed, the concept of differentiation therapy can be applied to both a chemotherapeutic and a chemopreventive setting.

In a chemotherapeutic setting, differentiating agents may be used to induce remission or to debulk the tumor mass, during the earliest stages of any given clinical protocol. Currently, the use of atRA in APL is the only documented example in this respect. In the APL paradigm, the retinoid induces the following sequence of ordinate events in the leukemic blast: granulocytic maturation, G1 cell cycle arrest, and secondary apoptosis (Gianni *et al.*, 2000). The three processes are at the basis of the complete remissions observed in APL patients during the early stages of the clinical development of atRA (Warrell *et al.*, 1991). In spite of the enthusiasm raised by the initial results obtained in APL, it is unlikely that monotherapy with retinoids and other classes of cytodifferentiating agents may find significant applications during the early phases of cancer or leukemia treatment. In this context, it is more likely that differentiating agents will find applications in combination with cytotoxic agents or targeted therapy. This type of orientation is clearly evident even in the case of APL, where the drug is no more used alone to induce remission, but is part of a regimen involving classical chemotherapeutic agents.

Retinoid-based differentiation therapy is likely to be reevaluated on the basis of the emerging concepts of cancer stem cells and transcriptional therapy. From a theoretical point of view, the recent developments of the "cancer stem cell hypothesis" suggest that combinations between retinoids and cytotoxic agents may be a rational way to manage the neoplastic disease. The hypothesis states that the growth and progression of a tumor mass or leukemic disease are mainly the result of a minor cell population endowed with stem cell characteristics. As normal organ-specific stem cells, cancer stem cells are defined by their capacity to undergo self-renewal and differentiation. Cancer stem cells comprise only a small fraction of the total number of tumor cells and seem to be quiescent or cycling in a relatively slow manner. There is growing evidence that the cancer stem cell pool is more resistant than the other fraction of more differentiated neoplastic cells to the cytotoxic activity of chemotherapeutics. Hence, it is believed that the debulking effect of classical cytotoxic agents is primarily the result of an effect on the differentiated cellular component of the tumor or leukemia. Guzman *et al.* demonstrated that leukemic stem cells are more resistant to chemotherapy than the more differentiated myeloblastic counterparts. Similarly there are reports indicating that myeloma stem cells are resistant to many therapies used to treat myeloma, including chemotherapy and the proteasome inhibitor, velcade (Guzman *et al.*, 2002). On the basis of this evidence, it is possible to envisage the use of atRA and other retinoids to target cancer and leukemic stem cells, forcing them to differentiate into a cell type that may be more sensitive to cytotoxic stimuli.

A major obstacle in the development of differentiation therapy is represented by the fact that the terminally differentiated phenotype in APL is largely known and can be easily studied with the use of recognized molecular markers of myeloid differentiation. Furthermore, it is well established that the terminally differentiated leukemic blast is poised to undergo programmed cell death. Hence, treatment with differentiating agents, like retinoids, automatically translates into decreased neoplastic cell burden and measurable clinical benefit. All this may not be the case for other types of leukemia and cancer. Indeed, maturation of leukemic blasts of non-M3 origin along the monocytic, erythroblastic, or thrombocytic pathway generally results in growth inhibition, but is not necessarily associated with an apoptotic response. From a therapeutic prospective, this may represent an important obstacle, which can be overcome only by the combined use of retinoid or other differentiating agents and cytotoxic compounds, as alluded to above. Another practical problem is the lack of established molecular, morphological, and biochemical markers of differentiation in most types of leukemia and solid tumors. The deficiency makes it difficult to study the therapeutic use of differentiating agents with established surrogate markers. These and other problems need to be resolved in the next future before the advent of a more generalized use of retinoids in oncology. However, the recent progress in the elucidation of the molecular mechanisms underlying the pharmacological activity of RA and derivatives is likely to indicate the most rational way to achieve this goal.

ACKNOWLEDGMENTS

Writing of this chapter was made possible by the financial support of the Associazione Italiana per la Ricerca contro il Cancro (AIRC), the Istituto Superiore di Sanità, "the Fondo d'Investimento per la Ricerca Biotecnologia" (FIRB). The Financial support of the Weizmann-Negri Foundation is also acknowledged. We thank Prof. Silvio Garattini and Dr. Maddalena Fratelli for critical reading of the manuscript.

REFERENCES

Abemayor, E., and Sidell, N. (1989). Human neuroblastoma cell lines as models for the *in vitro* study of neoplastic and neuronal cell differentiation. *Environ. Health Perspect.* **80**, 3–15.

Adam-Stitah, S., Penna, L., Chambon, P., and Rochette-Egly, C. (1999). Hyperphosphorylation of the retinoid X receptor alpha by activated c-Jun NH2-terminal kinases. *J. Biol. Chem.* **274**, 18932–18941.

Aggarwal, S., Kim, S. W., Cheon, K., Tabassam, F. H., Yoon, J. H., and Koo, J. S. (2006). Nonclassical action of retinoic acid on the activation of the cAMP response element-binding protein in normal human bronchial epithelial cells. *Mol. Biol. Cell* **17**, 566–575.

Allenby, G. (1995). The ying-yang of RAR and AP-1: Cancer treatment without overt toxicity. *Hum. Exp. Toxicol.* **14**, 226–230.

Alonso, A., Breuer, B., Steuer, B., and Fischer, J. (1991). The F9-EC cell line as a model for the analysis of differentiation. *Int. J. Dev. Biol.* **35**, 389–397.

Alsayed, Y., Uddin, S., Mahmud, N., Lekmine, F., Kalvakolanu, D. V., Minucci, S., Bokoch, G., and Platanias, L. C. (2001). Activation of Rac1 and the p38 mitogen-activated protein kinase pathway in response to all-trans-retinoic acid. *J. Biol. Chem.* **276**, 4012–4019.

Altucci, L., and Gronemeyer, H. (2002). Decryption of the retinoid death code in leukemia. *J. Clin. Immunol.* **22**, 117–123.

Altucci, L., Rossin, A., Hirsch, O., Nebbioso, A., Vitoux, D., Wilhelm, E., Guidez, F., De Simone, M., Schiavone, E. M., Grimwade, D., Zelent, A., and De The, H., *et al.* (2005). Rexinoid-triggered differentiation and tumor-selective apoptosis of acute myeloid leukemia by protein kinase A-mediated desubordination of retinoid X receptor. *Cancer Res.* **65**, 8754–8765.

Alvarez, R., Vega, M. J., Kammerer, S., Rossin, A., Germain, P., Gronemeyer, H., and de Lera, A. R. (2004). 9-cis-Retinoic acid analogues with bulky hydrophobic rings: New RXR-selective agonists. *Bioorg. Med. Chem. Lett.* **14**, 6117–6122.

Bain, G., Ray, W. J., Yao, M., and Gottlieb, D. I. (1994). From embryonal carcinoma cells to neurons: The P19 pathway. *Bioessays* **16**, 343–348.

Balendran, A., Hare, G. R., Kieloch, A., Williams, M. R., and Alessi, D. R. (2000). Further evidence that 3-phosphoinositide-dependent protein kinase-1 (PDK1) is required for the stability and phosphorylation of protein kinase C (PKC) isoforms. *FEBS Lett.* **484**, 217–223.

Bastie, J. N., Balitrand, N., Guillemot, I., Chomienne, C., and Delva, L. (2005). Cooperative action of 1alpha,25-dihydroxyvitamin D3 and retinoic acid in NB4 acute promyelocytic leukemia cell differentiation is transcriptionally controlled. *Exp. Cell Res.* **310**, 319–330.

Bastien, J., Plassat, J. L., Payrastre, B., and Rochette-Egly, C. (2006). The phosphoinositide 3-kinase/Akt pathway is essential for the retinoic acid-induced differentiation of F9 cells. *Oncogene* **25**, 2040–2047.
Beehler, B. C., Brinckerhoff, C. E., and Ostrowski, J. (2004). Selective retinoic acid receptor ligands for rheumatoid arthritis. *Curr. Opin. Investig. Drugs* **5**, 1153–1157.
Beere, H. M. (2005). Death versus survival: Functional interaction between the apoptotic and stress-inducible heat shock protein pathways. *J. Clin. Invest.* **115**, 2633–2639.
Benbrook, D. M., Madler, M. M., Spruce, L. W., Birckbichler, P. J., Nelson, E. C., Subramanian, S., Weerasekare, G. M., Gale, J. B., Patterson, M. K., Jr., Wang, B., Wang, W., Lu, S., *et al.* (1997). Biologically active heteroarotinoids exhibiting anticancer activity and decreased toxicity. *J. Med. Chem.* **40**, 3567–3583.
Benoit, G., Altucci, L., Flexor, M., Ruchaud, S., Lillehaug, J., Raffelsberger, W., Gronemeyer, H., and Lanotte, M. (1999). RAR-independent RXR signaling induces t(15;17) leukemia cell maturation. *EMBO J.* **18**, 7011–7018.
Bertolaso, L., Gibellini, D., Secchiero, P., Previati, M., Falgione, D., Visani, G., Rizzoli, R., Capitani, S., and Zauli, G. (1998). Accumulation of catalytically active PKC-zeta into the nucleus of HL-60 cell line plays a key role in the induction of granulocytic differentiation mediated by all-trans retinoic acid. *Br. J. Haematol.* **100**, 541–549.
Boehm, M. F., Zhang, L., Zhi, L., McClurg, M. R., Berger, E., Wagoner, M., Mais, D. E., Suto, C. M., Davies, J. A., Heyman, R. A., and Nadzan, A. M. (1995). Design and synthesis of potent retinoid X receptor selective ligands that induce apoptosis in leukemia cells. *J. Med. Chem.* **38**, 3146–3155.
Bogoyevitch, M. A. (2005). Therapeutic promise of JNK ATP-noncompetitive inhibitors. *Trends Mol. Med.* **11**, 232–239.
Bogoyevitch, M. A., Gillespie-Brown, J., Ketterman, A. J., Fuller, S. J., Ben-Levy, R., Ashworth, A., Marshall, C. J., and Sugden, P. H. (1996). Stimulation of the stress-activated mitogen-activated protein kinase subfamilies in perfused heart. p38/RK mitogen-activated protein kinases and c-Jun N-terminal kinases are activated by ischemia/reperfusion. *Circ. Res.* **79**, 162–173.
Bollag, W. (1994). Experimental basis of cancer combination chemotherapy with retinoids, cytokines, 1,25-dihydroxyvitamin D3, and analogs. *J. Cell. Biochem.* **56**, 427–435.
Bour, G., Gaillard, E., Bruck, N., Lalevee, S., Plassat, J. L., Busso, D., Samama, J. P., and Rochette-Egly, C. (2005). Cyclin H binding to the RARalpha activation function (AF)-2 domain directs phosphorylation of the AF-1 domain by cyclin-dependent kinase 7. *Proc. Natl. Acad. Sci. USA* **102**, 16608–16613.
Boylan, J. F., Lufkin, T., Achkar, C. C., Taneja, R., Chambon, P., and Gudas, L. J. (1995). Targeted disruption of retinoic acid receptor alpha (RAR alpha) and RAR gamma results in receptor-specific alterations in retinoic acid-mediated differentiation and retinoic acid metabolism. *Mol. Cell. Biol.* **15**, 843–851.
Breitman, T. R., Chen, Z. X., and Takahashi, N. (1994). Potential applications of cytodifferentiation therapy in hematologic malignancies. *Semin. Hematol.* **31**, 18–25.
Brtko, J., and Thalhamer, J. (2003). Renaissance of the biologically active vitamin A derivatives: Established and novel directed therapies for cancer and chemoprevention. *Curr. Pharm. Des.* **9**, 2067–2077.
Bruck, N., Bastien, J., Bour, G., Tarrade, A., Plassat, J. L., Bauer, A., Adam-Stitah, S., and Rochette-Egly, C. (2005). Phosphorylation of the retinoid x receptor at the omega loop, modulates the expression of retinoic-acid-target genes with a promoter context specificity. *Cell. Signal.* **17**, 1229–1239.
Caelles, C., Gonzalez-Sancho, J. M., and Munoz, A. (1997). Nuclear hormone receptor antagonism with AP-1 by inhibition of the JNK pathway. *Genes Dev.* **11**, 3351–3364.

Cao, Y., Wang, F., Liu, H. Y., Fu, Z. D., and Han, R. (2005). Resveratrol induces apoptosis and differentiation in acute promyelocytic leukemia (NB4) cells. *J. Asian Nat. Prod. Res.* **7**, 633–641.

Cao, Z., Flanders, K. C., Bertolette, D., Lyakh, L. A., Wurthner, J. U., Parks, W. T., Letterio, J. J., Ruscetti, F. W., and Roberts, A. B. (2003). Levels of phospho-Smad2/3 are sensors of the interplay between effects of TGF-beta and retinoic acid on monocytic and granulocytic differentiation of HL-60 cells. *Blood* **101**, 498–507.

Carpentier, A., Balitrand, N., Rochette-Egly, C., Shroot, B., Degos, L., and Chomienne, C. (1997). Distinct sensitivity of neuroblastoma cells for retinoid receptor agonists: Evidence for functional receptor heterodimers. *Oncogene* **15**, 1805–1813.

Carter, C. A., and Madden, V. J. (2000). A newly characterized human endometrial adenocarcinoma cell line (CAC-1) differentiates in response to retinoic acid treatment. *Exp. Mol. Pathol.* **69**, 175–191.

Carystinos, G. D., Alaoui-Jamali, M. A., Phipps, J., Yen, L., and Batist, G. (2001). Upregulation of gap junctional intercellular communication and connexin 43 expression by cyclic-AMP and all-trans-retinoic acid is associated with glutathione depletion and chemosensitivity in neuroblastoma cells. *Cancer Chemother. Pharmacol.* **47**, 126–132.

Castaigne, S., Chomienne, C., Daniel, M. T., Ballerini, P., Berger, R., Fenaux, P., and Degos, L. (1990). All-trans retinoic acid as a differentiation therapy for acute promyelocytic leukemia. I. Clinical results. *Blood* **76**, 1704–1709.

Castel, V., and Canete, A. (2004). A comparison of current neuroblastoma chemotherapeutics. *Expert Opin. Pharmacother.* **5**, 71–80.

Chambon, P. (1996). A decade of molecular biology of retinoic acid receptors. *FASEB J.* **10**, 940–954.

Chandraratna, R. A. (1997). Tazarotene: The first receptor-selective topical retinoid for the treatment of psoriasis. *J. Am. Acad. Dermatol.* **37**, S12–S17.

Chandraratna, R. A. (1998a). Future trends: A new generation of retinoids. *J. Am. Acad. Dermatol.* **39**, S149–S152.

Chandraratna, R. A. (1998b). Rational design of receptor-selective retinoids. *J. Am. Acad. Dermatol.* **39**, S124–S128.

Chang, L., and Karin, M. (2001). Mammalian MAP kinase signalling cascades. *Nature* **410**, 37–40.

Chelbi-Alix, M. K., and Pelicano, L. (1999). Retinoic acid and interferon signaling cross talk in normal and RA-resistant APL cells. *Leukemia* **13**, 1167–1174.

Chen, J. Y., Penco, S., Ostrowski, J., Balaguer, P., Pons, M., Starrett, J. E., Reczek, P., Chambon, P., and Gronemeyer, H. (1995). RAR-specific agonist/antagonists which dissociate transactivation and AP1 transrepression inhibit anchorage-independent cell proliferation. *EMBO J.* **14**, 1187–1197.

Chen, Z., Guidez, F., Rousselot, P., Agadir, A., Chen, S. J., Wang, Z. Y., Degos, L., Zelent, A., Waxman, S., and Chomienne, C. (1994). PLZF-RAR alpha fusion proteins generated from the variant t(11;17)(q23;q21) translocation in acute promyelocytic leukemia inhibit ligand-dependent transactivation of wild-type retinoic acid receptors. *Proc. Natl. Acad. Sci. USA* **91**, 1178–1182.

Cheung, B., Hocker, J. E., Smith, S. A., Norris, M. D., Haber, M., and Marshall, G. M. (1998). Favorable prognostic significance of high-level retinoic acid receptor beta expression in neuroblastoma mediated by effects on cell cycle regulation. *Oncogene* **17**, 751–759.

Cho, Y., Tighe, A. P., and Talmage, D. A. (1997). Retinoic acid induced growth arrest of human breast carcinoma cells requires protein kinase C alpha expression and activity. *J. Cell Physiol.* **172**, 306–313.

Cho, Y., Klein, M. G., and Talmage, D. A. (1998). Distinct functions of protein kinase Calpha and protein kinase Cbeta during retinoic acid-induced differentiation of F9 cells. *Cell Growth Differ.* **9**, 147–154.

Choi, Y., and Fuchs, E. (1990). TGF-beta and retinoic acid: Regulators of growth and modifiers of differentiation in human epidermal cells. *Cell Regul.* **1,** 791–809.

Chu, P. W., Cheung, W. M., and Kwong, Y. L. (2003). Differential effects of 9-cis, 13-cis and all-trans retinoic acids on the neuronal differentiation of human neuroblastoma cells. *Neuroreport* **14,** 1935–1939.

Collins, S. J. (2002). The role of retinoids and retinoic acid receptors in normal hematopoicsis. *Leukemia* **16,** 1896–1905.

Colotta, F., Re, F., Polentarutti, N., Sozzani, S., and Mantovani, A. (1992). Modulation of granulocyte survival and programmed cell death by cytokines and bacterial products. *Blood* **80,** 2012–2020.

Cote, S., and Momparler, R. L. (1997). Activation of the retinoic acid receptor beta gene by 5-aza-2'-deoxycytidine in human DLD-1 colon carcinoma cells. *Anticancer Drugs* **8,** 56–61.

Cote, S., Zhou, D., Bianchini, A., Nervi, C., Gallagher, R. E., and Miller, W. H., Jr. (2000). Altered ligand binding and transcriptional regulation by mutations in the PML/RARalpha ligand-binding domain arising in retinoic acid-resistant patients with acute promyelocytic leukemia. *Blood* **96,** 3200–3208.

Crowe, D. L. (2002). Receptor selective synthetic retinoids as potential cancer chemotherapy agents. *Curr. Cancer Drug Targets* **2,** 77–86.

Dai, X., Yamasaki, K., Shirakata, Y., Sayama, K., and Hashimoto, K. (2004). All-trans-retinoic acid induces interleukin-8 via the nuclear factor-kappaB and p38 mitogen-activated protein kinase pathways in normal human keratinocytes. *J. Invest. Dermatol.* **123,** 1078–1085.

De Benedetti, F., Falk, L. A., Ellingsworth, L. R., Ruscetti, F. W., and Faltynek, C. R. (1990). Synergy between transforming growth factor-beta and tumor necrosis factor-alpha in the induction of monocytic differentiation of human leukemic cell lines. *Blood* **75,** 626–632.

del Rincon, S. V., Rousseau, C., Samanta, R., and Miller, W. H., Jr. (2003). Retinoic acid-induced growth arrest of MCF-7 cells involves the selective regulation of the IRS-1/PI 3-kinase/AKT pathway. *Oncogene* **22,** 3353–3360.

Delmotte, M. H., Tahayato, A., Formstecher, P., and Lefebvre, P. (1999). Serine 157, a retinoic acid receptor alpha residue phosphorylated by protein kinase C *in vitro*, is involved in RXR. RARalpha heterodimerization and transcriptional activity. *J. Biol. Chem.* **274,** 38225–38231.

Demary, K., Wong, L., Liou, J. S., Faller, D. V., and Spanjaard, R. A. (2001). Redox control of retinoic acid receptor activity: A novel mechanism for retinoic acid resistance in melanoma cells. *Endocrinology* **142,** 2600–2605.

Desai, S. H., and Niles, R. M. (1997). Characterization of retinoic acid-induced AP-1 activity in B16 mouse melanoma cells. *J. Biol. Chem.* **272,** 12809–12815.

Di Croce, L., Raker, V. A., Corsaro, M., Fazi, F., Fanelli, M., Faretta, M., Fuks, F., Lo Coco, F., Kouzarides, T., Nervi, C., Minucci, S., Pelicci, P. G., *et al.* (2002). Methyltransferase recruitment and DNA hypermethylation of target promoters by an oncogenic transcription factor. *Science* **295,** 1079–1082.

Di Lorenzo, D., Gianni, M., Savoldi, G. F., Ferrari, F., Albertini, A., and Garattini, E. (1993). Progesterone induced expression of alkaline phosphatase is associated with a secretory phenotype in T47D breast cancer cells. *Biochem. Biophys. Res. Commun.* **192,** 1066–1072.

Dimberg, A., Nilsson, K., and Oberg, F. (2000). Phosphorylation-deficient Stat1 inhibits retinoic acid-induced differentiation and cell cycle arrest in U-937 monoblasts. *Blood* **96,** 2870–2878.

Dimberg, A., Karlberg, I., Nilsson, K., and Oberg, F. (2003). Ser727/Tyr701-phosphorylated Stat1 is required for the regulation of c-Myc, cyclins, and p27Kip1 associated with ATRA-induced G0/G1 arrest of U-937 cells. *Blood* **102,** 254–261.

Ding, W., Li, Y. P., Nobile, L. M., Grills, G., Carrera, I., Paietta, E., Tallman, M. S., Wiernik, P. H., and Gallagher, R. E. (1998). Leukemic cellular retinoic acid resistance and missense mutations in the PML-RARalpha fusion gene after relapse of acute promyelocytic leukemia from treatment with all-trans retinoic acid and intensive chemotherapy. *Blood* **92,** 1172–1183.

Dong, S., and Tweardy, D. J. (2002). Interactions of STAT5b-RARalpha, a novel acute promyelocytic leukemia fusion protein, with retinoic acid receptor and STAT3 signaling pathways. *Blood* **99**, 2637–2646.
Edwards, D. P. (1999). Coregulatory proteins in nuclear hormone receptor action. *Vitam. Horm.* **55**, 165–218.
Eferl, R., and Wagner, E. F. (2003). AP-1: A double-edged sword in tumorigenesis. *Nat. Rev. Cancer* **3**, 859–868.
Egyed, M., Kollar, B., Rumi, G., Keller, E., Vass, J., and Fekete, S. (2003). Effect of retinoic acid treatment on cytogenetic remission of chronic myeloid leukaemia. *Acta Haematol.* **109**, 84–89.
Eisenhauer, E. A., Lippman, S. M., Kavanagh, J. J., Parades-Espinoza, M., Arnold, A., Hong, W. K., Massimini, G., Schleuniger, U., Bollag, W., Holdener, E. E., and Krakoff, I. (1994). Combination 13-cis-retinoic acid and interferon alpha-2a in the therapy of solid tumors. *Leukemia* **8**, 1622–1625.
Ekmekci, C. G., Gutierrez, M. I., Siraj, A. K., Ozbek, U., and Bhatia, K. (2004). Aberrant methylation of multiple tumor suppressor genes in acute myeloid leukemia. *Am. J. Hematol.* **77**, 233–240.
Emionite, L., Galmozzi, F., Raffo, P., Vergani, L., and Toma, S. (2003). Retinoids and malignant melanoma: A pathway of proliferation inhibition on SK MEL28 cell line. *Anticancer Res.* **23**, 13–19.
Emionite, L., Galmozzi, F., Grattarola, M., Boccardo, F., Vergani, L., and Toma, S. (2004). Histone deacetylase inhibitors enhance retinoid response in human breast cancer cell lines. *Anticancer Res.* **24**, 4019–4024.
Fackler, M. J., McVeigh, M., Evron, E., Garrett, E., Mehrotra, J., Polyak, K., Sukumar, S., and Argani, P. (2003). DNA methylation of RASSF1A, HIN-1, RAR-beta, Cyclin D2 and Twist in *in situ* and invasive lobular breast carcinoma. *Int. J. Cancer* **107**, 970–975.
Faria, T. N., Mendelsohn, C., Chambon, P., and Gudas, L. J. (1999). The targeted disruption of both alleles of RARbeta(2) in F9 cells results in the loss of retinoic acid-associated growth arrest. *J. Biol. Chem.* **274**, 26783–26788.
Farias, E. F., Ong, D. E., Ghyselinck, N. B., Nakajo, S., Kuppumbatti, Y. S., and Miray Lopez, R. (2005). Cellular retinol-binding protein I, a regulator of breast epithelial retinoic acid receptor activity, cell differentiation, and tumorigenicity. *J. Natl. Cancer Inst.* **97**, 21–29.
Fazi, F., Travaglini, L., Carotti, D., Palitti, F., Diverio, D., Alcalay, M., McNamara, S., Miller, W. H., Jr., Lo Coco, F., Pelicci, P. G., and Nervi, C. (2005). Retinoic acid targets DNA-methyltransferases and histone deacetylases during APL blast differentiation *in vitro* and *in vivo*. *Oncogene* **24**, 1820–1830.
Ferrara, F. F., Fazi, F., Bianchini, A., Padula, F., Gelmetti, V., Minucci, S., Mancini, M., Pelicci, P. G., Lo Coco, F., and Nervi, C. (2001). Histone deacetylase-targeted treatment restores retinoic acid signaling and differentiation in acute myeloid leukemia. *Cancer Res.* **61**, 2–7.
Fisher, G. J., Talwar, H. S., Lin, J., Lin, P., McPhillips, F., Wang, Z., Li, X., Wan, Y., Kang, S., and Voorhees, J. J. (1998). Retinoic acid inhibits induction of c-Jun protein by ultraviolet radiation that occurs subsequent to activation of mitogen-activated protein kinase pathways in human skin *in vivo*. *J. Clin. Invest.* **101**, 1432–1440.
Fossa, S. D., Mickisch, G. H., De Mulder, P. H., Horenblas, S., van Oosterom, A. T., van Poppel, H., Fey, M., Croles, J. J., de Prijck, L., and Van Glabbeke, M. (2004). Interferon-alpha-2a with or without 13-cis retinoic acid in patients with progressive, measurable metastatic renal cell carcinoma. *Cancer* **101**, 533–540.
Gaines, P., and Berliner, N. (2003). Retinoids in myelopoiesis. *J. Biol. Regul. Homeost. Agents* **17**, 46–65.
Gajovic, S., St-Onge, L., Yokota, Y., and Gruss, P. (1997). Retinoic acid mediates Pax6 expression during *in vitro* differentiation of embryonic stem cells. *Differentiation* **62**, 187–192.

Garattini, E., and Gianni, M. (1996). Leukocyte alkaline phosphatase a specific marker for the post-mitotic neutrophilic granulocyte: Regulation in acute promyelocytic leukemia. *Leuk. Lymphoma* **23,** 493–503.

Garattini, E., and Terao, M. (2001). Cytodifferentiation: A novel approach to cancer treatment and prevention. *Curr. Opin. Pharmacol.* **1,** 358–363.

Garattini, E., Mologni, L., Ponzanelli, I., and Terao, M. (1998). Cross-talk between retinoic acid and interferons: Molecular mechanisms of interaction in acute promyelocytic leukemia cells. *Leuk. Lymphoma* **30,** 467–475.

Garattini, E., Gianni, M., and Terao, M. (2004). Retinoid related molecules an emerging class of apoptotic agents with promising therapeutic potential in oncology: Pharmacological activity and mechanisms of action. *Curr. Pharm. Des.* **10,** 433–448.

Germain, P., Kammerer, S., Perez, E., Peluso-Iltis, C., Tortolani, D., Zusi, F. C., Starrett, J., Lapointe, P., Daris, J. P., Marinier, A., de Lera, A. R., Rochel, N., et al. (2004). Rational design of RAR-selective ligands revealed by RARbeta crystal structure. *EMBO Rep.* **5,** 877–882.

Gery, S., Park, D. J., Vuong, P. T., Chih, D. Y., Lemp, N., and Koeffler, H. P. (2004). Retinoic acid regulates C/EBP homologous protein expression (CHOP), which negatively regulates myeloid target genes. *Blood* **104,** 3911–3917.

Gianni, M., Studer, M., Carpani, G., Terao, M., and Garattini, E. (1991). Retinoic acid induces liver/bone/kidney-type alkaline phosphatase gene expression in F9 teratocarcinoma cells. *Biochem. J.* **274**(Pt. 3), 673–678.

Gianni, M., Zanotta, S., Terao, M., Garattini, S., and Garattini, E. (1993). Effects of synthetic retinoids and retinoic acid isomers on the expression of alkaline phosphatase in F9 teratocarcinoma cells. *Biochem. Biophys. Res. Commun.* **196,** 252–259.

Gianni, M., Terao, M., Zanotta, S., Barbui, T., Rambaldi, A., and Garattini, E. (1994). Retinoic acid and granulocyte colony-stimulating factor synergistically induce leukocyte alkaline phosphatase in acute promyelocytic leukemia cells. *Blood* **83,** 1909–1921.

Gianni, M., Li Calzi, M., Terao, M., Rambaldi, A., and Garattini, E. (1995a). Tyrosine kinases but not cAMP-dependent protein kinase mediate the induction of leukocyte alkaline phosphatase by granulocyte-colony-stimulating factor and retinoic acid in acute promyelocytic leukemia cells. *Biochem. Biophys. Res. Commun.* **208,** 846–854.

Gianni, M., Norio, P., Terao, M., Falanga, A., Marchetti, M., Rambaldi, A., and Garattini, E. (1995b). Effects of dexamethasone on pro-inflammatory cytokine expression, cell growth and maturation during granulocytic differentiation of acute promyelocytic leukemia cells. *Eur. Cytokine Netw.* **6,** 157–165.

Gianni, M., Terao, M., Norio, P., Barbui, T., Rambaldi, A., and Garattini, E. (1995c). All-trans retinoic acid and cyclic adenosine monophosphate cooperate in the expression of leukocyte alkaline phosphatase in acute promyelocytic leukemia cells. *Blood* **85,** 3619–3635.

Gianni, M., Li Calzi, M., Terao, M., Guiso, G., Caccia, S., Barbui, T., Rambaldi, A., and Garattini, E. (1996a). AM580, a stable benzoic derivative of retinoic acid, has powerful and selective cyto-differentiating effects on acute promyelocytic leukemia cells. *Blood* **87,** 1520–1531.

Gianni, M., Zanotta, S., Terao, M., Rambaldi, A., and Garattini, E. (1996b). Interferons induce normal and aberrant retinoic-acid receptors type alpha in acute promyelocytic leukemia cells: Potentiation of the induction of retinoid-dependent differentiation markers. *Int. J. Cancer* **68,** 75–83.

Gianni, M., Terao, M., Fortino, I., LiCalzi, M., Viggiano, V., Barbui, T., Rambaldi, A., and Garattini, E. (1997). Stat1 is induced and activated by all-trans retinoic acid in acute promyelocytic leukemia cells. *Blood* **89,** 1001–1012.

Gianni, M., Ponzanelli, I., Mologni, L., Reichert, U., Rambaldi, A., Terao, M., and Garattini, E. (2000). Retinoid-dependent growth inhibition, differentiation and apoptosis in acute promyelocytic leukemia cells. Expression and activation of caspases. *Cell Death Differ.* **7,** 447–460.

Gianni, M., Kalac, Y., Ponzanelli, I., Rambaldi, A., Terao, M., and Garattini, E. (2001). Tyrosine kinase inhibitor STI571 potentiates the pharmacologic activity of retinoic acid in acute promyelocytic leukemia cells: Effects on the degradation of RARalpha and PML-RARalpha. *Blood* **97,** 3234–3243.

Gianni, M., Bauer, A., Garattini, E., Chambon, P., and Rochette-Egly, C. (2002a). Phosphorylation by p38MAPK and recruitment of SUG-1 are required for RA-induced RAR gamma degradation and transactivation. *EMBO J.* **21,** 3760–3769.

Gianni, M., Kopf, E., Bastien, J., Oulad-Abdelghani, M., Garattini, E., Chambon, P., and Rochette-Egly, C. (2002b). Down-regulation of the phosphatidylinositol 3-kinase/Akt pathway is involved in retinoic acid-induced phosphorylation, degradation, and transcriptional activity of retinoic acid receptor gamma 2. *J. Biol. Chem.* **277,** 24859–24862.

Gianni, M., Tarrade, A., Nigro, E. A., Garattini, E., and Rochette-Egly, C. (2003). The AF-1 and AF-2 domains of RAR gamma 2 and RXR alpha cooperate for triggering the transactivation and the degradation of RAR gamma 2/RXR alpha heterodimers. *J. Biol. Chem.* **278,** 34458–34466.

Gianni, M., Parrella, E., Raska, I., Jr., Gaillard, E., Nigro, E. A., Gaudon, C., Garattini, E., and Rochette-Egly, C. (2006). P38MAPK-dependent phosphorylation and degradation of SRC-3/AIB1 and RARalpha-mediated transcription. *EMBO J.* **25,** 739–751.

Glass, C. K., Rose, D. W., and Rosenfeld, M. G. (1997). Nuclear receptor coactivators. *Curr. Opin. Cell Biol.* **9,** 222–232.

Gomez-Santos, C., Ambrosio, S., Ventura, F., Ferrer, I., and Reiriz, J. (2002). TGF-beta1 increases tyrosine hydroxylase expression by a mechanism blocked by BMP-2 in human neuroblastoma SH-SY5Y cells. *Brain Res.* **958,** 152–160.

Gonzalez, M. V., Gonzalez-Sancho, J. M., Caelles, C., Munoz, A., and Jimenez, B. (1999). Hormone-activated nuclear receptors inhibit the stimulation of the JNK and ERK signalling pathways in endothelial cells. *FEBS Lett.* **459,** 272–276.

Gottardis, M. M., Lamph, W. W., Shalinsky, D. R., Wellstein, A., and Heyman, R. A. (1996). The efficacy of 9-cis retinoic acid in experimental models of cancer. *Breast Cancer Res. Treat.* **38,** 85–96.

Grabel, L., Becker, S., Lock, L., Maye, P., and Zanders, T. (1998). Using EC and ES cell culture to study early development: Recent observations on Indian hedgehog and Bmps. *Int. J. Dev. Biol.* **42,** 917–925.

Grande, A., Montanari, M., Manfredini, R., Tagliafico, E., Zanocco-Marani, T., Trevisan, F., Ligabue, G., Siena, M., Ferrari, S., and Ferrari, S. (2001). A functionally active RARalpha nuclear receptor is expressed in retinoic acid non responsive early myeloblastic cell lines. *Cell Death Differ.* **8,** 70–82.

Gravis, G., Pech-Gourgh, F., Viens, P., Alzieu, C., Camerlo, J., Oziel-Taieb, S., Jausseran, M., and Maraninchi, D. (1999). Phase II study of a combination of low-dose cisplatin with 13-cis-retinoic acid and interferon-alpha in patients with advanced head and neck squamous cell carcinoma. *Anticancer Drugs* **10,** 369–374.

Gronemeyer, H., and Miturski, R. (2001). Molecular mechanisms of retinoid action. *Cell. Mol. Biol. Lett.* **6,** 3–52.

Guidez, F., Huang, W., Tong, J. H., Dubois, C., Balitrand, N., Waxman, S., Michaux, J. L., Martiat, P., Degos, L., Chen, Z., and Chomienne, C. (1994). Poor response to all-trans retinoic acid therapy in a t(11;17) PLZF/RAR alpha patient. *Leukemia* **8,** 312–317.

Guillemin, M. C., Raffoux, E., Vitoux, D., Kogan, S., Soilihi, H., Lallemand-Breitenbach, V., Zhu, J., Janin, A., Daniel, M. T., Gourmel, B., Degos, L., Dombret, H., *et al.* (2002). In vivo activation of cAMP signaling induces growth arrest and differentiation in acute promyelocytic leukemia. *J. Exp. Med.* **196,** 1373–1380.

Guzman, M. L., Swiderski, C. F., Howard, D. S., Grimes, B. A., Rossi, R. M., Szilvassy, S. J., and Jordan, C. T. (2002). Preferential induction of apoptosis for primary human leukemic stem cells. *Proc. Natl. Acad. Sci. USA* **99,** 16220–16225.

Hanabata, T., Tsukuda, K., Toyooka, S., Yano, M., Aoe, M., Nagahiro, I., Sano, Y., Date, H., and Shimizu, N. (2004). DNA methylation of multiple genes and clinicopathological relationship of non-small cell lung cancers. *Oncol. Rep.* **12,** 177–180.

Harris, M. N., Ozpolat, B., Abdi, F., Gu, S., Legler, A., Mawuenyega, K. G., Tirado-Gomez, M., Lopez-Berestein, G., and Chen, X. (2004). Comparative proteomic analysis of all-trans-retinoic acid treatment reveals systematic posttranscriptional control mechanisms in acute promyelocytic leukemia. *Blood* **104,** 1314–1323.

He, L. Z., Tolentino, T., Grayson, P., Zhong, S., Warrell, R. P., Jr., Rifkind, R. A., Marks, P. A., Richon, V. M., and Pandolfi, P. P. (2001). Histone deacetylase inhibitors induce remission in transgenic models of therapy-resistant acute promyelocytic leukemia. *J. Clin. Invest.* **108,** 1321–1330.

Hideshima, T., Akiyama, M., Hayashi, T., Richardson, P., Schlossman, R., Chauhan, D., and Anderson, K. C. (2003). Targeting p38 MAPK inhibits multiple myeloma cell growth in the bone marrow milieu. *Blood* **101,** 703–705.

Higuchi, E., Chandraratna, R. A., Hong, W. K., and Lotan, R. (2003). Induction of TIG3, a putative class II tumor suppressor gene, by retinoic acid in head and neck and lung carcinoma cells and its association with suppression of the transformed phenotype. *Oncogene* **22,** 4627–4635.

Higuchi, T., Kizaki, M., and Omine, M. (2004). Induction of differentiation of retinoic acid-resistant acute promyelocytic leukemia cells by the combination of all-trans retinoic acid and granulocyte colony-stimulating factor. *Leuk. Res.* **28,** 525–532.

Hisatake, J., O'Kelly, J., Uskokovic, M. R., Tomoyasu, S., and Koeffler, H. P. (2001). Novel vitamin D(3) analog, 21-(3-methyl-3-hydroxy-butyl)-19-nor D(3), that modulates cell growth, differentiation, apoptosis, cell cycle, and induction of PTEN in leukemic cells. *Blood* **97,** 2427–2433.

Holtzer, H., Biehl, J., and Holtzer, S. (1985). Induction-dependent and lineage-dependent models for cell diversification are mutually exclusive. *Prog. Clin. Biol. Res.* **175,** 3–11.

Hong, W. K., Endicott, J., Itri, L. M., Doos, W., Batsakis, J. G., Bell, R., Fofonoff, S., Byers, R., Atkinson, E. N., Vaughan, C., Toth, B. B., Kramer, A., *et al.* (1986). 13-cis-Retinoic acid in the treatment of oral leukoplakia. *N. Engl. J. Med.* **315,** 1501–1505.

Hsu, S. L., Hsu, J. W., Liu, M. C., Chen, L. Y., and Chang, C. D. (2000). Retinoic acid-mediated G1 arrest is associated with induction of p27(Kip1) and inhibition of cyclin-dependent kinase 3 in human lung squamous carcinoma CH27 cells. *Exp. Cell Res.* **258,** 322–331.

Huang, M. E., Ye, Y. C., Chen, S. R., Chai, J. R., Lu, J. X., Zhoa, L., Gu, L. J., and Wang, Z. Y. (1988). Use of all-trans retinoic acid in the treatment of acute promyelocytic leukemia. *Blood* **72,** 567–572.

Huang, Y., Boskovic, G., and Niles, R. M. (2003). Retinoic acid-induced AP-1 transcriptional activity regulates B16 mouse melanoma growth inhibition and differentiation. *J. Cell Physiol.* **194,** 162–170.

Idres, N., Benoit, G., Flexor, M. A., Lanotte, M., and Chabot, G. G. (2001). Granulocytic differentiation of human NB4 promyelocytic leukemia cells induced by all-trans retinoic acid metabolites. *Cancer Res.* **61,** 700–705.

Imaizumi, M., and Breitman, T. R. (1987). Retinoic acid-induced differentiation of the human promyelocytic leukemia cell line, HL-60, and fresh human leukemia cells in primary culture: A model for differentiation inducing therapy of leukemia. *Eur. J. Haematol.* **38,** 289–302.

Imaizumi, M., Sato, A., Koizumi, Y., Inoue, S., Suzuki, H., Suwabe, N., Yoshinari, M., Ichinohasama, R., Endo, K., Sawai, T., and Tada, K. (1994). Potentiated maturation with a high proliferating activity of acute promyelocytic leukemia induced *in vitro* by granulocyte or granulocyte/macrophage colony-stimulating factors in combination with all-trans retinoic acid. *Leukemia* **8**, 1301–1308.

Ingraham, C. A., Cox, M. E., Ward, D. C., Fults, D. W., and Maness, P. F. (1989). c-src and other proto-oncogenes implicated in neuronal differentiation. *Mol. Chem. Neuropathol.* **10**, 1–14.

Ishaq, M., Fan, M., and Natarajan, V. (2000). Accumulation of RXR alpha during activation of cycling human T lymphocytes: Modulation of RXRE transactivation function by mitogen-activated protein kinase pathways. *J. Immunol.* **165**, 4217–4225.

Ishaq, M., Fan, M., Wigmore, K., Gaddam, A., and Natarajan, V. (2002). Regulation of retinoid X receptor responsive element-dependent transcription in T lymphocytes by Ser/Thr phosphatases: Functional divergence of protein kinase C (PKC)theta; and PKC alpha in mediating calcineurin-induced transactivation. *J. Immunol.* **169**, 732–738.

Ishida, S., Shudo, K., Takada, S., and Koike, K. (1994). Transcription from the P2 promoter of human protooncogene myc is suppressed by retinoic acid through an interaction between the E2F element and its binding proteins. *Cell Growth Differ.* **5**, 287–294.

Ivanova, T., Petrenko, A., Gritsko, T., Vinokourova, S., Eshilev, E., Kobzeva, V., Kisseljov, F., and Kisseljova, N. (2002). Methylation and silencing of the retinoic acid receptor-beta 2 gene in cervical cancer. *BMC Cancer* **2**, 4.

Jing, Y., Zhang, J., Waxman, S., and Mira-y-Lopez, R. (1996). Upregulation of cytokeratins 8 and 18 in human breast cancer T47D cells is retinoid-specific and retinoic acid receptor-dependent. *Differentiation* **60**, 109–117.

Jones, L. C., Tefferi, A., Idos, G. E., Kumagai, T., Hofmann, W. K., and Koeffler, H. P. (2004). RARbeta2 is a candidate tumor suppressor gene in myelofibrosis with myeloid metaplasia. *Oncogene* **23**, 7846–7853.

Joshi, S., Guleria, R., Pan, J., DiPette, D., and Singh, U. S. (2006). Retinoic acid receptors and tissue-transglutaminase mediate short-term effect of retinoic acid on migration and invasion of neuroblastoma SH-SY5Y cells. *Oncogene* **25**, 240–247.

Kamashev, D., Vitoux, D., and De The, H. (2004). PML-RARA-RXR oligomers mediate retinoid and rexinoid/cAMP cross-talk in acute promyelocytic leukemia cell differentiation. *J. Exp. Med.* **199**, 1163–1174.

Kambhampati, S., Li, Y., Verma, A., Sassano, A., Majchrzak, B., Deb, D. K., Parmar, S., Giafis, N., Kalvakolanu, D. V., Rahman, A., Uddin, S., Minucci, S., *et al.* (2003). Activation of protein kinase C delta by all-trans-retinoic acid. *J. Biol. Chem.* **278**, 32544–32551.

Kambhampati, S., Verma, A., Li, Y., Parmar, S., Sassano, A., and Platanias, L. C. (2004). Signalling pathways activated by all-trans-retinoic acid in acute promyelocytic leukemia cells. *Leuk. Lymphoma* **45**, 2175–2185.

Kang, S., Li, X. Y., Duell, E. A., and Voorhees, J. J. (1997). The retinoid X receptor agonist 9-cis-retinoic acid and the 24-hydroxylase inhibitor ketoconazole increase activity of 1,25-dihydroxyvitamin D3 in human skin *in vivo*. *J. Invest. Dermatol.* **108**, 513–518.

Karin, M., and Shaulian, E. (2001). AP-1: Linking hydrogen peroxide and oxidative stress to the control of cell proliferation and death. *IUBMB Life* **52**, 17–24.

Kastner, P., Mark, M., and Chambon, P. (1995). Nonsteroid nuclear receptors: What are genetic studies telling us about their role in real life? *Cell* **83**, 859–869.

Khuri, F. R., Cho, Y., and Talmage, D. A. (1996). Retinoic acid-induced transition from protein kinase C beta to protein kinase C alpha in differentiated F9 cells: Correlation with altered regulation of proto-oncogene expression by phorbol esters. *Cell Growth Differ.* **7**, 595–602.

Klaassen, I., Brakenhoff, R. H., Smeets, S. J., Snow, G. B., and Braakhuis, B. J. (2001). Expression of retinoic acid receptor gamma correlates with retinoic acid sensitivity and metabolism in head and neck squamous cell carcinoma cell lines. *Int. J. Cancer* **92**, 661–665.

Kolla, V., Lindner, D. J., Xiao, W., Borden, E. C., and Kalvakolanu, D. V. (1996). Modulation of interferon (IFN)-inducible gene expression by retinoic acid. Up-regulation of STAT1 protein in IFN-unresponsive cells. *J. Biol. Chem.* **271,** 10508–10514.

Kolla, V., Weihua, X., and Kalvakolanu, D. V. (1997). Modulation of interferon action by retinoids. Induction of murine STAT1 gene expression by retinoic acid. *J. Biol. Chem.* **272,** 9742–9748.

Koskela, K., Pelliniemi, T. T., Pulkki, K., and Remes, K. (2004). Treatment of multiple myeloma with all-trans retinoic acid alone and in combination with chemotherapy: A phase I/II trial. *Leuk. Lymphoma* **45,** 749–754.

Kovarik, P., Stoiber, D., Eyers, P. A., Menghini, R., Neininger, A., Gaestel, M., Cohen, P., and Decker, T. (1999). Stress-induced phosphorylation of STAT1 at Ser727 requires p38 mitogen-activated protein kinase whereas IFN-gamma uses a different signaling pathway. *Proc. Natl. Acad. Sci. USA* **96,** 13956–13961.

Kraemer, K. H., DiGiovanna, J. J., Moshell, A. N., Tarone, R. E., and Peck, G. L. (1988). Prevention of skin cancer in xeroderma pigmentosum with the use of oral isotretinoin. *N. Engl. J. Med.* **318,** 1633–1637.

Kuendgen, A., Strupp, C., Aivado, M., Bernhardt, A., Hildebrandt, B., Haas, R., Germing, U., and Gattermann, N. (2004). Treatment of myelodysplastic syndromes with valproic acid alone or in combination with all-trans retinoic acid. *Blood* **104,** 1266–1269.

Kunisada, M., Budiyanto, A., Bito, T., Nishigori, C., and Ueda, M. (2005). Retinoic acid suppresses telomerase activity in HSC-1 human cutaneous squamous cell carcinoma. *Br. J. Dermatol.* **152,** 435–443.

Lanotte, M., Martin-Thouvenin, V., Najman, S., Balerini, P., Valensi, F., and Berger, R. (1991). NB4, a maturation inducible cell line with t(15;17) marker isolated from a human acute promyelocytic leukemia (M3). *Blood* **77,** 1080–1086.

Lauritsen, K. J., List, H. J., Reiter, R., Wellstein, A., and Riegel, A. T. (2002). A role for TGF-beta in estrogen and retinoid mediated regulation of the nuclear receptor coactivator AIB1 in MCF-7 breast cancer cells. *Oncogene* **21,** 7147–7155.

Le Good, J. A., Ziegler, W. H., Parekh, D. B., Alessi, D. R., Cohen, P., and Parker, P. J. (1998). Protein kinase C isotypes controlled by phosphoinositide 3-kinase through the protein kinase PDK1. *Science* **281,** 2042–2045.

Lee, H. Y., Sueoka, N., Hong, W. K., Mangelsdorf, D. J., Claret, F. X., and Kurie, J. M. (1999). All-trans-retinoic acid inhibits Jun N-terminal kinase by increasing dual-specificity phosphatase activity. *Mol. Cell. Biol.* **19,** 1973–1980.

Lee, J. J., Hong, W. K., Hittelman, W. N., Mao, L., Lotan, R., Shin, D. M., Benner, S. E., Xu, X. C., Lee, J. S., Papadimitrakopoulou, V. M., Geyer, C., Perez, C., *et al.* (2000). Predicting cancer development in oral leukoplakia: Ten years of translational research. *Clin. Cancer Res.* **6,** 1702–1710.

Lehtonen, E., Laasonen, A., and Tienari, J. (1989). Teratocarcinoma stem cells as a model for differentiation in the mouse embryo. *Int. J. Dev. Biol.* **33,** 105–115.

Lembo, D., Gaboli, M., Caliendo, A., Falciani, F., Garattini, E., and Landolfo, S. (1992). Regulation of the 202 gene expression by interferons in L929 cells. *Biochem. Biophys. Res. Commun.* **187,** 628–634.

Leone, G., Sica, S., Ortu La Barbera, E., Testa, U., Riccioni, R., Labbaye, C., Peschle, C., and Zollino, M. (1998). Secondary leukemia responsive to retinoic acid with abnormal localization of RARalpha protein: A report of two cases. *Blood* **91,** 4811–4812.

Li, X. Y., Xiao, J. H., Feng, X., Qin, L., and Voorhees, J. J. (1997). Retinoid X receptor-specific ligands synergistically upregulate 1, 25-dihydroxyvitamin D3-dependent transcription in epidermal keratinocytes *in vitro* and *in vivo*. *J. Invest. Dermatol.* **108,** 506–512.

Li, Y., Hashimoto, Y., Agadir, A., Kagechika, H., and Zhang, X. (1999). Identification of a novel class of retinoic acid receptor beta-selective retinoid antagonists and their inhibitory effects on AP-1 activity and retinoic acid-induced apoptosis in human breast cancer cells. *J. Biol. Chem.* **274,** 15360–15366.

Lindner, D. J., Borden, E. C., and Kalvakolanu, D. V. (1997). Synergistic antitumor effects of a combination of interferons and retinoic acid on human tumor cells *in vitro* and *in vivo*. *Clin. Cancer Res.* **3,** 931–937.

Lippman, S. M., Kavanagh, J. J., Paredes-Espinoza, M., Delgadillo-Madrueno, F., Paredes-Casillas, P., Hong, W. K., Massimini, G., Holdener, E. E., and Krakoff, I. H. (1993). 13-cis-retinoic acid plus interferon-alpha 2a in locally advanced squamous cell carcinoma of the cervix. *J. Natl. Cancer Inst.* **85,** 499–500.

Lohnes, D., Dierich, A., Ghyselinck, N., Kastner, P., Lampron, C., LeMeur, M., Lufkin, T., Mendelsohn, C., Nakshatri, H., and Chambon, P. (1992). Retinoid receptors and binding proteins. *J. Cell Sci. Suppl.* **16,** 69–76.

Lomo, J., Smeland, E. B., Ulven, S., Natarajan, V., Blomhoff, R., Gandhi, U., Dawson, M. I., and Blomhoff, H. K. (1998). RAR-, not RXR, ligands inhibit cell activation and prevent apoptosis in B-lymphocytes. *J. Cell Physiol.* **175,** 68–77.

Lopez-Carballo, G., Moreno, L., Masia, S., Perez, P., and Barettino, D. (2002). Activation of the phosphatidylinositol 3-kinase/Akt signaling pathway by retinoic acid is required for neural differentiation of SH-SY5Y human neuroblastoma cells. *J. Biol. Chem.* **277,** 25297–25304.

Lotan, R. (1996). Retinoids and their receptors in modulation of differentiation, development, and prevention of head and neck cancers. *Anticancer Res.* **16,** 2415–2419.

Lotan, R. (1997). Roles of retinoids and their nuclear receptors in the development and prevention of upper aerodigestive tract cancers. *Environ. Health Perspect.* **105**(Suppl. 4), 985–988.

Lovat, P. E., Irving, H., Annicchiarico-Petruzzelli, M., Bernassola, F., Malcolm, A. J., Pearson, A. D., Melino, G., and Redfern, C. P. (1997). Retinoids in neuroblastoma therapy: Distinct biological properties of 9-cis- and all-trans-retinoic acid. *Eur. J. Cancer* **33,** 2075–2080.

Manfredini, R., Trevisan, F., Grande, A., Tagliafico, E., Montanari, M., Lemoli, R., Visani, G., Tura, S., Ferrari, S., and Ferrari, S. (1999). Induction of a functional vitamin D receptor in all-trans-retinoic acid-induced monocytic differentiation of M2-type leukemic blast cells. *Cancer Res.* **59,** 3803–3811.

Mangelsdorf, D. J., and Evans, R. M. (1995). The RXR heterodimers and orphan receptors. *Cell* **83,** 841–850.

Mangelsdorf, D. J., Kliewer, S. A., Kakizuka, A., Umesono, K., and Evans, R. M. (1993). Retinoid receptors. *Recent Prog. Horm. Res.* **48,** 99–121.

Mann, K. K., Padovani, A. M., Guo, Q., Colosimo, A. L., Lee, H. Y., Kurie, J. M., and Miller, W. H., Jr. (2005). Arsenic trioxide inhibits nuclear receptor function via SEK1/JNK-mediated RXRalpha phosphorylation. *J. Clin. Invest.* **115,** 2924–2933.

Martelli, A. M., Tazzari, P. L., Tabellini, G., Bortul, R., Billi, A. M., Manzoli, L., Ruggeri, A., Conte, R., and Cocco, L. (2003). A new selective AKT pharmacological inhibitor reduces resistance to chemotherapeutic drugs, TRAIL, all-trans-retinoic acid, and ionizing radiation of human leukemia cells. *Leukemia* **17,** 1794–1805.

Masuda, M., Toh, S., Koike, K., Kuratomi, Y., Suzui, M., Deguchi, A., Komiyama, S., and Weinstein, I. B. (2002). The roles of JNK1 and Stat3 in the response of head and neck cancer cell lines to combined treatment with all-trans-retinoic acid and 5-fluorouracil. *Jpn. J. Cancer Res.* **93,** 329–339.

Matikainen, S., Ronni, T., Hurme, M., Pine, R., and Julkunen, I. (1996). Retinoic acid activates interferon regulatory factor-1 gene expression in myeloid cells. *Blood* **88,** 114–123.

Matikainen, S., Ronni, T., Lehtonen, A., Sareneva, T., Melen, K., Nordling, S., Levy, D. E., and Julkunen, I. (1997). Retinoic acid induces signal transducer and activator of transcription (STAT) 1, STAT2, and p48 expression in myeloid leukemia cells and enhances their responsiveness to interferons. *Cell Growth Differ.* **8,** 687–698.

Matikainen, S., Lehtonen, A., Sareneva, T., and Julkunen, I. (1998). Regulation of IRF and STAT gene expression by retinoic acid. *Leuk. Lymphoma* **30,** 63–71.
McBurney, M. W., Reuhl, K. R., Ally, A. I., Nasipuri, S., Bell, J. C., and Craig, J. (1988). Differentiation and maturation of embryonal carcinoma-derived neurons in cell culture. *J. Neurosci.* **8,** 1063–1073.
Mehta, K. D., and Miller, L. (1999). Inhibition of stress-activated p38 mitogen-activated protein kinase induces low-density lipoprotein receptor expression. *Trends Cardiovasc. Med.* **9,** 201–205.
Melino, G., Draoui, M., Bellincampi, L., Bernassola, F., Bernardini, S., Piacentini, M., Reichert, U., and Cohen, P. (1997). Retinoic acid receptors alpha and gamma mediate the induction of "tissue" transglutaminase activity and apoptosis in human neuroblastoma cells. *Exp. Cell Res.* **235,** 55–61.
Melnick, A., and Licht, J. D. (1999). Deconstructing a disease: RARalpha, its fusion partners, and their roles in the pathogenesis of acute promyelocytic leukemia. *Blood* **93,** 3167–3215.
Miloso, M., Villa, D., Crimi, M., Galbiati, S., Donzelli, E., Nicolini, G., and Tredici, G. (2004). Retinoic acid-induced neuritogenesis of human neuroblastoma SH-SY5Y cells is ERK independent and PKC dependent. *J. Neurosci. Res.* **75,** 241–252.
Miranda, M. B., McGuire, T. F., and Johnson, D. E. (2002). Importance of MEK-1/-2 signaling in monocytic and granulocytic differentiation of myeloid cell lines. *Leukemia* **16,** 683–692.
Mologni, L., Ponzanelli, I., Bresciani, F., Sardiello, G., Bergamaschi, D., Gianni, M., Reichert, U., Rambaldi, A., Terao, M., and Garattini, E. (1999). The novel synthetic retinoid 6-[3-adamantyl-4-hydroxyphenyl]-2-naphthalene carboxylic acid (CD437) causes apoptosis in acute promyelocytic leukemia cells through rapid activation of caspases. *Blood* **93,** 1045–1061.
Moore, D. M., Kalvakolanu, D. V., Lippman, S. M., Kavanagh, J. J., Hong, W. K., Borden, E. C., Paredes-Espinoza, M., and Krakoff, I. H. (1994). Retinoic acid and interferon in human cancer: Mechanistic and clinical studies. *Semin. Hematol.* **31,** 31–37.
Morita, K., Nishikawa, M., Kobayashi, K., Deguchi, K., Ito, M., Nakano, T., Shima, H., Nagao, M., Kuno, T., and Tanaka, C. (1992). Augmentation of retinoic acid-induced granulocytic differentiation in HL-60 leukemia cells by serine/threonine protein phosphatase inhibitors. *FEBS Lett.* **314,** 340–344.
Muindi, J., Frankel, S. R., Miller, W. H., Jr., Jakubowski, A., Scheinberg, D. A., Young, C. W., Dmitrovsky, E., and Warrell, R. P., Jr. (1992). Continuous treatment with all-trans retinoic acid causes a progressive reduction in plasma drug concentrations: Implications for relapse and retinoid "resistance" in patients with acute promyelocytic leukemia. *Blood* **79,** 299–303.
Musto, P., Falcone, A., Sajeva, M. R., D'Arena, G., Bonini, A., and Carotenuto, M. (1995). All-trans retinoic acid for advanced multiple myeloma. *Blood* **85,** 3769–3770.
Nagy, L., Thomazy, V. A., Shipley, G. L., Fesus, L., Lamph, W., Heyman, R. A., Chandraratna, R. A., and Davies, P. J. (1995). Activation of retinoid X receptors induces apoptosis in HL-60 cell lines. *Mol. Cell. Biol.* **15,** 3540–3551.
Nakagawa, S., Fujii, T., Yokoyama, G., Kazanietz, M. G., Yamana, H., and Shirouzu, K. (2003). Cell growth inhibition by all-trans retinoic acid in SKBR-3 breast cancer cells: Involvement of protein kinase Calpha and extracellular signal-regulated kinase mitogen-activated protein kinase. *Mol. Carcinog.* **38,** 106–116.
Neri, L. M., Borgatti, P., Tazzari, P. L., Bortul, R., Cappellini, A., Tabellini, G., Bellacosa, A., Capitani, S., and Martelli, A. M. (2003). The phosphoinositide 3-kinase/AKT1 pathway involvement in drug and all-trans-retinoic acid resistance of leukemia cells. *Mol. Cancer Res.* **1,** 234–246.
Newton, A. C. (1997). Regulation of protein kinase C. *Curr. Opin. Cell Biol.* **9,** 161–167.
Nguyen, T., Hocker, J. E., Thomas, W., Smith, S. A., Norris, M. D., Haber, M., Cheung, B., and Marshall, G. M. (2003). Combined RAR alpha- and RXR-specific ligands overcome N-myc-associated retinoid resistance in neuroblastoma cells. *Biochem. Biophys. Res. Commun.* **302,** 462–468.

Niles, R. M. (2000). Recent advances in the use of vitamin A (retinoids) in the prevention and treatment of cancer. *Nutrition* **16,** 1084–1089.

Niles, R. M. (2003). Vitamin A (retinoids) regulation of mouse melanoma growth and differentiation. *J. Nutr.* **133,** 282S–286S.

O'Connor, T. J., and Fujita, D. J. (1995). Differentiation of B16 murine melanoma cells is associated with an increased level of c-SRC. *Melanoma Res.* **5,** 5–13.

Okuse, K., Mizuno, N., Matsuoka, I., and Kurihara, K. (1993). Induction of cholinergic and adrenergic differentiation in N-18 cells by differentiation agents and DNA demethylating agents. *Brain Res.* **626,** 225–233.

Ouyang, D. Y., Wang, Y. Y., and Zheng, Y. T. (2005). Activation of c-Jun N-terminal kinases by ribotoxic stresses. *Cell. Mol. Immunol.* **2,** 419–425.

Paik, J., Blaner, W. S., Sommer, K. M., Moe, R., and Swisshlem, K. (2003). Retinoids, retinoic acid receptors, and breast cancer. *Cancer Invest.* **21,** 304–312.

Palm-Leis, A., Singh, U. S., Herbelin, B. S., Olsovsky, G. D., Baker, K. M., and Pan, J. (2004). Mitogen-activated protein kinases and mitogen-activated protein kinase phosphatases mediate the inhibitory effects of all-trans retinoic acid on the hypertrophic growth of cardiomyocytes. *J. Biol. Chem.* **279,** 54905–54917.

Pandolfi, P. P. (2001a). Histone deacetylases and transcriptional therapy with their inhibitors. *Cancer Chemother. Pharmacol.* **48**(Suppl. 1), S17–S19.

Pandolfi, P. P. (2001b). *In vivo* analysis of the molecular genetics of acute promyelocytic leukemia. *Oncogene* **20,** 5726–5735.

Parnas, D., and Linial, M. (1997). Acceleration of neuronal maturation of P19 cells by increasing culture density. *Brain Res. Dev. Brain Res.* **101,** 115–124.

Parrella, E., Gianni, M., Cecconi, V., Nigro, E., Barzago, M. M., Rambaldi, A., Rochette-Egly, C., Terao, M., and Garattini, E. (2004). Phosphodiesterase IV inhibition by piclamilast potentiates the cytodifferentiating action of retinoids in myeloid leukemia cells. Cross-talk between the cAMP and the retinoic acid signaling pathways. *J. Biol. Chem.* **279,** 42026–42040.

Pelicano, L., Li, F., Schindler, C., and Chelbi-Alix, M. K. (1997). Retinoic acid enhances the expression of interferon-induced proteins: Evidence for multiple mechanisms of action. *Oncogene* **15,** 2349–2359.

Pellegrini, R., Mariotti, A., Tagliabue, E., Bressan, R., Bunone, G., Coradini, D., Della Valle, G., Formelli, F., Cleris, L., and Radice, P. (1995). Modulation of markers associated with tumor aggressiveness in human breast cancer cell lines by N-(4-hydroxyphenyl) retinamide. *Cell Growth Differ.* **6,** 863–869.

Percario, Z. A., Giandomenico, V., Fiorucci, G., Chiantore, M. V., Vannucchi, S., Hiscott, J., Affabris, E., and Romeo, G. (1999). Retinoic acid is able to induce interferon regulatory factor 1 in squamous carcinoma cells via a STAT-1 independent signalling pathway. *Cell Growth Differ.* **10,** 263–270.

Petersen, O. W., Ronnov-Jessen, L., Weaver, V. M., and Bissell, M. J. (1998). Differentiation and cancer in the mammary gland: Shedding light on an old dichotomy. *Adv. Cancer Res.* **75,** 135–161.

Pettersson, F., Couture, M. C., Hanna, N., and Miller, W. H. (2004). Enhanced retinoid-induced apoptosis of MDA-MB-231 breast cancer cells by PKC inhibitors involves activation of ERK. *Oncogene* **23,** 7053–7066.

Petti, M. C., Fazi, F., Gentile, M., Diverio, D., De Fabritiis, P., De Propris, P., Fiorini, M. S., Spiriti, M. A., Padula, F., Pelicci, P. G., Nervi, C., and Coco, F. L. (2002). Complete remission through blast cell differentiation in PLZF/RARalpha-positive acute promyelocytic leukemia: *In vitro* and *in vivo* studies. *Blood* **100,** 1065–1067.

Pili, R., Kruszewski, M. P., Hager, B. W., Lantz, J., and Carducci, M. A. (2001). Combination of phenylbutyrate and 13-cis retinoic acid inhibits prostate tumor growth and angiogenesis. *Cancer Res.* **61,** 1477–1485.

Pisano, C., Kollar, P., Gianni, M., Kalac, Y., Giordano, V., Ferrara, F. F., Tancredi, R., Devoto, A., Rinaldi, A., Rambaldi, A., Penco, S., Marzi, M., *et al.* (2002). Bis-indols: A novel class of molecules enhancing the cytodifferentiating properties of retinoids in myeloid leukemia cells. *Blood* **100,** 3719–3730.

Ponzanelli, I., Gianni, M., Giavazzi, R., Garofalo, A., Nicoletti, I., Reichert, U., Erba, E., Rambaldi, A., Terao, M., and Garattini, E. (2000). Isolation and characterization of an acute promyelocytic leukemia cell line selectively resistant to the novel antileukemic and apoptogenic retinoid 6-[3-adamantyl-4-hydroxyphenyl]-2-naphthalene carboxylic acid. *Blood* **95,** 2672–2682.

Pyle, S. J., Roberts, K. G., and Reuhl, K. R. (2001). Delayed expression of the NFH subunit in differentiating P19 cells. *Brain Res. Dev. Brain Res.* **132,** 103–106.

Radominska-Pandya, A., Chen, G., Czernik, P. J., Little, J. M., Samokyszyn, V. M., Carter, C. A., and Nowak, G. (2000). Direct interaction of all-trans-retinoic acid with protein kinase C (PKC). Implications for PKC signaling and cancer therapy. *J. Biol. Chem.* **275,** 22324–22330.

Ramana, C. V., Chatterjee-Kishore, M., Nguyen, H., and Stark, G. R. (2000). Complex roles of Stat1 in regulating gene expression. *Oncogene* **19,** 2619–2627.

Ramana, C. V., Gil, M. P., Schreiber, R. D., and Stark, G. R. (2002). Stat1-dependent and -independent pathways in IFN-gamma-dependent signaling. *Trends Immunol.* **23,** 96–101.

Ramirez, C. J., Haberbusch, J. M., Soprano, D. R., and Soprano, K. J. (2005). Retinoic acid induced repression of AP-1 activity is mediated by protein phosphatase 2A in ovarian carcinoma cells. *J. Cell Biochem.* **96,** 170–182.

Renoncourt, Y., Carroll, P., Filippi, P., Arce, V., and Alonso, S. (1998). Neurons derived *in vitro* from ES cells express homeoproteins characteristic of motoneurons and interneurons. *Mech. Dev.* **79,** 185–197.

Reynolds, C. P., Matthay, K. K., Villablanca, J. G., and Maurer, B. J. (2003). Retinoid therapy of high-risk neuroblastoma. *Cancer Lett.* **197,** 185–192.

Roberts, A. B., and Sporn, M. B. (1992). Mechanistic interrelationships between two superfamilies: The steroid/retinoid receptors and transforming growth factor-beta. *Cancer Surv.* **14,** 205–220.

Rochette-Egly, C. (2003). Nuclear receptors: Integration of multiple signalling pathways through phosphorylation. *Cell. Signal.* **15,** 355–366.

Rochette-Egly, C., and Chambon, P. (2001). F9 embryocarcinoma cells: A cell autonomous model to study the functional selectivity of RARs and RXRs in retinoid signaling. *Histol. Histopathol.* **16,** 909–922.

Rochette-Egly, C., Oulad-Abdelghani, M., Staub, A., Pfister, V., Scheuer, I., Chambon, P., and Gaub, M. P. (1995). Phosphorylation of the retinoic acid receptor-alpha by protein kinase A. *Mol. Endocrinol.* **9,** 860–871.

Rousseau, C., Nichol, J. N., Pettersson, F., Couture, M. C., and Miller, W. H., Jr. (2004). ERbeta sensitizes breast cancer cells to retinoic acid: Evidence of transcriptional crosstalk. *Mol. Cancer Res.* **2,** 523–531.

Roux, P. P., and Blenis, J. (2004). ERK and p38 MAPK-activated protein kinases: A family of protein kinases with diverse biological functions. *Microbiol. Mol. Biol. Rev.* **68,** 320–344.

Salomoni, P., and Pandolfi, P. P. (2000). Transcriptional regulation of cellular transformation. *Nat. Med.* **6,** 742–744.

Sankowski, A., Janik, P., Jeziorska, M., Swietochowska, B., Ciesla, W., Malek, A., and Przybyszewska, M. (1987). The results of topical application of 13-cis-retinoic acid on basal cell carcinoma. A correlation of the clinical effect with histopathological examination and serum retinol level. *Neoplasma* **34,** 485–489.

Sanz, M. A., Martin, G., Gonzalez, M., Leon, A., Rayon, C., Rivas, C., Colomer, D., Amutio, E., Capote, F. J., Milone, G. A., de la Serna, J., Roman, J., *et al.* (2004). Risk-adapted treatment of acute promyelocytic leukemia with all-trans-retinoic acid and anthracycline monochemotherapy: A multicenter study by the PETHEMA group. *Blood* **103,** 1237–1243.

Satake, K., Takagi, E., Ishii, A., Kato, Y., Imagawa, Y., Kimura, Y., and Tsukuda, M. (2003). Anti-tumor effect of vitamin A and D on head and neck squamous cell carcinoma. *Auris Nasus Larynx* **30,** 403–412.

Schneider, S. M., Offterdinger, M., Huber, H., and Grunt, T. W. (2000). Activation of retinoic acid receptor alpha is sufficient for full induction of retinoid responses in SK-BR-3 and T47D human breast cancer cells. *Cancer Res.* **60,** 5479–5487.

Seewaldt, V. L., Kim, J. H., Parker, M. B., Dietze, E. C., Srinivasan, K. V., and Caldwell, L. E. (1999). Dysregulated expression of cyclin D1 in normal human mammary epithelial cells inhibits all-trans-retinoic acid-mediated G0/G1-phase arrest and differentiation *in vitro*. *Exp. Cell Res.* **249,** 70–85.

Segura-Pacheco, B., Trejo-Becerril, C., Perez-Cardenas, E., Taja-Chayeb, L., Mariscal, I., Chavez, A., Acuna, C., Salazar, A. M., Lizano, M., and Duenas-Gonzalez, A. (2003). Reactivation of tumor suppressor genes by the cardiovascular drugs hydralazine and procainamide and their potential use in cancer therapy. *Clin. Cancer Res.* **9,** 1596–1603.

Shaulian, E., and Karin, M. (2002). AP-1 as a regulator of cell life and death. *Nat. Cell Biol.* **4,** E131–E136.

Shen, H. M., and Liu, Z. G. (2006). JNK signaling pathway is a key modulator in cell death mediated by reactive oxygen and nitrogen species. *Free Radic. Biol. Med.* **40,** 928–939.

Shen, Y., Mani, S., and Meiri, K. F. (2004). Failure to express GAP-43 leads to disruption of a multipotent precursor and inhibits astrocyte differentiation. *Mol. Cell. Neurosci.* **26,** 390–405.

Shibata, H., Spencer, T. E., Onate, S. A., Jenster, G., Tsai, S. Y., Tsai, M. J., and O'Malley, B. W. (1997). Role of co-activators and co-repressors in the mechanism of steroid/thyroid receptor action. *Recent Prog. Horm. Res.* **52,** 141–164; discussion 164–165.

Shimada, K., Nakamura, M., Ishida, E., and Konishi, N. (2006). Molecular roles of MAP kinases and FADD phosphorylation in prostate cancer. *Histol. Histopathol.* **21,** 415–422.

Shimodaira, S., Kitano, K., Nishizawa, Y., Ichikawa, N., Ishida, F., Kamino, I., Matsui, H., and Kiyosawa, K. (1999). Acute myelogenous leukemia with a t(2;17;4)(p13;q21;p16) aberration: Effective treatment with all-trans retinoic acid and granulocyte colony-stimulating factor. *Intern. Med.* **38,** 150–154.

Shin, D. M., Xu, X. C., Lippman, S. M., Lee, J. J., Lee, J. S., Batsakis, J. G., Ro, J. Y., Martin, J. W., Hittelman, W. N., Lotan, R., and Hong, W. K. (1997). Accumulation of p53 protein and retinoic acid receptor beta in retinoid chemoprevention. *Clin. Cancer Res.* **3,** 875–880.

Shin, D. M., Khuri, F. R., Murphy, B., Garden, A. S., Clayman, G., Francisco, M., Liu, D., Glisson, B. S., Ginsberg, L., Papadimitrakopoulou, V., Myers, J., Morrison, W., *et al.* (2001). Combined interferon-alfa, 13-cis-retinoic acid, and alpha-tocopherol in locally advanced head and neck squamous cell carcinoma: Novel bioadjuvant phase II trial. *J. Clin. Oncol.* **19,** 3010–3017.

Sidell, N., Taga, T., Hirano, T., Kishimoto, T., and Saxon, A. (1991). Retinoic acid-induced growth inhibition of a human myeloma cell line via down-regulation of IL-6 receptors. *J. Immunol.* **146,** 3809–3814.

Sirulnik, A., Melnick, A., Zelent, A., and Licht, J. D. (2003). Molecular pathogenesis of acute promyelocytic leukaemia and APL variants. *Best Pract. Res. Clin. Haematol.* **16,** 387–408.

Siwak, D. R., Mendoza-Gamboa, E., and Tari, A. M. (2003). HER2/neu uses Akt to suppress retinoic acid response element binding activity in MDA-MB-453 breast cancer cells. *Int. J. Oncol.* **23,** 1739–1745.

Smith, M. A., Parkinson, D. R., Cheson, B. D., and Friedman, M. A. (1992). Retinoids in cancer therapy. *J. Clin. Oncol.* **10,** 839–864.

Sparatore, B., Patrone, M., Passalacqua, M., Pedrazzi, M., Pontremoli, S., and Melloni, E. (2000). Human neuroblastoma cell differentiation requires protein kinase C-theta. *Biochem. Biophys. Res. Commun.* **279,** 589–594.

Sporn, M. B., and Roberts, A. B. (1985). Suppression of carcinogenesis by retinoids: Interactions with peptide growth factors and their receptors as a key mechanism. *Princess Takamatsu Symp.* **16,** 149–158.

Sporn, M. B., Roberts, A. B., Wakefield, L. M., Glick, A. B., and Danielpour, D. (1989). Transforming growth factor-beta and suppression of carcinogenesis. *Princess Takamatsu Symp.* **20,** 259–266.

Srinivas, H., Juroske, D. M., Kalyankrishna, S., Cody, D. D., Price, R. E., Xu, X. C., Narayanan, R., Weigel, N. L., and Kurie, J. M. (2005). c-Jun N-terminal kinase contributes to aberrant retinoid signaling in lung cancer cells by phosphorylating and inducing proteasomal degradation of retinoic acid receptor alpha. *Mol. Cell. Biol.* **25,** 1054–1069.

Srinivas, H., Xia, D., Moore, N. L., Uray, I. P., Kim, H., Ma, L., Weigel, N. L., Brown, P. H., and Kurie, J. M. (2006). Akt phosphorylates and suppresses the transactivation of retinoic acid receptor alpha. *Biochem. J.* **395,** 653–662.

Stark, G. R., Kerr, I. M., Williams, B. R., Silverman, R. H., and Schreiber, R. D. (1998). How cells respond to interferons. *Annu. Rev. Biochem.* **67,** 227–264.

Studer, U. E., Biedermann, C., Chollet, D., Karrer, P., Kraft, R., Toggenburg, H., and Vonbank, F. (1984). Prevention of recurrent superficial bladder tumors by oral etretinate: Preliminary results of a randomized, double blind multicenter trial in Switzerland. *J. Urol.* **131,** 47–49.

Studer, U. E., Jenzer, S., Biedermann, C., Chollet, D., Kraft, R., von Toggenburg, H., and Vonbank, F. (1995). Adjuvant treatment with a vitamin A analogue (etretinate) after transurethral resection of superficial bladder tumors. Final analysis of a prospective, randomized multicenter trial in Switzerland. *Eur. Urol.* **28,** 284–290.

Sun, S. Y., Kurie, J. M., Yue, P., Dawson, M. I., Shroot, B., Chandraratna, R. A., Hong, W. K., and Lotan, R. (1999). Differential responses of normal, premalignant, and malignant human bronchial epithelial cells to receptor-selective retinoids. *Clin. Cancer Res.* **5,** 431–437.

Sun, S. Y., Yue, P., Mao, L., Dawson, M. I., Shroot, B., Lamph, W. W., Heyman, R. A., Chandraratna, R. A., Shudo, K., Hong, W. K., and Lotan, R. (2000). Identification of receptor-selective retinoids that are potent inhibitors of the growth of human head and neck squamous cell carcinoma cells. *Clin. Cancer Res.* **6,** 1563–1573.

Tahayato, A., Lefebvre, P., Formstecher, P., and Dautrevaux, M. (1993). A protein kinase C-dependent activity modulates retinoic acid-induced transcription. *Mol. Endocrinol.* **7,** 1642–1653.

Taimi, M., Breitman, T. R., and Takahashi, N. (2001). Cyclic AMP-dependent protein kinase isoenzymes in human myeloid leukemia (HL60) and breast tumor (MCF-7) cells. *Arch. Biochem. Biophys.* **392,** 137–144.

Tallman, M. S., Nabhan, C., Feusner, J. H., and Rowe, J. M. (2002). Acute promyelocytic leukemia: Evolving therapeutic strategies. *Blood* **99,** 759–767.

Tamura, K., Kagechika, H., Hashimoto, Y., Shudo, K., Ohsugi, K., and Ide, H. (1990). Synthetic retinoids, retinobenzoic acids, Am80, Am580 and Ch55 regulate morphogenesis in chick limb bud. *Cell Differ. Dev.* **32,** 17–26.

Taneja, R., Rochette-Egly, C., Plassat, J. L., Penna, L., Gaub, M. P., and Chambon, P. (1997). Phosphorylation of activation functions AF-1 and AF-2 of RAR alpha and RAR gamma is indispensable for differentiation of F9 cells upon retinoic acid and cAMP treatment. *EMBO J.* **16,** 6452–6465.

Tari, A. M., Lim, S. J., Hung, M. C., Esteva, F. J., and Lopez-Berestein, G. (2002). Her2/neu induces all-trans retinoic acid (ATRA) resistance in breast cancer cells. *Oncogene* **21,** 5224–5232.

Tawara, I., Nishikawa, M., Morita, K., Kobayashi, K., Toyoda, H., Omay, S. B., Shima, H., Nagao, M., Kuno, T., Tanaka, C., and Shirakawa, S. (1993). Down-regulation by retinoic acid of the catalytic subunit of protein phosphatase type 2A during granulocytic differentiation of HL-60 cells. *FEBS Lett.* **321**, 224–228.

Tenen, D. G. (2001). Abnormalities of the CEBP alpha transcription factor: A major target in acute myeloid leukemia. *Leukemia* **15**, 688–689.

Teruel, T., Hernandez, R., Benito, M., and Lorenzo, M. (2003). Rosiglitazone and retinoic acid induce uncoupling protein-1 (UCP-1) in a p38 mitogen-activated protein kinase-dependent manner in fetal primary brown adipocytes. *J. Biol. Chem.* **278**, 263–269.

Tibbles, L. A., and Woodgett, J. R. (1999). The stress-activated protein kinase pathways. *Cell. Mol. Life Sci.* **55**, 1230–1254.

Tkatch, L. S., Rubin, K. A., Ziegler, S. F., and Tweardy, D. J. (1995). Modulation of human G-CSF receptor mRNA and protein in normal and leukemic myeloid cells by G-CSF and retinoic acid. *J. Leukoc. Biol.* **57**, 964–971.

Turley, J. M., Falk, L. A., Ruscetti, F. W., Kasper, J. J., Francomano, T., Fu, T., Bang, O. S., and Birchenall-Roberts, M. C. (1996). Transforming growth factor beta 1 functions in monocytic differentiation of hematopoietic cells through autocrine and paracrine mechanisms. *Cell Growth Differ.* **7**, 1535–1544.

Valette, A., and Botanch, C. (1990). Transforming growth factor beta (TGF-beta) potentiates the inhibitory effect of retinoic acid on human breast carcinoma (MCF-7) cell proliferation. *Growth Factors* **2**, 283–287.

Veal, G. J., Errington, J., Redfern, C. P., Pearson, A. D., and Boddy, A. V. (2002). Influence of isomerisation on the growth inhibitory effects and cellular activity of 13-cis and all-trans retinoic acid in neuroblastoma cells. *Biochem. Pharmacol.* **63**, 207–215.

Verbeek, W., Gombart, A. F., Chumakov, A. M., Muller, C., Friedman, A. D., and Koeffler, H. P. (1999). C/EBPε directly interacts with the DNA binding domain of c-myb and cooperatively activates transcription of myeloid promoters. *Blood* **93**, 3327–3337.

Verheijen, M. H., Wolthuis, R. M., Bos, J. L., and Defize, L. H. (1999). The Ras/Erk pathway induces primitive endoderm but prevents parietal endoderm differentiation of F9 embryonal carcinoma cells. *J. Biol. Chem.* **274**, 1487–1494.

Veronesi, U., De Palo, G., Marubini, E., Costa, A., Formelli, F., Mariani, L., Decensi, A., Camerini, T., Del Turco, M. R., Di Mauro, M. R., Muraca, M. G., Del Vecchio, M., *et al.* (1999). Randomized trial of fenretinide to prevent second breast malignancy in women with early breast cancer. *J. Natl. Cancer Inst.* **91**, 1847–1856.

Wakefield, L., Kim, S. J., Glick, A., Winokur, T., Colletta, A., and Sporn, M. (1990). Regulation of transforming growth factor-beta subtypes by members of the steroid hormone superfamily. *J. Cell Sci. Suppl.* **13**, 139–148.

Walz, T. M., Malm, C., and Wasteson, A. (1993). Expression of the transforming growth factor alpha protooncogene in differentiating human promyelocytic leukemia (HL-60) cells. *Cancer Res.* **53**, 191–196.

Wang, Q., and Wieder, R. (2004). All-trans retinoic acid potentiates Taxotere-induced cell death mediated by Jun N-terminal kinase in breast cancer cells. *Oncogene* **23**, 426–433.

Wang, X., and Studzinski, G. P. (2001). Activation of extracellular signal-regulated kinases (ERKs) defines the first phase of 1,25-dihydroxyvitamin D3-induced differentiation of HL60 cells. *J. Cell Biochem.* **80**, 471–482.

Wang, X. Z., and Ron, D. (1996). Stress-induced phosphorylation and activation of the transcription factor CHOP (GADD153) by p38 MAP Kinase. *Science* **272**, 1347–1349.

Warrell, R. P., Jr., Frankel, S. R., Miller, W. H., Jr., Scheinberg, D. A., Itri, L. M., Hittelman, W. N., Vyas, R., Andreeff, M., Tafuri, A., Jakubowski, A., Gabrilove, J., Gordon, M. S., *et al.* (1991). Differentiation therapy of acute promyelocytic leukemia with tretinoin (all-trans-retinoic acid). *N. Engl. J. Med.* **324**, 1385–1393.

Warrell, R. P., Jr., He, L. Z., Richon, V., Calleja, E., and Pandolfi, P. P. (1998). Therapeutic targeting of transcription in acute promyelocytic leukemia by use of an inhibitor of histone deacetylase. *J. Natl. Cancer Inst.* **90**, 1621–1625.

Wei, L. N. (2004). Retinoids and receptor interacting protein 140 (RIP140) in gene regulation. *Curr. Med. Chem.* **11**, 1527–1532.

Westin, S., Rosenfeld, M. G., and Glass, C. K. (2000). Nuclear receptor coactivators. *Adv. Pharmacol.* **47**, 89–112.

Weston, A. D., Blumberg, B., and Underhill, T. M. (2003). Active repression by unliganded retinoid receptors in development: Less is sometimes more. *J. Cell Biol.* **161**, 223–228.

Widschwendter, M., Berger, J., Daxenbichler, G., Muller-Holzner, E., Widschwendter, A., Mayr, A., Marth, C., and Zeimet, A. G. (1997). Loss of retinoic acid receptor beta expression in breast cancer and morphologically normal adjacent tissue but not in the normal breast tissue distant from the cancer. *Cancer Res.* **57**, 4158–4161.

Widschwendter, M., Berger, J., Hermann, M., Muller, H. M., Amberger, A., Zeschnigk, M., Widschwendter, A., Abendstein, B., Zeimet, A. G., Daxenbichler, G., and Marth, C. (2000). Methylation and silencing of the retinoic acid receptor-beta2 gene in breast cancer. *J. Natl. Cancer Inst.* **92**, 826–832.

Witcher, M., Shiu, H. Y., Guo, Q., and Miller, W. H., Jr. (2004). Combination of retinoic acid and tumor necrosis factor overcomes the maturation block in a variety of retinoic acid-resistant acute promyelocytic leukemia cells. *Blood* **104**, 3335–3342.

Xu, L., Glass, C. K., and Rosenfeld, M. G. (1999a). Coactivator and corepressor complexes in nuclear receptor function. *Curr. Opin. Genet. Dev.* **9**, 140–147.

Xu, X. C., Liu, X., Tahara, E., Lippman, S. M., and Lotan, R. (1999b). Expression and up-regulation of retinoic acid receptor-beta is associated with retinoid sensitivity and colony formation in esophageal cancer cell lines. *Cancer Res.* **59**, 2477–2483.

Yang, K. D., Mizobuchi, T., Kharbanda, S. M., Datta, R., Huberman, E., Kufe, D. W., and Stone, R. M. (1994). All-trans retinoic acid reverses phorbol ester resistance in a human myeloid leukemia cell line. *Blood* **83**, 490–496.

Yang, K. D., Chao, C. Y., and Shaio, M. F. (1998). Pentoxifylline synergizes with all-trans retinoic acid to induce differentiation of HL-60 myelocytic cells, but suppresses tRA-augmented clonal growth of normal CFU-GM. *Acta Haematol.* **99**, 191–199.

Yang, Q., Sakurai, T., Yoshimura, G., Mori, I., Nakamura, M., Nakamura, Y., Suzuma, T., Tamaki, T., Umemura, T., and Kakudo, K. (2001). Hypermethylation does not account for the frequent loss of the retinoic acid receptor beta2 in breast carcinoma. *Anticancer Res.* **21**, 1829–1833.

Yang, Q., Sakurai, T., and Kakudo, K. (2002). Retinoid, retinoic acid receptor beta and breast cancer. *Breast Cancer Res. Treat.* **76**, 167–173.

Yen, A., Roberson, M. S., Varvayanis, S., and Lee, A. T. (1998). Retinoic acid induced mitogen-activated protein (MAP)/extracellular signal-regulated kinase (ERK) kinase-dependent MAP kinase activation needed to elicit HL-60 cell differentiation and growth arrest. *Cancer Res.* **58**, 3163–3172.

Yen, A., Roberson, M. S., and Varvayanis, S. (1999). Retinoic acid selectively activates the ERK2 but not JNK/SAPK or p38 MAP kinases when inducing myeloid differentiation. *In Vitro Cell. Dev. Biol. Anim.* **35**, 527–532.

Youssef, E. M., Lotan, D., Issa, J. P., Wakasa, K., Fan, Y. H., Mao, L., Hassan, K., Feng, L., Lee, J. J., Lippman, S. M., Hong, W. K., Lotan, R., *et al.* (2004). Hypermethylation of the retinoic acid receptor-beta(2) gene in head and neck carcinogenesis. *Clin. Cancer Res.* **10**, 1733–1742.

Yu, K. H., Weng, L. J., Fu, S., Piantadosi, S., and Gore, S. D. (1999). Augmentation of phenylbutyrate-induced differentiation of myeloid leukemia cells using all-trans retinoic acid. *Leukemia* **13**, 1258–1265.

Yung, B. Y., and Hui, E. K. (1995). Differential cellular distribution of retinoic acid during staurosporine potentiation of retinoic acid-induced granulocytic differentiation in human leukemia HL-60 cells. *J. Biomed. Sci.* **2,** 154–159.

Zhao, H. H., Herrera, R. E., Coronado-Heinsohn, E., Yang, M. C., Ludes-Meyers, J. H., Seybold-Tilson, K. J., Nawaz, Z., Yee, D., Barr, F. G., Diab, S. G., Brown, P. H., Fuqua, S. A., *et al.* (2001). Forkhead homologue in rhabdomyosarcoma functions as a bifunctional nuclear receptor-interacting protein with both coactivator and corepressor functions. *J. Biol. Chem.* **276,** 27907–27912.

13

Effects of Vitamins, Including Vitamin A, on HIV/AIDS Patients

Saurabh Mehta*,† and Wafaie Fawzi*,†

*Department of Epidemiology, Harvard School of Public Health
677 Huntington Avenue, Boston, Massachusetts 02115
†Department of Nutrition, Harvard School of Public Health
677 Huntington Avenue, Boston, Massachusetts 02115

I. Introduction
II. Vitamins and Immune Function
III. Vitamins, HIV Transmission, and Pregnancy Outcomes
   A. Evidence from Observational Studies
   B. Evidence from Trials
IV. Vitamins and HIV Disease Progression in Adults
   A. Evidence from Observational Studies
   B. Evidence from Trials
V. Vitamins, Growth, and Disease Progression in HIV-Infected Children and HIV-Negative Children Born to HIV-Infected Mothers
VI. Comment
VII. Future Research
   References

An estimated 25 million lives have been lost to acquired immune-deficiency syndrome (AIDS) since the immunodeficiency syndrome was first described in 1981. The progress made in the field of treatment in the form of antiretroviral therapy (ART) for HIV disease/AIDS has prolonged as well as improved the quality of life of HIV-infected individuals. However, access to such treatment remains a major concern in most parts of the world, especially in the developing countries. Hence, there is a constant need to find low-cost interventions to complement the role of ART in prevention of HIV infection and slowing clinical disease progression.

Nutritional interventions, particularly vitamin supplementation, have the potential to be a low-cost method for being such an intervention by virtue of their modulation of the immune system. Among all the vitamins, the role of vitamin A has been studied most extensively; most observational studies have found that low vitamin A levels are associated with increased risk of transmission of HIV from mother to child. This finding has not been supported by large randomized trials of vitamin A supplementation; on the contrary, these trials have found that vitamin A supplementation increases the risk of mother-to-child transmission (MTCT). There are a number of potential mechanisms that might explain these contradictory findings. One is the issue of reverse causality in observational studies—for instance, advanced HIV disease may suppress release of vitamin A from the liver. This would lead to low levels of vitamin A in the plasma despite the body having enough vitamin A liver stores. Further, advanced HIV disease is likely to increase the risk of MTCT, and hence it would appear that low serum vitamin A levels are associated with increased MTCT. The HIV genome also has a retinoic acid receptor element—hence, vitamin A may increase HIV replication via interacting with this element, thus increasing risk of MTCT. Finally, vitamin A is known to increase lymphoid cell differentiation, which leads to an increase in CCR5 receptors. These receptors are essential for attachment of HIV to the lymphocytes and therefore, an increase in their number is likely to increase HIV replication.

Vitamin A supplementation in HIV-infected children, on the other hand, has been associated with protective effects against mortality and morbidity, similar to that seen in HIV-negative children. The risk for lower respiratory tract infection and severe watery diarrhea has been shown to be lower in HIV-infected children supplemented with vitamin A. All-cause mortality and AIDS-related deaths have also been found to be lower in vitamin A-supplemented HIV-infected children.

The benefits of multivitamin supplementation, particularly vitamins B, C, and E, have been more consistent across studies. Multivitamin supplementation in HIV-infected pregnant mothers has been shown to reduce the incidence of adverse pregnancy outcomes such as fetal loss and low birth weight. It also has been shown to decrease rates of MTCT among women who have poor nutritional or immunologic status. Further, multivitamin supplementation reduces the rate of HIV disease progression among patients in early stage of disease, thus delaying the need for ART by prolonging the pre-ART stage.

In brief, there is no evidence to recommend vitamin A supplementation of HIV-infected pregnant women; however, periodic vitamin A supplementation of HIV-infected infants and children is beneficial in reducing all-cause mortality and morbidity and is recommended. Similarly, multivitamin supplementation of people infected with HIV, particularly pregnant women, is strongly suggested. © 2007 Elsevier Inc.

I. INTRODUCTION

June 5, 1981. The Centers for Disease Control and Prevention (CDC, 1981) released a report entitled "Pneumocystis Pneumonia—Los Angeles," in which, five case studies were presented—all with a common theme of homosexuality, male gender, and some unknown cellular-immune dysfunction. This dysfunction is what we today know as the acquired immunedeficiency syndrome (AIDS). Since then, a huge body of literature on AIDS has been amassed in a relatively short span of time. Extensive academic as well as economic resources have been invested to study AIDS at a scale and speed unparalleled by any other disease or syndrome. In spite of all these efforts, an estimated 25 million people have died from AIDS in the last 25 years.

December 31, 2005. In the latest UNAIDS update on the AIDS epidemic, ~40.3 million people are believed to be living with the human immunodeficiency virus (HIV) infection or AIDS worldwide. Over 60% of the people living with HIV/AIDS (PLWHA) are in sub-Saharan Africa, the most affected geographical region, which is home to just about 10% of the world's population. This region also has 77% of women living with HIV/AIDS globally and over 89% of the world's total number of new infections among children. In recent years, this epidemic has rapidly expanded in other areas of the world. Notably in 2005, South/Southeast Asia region had an estimated 7.4 million people living with HIV/AIDS (UNAIDS, 2005).

AIDS is characterized by a progressive deterioration in immune function. Interventions that may offset this impairment have the potential to decrease the viral load and consequently the risk of transmission—both from motherto-child as well as horizontal partner-to-partner transmission. Such an improvement in immune function is also likely to slow down HIV disease progression, apart from improving the quality of life. The latter has become increasingly important in the face of HIV infection becoming a chronic disease in those with access to adequate antiretroviral therapy.

By virtue of its ability to influence humoral and cell-mediated immune function, nutritional status has been extensively studied and researched as an adjunct therapy for decreasing transmission and also for reducing morbidity in HIV-infected patients. There are two other important points in the favor of pursuit of nutritional interventions. First, they are relatively inexpensive;

second, they may potentially demonstrate benefit regardless of HIV status, especially in pregnant women with reference to improvement in pregnancy outcomes and child health. Hence, a nutrition program is likely to cover a relatively greater proportion of the population, a feature which is important in regions such as sub-Saharan Africa where the prevention of mother-to-child transmission (PMTCT) services coverage was a mere 5% in 2003 (UNAIDS, 2005).

Vitamin supplementation is one of the most commonly practiced methods of improving nutritional status in HIV-infected patients. This practice is based on the vast literature from laboratory-based and epidemiological studies implicating various vitamins in improving immune responses. In this chapter, we will examine HIV disease progression and HIV transmission as separate and outline the evidence for use or nonuse of various vitamins for these outcomes, with a special focus on vitamin A. HIV disease itself has varying effects on different body systems and is likely to alter nutritional status by inducing abnormalities in absorption, utilization, and excretion of nutrients as well as by reducing appetite and consequently intake of various nutrients (Keusch and Farthing, 1990). For example, a review examining the etiologies for HIV-associated wasting and weight loss found that more than 25% of women in the nutrition for healthy living (NFHL) cohort had inadequate dietary intakes of vitamins A, C, E, and B6 (Mangili *et al.*, 2006). However, for the purposes of this chapter, we would restrict ourselves to description of the effects of various vitamins on HIV-related outcomes.

II. VITAMINS AND IMMUNE FUNCTION

A number of reviews available in the literature detail the cellular and molecular mechanisms by which vitamin A may influence immune function (Semba, 1998, 1999; Stephensen, 2001); to complement these, there is a comprehensive review of the effects of "vitamin A supplementation" on immune responses as well (Villamor and Fawzi, 2005). Briefly, vitamin A is believed to be important at all levels of the immune system (Ross and Stephensen, 1996; Semba, 1998)—its various functions include maintaining the integrity of the epithelia, increasing the levels of acute phase reactants in response to infection, regulating monocyte differentiation and function, improving the cytotoxicity of natural killer cells, enhancing the antibody responses to tetanus toxoid (Semba *et al.*, 1992) and measles vaccines (Coutsoudis *et al.*, 1992), and increasing the total lymphocyte count, especially the CD4 subset. Similarly, various other vitamins regulate cellular and humoral immune function at a variety of levels. A summary of the possible roles of various vitamins in immune function is provided in Table I.

TABLE I. Vitamins and Immune Function

Vitamin A (Coutsoudis et al., 1992; Ross and Stephensen, 1996; Semba, 1998; Semba et al., 1992)
  Innate Immunity
    Maintenance of epithelial integrity
    Acute phase response—increase in serum amyloid A and C reactive protein during infection
    Enhanced monocyte differentiation and function
    Increased cytotoxicity of natural killer cells
    Improved neutrophil function
  Adaptive Immunity
    Increase in T-cell counts, particularly of the CD4 cells
    Increase in the antibody response to tetanus toxoid and measles vaccine

Vitamin B6 (Meydani et al., 1991)
  Increased lymphocyte production
  Increased cell-mediated cytotoxicity
  Increased delayed-type hypersensitivity (DTH) responses
  Increased antibody production

Folic Acid (Bendich and Cohen, 1988)
  Improved neutrophil phagocytosis and activity

Vitamin B12 (Bendich and Cohen, 1988)
  Improved antibody immunity
  Impaired neutrophil function

Vitamin E (Bendich, 1988; Meydani et al., 1990, 1997; Wang et al., 1994, 1995)
  Improved DTH skin response, neutrophil phagocytosis, lymphocyte proliferation
  Increased IL-2 production
  Increased natural killer cell cytotoxicity
  Reduced production of inflammatory cytokines such as TNF, IL-6
  Improved antibody response to T-cell-dependent vaccines

Vitamin C (Bendich, 1988; Hemila, 1997; Winklhofer-Roob et al., 1997)
  Improved T- and B-lymphocyte proliferative responses
  Reduced concentration of proinflammatory cytokines, including IL-6 and 1-acid glycoprotein

Vitamin D (Abe et al., 1984; Bar-Shavit et al., 1981; Mariani et al., 1999; Yang et al., 1993)
  Improved phagocytic capacity of macrophages
  Improved cell-mediated immunity
  Increase in natural killer cell number and their cytolytic activity

III. VITAMINS, HIV TRANSMISSION, AND PREGNANCY OUTCOMES

HIV transmission can occur via many routes, with person-to-person (horizontal) transmission and mother-to-child (vertical) transmission, constituting the major transmission pathways. Mother-to-child transmission (MTCT) can be further segmented into three time periods with varying degrees of risk of transmission—intrauterine (during pregnancy), intrapartum (during passage through the birth canal), and during breast-feeding. In the absence of preventive measures, HIV is transmitted from an HIV-infected mother to her infant in $\sim$25–48% of the cases in developing countries (15–25% in the setting of a developed country) (Dabis *et al.*, 1993; UNAIDS, 1998). In addition to the risk of MTCT of HIV, infants born to HIV-infected mothers experience increased risk of prematurity, low birth weight, and small-for-gestational age (Brocklehurst and French, 1998). A number of strategies have been implemented and studied to reduce transmission as well as improve pregnancy outcomes—the most effective of which is the use of antiretroviral therapy, primarily with the drugs Zidovudine and Nevirapine (Connor *et al.*, 1994; Guay *et al.*, 1999; Jackson *et al.*, 2003; Wiktor *et al.*, 1999). Complementary approaches that have been or are being evaluated include delivery via caesarean section, avoidance of breast-feeding, inactivation of HIV through expression and heat treatment of breast milk, and a variety of nutritional interventions. Among the latter, supplementation with vitamins has been an obvious choice, prompted by the fact that the regions with the highest HIV burden often have high prevalence of vitamin deficiencies, as indicated by intake as well as serum levels of these vitamins. Thus, vitamin supplementation may have benefits even among the HIV-negative population. However, to reiterate, nutritional interventions are at best a complementary therapy and not a substitute for antiretroviral treatment. Treatment courses with antiretroviral drugs should be the primary mode of therapy for all HIV-infected pregnant women who are advanced in their disease and are eligible for such therapy per national and international guidelines.

There are many biologically plausible mechanisms by which vitamins may reduce HIV transmission. As mentioned in the earlier section, most vitamins are modulators or regulators of the immune system; hence, they might influence HIV transmission by a nonspecific immunostimulatory effect and a subsequent decrease in the systemic viral load. Vitamins may specifically reduce the risk of intrapartum transmission by strengthening the epithelial integrity of the placenta and the lower genital tract to decrease fetal exposure to HIV. Moreover, improved nutritional status may reduce inflammation of the breast tissue, consequently decreasing viral shedding in breast milk and protecting against HIV transmission via breast-feeding. Maternal vitamin supplementation may also lead to more robust immune and gastrointestinal systems in the newborn, which may provide additional defense against transmission through breast-feeding (Dreyfuss and Fawzi, 2002).

A. EVIDENCE FROM OBSERVATIONAL STUDIES

A few observational studies conducted in Malawi and Rwanda in the 1990s suggested that low serum vitamin A concentrations in HIV-infected pregnant women are associated with an increased risk of MTCT of HIV. In Malawi (Semba et al., 1994), the investigators enrolled 338 HIV-infected women and measured their serum vitamin A concentrations and assessed the HIV status of the newborn infant. Overall, the rate of MTCT was 21.9% among mothers whose infants were alive at 12 months postpartum. The investigators divided the mothers into four groups based on their vitamin A levels—less than 0.70, between 0.70 and 1.05, between 1.05 and 1.40, and greater than or equal to 1.40 μmol/liter. The rates of MTCT in each group were 32.4%, 26.2%, 16.0%, and 7.2%, respectively. The odds ratio (OR) for vertical transmission was 1.78 [95% confidence interval (CI): 1.18, 2.70] for every 0.45 μmol/liter decrease in serum retinol levels, adjusted for maternal age, body mass index (BMI), CD4 cell count, birth weight, and gestational age. In another study in Malawi, the same investigators found that low maternal plasma vitamin A during HIV infection also predicted increased infant mortality (Semba et al., 1995b). Infant mortality rate was 14.2% in the children born to women with vitamin A levels greater than 1.75 μmol/liter, as compared to 93.3% in children born to women with vitamin A levels less than 0.35 μmol/liter. The study in Rwanda (Graham et al., 1993) followed 302 HIV-infected women during pregnancy and characterized their vitamin A status as normal if serum retinol levels were greater than or equal to 0.70 μmol/liter and as low for levels below 0.70 μmol/liter. The OR for infant HIV infection, adjusted for percent CD4 cells and hematocrit, was reported to be 1.96 (95% CI: 1.11, 3.45) for the low vitamin A group, as compared to the normal vitamin A group.

However, the results of studies examining the relationship between vitamin A levels and MTCT in the United States have been more equivocal. A nonsignificant protective trend against transmission was seen with increasing vitamin A levels in a study of 449 HIV-infected pregnant women enrolled from across the US [OR of 1.78 (95% CI: 0.69, 4.54), comparing women with levels below 0.70 μmol/liter to women with levels greater than or equal to 1.05 μmol/liter, adjusted for plasma viral RNA, %CD4+ cells, birth weight, and hard drug use] (Burns et al., 1999). However, the incidence of low birth weight (<2500 g) in infants of mothers with low (20 to <30 μg/dl) and very low (<20 μg/dl) vitamin A levels was significantly higher than in those infants born to mothers with higher levels of vitamin A (OR of 4.58; 95% CI: 1.57, 13.4; and OR of 6.99; 95% CI: 1.09, 45.0, respectively). In another study, 133 women were enrolled from 2 urban areas of the eastern United States and an OR of 4.54 (95% CI: 1.08, 20.0) for vertical transmission of HIV infection was observed, comparing women with levels below 0.70 μmol/liter to women with levels greater than or equal to 1.05 μmol/liter, adjusted for percent CD4 cells, mode of delivery, gestational age, duration of membrane rupture, and race (Greenberg et al., 1997). A small cohort study of 95 HIV+ women in the US found that the levels of serum retinol, β-carotene, and vitamin E during pregnancy were not associated with vertical HIV transmission (Burger et al., 1997).

Some studies have assessed the role of vitamins in horizontal transmission of HIV. A nested case-control study of sexually active, adult women in Kigali, Rwanda compared the baseline serum levels of various nutrients in 45 women who seroconverted during the 24-month study period with serum levels in 74 women who remained seronegative throughout the study. No differences in the baseline serum levels of vitamin A, carotenoids, and vitamin E were observed (Moore et al., 1993). In a nested case-control study in India that enrolled individuals attending 2 sexually transmitted disease (STD) clinics in Pune, subjects having a β-carotene concentration of less than 0.075 μmol/liter were 21 times more likely to acquire HIV infection than those with higher levels (adjusted OR of 21.1; $p=0.01$) (Mehendale et al., 2001). There was no association between HIV infection and other non-provitamin A carotenoids. These results contrast those from a study in Kenya that enrolled men with concurrent genital ulcers and found that a low serum retinol concentration (less than 20 μg/dl) was associated with a lower probability of seroconversion. Seroconversion was found to be independently associated with a retinol level greater than 20 μg/dl [hazard ratio (HR) of 2.43; 95% CI: 1.25, 4.70] (MacDonald et al., 2001). Another nested case-control study in Tanzania found no significant relationship between serum vitamin A concentrations and the risk of HIV seroconversion among women attending family planning clinics in Dar es Salaam (Villamor et al., 2006.)

The evidence from observational studies of vitamins and HIV transmission is summarized in Table II.

B. EVIDENCE FROM TRIALS

The results from the observational studies motivated the initiation of randomized, placebo-controlled intervention trials to assess the role of micronutrient supplementation in vertical HIV transmission and pregnancy outcomes. All of these trials were conducted on antiretroviral-naive pregnant women. In Malawi, 697 HIV-infected pregnant women were randomly assigned to receive daily doses of iron and folate, either alone or with vitamin A (3-mg retinol equivalent = 10,000 IU preformed vitamin A) from

TABLE II. Vitamins and HIV Transmission

Study	Exposure	Risk of transmission and other results
Observational studies		
Vertical transmission		
Malawi (Semba et al., 1994, 1995b)	Low maternal vitamin A	↑; ↑ Infant mortality
Rwanda (Graham et al., 1993)	Low maternal vitamin A	↑
United States (Burns et al., 1999)	Low maternal vitamin A	↔; ↑ Risk of low birth weight
United States (Greenberg et al., 1997)	Low maternal vitamin A	↑
United States (Burger et al., 1997)	Low maternal vitamin A	↔
	Low maternal β-carotene	↔
	Low maternal vitamin E	↔
Horizontal transmission		
Rwanda (Moore et al., 1993)	Low serum vitamin A	↔
	Low serum vitamin E	↔
	Low serum carotenoids	↔
India (Mehendale et al., 2001)	Low serum β-carotene	↑
Kenya (MacDonald et al., 2001)	Low serum vitamin A	↓
Tanzania (Villamor et al., 2006)	Low serum vitamin A	↔
Trials		
Malawi (Kumwenda et al., 2002)	Vitamin A supplementation	↔; ↓ Risk of low birth weight and anemia in the newborn; no association with prematurity
South Africa (Coutsoudis et al., 1999)	Vitamin A plus β-carotene supplementation	↔; No association with low birth weight, infant mortality; ↓ risk of preterm delivery
Tanzania (Fawzi et al., 1998, 2000b, 2002, 2004a; Villamor et al., 2002b)	Vitamin A plus β-carotene supplementation	↑; No association with low birth weight, fetal mortality, severe preterm (<34 weeks), small SGA; ↑ risk of HIV shedding through the lower genital tract
	Multivitamin (vitamins B, C, and E) supplementation	↓ Only among children of women who had a poorer nutritional and immunologic status at baseline; ↓ risk of low birth weight, fetal mortality, severe preterm (<34 weeks), small SGA
Kenya (Baeten et al., 2002; McClelland et al., 2004)	Vitamin A supplementation	↔; No effect on HIV shedding in vaginal secretions
Kenya (McClelland et al., 2004)	Multivitamin + Se supplementation	↑ Risk of HIV shedding in vaginal secretions
Zimbabwe (Humphrey et al., 2006)	Vitamin A supplementation to either mother or infant	↑; ↑ Risk of infant mortality

↑, Increased risk; ↓, decreased risk; ↔, no association; SGA, small for gestational age; Se, selenium.

18 to 28 weeks of gestation until delivery. Supplementation with vitamin A had no effect on prematurity or on HIV transmission from mother to child assessed at 6 weeks or at 24 months postpartum; however, the vitamin A group had significantly lower proportion of low birth weight infants (14% versus 21.1% in the control group; $p = 0.03$) and a reduced incidence of anemia in the newborn children at 6 weeks postpartum (23.4% versus 40.6%; $p < 0.001$) (Kumwenda et al., 2002).

In South Africa, 728 HIV-infected pregnant women were randomized to receive either 5000 IU vitamin A (1.667-mg retinol equivalent) plus 30-mg β-carotene (5-mg retinol equivalent) daily during the third trimester of pregnancy plus 200,000 IU vitamin A at delivery or placebo. No difference was found either in the risk of HIV infection or in birth weights or in the fetal/infant mortality rates in the two groups at 3 months of age. However, women in the vitamin A group were significantly less likely to have a preterm delivery (11.4% versus 17.4% in the placebo group; $p = 0.03$) (Coutsoudis et al., 1999).

In Tanzania, the Trial of Vitamins (TOV) study randomized 1078 HIV-infected pregnant women at 12–27 weeks of gestation to receive either daily vitamin A supplements [30-mg β-carotene (5-mg retinol equivalent) and 5000 IU of preformed vitamin A (1.667-mg retinol equivalent)], multivitamins (20-mg thiamine, 20-mg riboflavin, 25-mg B6, 100-mg niacin, 50-μg B12, 500-mg C, 30-mg E, and 0.8-mg folic acid), both, or neither using a 2×2 factorial design. At delivery, women in vitamin A groups received an additional oral dose of vitamin A (200,000 IU), whereas women in nonvitamin A groups received a placebo. In addition, all women received 120 mg of ferrous iron and 5 mg of folate tablets daily and 300 mg of Chloroquine as malaria prophylaxis weekly. Neither multivitamins [relative risk (RR) of 0.95; 95% CI: 0.73, 1.24] nor vitamin A (RR of 1.06; 95% CI: 0.81, 1.39) had an effect on the risk of HIV transmission or survival through 6 weeks postpartum. Babies born HIV-negative to the mothers supplemented with multivitamins, but not vitamin A, had higher birth weights as compared to those in the no-multivitamin arm (94 g difference in birth weights, $p = 0.02$) (Fawzi et al., 2000b). Moreover, the multivitamin supplements, but not vitamin A, decreased the risk of fetal death by 39% [30 fetal deaths in the multivitamin group, compared to 49 among those not on multivitamins (RR of 0.61; 95% CI: 0.39, 0.94)]. Multivitamin supplementation decreased the risk of low birth weight (<2500 g) by 44% (RR of 0.56; 95% CI: 0.38, 0.82), severe preterm birth (<34 weeks of gestation) by 39% (RR of 0.61; 95% CI: 0.38, 0.96), and small size for gestational age at birth by 43% (RR of 0.57; 95% CI: 0.39, 0.82). Vitamin A supplementation had no significant effect on these endpoints. Multivitamins, but not vitamin A, resulted in a significant increase in CD4, CD8, and CD3 counts in the women (Fawzi et al., 1998). Maternal multivitamin supplements, but not vitamin A, also led to a significant increase in the weight gain during pregnancy, along with reducing the risk of developing hypertension during pregnancy (RR of 0.62; 95% CI: 0.40, 0.94) (Merchant *et al.*, 2005; Villamor *et al.*, 2002b). After assessment of HIV transmission through breast-feeding, vitamin A/β-carotene resulted in a significant increase of 38% in the risk of overall vertical HIV transmission, whereas multivitamins excluding A had no effect on the risk of transmission (Fawzi *et al.*, 2002). Children who were HIV-negative at 6 weeks of age and were born to women who were in relatively poorer nutritional or immunologic conditions at baseline and whose mothers received multivitamins experienced significant reductions in HIV transmission through breast-feeding and improvement in HIV-free survival. Vitamin A/β-carotene supplementation also led to a significant increased in shedding of the virus through the lower genital tract (Fawzi *et al.*, 2004a). There are two important differences to note between TOV and the previously mentioned vitamin A trials—first, the supplementation of the mothers continued beyond the antenatal period, unlike the other trials; second, the effect of the supplementation regimens on the various routes of vertical transmission (intrauterine, intrapartum, and breast-feeding) could be assessed.

Another trial in Mombasa in Kenya randomized 400 HIV-infected women to receive either vitamin A (10,000 IU) supplementation daily for 6 weeks or placebo to examine difference in viral shedding in vaginal secretions between the 2 groups, if any. No such difference was seen at the end of follow-up; moreover, there was no effect observed on plasma viral load or on CD4 or CD8 counts with vitamin A supplementation in these women (Baeten *et al.*, 2002). The same investigators then conducted a similar trial with multivitamin plus selenium supplementation versus placebo for 6 weeks to examine the same outcome. They found that micronutrient supplementation led to a 2.5 times higher risk ($p = 0.001$) of viral shedding and this effect was greatest among women who had normal selenium levels at baseline (McClelland *et al.*, 2004). Moreover, this trial used a similar dose of vitamins B, C, and E as the TOV in Tanzania. These observations appear to indicate selenium as a potential risk factor for HIV transmission, an issue that needs to be examined further.

In Zimbabwe, the investigators assessed the efficacy of a single large dose of vitamin A given to women (400,000 IU) early during the postpartum period and/or to infants (50,000 IU). The authors used the data from a total of 4495 infants born to HIV-infected women and found that vitamin A supplementation to either mothers or infants resulted in an increased risk of infant HIV infection or death, although the effect of providing the supplement to both mother and infant was not different from the effect of providing a placebo (Humphrey *et al.*, 2006).

The findings from these trials assessing the role of vitamin supplementation in the prevention of HIV transmission are summarized in Table II.

IV. VITAMINS AND HIV DISEASE PROGRESSION IN ADULTS

The benefit of slowing HIV disease progression is likely to be twofold—one is that it will improve the quality of life of the infected patient, and second is the potential to decrease transmission of HIV, as the stage of HIV disease is a key determinant of probability of transmission. There have been a few studies, which have been reviewed below, exploring the role of vitamin status or vitamin supplementation in slowing clinical, immunologic, and virological disease progression.

A. EVIDENCE FROM OBSERVATIONAL STUDIES

In the United States, two long-term prospective studies have followed cohorts of HIV-infected homosexual or bisexual men. One was titled the San Francisco Men's Health Study (SFMHS) and it enrolled 1034 single men between 25 and 54 years of age (Winkelstein et al., 1987). The other was named Multicenter AIDS Cohort study (MACS) and it enrolled about 5000 homosexual male volunteers for participation in a semiannual interview, physical examination, and laboratory testing in 4 metropolitan areas (Kaslow et al., 1987).

In the SFMHS cohort, nutrient intake was assessed for 296 HIV-infected men using food frequency questionnaires (FFQs), and progression to clinical AIDS was determined. The authors found that vitamin A intake was positively associated with the CD4 cell count at baseline but not with progression to AIDS. A doubling of daily intake of carotenoids and vitamin A were associated with RRs of 0.93 and 0.98, respectively, for progression to AIDS, neither of which were statistically significant [adjusted for age, smoking, energy intake, symptoms, and CD4 cell count; 95% CI: (0.78, 1.12) for carotenoids and (0.78, 1.23) for vitamin A]. However, intakes of vitamins B1 (thiamine), B2 (riboflavin), folate, niacin, C, and E were all associated with a decreased risk of progression to AIDS; only B2 and E intakes remained significant after adjustment for age, smoking, energy intake, HIV symptoms, and CD4 cell count (Abrams et al., 1993). Overall, the use of a daily multivitamin was associated with a significantly reduced risk for low CD4 counts as well as a significant reduction in the risk of progression to AIDS by 30% in this cohort.

In the MACS cohort, nutrient intakes were also assessed using FFQs for 281 HIV-infected men and the relationships of various micronutrients with 2 endpoints were examined—progression to clinical AIDS (Tang et al., 1993) and death (Tang et al., 1996). A U-shaped relationship was found between vitamin A intake (from diet and supplements) and risks of progression to clinical AIDS and death; the middle two quartiles of intake were associated with significantly slower progression to AIDS (RR of 0.55; 95% CI: 0.35, 0.88), as compared to the lowest and the highest quartiles. No association was noted between serum levels of vitamin A and progression to AIDS in this cohort; however, most of the participants had vitamin A levels in the normal to high range (median = 2.44 μmol/liter) and only 2% of them had serum vitamin A concentrations indicative of vitamin A deficiency (Tang et al., 1997b). Higher intakes of vitamins B1, B2, niacin, and C were associated with a decreased risk of progression to clinical AIDS (Tang et al., 1993); higher intakes of B vitamins (B1, B2, B6, and niacin) as well as the use of vitamin B supplements (B1, B2, and B6) were also associated with increased survival in this cohort (Tang et al., 1996). It is important to note that ~40% reduction in mortality was seen with the supplemental use of vitamins B1 and B2 at levels that were five times the recommended dietary allowance (RDA) and of B6 at levels that were two times the RDA. No effect on survival was seen with vitamins C or E in this cohort. However, men in the highest quartile of serum vitamin E levels ($\geq$23.5 μmol/liter) had a 34% decrease in the risk of progression to AIDS, when compared to men in the lowest quartile (Tang et al., 1997b). In a subsample of the MACS cohort, the investigators found that men with low serum vitamin B12 concentrations (<120 pmol/liter) progressed to AIDS much faster than those with higher vitamin B12 levels (median AIDS-free time = 4 versus 8 years, respectively, $p = 0.004$) (Tang et al., 1997a). This relationship persisted after adjusting for HIV symptoms, CD4 cell counts, age, serum albumin, use of antiretroviral treatment before AIDS, frequency of alcohol consumption, and serum folate concentration as well as after excluding men with advanced disease at baseline.

In a cross-sectional study of 132 adults attending an HIV clinic in Cape Town, South Africa, vitamin A levels were found to be low (<1.05 μmol/liter) in 39% of patients in early stages (WHO stages I and II) compared to being low in 48% and 79% of the patients in WHO stage III and IV, respectively ($p < 0.01$). There was a weak positive association of serum retinol seen with CD4 cell count ($r = 0.27$, 95% CI: 0.1, 0.43). A novel aspect of this study was that they tried to eliminate confounding by the acute phase response by controlling for C-reactive protein in multivariate analysis and by excluding active opportunistic infections (Visser et al., 2003).

A smaller study enrolled 108 HIV-infected homosexual men in the United States and followed them for 18 months; biochemical measurements of nutrients were made at baseline and after every 6 months. Low plasma vitamin A and B12 concentrations were associated with a significant decline in CD4 count, and normalization of these concentrations over time led to a significant increase in the CD4 cell counts (Baum et al., 1995). Low vitamin B12 at baseline also significantly predicted faster HIV disease progression. Importantly, these results were unaffected by the use of Zidovudine, indicating a benefit regardless of antiretroviral treatment. In another study by the same investigators, 125 HIV-infected drug-using men and women were followed for 3.5 years in Miami, Florida. Various immunologic parameters and nutrient levels were assessed every 6 months. Low serum vitamin A concentration was not found to predict mortality after multivariate adjustment for CD4 cell count less than 200 cells/mm^3 at baseline, CD4 cell counts over time, and for other nutritional deficiencies (Baum et al., 1997).

However, in a trial of micronutrient supplements for HIV-infected patients with diarrhea in Zambia, serum vitamin A and E levels before treatment were found to be a significant predictor of early mortality (Kelly et al., 1999).

A team of investigators at Johns Hopkins University followed a cohort of 2000 intravenous drug users in Baltimore. In one study, they analyzed blood samples for vitamin A levels in a random subsample of 179 subjects and measured their survival. They found that low plasma vitamin A (vitamin A levels less than 1.05 μmol/liter) was associated with lower CD4 levels as well as with increased mortality (RR of 6.3; 95% CI: 2.1, 18.6) among the HIV-infected subjects (Semba et al., 1993). The investigators also did a nested case-control study in which 50 HIV-infected adult subjects who died from AIDS and infections were matched with 235 HIV-infected controls that survived. They found that low plasma vitamin A was associated with a higher risk of mortality (OR of 4.6; 95% CI: 1.8, 11.3) (Semba et al., 1995a).

In South Africa, a case-control study of 175 pairs of black HIV-infected subjects, matched by age, gender, CD4 cell counts, and year of HIV diagnosis, found that the use of B vitamins was associated with an increase in median duration to AIDS by about 40 weeks (72.7 weeks with B vitamin use versus 32.0 weeks with no vitamin use, $p = 0.004$) and an increase in the survival time by more than 2 years (264.6 weeks with B vitamin use versus 144.8 weeks with no vitamin use, $p = 0.001$) (Kanter et al., 1999).

In a study of 50 HIV-infected female outpatients in Germany examining osteopenia, serum levels of vitamin D (1,25-dihydroxyvitamin D) were significantly reduced compared to 50 age-matched healthy controls and there was a significant positive correlation between the CD4 counts and vitamin D (correlation of 0.45, $p < 0.05$) (Teichmann et al., 2003). Another study found significantly lower serum levels of 1,25-vitamin D in HIV-infected symptomatic patients compared with controls as well as with HIV-infected asymptomatic patients. Low serum level of 1,25-vitamin D was correlated with a low CD4 cell count and also predicted a shorter survival time as compared to HIV-infected subjects with levels above 25 pg/ml (Haug et al., 1994).

A summary of the results of observational studies examining the relationship between vitamins and HIV disease progression is provided in Table III.

TABLE III. Vitamins and HIV Disease Progression

Study	Exposure	Main results
Observational studies		
SFMHS, United States (Abrams et al., 1993)	Vitamin A intake	↑ CD4; no effect on clinical progression
	Daily multivitamin intake	↑ CD4; ↓ clinical progression
MACS, United States (Tang et al., 1993, 1996, 1997a,b)	Vitamin A intake	U-shaped relationship with clinical progression and mortality
	Serum vitamin A	No effect on clinical progression
	Intake of B vitamins (B1, B2, B6, niacin)/use of supplements (B1, B2, B6)	↓ Clinical progression and mortality
	Low serum vitamin E	↑ Clinical progression
	Low serum vitamin B12	↑ Clinical progression
South Africa (Visser et al., 2003)	Low serum vitamin A	↓ CD4; ↑ clinical progression
United States (Baum et al., 1995)	Low serum vitamin A	↓ CD4
	Low serum vitamin B12	↓ CD4; ↑ clinical progression
United States (Baum et al., 1997)	Low serum vitamin A	No association with mortality
Zambia (Kelly et al., 1999)	Low serum vitamin A and E	↑ Mortality
United States (Semba et al., 1993, 1995a)	Low serum vitamin A	↓ CD4; ↑ mortality
South Africa (Kanter et al., 1999)	Use of B-vitamins	↓ Clinical progression and mortality
Germany (Teichmann et al., 2003)	Low serum vitamin D*	↓ CD4
Norway (Haug et al., 1994)	Low serum vitamin D*	↓ CD4; ↑ mortality
Trials		
United States (Coodley et al., 1993)	β-Carotene supplementation	↑ CD4
United States (Coodley et al., 1996)	β-Carotene supplementation	No association with CD4 cell counts
France (Constans et al., 1996)	β-Carotene supplementation	No association with CD4 cell counts
Kenya (Baeten et al., 2002)	Vitamin A supplementation	No association with CD4 cell counts or viral load

(*Continues*)

TABLE III. (*Continued*)

Study	Exposure	Main results
Kenya (McClelland et al., 2004)	Multivitamins plus Se supplementation	↑ CD4; no association with viral load
United States (Semba et al., 1998)	Vitamin A supplementation	No association with CD4 cell counts or viral load
United States (Humphrey et al., 1999)	Vitamin A supplementation	No association with CD4 cell counts or viral load
Thailand (Jiamton et al., 2003)	Multiple micronutrient supplementation (vitamins A, B1, B2, B6, B12, C, D, E, K, β-carotene, folate, iron, zinc, and selenium)	No association with CD4 cell counts or viral load; ↓ mortality only among individuals with baseline CD4 < 100 cells/mm^3
Zambia (Kelly et al., 1999)	Vitamins A, C, E, Se, Zn	No association with CD4 cell counts
Canada (Allard et al., 1998)	Vitamins C and E	↓ Viral load
Tanzania (Fawzi et al., 2004b)	Vitamin A supplementation	No association with CD4 cell count, viral load, or clinical progression
	Multivitamin (vitamins B, C, and E) supplementation	↑ CD4; ↓ viral load; ↓ clinical progression
South Africa (Coutsoudis et al., 1997)	Vitamin A and β-carotene supplementation	No association with viral load

SFMHS, San Francisco Men's Health Study; MACS, Multicenter AIDS Cohort Study; ↑, increase in; ↓, decrease in; *, 1,25-dihydroxyvitamin D; Se, selenium; Zn, zinc.

B. EVIDENCE FROM TRIALS

A number of intervention studies have examined the effect of vitamin A supplementation on HIV disease progression. Some of these studies used β-carotene as the supplement because it has additional antioxidant properties, apart from contributing to body stores of vitamin A. A double-blind placebo-controlled clinical trial enrolled 21 HIV-infected patients and randomized them to receive either β-carotene (180 mg/day) or placebo for 4 weeks, and then crossed them over so that participants received the alternative treatment for the ensuing 4 weeks. β-Carotene supplementation resulted in significant increases in the leucocyte counts, the percentage change in CD4 cell counts, and the percentage change in the ratio of CD4 to CD8 cells.

A decrease in these parameters was observed when the subjects were crossed over to the placebo regimen (Coodley et al., 1993). In a larger trial, the same investigators randomly assigned 72 HIV-infected patients to receive either β-carotene (180 mg/day) or placebo for a period of 3 months. No effect on immune cells was seen in this study. However, a multivitamin supplement (containing 5000 IU of vitamin A) was provided to all study subjects, which in the authors' opinion may have masked the difference between the two groups (Coodley et al., 1996). In another study, 52 HIV-infected patients with CD4 counts below 400 cells/mm^3 were randomized to receive either 60-mg β-carotene (10-mg retinol equivalent) or 0.25-mg selenium/day for 1 year. An improvement in measures of oxidative stress was reported but no effect on CD4 cell counts was observed in either arm (Constans et al., 1996).

In the trial from Mombasa, Kenya referred to earlier, investigators enrolled 400 HIV-infected women and randomized them to receive either 10,000 IU of vitamin A daily for 6 weeks or placebo. No difference in plasma viral load or CD4 or CD8 cell counts was observed in the two groups. Moreover, no effect was noted even among the women who were deficient at baseline (Baeten et al., 2002). The same investigators then conducted a similar trial using micronutrient (multivitamins and selenium) supplementation instead of vitamin A. Micronutrient-supplemented patients had higher CD4 (+23 cells/μl) and CD8 (+74 cells/μl) counts than placebo patients after adjustment for baseline CD4 and CD8 count, respectively. There was no difference in plasma viral load between the two groups (McClelland et al., 2004).

In a short-term trial, 120 HIV-infected injection drug users were randomized to receive either a single high-dose of vitamin A (200,000 IU) or placebo. No significant effect was observed on HIV load or on CD4 lymphocyte count at 2 and 4 weeks after treatment (Semba et al., 1998). Similar results were obtained in another trial, which recruited 40 HIV-infected women of reproductive age and randomized them to receive either a single high-dose of vitamin A (300,000 IU) or a placebo (Humphrey et al., 1999). The South African trial on vertical transmission described earlier also did not find any significant effect of vitamin A and β-carotene supplementation on HIV load or on the immunologic progression of HIV disease (Coutsoudis et al., 1997).

There have been very few trials examining the role of vitamins, other than A, in HIV disease progression. One randomized, double-blind, placebo-controlled trial in Thailand provided daily micronutrient supplementation (including vitamin A, β-carotene, vitamins D, E, K, C, B1, B2, B6, B12, folate, iron, zinc, and selenium) for 48 weeks to 481 HIV-infected asymptomatic men and women. The supplements had no effect on CD4 cell counts or viral load, but were associated with a reduction in mortality, which was significant only among individuals with baseline CD4 cell counts below 100 cells/mm^3 (Jiamton et al., 2003). Though this study indicates that there is a benefit of micronutrient supplementation, a few aspects of the study deserve mention. First, the supplementation was provided for a limited duration of time (48 weeks) to subjects who were asymptomatic at baseline; and second, the study had limited statistical power as the sample size was relatively small ($n = 481$).

In Zambia, 141 HIV-infected patients with persistent diarrhea were randomized to albendazole plus vitamins A, C, and E, selenium, and zinc or to albendazole plus placebo for 2 weeks. There was no effect of the supplementation on CD4 cell counts or on any clinical markers of illness severity (Kelly et al., 1999). Again, a limitation of this study was its short duration of 2 weeks, in which a beneficial benefit of micronutrient supplementation, if any, may not be seen.

In a randomized, placebo-controlled trial in Canada, 49 HIV-infected patients were randomized to receive large daily doses of vitamin E (800 IU) and vitamin C (1000 mg) or matched placebo for a duration of 3 months. A general trend toward reduced viral load was observed, along with a significant reduction in the measures of oxidative stress (breath pentane output, plasma malondialdehyde, and lipid peroxides) (Allard et al., 1998).

The TOV study in Tanzania also looked at various clinical and laboratory markers of disease progression in the HIV-infected pregnant women. One of the chief findings was a 30% reduction in the risk of progression to WHO stage IV or AIDS-related death with multivitamin use (vitamins B, C, and E; RR of 0.71; 95% CI: 0.51, 0.98) (Fawzi et al., 2004b). Multivitamins significantly reduced oral and gastrointestinal manifestations of HIV disease such as oral thrush, oral ulcers, and difficulty in swallowing, along with significantly decreasing incidence of reported fatigue, rash, and acute upper respiratory infections. Vitamin A alone had no significant beneficial effect on any of these symptoms/signs. Multivitamin supplements also led to a significant increase in the CD4 and CD8 cell counts, apart from significantly lowering viral load (Fawzi et al., 2004b). Multivitamins were also found to have a protective effect on wasting in these HIV-infected women (Villamor et al., 2005). Vitamin A supplementation had no beneficial effect on CD4, CD3, and CD8 counts in the HIV-infected pregnant women (Fawzi et al., 1998).

The findings of the trials assessing the role of vitamin supplementation in HIV disease progression are summarized in Table III.

V. VITAMINS, GROWTH, AND DISEASE PROGRESSION IN HIV-INFECTED CHILDREN AND HIV-NEGATIVE CHILDREN BORN TO HIV-INFECTED MOTHERS

HIV infection in children appears to be bimodal in the developed countries—10–20% of infected children progress rapidly to AIDS and usually die before 4 years of age; whereas the remaining 80–90% have a mean survival of ~9–10 years (Rogers et al., 1994). However, survival in African countries is shorter than what is seen in the more developed world. Median survival in studies in Uganda (Marum *et al.*, 1996), Rwanda (Spira *et al.*, 1999), West Africa (Dabis *et al.*, 2001), and Gambia (Schim van der Loeff *et al.*, 2003) have ranged from a little over a year to 2.5 years. HIV-infected children are more likely to suffer from nutritional deficiencies and related diseases, especially in developing areas of the world. Failure to thrive, weight loss, delayed milestones, persistent diarrhea, persistent fever, and severe pneumonia are all expected to occur at a greater frequency in HIV-infected children. In a study in Zaire, the incidence rates of acute diarrhea was 1.7 times higher in HIV-infected children as compared to uninfected children (170 per 100 child years versus 100 per 100 child years) (Thea *et al.*, 1993). In the same study, HIV-infected children were twice as likely die of diarrheal disease as compared to uninfected children. In this context, the need and relevance of nutritional interventions is perhaps greater in HIV-infected children.

A few trials have directly examined the effects of vitamin A supplementation in children with HIV infection or children born to HIV-infected women. The first one in Durban, South Africa, randomized 118 infants born to HIV-infected women to either a vitamin A supplementation arm or a placebo arm. Vitamin A supplementation was provided at 1 and 3 months of age (50,000 IU), at 6 and 9 months (100,000 IU), and at 12 and 15 months (200,000 IU). Morbidity associated with diarrhea in vitamin A supplemented HIV-infected children went down by almost 50% (OR of 0.51; 95% CI: 0.27, 0.99), whereas there was no effect of supplementation seen in uninfected children (Coutsoudis *et al.*, 1995). In another trial in South Africa among HIV-infected children, supplementation with vitamin A before influenza vaccination decreased the increase in HIV viral load typically observed postimmunization (Hanekom *et al.*, 2000).

Another trial in Tanzania examined whether vitamin A supplementation could ameliorate the adverse effect of various infections, such as HIV, malaria, and diarrheal disease on child growth. 687 children between 6 and 60 months of age, who had been admitted to the hospital with pneumonia, were randomized to either receive vitamin A supplementation or placebo. The vitamin A regimen included an oral dose of 200,000 IU (100,000 IU if age <12 months) at day of admission, a second dose on the following day, and third and fourth doses at 4 and 8 months after discharge from the hospital. HIV infection was found in 9% of the children. Among these HIV-infected children, the average gain in length was 2.8 cm (95% CI: 1.0–4.6) more in the vitamin A supplemented group than in the children who received placebo, whereas no effect was seen in infants who were HIV-negative (Villamor *et al.*, 2002a). HIV-infected children, who were supplemented with vitamin A, were at lower risk for a respiratory tract infection. Vitamin A supplementation also significantly decreased the risk of severe watery diarrhea (multivariate OR of 0.56; 95% CI: 0.32, 0.99) (Fawzi *et al.*, 2000a). Vitamin A supplements reduced all-cause mortality by 63% among these HIV-infected children (RR of 0.37; 95% CI: 0.14, 0.95) and by 42% among uninfected children

(RR of 0.58; 95% CI: 0.28, 1.19). Vitamin A supplements were also associated with a 68% reduction in AIDS-related deaths ($p = 0.05$) and a 92% reduction in diarrhea-related deaths ($p = 0.01$) (Fawzi et al., 1999).

A randomized trial in Uganda enrolled 181 HIV-infected children at 6 months and randomized them to receive vitamin A supplementation (60-mg retinol equivalent = 200,000 IU) or placebo every 3 months from ages 15 to 36 months. Children were followed for a median duration of 17.8 months after reaching 15 months of age and the trial was stopped when a new national policy for mass supplementation of vitamin A was implemented in Uganda. The investigators found that the children in vitamin A group had significantly lower risk of mortality as compared to the children in placebo group (RR of 0.54; 95% CI: 0.30, 0.98) (Semba et al., 2005).

In another study, the Bayley Scales of infant development were administered at 6, 12, and 18 months of age to a subset ($n = 327$) of the children born to women who participated in the TOV trial in Tanzania. The effect of maternal vitamin A and multivitamin (vitamins B, C, and E) supplementation was assessed using linear regression models and Cox proportional hazard models for the Mental Development Index, the Psychomotor Development Index, and raw scores separately. Multivitamin supplementation was associated significantly with a mean increase in Psychomotor Development Index score of 2.6 (95% CI: 0.1, 5.1). Multivitamins were also significantly protective against the risk for developmental delay on the motor scale (RR of 0.4; 95% CI: 0.2, 0.7) but not on the Mental Development Index. Vitamin A supplementation had no significant effect on these outcomes (McGrath et al., 2006).

The efficacy of a single large dose of vitamin A given to women early during the postpartum period and/or to infants was assessed in a trial in Zimbabwe, referred to earlier. Vitamin A supplementation had no effect on mortality in infants who were positive by polymerase chain reaction (PCR) for HIV at baseline; however, neonatal supplementation decreased mortality by 28% ($p = 0.01$) in infants who were PCR negative at baseline but PCR positive at 6 weeks, but maternal supplementation had no effect; in the majority of infants, that is, those who were PCR negative at 6 weeks—all three vitamin A regimens were significantly associated with twofold higher mortality ($p \leq 0.05$) (Humphrey et al., 2006).

Table IV presents a brief summary of these trials examining HIV disease progression and mortality in HIV-infected children.

VI. COMMENT

Most observational studies have demonstrated a harmful association between low serum vitamin A and MTCT; however, results from these studies need to be interpreted cautiously. It is well established that a state

TABLE IV. Vitamins and HIV Disease in Children

Trial	Exposure	Main results
South Africa (Coutsoudis et al., 1995)	Vitamin A supplementation	↓ Morbidity associated with diarrhea in vitamin A supplemented HIV-infected children; no effect in HIV-uninfected children
South Africa (Hanekom et al., 2000)	Vitamin A supplementation	Decreased the increase in HIV viral load observed after influenza vaccination
Tanzania (Fawzi et al., 1999, 2000a; Villamor et al., 2002a)	Vitamin A supplementation	↓ All-cause mortality; ↓ risk of severe watery diarrhea; ↑ average length gain in HIV-infected children; ↓ risk of respiratory infection in HIV-infected children
Uganda (Semba et al., 2005)	Vitamin A supplementation	↓ Mortality
Tanzania (McGrath et al., 2006)	Maternal vitamin A supplementation	No effect on psychomotor development index or on the risk for developmental delay on the motor scale
	Maternal multivitamin (vitamins B, C, and E) supplementation	↑ Psychomotor development index score; ↓ risk of developmental delay on the motor scale
Zimbabwe (Humphrey et al., 2006)	Vitamin A supplementation	No effect on mortality in infants who were positive by PCR for HIV at baseline; ↓ mortality with neonatal supplementation in infants PCR negative at baseline but PCR positive at 6 weeks; ↑ mortality in infants who were PCR negative at 6 weeks

↑, Increase in; ↓, decrease in.

of infection may lead to decreased mobilization of vitamin A stores from the liver, resulting in a low serum level (Baeten et al., 2004; Filteau et al., 1993). Hence, low vitamin A levels may be a marker of advanced HIV disease stage, and this disease progression may be responsible for the increased risk of transmission observed. Reverse causality in the form of HIV disease itself adversely affecting absorption and metabolism of nutrients leading to biochemical deficiency might bias the observed relationship as well. Additionally, there may be residual confounding due to factors such as access to care, opportunistic infections, and other nutritional deficiencies, which were not assessed in these studies. It is also likely that the follow-up time and/or the stage of disease of vitamin A-deficient women differed from vitamin A-sufficient women and consequently they were more likely to develop AIDS and be excluded from the cohort (Brookmeyer et al., 1987). Another limitation of these studies is that they could not evaluate the relationship between maternal vitamin A status and the different routes of HIV transmission from mother-to-child, as the HIV status of the infant was determined at different times postnatally. Moreover, the women in the studies carried out in the United States were not breast-feeding—hence, the association between this route of transmission and maternal vitamin A levels could not be assessed (Dreyfuss and Fawzi, 2002).

The trials, on the other hand, have either found no effect or an increase in the risk of transmission of HIV with vitamin A supplementation. The findings from these trials raise concerns about the safety of maternal vitamin A supplementation programs as recommended by the World Health Organization (WHO/UNICEF/IVACG, 1997)—which are based on large trials such as the one in Nepal where investigators found that vitamin A supplementation halved maternal mortality in presumably HIV-negative pregnant women (West et al., 1999)—especially in areas where HIV infection is endemic (Fawzi, 2006). These findings also raise the question of how vitamin A may be leading to such an adverse effect. As mentioned earlier, vitamin A leads to an increased shedding of the virus from the lower genital tract (Fawzi et al., 2004a); however, the mechanism of this is not clear. It has also been hypothesized that vitamin A may lead to increased density of CCR5 receptors by increasing the multiplication and differentiation of lymphoid and myeloid cells; the CCR5 receptors are critical for the attachment of the virus to the lymphocytes as well as for subsequent replication of the virus (MacDonald et al., 2001). Vitamin A could also potentially modulate HIV replication because the virus genome contains a retinoic acid response element (Semba et al., 1998). It is also possible that the adverse effect noted in the Tanzania trial, at least, is due to the β-carotene component of the regimen. β-Carotene has been found to be relatively safe for short periods in HIV-infected individuals (Coodley et al., 1993; Nimmagadda et al., 1998); however, there has been no study of the safety of prolonged supplementation.

Unlike vitamin A, the evidence for a beneficial effect of multivitamin supplementation is consistent across observational studies and randomized trials. To reemphasize though, multivitamins are at best a complementary intervention in the care of the HIV-infected patient and not an alternative to antiretroviral therapy. A daily multivitamin is likely to reduce HIV disease progression and thus prolong the time before initiation antiretroviral therapy is recommended. This is likely to help preserve the antiretroviral drugs for later stages of disease, avert their adverse effects, and thus enhance compliance and quality of life of the HIV-infected patient.

Overall, the evidence reviewed suggests that daily supplementation with vitamin A or β-carotene to HIV-infected women is not to be advised.

However, a strong case can be made for supplementation of vitamins B, C, and E at multiples of the RDA among HIV-infected, pregnant, or lactating women. The results also support periodic vitamin A supplementation of HIV-positive infants and children starting at 6 months of age, as it is likely to prolong their survival and decrease morbidity associated with various childhood illnesses.

VII. FUTURE RESEARCH

Future research should address the role of other vitamins, for example, vitamin D, as well as determine the safety and the efficacy of multivitamin supplements among adults who are advanced in their disease and are receiving antiretroviral therapy, and among children. The protective effects of multivitamins on pregnancy and child outcomes among HIV-infected women may not be generalizable to HIV-negative women, and research among the HIV-negative pregnant and lactating women is needed. Similarly, more studies need to be conducted among men and children with multivitamin supplementation to assess any difference in response with reference to disease progression. The issue of whether single versus multiple RDAs of multivitamins have similar benefits is already being examined in a trial setting. Moreover, the conflicting findings from the various studies reinforce the need for international collaboration to define the baseline nutritional profiles and prevailing burden of infections in each area of the world to help us interpret the results of interventions in a region-relevant manner.

ACKNOWLEDGMENTS

We would like to express our gratitude to Molly Franke and Vasanti Malik for reviewing an earlier version of this chapter and for providing feedback on the same.

REFERENCES

Abe, E., Shiina, Y., Miyaura, C., Tanaka, H., Hayashi, T., Kanegasaki, S., Saito, M., Nishii, Y., DeLuca, H. F., and Suda, T. (1984). Activation and fusion induced by 1 alpha, 25-dihydroxyvitamin D3 and their relation in alveolar macrophages. *Proc. Natl. Acad. Sci. USA* **81**(22), 7112–7116.

Abrams, B., Duncan, D., and Hertz-Picciotto, I. (1993). A prospective study of dietary intake and acquired immune deficiency syndrome in HIV-seropositive homosexual men. *J. Acquir. Immune Defic. Syndr.* **6**(8), 949–958.

Allard, J. P., Aghdassi, E., Chau, J., Tam, C., Kovacs, C. M., Salit, I. E., and Walmsley, S. L. (1998). Effects of vitamin E and C supplementation on oxidative stress and viral load in HIV-infected subjects. *AIDS* **12**(13), 1653–1659.

Baeten, J. M., McClelland, R. S., Overbaugh, J., Richardson, B. A., Emery, S., Lavreys, L., Mandaliya, K., Bankson, D. D., Ndinya-Achola, J. O., Bwayo, J. J., and Kreiss, J. K. (2002). Vitamin A supplementation and human immunodeficiency virus type 1 shedding in women: Results of a randomized clinical trial. *J. Infect. Dis.* **185**(8), 1187–1191.

Baeten, J. M., Richardson, B. A., Bankson, D. D., Wener, M. H., Kreiss, J. K., Lavreys, L., Mandaliya, K., Bwayo, J. J., and McClelland, R. S. (2004). Use of serum retinol-binding protein for prediction of vitamin A deficiency: Effects of HIV-1 infection, protein malnutrition, and the acute phase response. *Am. J. Clin. Nutr.* **79**(2), 218–225.

Bar-Shavit, Z., Noff, D., Edelstein, S., Meyer, M., Shibolet, S., and Goldman, R. (1981). 1,25-Dihydroxyvitamin D3 and the regulation of macrophage function. *Calcif. Tissue Int.* **33**(6), 673–676.

Baum, M. K., Shor-Posner, G., Lu, Y., Rosner, B., Sauberlich, H. E., Fletcher, M. A., Szapocznik, J., Eisdorfer, C., Buring, J. E., and Hennekens, C. H. (1995). Micronutrients and HIV-1 disease progression. *AIDS* **9**(9), 1051–1056.

Baum, M. K., Shor-Posner, G., Lai, S., Zhang, G., Lai, H., Fletcher, M. A., Sauberlich, H., and Page, J. B. (1997). High risk of HIV-related mortality is associated with selenium deficiency. *J. Acquir. Immune. Defic. Syndr. Hum. Retrovirol.* **15**(5), 370–374.

Bendich, A. (1988). Antioxidant vitamins and immune responses. *In* "Nutrition and Immunology" (R. Chandra, Ed.), pp. 125–147. Liss, New York.

Bendich, A., and Cohen, M. (1988). B vitamins: Effects on specific and nonspecific immune responses. *In* "Nutrition and Immunology" (R. Chandra, Ed.), pp. 101–123. Liss, New York.

Brocklehurst, P., and French, R. (1998). The association between maternal HIV infection and perinatal outcome: A systematic review of the literature and meta-analysis. *Br. J. Obstet. Gynaecol.* **105**(8), 836–848.

Brookmeyer, R., Gail, M. H., and Polk, B. F. (1987). The prevalent cohort study and the acquired immunodeficiency syndrome. *Am. J. Epidemiol.* **126**(1), 14–24.

Burger, H., Kovacs, A., Weiser, B., Grimson, R., Nachman, S., Tropper, P., van Bennekum, A. M., Elie, M. C., and Blaner, W. S. (1997). Maternal serum vitamin A levels are not associated with mother-to-child transmission of HIV-1 in the United States. *J. Acquir. Immune Defic. Syndr. Hum. Retrovirol.* **14**(4), 321–326.

Burns, D. N., FitzGerald, G., Semba, R., Hershow, R., Zorrilla, C., Pitt, J., Hammill, H., Cooper, E. R., Fowler, M. G., and Landesman, S. (1999). Vitamin A deficiency and other nutritional indices during pregnancy in human immunodeficiency virus infection: Prevalence, clinical correlates, and outcome. Women and Infants Transmission Study Group. *Clin. Infect. Dis.* **29**(2), 328–334.

CDC (1981). Pneumocystis pneumonia—Los Angeles. *MMWR Morb. Mortal. Wkly. Rep.* **30**(21), 250–252.

Connor, E. M., Sperling, R. S., Gelber, R., Kiselev, P., Scott, G., O'Sullivan, M. J., VanDyke, R., Bey, M., Shearer, W., Jacobson, R. L., Jimenz, E., O'Neill, E., *et al.* (1994). Reduction of maternal-infant transmission of human immunodeficiency virus type 1 with zidovudine treatment. Pediatric AIDS Clinical Trials Group Protocol 076 Study Group. *N. Engl. J. Med.* **331**(18), 1173–1180.

Constans, J., Delmas-Beauvieux, M. C., Sergeant, C., Peuchant, E., Pellegrin, J. L., Pellegrin, I., Clerc, M., Fleury, H., Simonoff, M., Leng, B., and Conri, C. (1996). One-year antioxidant supplementation with beta-carotene or selenium for patients infected with human immunodeficiency virus: A pilot study. *Clin. Infect. Dis.* **23**(3), 654–656.

Coodley, G. O., Nelson, H. D., Loveless, M. O., and Folk, C. (1993). Beta-carotene in HIV infection. *J. Acquir. Immune Defic. Syndr.* **6**(3), 272–276.

Coodley, G. O., Coodley, M. K., Lusk, R., Green, T. R., Bakke, A. C., Wilson, D., Wachenheim, D., Sexton, G., and Salveson, C. (1996). Beta-carotene in HIV infection: An extended evaluation. *AIDS* **10**(9), 967–973.

Coutsoudis, A., Kiepiela, P., Coovadia, H. M., and Broughton, M. (1992). Vitamin A supplementation enhances specific IgG antibody levels and total lymphocyte numbers while improving morbidity in measles. *Pediatr. Infect. Dis. J.* **11**(3), 203–209.

Coutsoudis, A., Bobat, R. A., Coovadia, H. M., Kuhn, L., Tsai, W. Y., and Stein, Z. A. (1995). The effects of vitamin A supplementation on the morbidity of children born to HIV-infected women. *Am. J. Public Health* **85**(8 Pt. 1), 1076–1081.

Coutsoudis, A., Moodley, D., Pillay, K., Harrigan, R., Stone, C., Moodley, J., and Coovadia, H. M. (1997). Effects of vitamin A supplementation on viral load in HIV-1-infected pregnant women. *J. Acquir. Immune Defic. Syndr. Hum. Retrovirol.* **15**(1), 86–87.

Coutsoudis, A., Pillay, K., Spooner, E., Kuhn, L., and Coovadia, H. M. (1999). Randomized trial testing the effect of vitamin A supplementation on pregnancy outcomes and early mother-to-child HIV-1 transmission in Durban, South Africa. South African Vitamin A Study Group. *AIDS* **13**(12), 1517–1524.

Dabis, F., Msellati, P., Dunn, D., Lepage, P., Newell, M. L., Peckham, C., and Van de Perre, P. (1993). Estimating the rate of mother-to-child transmission of HIV. Report of a workshop on methodological issues Ghent (Belgium), 17–20 February 1992. The Working Group on Mother-to-Child Transmission of HIV. *AIDS* **7**(8), 1139–1148.

Dabis, F., Elenga, N., Meda, N., Leroy, V., Viho, I., Manigart, O., Dequae-Merchadou, L., Msellati, P., and Sombie, I. (2001). 18-Month mortality and perinatal exposure to zidovudine in West Africa. *AIDS* **15**(6), 771–779.

Dreyfuss, M. L., and Fawzi, W. W. (2002). Micronutrients and vertical transmission of HIV-1. *Am. J. Clin. Nutr.* **75**(6), 959–970.

Fawzi, W., Msamanga, G., Antelman, G., Xu, C., Hertzmark, E., Spiegelman, D., Hunter, D., and Anderson, D. (2004a). Effect of prenatal vitamin supplementation on lower-genital levels of HIV type 1 and interleukin type 1 beta at 36 weeks of gestation. *Clin. Infect. Dis.* **38**(5), 716–722.

Fawzi, W. W. (2006). The benefits and concerns related to vitamin a supplementation. *J. Infect. Dis.* **193**(6), 756–769.

Fawzi, W. W., Msamanga, G. I., Spiegelman, D., Urassa, E. J., McGrath, N., Mwakagile, D., Antelman, G., Mbise, R., Herrera, G., Kapiga, S., Willett, W., and Hunter, D. J. (1998). Randomised trial of effects of vitamin supplements on pregnancy outcomes and T cell counts in HIV-1-infected women in Tanzania. *Lancet* **351**(9114), 1477–1482.

Fawzi, W. W., Mbise, R. L., Hertzmark, E., Fataki, M. R., Herrera, M. G., Ndossi, G., and Spiegelman, D. (1999). A randomized trial of vitamin A supplements in relation to mortality among human immunodeficiency virus-infected and uninfected children in Tanzania. *Pediatr. Infect. Dis. J.* **18**(2), 127–133.

Fawzi, W. W., Mbise, R., Spiegelman, D., Fataki, M., Hertzmark, E., and Ndossi, G. (2000a). Vitamin A supplements and diarrheal and respiratory tract infections among children in Dar es Salaam, Tanzania. *J. Pediatr.* **137**(5), 660–667.

Fawzi, W. W., Msamanga, G., Hunter, D., Urassa, E., Renjifo, B., Mwakagile, D., Hertzmark, E., Coley, J., Garland, M., Kapiga, S., Antelman, G., Essex, M., *et al.* (2000b). Randomized trial of vitamin supplements in relation to vertical transmission of HIV-1 in Tanzania. *J. Acquir. Immune Defic. Syndr.* **23**(3), 246–254.

Fawzi, W. W., Msamanga, G. I., Hunter, D., Renjifo, B., Antelman, G., Bang, H., Manji, K., Kapiga, S., Mwakagile, D., Essex, M., and Spiegelman, D. (2002). Randomized trial of vitamin supplements in relation to transmission of HIV-1 through breastfeeding and early child mortality. *AIDS* **16**(14), 1935–1944.

Fawzi, W. W., Msamanga, G. I., Spiegelman, D., Wei, R., Kapiga, S., Villamor, E., Mwakagile, D., Mugusi, F., Hertzmark, E., Essex, M., and Hunter, D. J. (2004b). A randomized trial of multivitamin supplements and HIV disease progression and mortality. *N. Engl. J. Med.* **351**(1), 23–32.

Filteau, S. M., Morris, S. S., Abbott, R. A., Tomkins, A. M., Kirkwood, B. R., Arthur, P., Ross, D. A., Gyapong, J. O., and Raynes, J. G. (1993). Influence of morbidity on serum retinol of children in a community-based study in northern Ghana. *Am. J. Clin. Nutr.* **58**(2), 192–197.

Graham, N. M., Bulterys, M., Chao, A., Humphrey, J., Clement, L., Dushimimana, A., Kurawige, J. B., Flynn, C., and Saah, A. (1993). Effect of maternal vitamin A deficiency on infant mortality and perinatal HIV transmission. National Conference on Human Retroviruses and Related Infection. Johns Hopkins University, Baltimore.

Greenberg, B. L., Semba, R. D., Vink, P. E., Farley, J. J., Sivapalasingam, M., Steketee, R. W., Thea, D. M., and Schoenbaum, E. E. (1997). Vitamin A deficiency and maternal-infant transmissions of HIV in two metropolitan areas in the United States. *AIDS* **11**(3), 325–332.

Guay, L. A., Musoke, P., Fleming, T., Bagenda, D., Allen, M., Nakabiito, C., Sherman, J., Bakaki, P., Ducar, C., Deseyve, M., Emel, L., Mirochnick, M., et al. (1999). Intrapartum and neonatal single-dose nevirapine compared with zidovudine for prevention of mother-to-child transmission of HIV-1 in Kampala, Uganda: HIVNET 012 randomised trial. *Lancet* **354**(9181), 795–802.

Hanekom, W. A., Yogev, R., Heald, L. M., Edwards, K. M., Hussey, G. D., and Chadwick, E. G. (2000). Effect of vitamin A therapy on serologic responses and viral load changes after influenza vaccination in children infected with the human immunodeficiency virus. *J. Pediatr.* **136**(4), 550–552.

Haug, C., Muller, F., Aukrust, P., and Froland, S. S. (1994). Subnormal serum concentration of 1,25-vitamin D in human immunodeficiency virus infection: Correlation with degree of immune deficiency and survival. *J. Infect. Dis.* **169**(4), 889–893.

Hemila, H. (1997). Vitamin C and infectious diseases. In "Vitamin C in Health and Disease" (L. Pacler and J. Fuchs, Eds.), pp. 471–504. Marcel Dekker, Inc., New York.

Humphrey, J. H., Quinn, T., Fine, D., Lederman, H., Yamini-Roodsari, S., Wu, L. S., Moeller, S., and Ruff, A. J. (1999). Short-term effects of large-dose vitamin A supplementation on viral load and immune response in HIV-infected women. *J. Acquir. Immune. Defic. Syndr. Hum. Retrovirol.* **20**(1), 44–51.

Humphrey, J. H., Iliff, P. J., Marinda, E. T., Mutasa, K., Moulton, L. H., Chidawanyika, H., Ward, B. J., Nathoo, K. J., Malaba, L. C., Zijenah, L. S., Zvandasara, P., Ntozini, R., et al. (2006). Effects of a single large dose of vitamin A, given during the postpartum period to HIV-positive women and their infants, on child HIV infection, HIV-free survival, and mortality. *J. Infect. Dis.* **193**(6), 860–871.

Jackson, J. B., Musoke, P., Fleming, T., Guay, L. A., Bagenda, D., Allen, M., Nakabiito, C., Sherman, J., Bakaki, P., Owor, M., Ducar, C., Deseyve, M., et al. (2003). Intrapartum and neonatal single-dose nevirapine compared with zidovudine for prevention of mother-to-child transmission of HIV-1 in Kampala, Uganda: 18-month follow-up of the HIVNET 012 randomised trial. *Lancet* **362**(9387), 859–868.

Jiamton, S., Pepin, J., Suttent, R., Filteau, S., Mahakkanukrauh, B., Hanshaoworakul, W., Chaisilwattana, P., Suthipinittharm, P., Shetty, P., and Jaffar, S. (2003). A randomized trial of the impact of multiple micronutrient supplementation on mortality among HIV-infected individuals living in Bangkok. *AIDS* **17**(17), 2461–2469.

Kanter, A. S., Spencer, D. C., Steinberg, M. H., Soltysik, R., Yarnold, P. R., and Graham, N. M. (1999). Supplemental vitamin B and progression to AIDS and death in black South African patients infected with HIV. *J. Acquir. Immune Defic. Syndr.* **21**(3), 252–253.

Kaslow, R. A., Ostrow, D. G., Detels, R., Phair, J. P., Polk, B. F., and Rinaldo, C. R., Jr. (1987). The Multicenter AIDS Cohort Study: Rationale, organization, and selected characteristics of the participants. *Am. J. Epidemiol.* **126**(2), 310–318.

Kelly, P., Musonda, R., Kafwembe, E., Kaetano, L., Keane, E., and Farthing, M. (1999). Micronutrient supplementation in the AIDS diarrhoea-wasting syndrome in Zambia: A randomized controlled trial. *AIDS* **13**(4), 495–500.

Keusch, G. T., and Farthing, M. J. (1990). Nutritional aspects of AIDS. *Annu. Rev. Nutr.* **10**, 475–501.

Kumwenda, N., Miotti, P. G., Taha, T. E., Broadhead, R., Biggar, R. J., Jackson, J. B., Melikian, G., and Semba, R. D. (2002). Antenatal vitamin A supplementation increases birth weight and decreases anemia among infants born to human immunodeficiency virus-infected women in Malawi. *Clin. Infect. Dis.* **35**(5), 618–624.

MacDonald, K. S., Malonza, I., Chen, D. K., Nagelkerke, N. J., Nasio, J. M., Ndinya-Achola, J., Bwayo, J. J., Sitar, D. S., Aoki, F. Y., and Plummer, F. A. (2001). Vitamin A and risk of HIV-1 seroconversion among Kenyan men with genital ulcers. *AIDS* **15**(5), 635–639.

Mangili, A., Murman, D. H., Zampini, A. M., and Wanke, C. A. (2006). Nutrition and HIV infection: Review of weight loss and wasting in the era of highly active antiretroviral therapy from the nutrition for healthy living cohort. *Clin. Infect. Dis.* **42**(6), 836–842.

Mariani, E., Ravaglia, G., Forti, P., Meneghetti, A., Tarozzi, A., Maioli, F., Boschi, F., Pratelli, L., Pizzoferrato, A., Piras, F., and Facchini, A. (1999). Vitamin D, thyroid hormones and muscle mass influence natural killer (NK) innate immunity in healthy nonagenarians and centenarians. *Clin. Exp. Immunol.* **116**(1), 19–27.

Marum, L., Bagenda, D., Guay, L. A., Aceng, E., Kalyesubula, I., Tindyebwa, D., Ndugwa, C., and Olness, K. (1996). Three-year mortality in a cohort of HIV-1 infected and uninfected Ugandan children. XI International Conference on AIDS & STDs, Vancouver.

McClelland, R. S., Baeten, J. M., Overbaugh, J., Richardson, B. A., Mandaliya, K., Emery, S., Lavreys, L., Ndinya-Achola, J. O., Bankson, D. D., Bwayo, J. J., and Kreiss, J. K. (2004). Micronutrient supplementation increases genital tract shedding of HIV-1 in women: Results of a randomized trial. *J. Acquir. Immune Defic. Syndr.* **37**(5), 1657–1663.

McGrath, N., Bellinger, D., Robins, J., Msamanga, G. I., Tronick, E., and Fawzi, W. W. (2006). Effect of maternal multivitamin supplementation on the mental and psychomotor development of children who are born to HIV-1-infected mothers in Tanzania. *Pediatrics* **117**(2), e216–e225.

Mehendale, S. M., Shepherd, M. E., Brookmeyer, R. S., Semba, R. D., Divekar, A. D., Gangakhedkar, R. R., Joshi, S., Risbud, A. R., Paranjape, R. S., Gadkari, D. A., and Bollinger, R. C. (2001). Low carotenoid concentration and the risk of HIV seroconversion in Pune, India. *J. Acquir. Immune Defic. Syndr.* **26**(4), 352–359.

Merchant, A. T., Msamanga, G., Villamor, E., Saathoff, E., O'Brien, M., Hertzmark, E., Hunter, D. J., and Fawzi, W. W. (2005). Multivitamin supplementation of HIV-positive women during pregnancy reduces hypertension. *J. Nutr.* **135**(7), 1776–1781.

Meydani, S. N., Barklund, M. P., Liu, S., Meydani, M., Miller, R. A., Cannon, J. G., Morrow, F. D., Rocklin, R., and Blumberg, J. B. (1990). Vitamin E supplementation enhances cell-mediated immunity in healthy elderly subjects. *Am. J. Clin. Nutr.* **52**(3), 557–563.

Meydani, S. N., Ribaya-Mercado, J. D., Russell, R. M., Sahyoun, N., Morrow, F. D., and Gershoff, S. N. (1991). Vitamin B-6 deficiency impairs interleukin 2 production and lymphocyte proliferation in elderly adults. *Am. J. Clin. Nutr.* **53**(5), 1275–1280.

Meydani, S. N., Meydani, M., Blumberg, J. B., Leka, L. S., Siber, G., Loszewski, R., Thompson, C., Pedrosa, M. C., Diamond, R. D., and Stollar, B. D. (1997). Vitamin E supplementation and *in vivo* immune response in healthy elderly subjects. A randomized controlled trial. *JAMA* **277**(17), 1380–1386.

Moore, P. S., Allen, S., Sowell, A. L., Van de Perre, P., Huff, D. L., Serufilira, A., Nsengumuremyi, F., and Hulley, S. B. (1993). Role of nutritional status and weight loss in HIV seroconversion among Rwandan women. *J. Acquir. Immune Defic Syndr.* **6**(6), 611–616.

Nimmagadda, A. P., Burri, B. J., Neidlinger, T., O'Brien, W. A., and Goetz, M. B. (1998). Effect of oral beta-carotene supplementation on plasma human immunodeficiency virus (HIV) RNA levels and CD4 + cell counts in HIV-infected patients. *Clin. Infect. Dis.* **27**(5), 1311–1313.

Rogers, M. F., Caldwell, M. B., Gwinn, M. L., and Simonds, R. J. (1994). Epidemiology of pediatric human immunodeficiency virus infection in the United States. *Acta Paediatr. Suppl.* **400**, 5–7.

Ross, A. C., and Stephensen, C. B. (1996). Vitamin A and retinoids in antiviral responses. *FASEB J.* **10**(9), 979–985.
Schim van der Loeff, M. F., Hansmann, A., Awasana, A. A., Ota, M. O., O'Donovan, D., Sarge-Njie, R., Ariyoshi, K., Milligan, P., and Whittle, H. (2003). Survival of HIV-1 and HIV-2 perinatally infected children in The Gambia. *AIDS* **17**(16), 2389–2394.
Semba, R. D. (1998). The role of vitamin A and related retinoids in immune function. *Nutr. Rev.* **56**(1 Pt. 2), S38–S48.
Semba, R. D. (1999). Vitamin A and immunity to viral, bacterial and protozoan infections. *Proc. Nutr. Soc.* **58**(3), 719–727.
Semba, R. D., Muhilal, Scott, A. L., Natadisastra, G., Wirasasmita, S., Mele, L., Ridwan, E., West, K. P., Jr., and Sommer, A. (1992). Depressed immune response to tetanus in children with vitamin A deficiency. *J. Nutr.* **122**(1), 101–107.
Semba, R. D., Graham, N. M., Caiaffa, W. T., Margolick, J. B., Clement, L., and Vlahov, D. (1993). Increased mortality associated with vitamin A deficiency during human immunodeficiency virus type 1 infection. *Arch. Intern. Med.* **153**(18), 2149–2154.
Semba, R. D., Miotti, P. G., Chiphangwi, J. D., Saah, A. J., Canner, J. K., Dallabetta, G. A., and Hoover, D. R. (1994). Maternal vitamin A deficiency and mother-to-child transmission of HIV-1. *Lancet* **343**(8913), 1593–1597.
Semba, R. D., Caiaffa, W. T., Graham, N. M., Cohn, S., and Vlahov, D. (1995a). Vitamin A deficiency and wasting as predictors of mortality in human immunodeficiency virus-infected injection drug users. *J. Infect. Dis.* **171**(5), 1196–1202.
Semba, R. D., Miotti, P. G., Chiphangwi, J. D., Liomba, G., Yang, L. P., Saah, A. J., Dallabetta, G. A., and Hoover, D. R. (1995b). Infant mortality and maternal vitamin A deficiency during human immunodeficiency virus infection. *Clin. Infect. Dis.* **21**(4), 966–972.
Semba, R. D., Lyles, C. M., Margolick, J. B., Caiaffa, W. T., Farzadegan, H., Cohn, S., and Vlahov, D. (1998). Vitamin A supplementation and human immunodeficiency virus load in injection drug users. *J. Infect. Dis.* **177**(3), 611–616.
Semba, R. D., Ndugwa, C., Perry, R. T., Clark, T. D., Jackson, J. B., Melikian, G., Tielsch, J., and Mmiro, F. (2005). Effect of periodic vitamin A supplementation on mortality and morbidity of human immunodeficiency virus-infected children in Uganda: A controlled clinical trial. *Nutrition* **21**(1), 25–31.
Spira, R., Lepage, P., Msellati, P., Van De Perre, P., Leroy, V., Simonon, A., Karita, E., and Dabis, F. (1999). Natural history of human immunodeficiency virus type 1 infection in children: A five-year prospective study in Rwanda. Mother-to-Child HIV-1 Transmission Study Group. *Pediatrics* **104**(5), e56.
Stephensen, C. B. (2001). Vitamin A, infection, and immune function. *Annu. Rev. Nutr.* **21**, 167–192.
Tang, A. M., Graham, N. M., Kirby, A. J., McCall, L. D., Willett, W. C., and Saah, A. J. (1993). Dietary micronutrient intake and risk of progression to acquired immunodeficiency syndrome (AIDS) in human immunodeficiency virus type 1 (HIV-1)-infected homosexual men. *Am. J. Epidemiol.* **138**(11), 937–951.
Tang, A. M., Graham, N. M., and Saah, A. J. (1996). Effects of micronutrient intake on survival in human immunodeficiency virus type 1 infection. *Am. J. Epidemiol.* **143**(12), 1244–1256.
Tang, A. M., Graham, N. M., Chandra, R. K., and Saah, A. J. (1997a). Low serum vitamin B-12 concentrations are associated with faster human immunodeficiency virus type 1 (HIV-1) disease progression. *J. Nutr.* **127**(2), 345–351.
Tang, A. M., Graham, N. M., Semba, R. D., and Saah, A. J. (1997b). Association between serum vitamin A and E levels and HIV-1 disease progression. *AIDS* **11**(5), 613–620.
Teichmann, J., Stephan, E., Lange, U., Discher, T., Friese, G., Lohmeyer, J., Stracke, H., and Bretzel, R. G. (2003). Osteopenia in HIV-infected women prior to highly active antiretroviral therapy. *J. Infect.* **46**(4), 221–227.

Thea, D. M., St. Louis, M. E., Atido, U., Kanjinga, K., Kembo, B., Matondo, M., Tshiamala, T., Kamenga, C., Davachi, F., Brown, C., Rand, W. M., and Keusch, G. T. (1993). A prospective study of diarrhea and HIV-1 infection among 429 Zairian infants. *N. Engl. J. Med.* **329**(23), 1696–1702.

UNAIDS (1998). Mother to child transmission of HIV: UNAIDS technical update. UNAIDS, Geneva.

UNAIDS (2005). AIDS Epidemic Update: 2005. World Health Organization, Geneva.

Villamor, E., and Fawzi, W. W. (2005). Effects of vitamin a supplementation on immune responses and correlation with clinical outcomes. *Clin. Microbiol. Rev.* **18**(3), 446–464.

Villamor, E., Mbise, R., Spiegelman, D., Hertzmark, E., Fataki, M., Peterson, K. E., Ndossi, G., and Fawzi, W. W. (2002a). Vitamin A supplements ameliorate the adverse effect of HIV-1, malaria, and diarrheal infections on child growth. *Pediatrics* **109**(1), E6.

Villamor, E., Msamanga, G., Spiegelman, D., Antelman, G., Peterson, K. E., Hunter, D. J., and Fawzi, W. W. (2002b). Effect of multivitamin and vitamin A supplements on weight gain during pregnancy among HIV-1-infected women. *Am. J. Clin. Nutr.* **76**(5), 1082–1090.

Villamor, E., Saathoff, E., Manji, K., Msamanga, G., Hunter, D. J., and Fawzi, W. W. (2005). Vitamin supplements, socioeconomic status, and morbidity events as predictors of wasting in HIV-infected women from Tanzania. *Am. J. Clin. Nutr.* **82**(4), 857–865.

Villamor, E., Kapiga, S. H., and Fawzi, W. W. (2006). Vitamin A serostatus and heterosexual transmission of HIV: Case-control study in Tanzania and review of the evidence. *Int. J. Vitam. Nutr. Res.* **76**(2), 81–85.

Visser, M. E., Maartens, G., Kossew, G., and Hussey, G. D. (2003). Plasma vitamin A and zinc levels in HIV-infected adults in Cape Town, South Africa. *Br. J. Nutr.* **89**(4), 475–482.

Wang, Y., Huang, D. S., Liang, B., and Watson, R. R. (1994). Nutritional status and immune responses in mice with murine AIDS are normalized by vitamin E supplementation. *J. Nutr.* **124**(10), 2024–2032.

Wang, Y., Huang, D. S., Wood, S., and Watson, R. R. (1995). Modulation of immune function and cytokine production by various levels of vitamin E supplementation during murine AIDS. *Immunopharmacology* **29**(3), 225–233.

West, K. P., Jr., Katz, J., Khatry, S. K., LeClerq, S. C., Pradhan, E. K., Shrestha, S. R., Connor, P. B., Dali, S. M., Christian, P., Pokhrel, R. P., and Sommer, A. (1999). Double blind, cluster randomised trial of low dose supplementation with vitamin A or beta carotene on mortality related to pregnancy in Nepal. The NNIPS-2 Study Group. *BMJ* **318**(7183), 570–575.

WHO/UNICEF/IVACG (1997). Vitamin A supplements: A guide to their use in the treatment and prevention of vitamin A deficiency and xerophthalmia. WHO, Geneva.

Wiktor, S. Z., Ekpini, E., Karon, J. M., Nkengasong, J., Maurice, C., Severin, S. T., Roels, T. H., Kouassi, M. K., Lackritz, E. M., Coulibaly, I. M., and Greenberg, A. E. (1999). Short-course oral zidovudine for prevention of mother-to-child transmission of HIV-1 in Abidjan, Cote d'Ivoire: A randomised trial. *Lancet* **353**(9155), 781–785.

Winkelstein, W., Jr., Lyman, D. M., Padian, N., Grant, R., Samuel, M., Wiley, J. A., Anderson, R. E., Lang, W., Riggs, J., and Levy, J. A. (1987). Sexual practices and risk of infection by the human immunodeficiency virus. The San Francisco Men's Health Study. *JAMA* **257**(3), 321–325.

Winklhofer-Roob, B. M., Ellemunter, H., Fruhwirth, M., Schlegel-Haueter, S. E., Khoschsorur, G., van't Hof, M. A., and Shmerling, D. H. (1997). Plasma vitamin C concentrations in patients with cystic fibrosis: Evidence of associations with lung inflammation. *Am. J. Clin. Nutr.* **65**(6), 1858–1866.

Yang, S., Smith, C., Prahl, J. M., Luo, X., and DeLuca, H. F. (1993). Vitamin D deficiency suppresses cell-mediated immunity *in vivo*. *Arch. Biochem. Biophys.* **303**(1), 98–106.

14

Vitamin A and Emphysema

Richard C. Baybutt* and Agostino Molteni[†]

*Department of Human Nutrition, Kansas State University
Manhattan, Kansas 66506
[†]Department of Pathology and Pharmacology, University of Missouri
Kansas City Medical School, Kansas City, Missouri 64108

I. Does Vitamin A Protect Against
   Pulmonary Emphysema?
   A. Vitamin A and Its Functions
   B. Vitamin A and the Lung
   C. Vitamin A and the Type II Pneumocyte
   D. Emphysema
   E. Emphysema and Vitamin A
   F. Cigarette Smoking, Vitamin A Deficiency, and the
      Development of Emphysema
   G. Emphysema, Vitamin A, and Elastin Metabolism
   H. Mechanisms for Protection Against Emphysema by
      Vitamin A
   I. Vitamin A, Lung Inflammation, and Emphysema
   J. Evidence for Lung Restoration by Vitamin A
II. Conclusions
   References

Within the last several years, research scientists and clinicians have been intrigued with the potential use of an active form of vitamin A, retinoic acid (RA), for the treatment and prevention of emphysema. The interest in this area can be largely attributed to the work of Massaro and Massaro (1996, 1997, 2000) in which they presented evidence that RA partially protects against and to some degree restores elastase-induced emphysema in rats. The mechanism for this protective effect of RA is in part related to elastin metabolism. RA also inhibits inflammation, an upstream event that may lead to the development of emphysema. Although there is evidence of this protective effect in young rats and a mechanistic explanation, more studies are needed in humans in order to establish a role for vitamin A in protecting against emphysema. Too many unanswered questions remain to definitively state that vitamin A protects against this disease in humans. Nevertheless, the potential for this novel approach in prevention and treatment of emphysema is an exciting area of research. © 2007 Elsevier Inc.

I. DOES VITAMIN A PROTECT AGAINST PULMONARY EMPHYSEMA?

A. VITAMIN A AND ITS FUNCTIONS

Vitamin A is a fat-soluble nutrient, and its derivatives have a number of already established functions, as well as new functions currently being discovered. The name vitamin A is synonymous with the alcohol form of the vitamin retinol. It is oxidized into retinal and further oxidized to RA which can exist as all-*trans* or several *cis* isomers. Besides the well-established role of retinal for vision, vitamin A in the form of RA exerts its effects via nuclear receptors that act as transcription factors for changing gene expression.

Carotenoids are a family of compounds of which some can form vitamin A. One of the carotenoids, β-carotene, is converted into retinal and eventually retinol by an intestinal central cleavage enzyme. As the activity of this enzyme is very low in humans, very small amounts are converted into vitamin A. The conversion ratio is about 12:1 (β-carotene:retinol). An appropriate animal model for the study of β-carotene metabolism in humans is the ferret. Thus, if one wants to test the effect of vitamin A in humans, then retinol or its derivatives should be used as opposed to β-carotene. Orally supplemented, purified β-carotene cannot be used in place of the dietary retinoid form for human studies because of the potential deleterious side products that are formed with excess β-carotene. When isolated β-carotene is ingested in excess by ferrets, abnormal cleavage products are formed, along with the typical intestinal central cleavage enzyme product, retinal, a vitamin A metabolite (Wang *et al.*, 1999). The atypical acentric products that are also formed have been found to interfere with normal retinol metabolism and potentially cause a functional deficiency. In the ferret model, cigarette smoking increased the production of these abnormal metabolites and decreased retinoid signaling.

There are many functions of vitamin A. Along with vision, vitamin A is important for reproduction, immune function, growth, bone development, cellular differentiation, proliferation, and cell signaling. Vitamin A is particularly important for normal epithelial cell (surface cell) function and integrity. These epithelial cells affected by vitamin A include the intestinal cells, skin cells, urogenital cells, and lung cells. Within the last 20 years, there has been an increased interest in vitamin A, as discoveries have highlighted the importance of vitamin A in maintenance of normal lung function and prevention of injury.

B. VITAMIN A AND THE LUNG

Adequate dietary intake of vitamin A is essential in preserving the integrity of the lung epithelium. Pulmonary anomalies observed in vitamin A-deficient rats resemble those found in premature infants that have inadequate stores of the vitamin (McMenamy and Zachman, 1993). Several studies have attempted to define the role of vitamin A in the tracheal epithelial cells (Klann and Marchok, 1982; Lancillotti *et al.*, 1992; McDowell *et al.*, 1984, 1987). These studies suggest that regeneration of the tracheal epithelium after injury requires vitamin A-dependent mucous cell proliferation. In vitamin A-deficient rats, the proliferation of the mucous cells is inhibited, which leads to the development of squamous metaplasia (Lancillotti *et al.*, 1992; McDowell *et al.*, 1984, 1987). More information is needed regarding the role of vitamin A in protecting against tissue pathologies in the upper and lower respiratory tract.

C. VITAMIN A AND THE TYPE II PNEUMOCYTE

There are critical cell types within the lung that facilitate and mediate lung cell repair. RA, an active metabolite of vitamin A, stimulates cell growth and proliferation in cultured tracheal cells (Klann and Marchok, 1982; Lancillotti *et al.*, 1992). In the lower part of the respiratory tract, the mechanism responsible for maintaining alveolar integrity stems in part from vitamin A-directed proliferation of type II pneumocytes in response to injury (Takahashi *et al.*, 1993). The type II pneumocyte not only plays a critical role in the lung by producing surfactant, a mixture of phospholipids and proteins that reduces surface tension in the air spaces and maintains alveolar patency, but it also serves as a progenitor for the type I pneumocyte, which is the major resident of the alveolar wall and therefore important for normal lung maintenance. Because of its large surface area, the type I pneumocyte is more susceptible to injury and death by environmental toxins. When the type I pneumocyte dies, the type II pneumocyte is thought to proliferate. Then one of its daughter cells differentiates into a type I pneumocyte and replaces the injured cell. In vitamin A deficiency, type II pneumocyte proliferation is inhibited and lung cell repair is obstructed in ozone-injured lung cells (Takahashi et al., 1993). When type II pneumocytes are cultured, RA stimulates at least two conditions necessary for cell repair: that of cell proliferation (Nabeyrat et al., 1998) and increased availability of polyamines through enhanced synthesis and uptake (Heger and Baybutt, 1999). Polyamines are organic polycations that are thought to act as a scaffolding to hold negatively charged DNA in such a way as to facilitate and enable cell proliferation (Pegg, 1986). Thus, type II pneumocytes depend on vitamin A in maintenance of lung tissue integrity and in mediating lung cell repair.

D. EMPHYSEMA

When there is a defect in the cell repair mechanisms of the lung, large air spaces may be created and lung function compromised. Emphysema is a destructive lung disease of the pulmonary parenchyma characterized by large air spaces in the lungs, which are associated with a small number of alveoli. In 1964, the US Surgeon General first reported a potential relationship between smoking and emphysema, and this connection was strengthened by numerous epidemiological and animal studies (US Surgeon-General, 1984). Having now established that cigarette smoking is the leading cause of emphysema, the focus has shifted to determining how smoke-induced emphysema develops.

Among the proposed mechanisms for the development of emphysema, the protease/antiprotease and oxidant/antioxidant hypotheses are popular. The major tenant of the protease/antiprotease theory is that cigarette smoking produces an imbalance between antiprotease, α-1 antitrypsin, and elastase, favoring elastase activity. It is thought that the α-1 antitrypsin protein is more susceptible to destruction by the cigarette smoke components than is the elastase. The active elastase catabolizes the matrix protein elastin that eventually leads to a reduced number of alveoli and the creation of large air spaces or areas of emphysema.

The basis for the oxidant/antioxidant hypothesis is that the oxidants in the cigarette smoke deplete the antioxidant supply in the lung and cause oxidative injury to the tissues, leading to emphysema. During exposure to cigarette smoke, large amounts of oxygen free radicals are generated which could damage lipid components of cell membranes as well as matrix components of the lung (Janoff et al., 1987). Destruction of lung matrix, especially elastin, leads to emphysema. The precise chemical and biochemical mechanism for the injuries induced by cigarette smoke are not yet thoroughly explored.

E. EMPHYSEMA AND VITAMIN A

An alternative hypothesis for the development of cigarette smoke-induced emphysema attributes the disease to consequences of vitamin A deficiency. There is evidence that suggests that cigarette smoke induces vitamin A deficiency and that this deficiency leads to emphysema. Support for this hypothesis is based on a series of related experiments. Weanling rats fed a vitamin A-deficient diet for 6 weeks with no smoke exposure develop emphysema (Baybutt *et al.*, 2000). These pathological effects of vitamin A deficiency did not appear to be due to malnutrition because there was no difference in the average food intake or average body weight between rats fed the vitamin A-deficient diet and those fed the vitamin A-adequate diet for the length of this study. These results suggest that vitamin A deficiency per se creates emphysemic lungs with the histological data presented in Fig. 1. Another novel and interesting result from this study was that within the same lung of the vitamin A-deficient rat, there were areas of inflammation along with areas of emphysema. It is likely that the areas of inflammation precede and eventually become the areas of emphysema.

The development of emphysema observed in the lungs of vitamin A-deficient rats is consistent with other studies. In an elastase-induced model of emphysema in rats, Massaro and Massaro (1996, 1997) found that development of emphysema could be prevented and to some extent reversed by the administration of all-*trans* RA. They also have shown an important role for RA in alveolar formation during development (Massaro and Massaro, 2001). RA initiates septation or formation of smaller and more numerous gas-exchange saccules or alveoli and increases the total number of alveoli relative to untreated rats (Massaro and Massaro, 2000). These studies suggest that vitamin A plays a critical role in lung development, maintenance of lung epithelium, and likely prevention or possible treatment of emphysema.

F. CIGARETTE SMOKING, VITAMIN A DEFICIENCY, AND THE DEVELOPMENT OF EMPHYSEMA

The overwhelming majority of cases of emphysema have been linked to cigarette smoking. Because vitamin A deficiency induces emphysema (Baybutt *et al.*, 2000), the emphysema resulting from cigarette smoke could be the consequence of a localized vitamin A deficiency of the lungs. In support of this hypothesis, there is evidence that cigarette smoking leads to vitamin A deficiency. Edes and Gysbers (1993) reported that feeding rats benzopyrene, a constituent in cigarette smoke, depleted the vitamin A content of the lungs and liver. As previously noted, vitamin A-deficient lungs produce areas of emphysema (Baybutt *et al.*, 2000). Therefore, the emphysema incurred from cigarette smoke could be the consequence of a localized vitamin A deficiency in the lungs. Further support for this hypothesis is reported in a study by

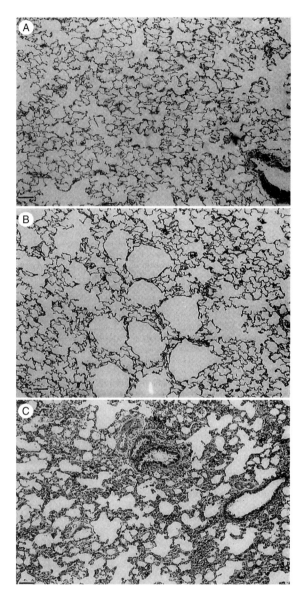

FIGURE 1. Vitamin A deficiency induces emphysema and lung inflammation. Representative section of the lung (A) of a rat receiving a diet with an adequate intake of vitamin A, and different areas from the lung of a rat fed a vitamin A-deficient diet (B and C). (A) shows a normal appearance, and (B) shows emphysematous areas of dilation of many alveolar spaces, extensive destruction of the septal walls, and thinning of the walls in other areas. Within the same lungs at different locations were areas of interstitial pneumonia (C), with distortion and/or reduction of the alveolar spaces. Small bronchi show presence of necrotic material and inflammatory cells in their lumen with partial disepithelization. Staining: H&E Magnification: 100× Bar (lower left corner) = 100 μm. Adapted from Baybutt et al. (2000).

Li *et al.* (2003) in which rats were exposed to cigarette smoke from 20 nonfiltered commercial cigarettes/day in a smoke chamber for 5 days/week while the control group was exposed to air. After 6 weeks, vitamin A levels significantly decreased in the serum, lung, and liver of smoke-treated rats and produced areas of emphysema in the deficient rats (Fig. 2). It is interesting to note the similarities of the lung pathologies between the vitamin A-deficient lungs caused by a deficient diet and the vitamin A-deficient lungs caused by cigarette smoke. Both conditions produced lungs with areas of inflammation and areas of emphysema (compare Figs. 1 and 2).

There also appears to be a connection between poor vitamin A status and emphysema. For smoke-induced emphysema, the degree of emphysema is inversely related to the vitamin A content of the lung (Li *et al.*, 2003). This study provided the first evidence that cigarette smoke inhalation by rats decreases lung and liver retinol levels, for there was no significant difference in mean daily food intake. Therefore, the depletion appeared to be largely attributed to the cigarette smoke. Others have found that intraperitoneal administration of tobacco extract or a tobacco constituent N'-nitrosonornicotine decreased the hepatic pool of vitamin A (Ammigan *et al.*, 1990). When adult ferrets are exposed to cigarette smoke, retinoid catabolism increases, with the result of significantly lower levels of RA in the lung (Liu *et al.*, 2000; Wang *et al.*, 1999); however, lung retinol concentrations are not significantly altered in this study. This apparent discrepancy can be explained by the age of the animal. The adult ferret used in this study would have ample supply of vitamin A or retinol stored, which could prevent the adult ferret from becoming vitamin A deficient.

It should be noted that vitamin A deficiency-induced emphysema was observed in growing rats (Baybutt *et al.*, 2000) when the development of the lung was still occurring. Whether the induction of vitamin A deficiency occurs in the adult animal or in humans in response to cigarette smoke remains to be determined. The precise role of vitamin A in smoke-induced emphysema is not known.

The amount of smoke exposure may be another important determinant for the smoke-induced vitamin A depletion. When weanling guinea pigs were exposed to a low dose of six cigarettes per day for 6 weeks, the levels of retinol in the lung increased (Mukherjee *et al.*, 1995). In contrast, when weanling rats were exposed to a higher dose of 20 cigarettes per day, lung retinol decreased after 6 weeks.

The effects of cigarette smoke on vitamin A status in humans is less clear and more difficult to determine. Results from cross-sectional studies indicated an inverse relationship between plasma retinol and degree of airway obstruction (Morabia *et al.*, 1990; Paiva *et al.*, 1996). Some studies with smoking pregnant women have reported decreased serum levels of retinol (Chelchowska *et al.*, 2001; Laskowska-Klita *et al.*, 1999); while in other studies, cigarette smoking did not affect serum retinol levels

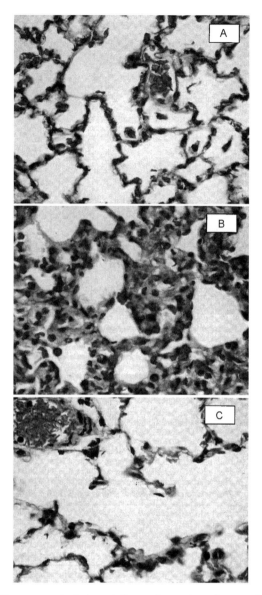

FIGURE 2. Cigarette smoke induces inflammation and emphysema. (A) Representative histological section of the lung of a control rat. No abnormalities were found. (B) Representative section of an area in the lung of a rat in response to cigarette smoke exposure for 6 weeks. The alveolar septa were significantly thickened with infiltration of inflammatory cells. Diffuse interstitial pneumonia was present. Large number of red blood cells are spread in the septa and the alveolar lumen. (C) Histological section of another area of the lung of the same cigarette smoke-exposed rat depicted in Fig. 2B. The alveolar wall was thinner than normal. Emphysematous area of dilation of many alveolar spaces, destruction of the septa wall was evident. Red blood cell and few chronic inflammatory cells were also present. Staining: Trichrome; Magnification: 400×. Adapted from Li et al. (2003).

(al Senaidy *et al.*, 1997; Lim *et al.*, 2001), or even increased serum retinol levels (Pamuk *et al.*, 1994). It is important to note that serum retinol is not an accurate measure for vitamin A status because most of this vitamin is stored in the liver. Methods have been developed for determining body content of vitamin A in humans using the dilution of an injected deuterated retinol (Furr *et al.*, 1989; Haskell *et al.*, 1997). Such methods have not been used for smokers and may help define the relationship between cigarette smokers and vitamin A status.

The mechanism for the cigarette smoke-induced vitamin A depletion and/or impaired function of vitamin A is not known. It may be due to induction of the cytochrome P450 system. Exposure to cigarette smoke increases lung and liver cytochrome P450 isoforms CYP1A1 and CYP1A2 (Liu *et al.*, 2003; Villard *et al.*, 1988), which increases catabolism of RA (Liu *et al.*, 2003; McSorley and Daly, 2000) and may lead to vitamin A depletion. Another mechanism for the cigarette smoke-induced vitamin A depletion may be that the cigarette smoke constituents block cellular uptake of retinol and create a functional deficiency. The cigarette smoke constituent benzopyrene inhibits uptake of retinol by tracheal cells and therefore could block its function (Biesalski and Engel, 1996).

G. EMPHYSEMA, VITAMIN A, AND ELASTIN METABOLISM

Another characteristic of emphysema is a decreased elastic recoil after exhalation because of a decrease in the amount of the matrix protein elastin, which provides structural framework and elasticity for the lung. There is less elastin staining detected in the lungs of vitamin A-deficient rats (Baybutt *et al.*, 2000). Furthermore, cigarette smoking that has been found to induce vitamin A deficiency (Li *et al.*, 2003) impairs elastin synthesis in the lungs of hamsters in an elastase-induced emphysema model (Osman *et al.*, 1985). Decreased elastin staining can also be observed in the lungs of rat fetuses from vitamin A-deficient mothers and is associated with a decreased elastin mRNA in the fetal lung (Antipatis *et al.*, 1998). In contrast, the levels of RA, RA receptor, and cellular retinol-binding protein are highest in lung interstitial fibroblasts during the time of maximal elastin synthesis, when extensive enlargement of the alveolar surface area occurs (McGowan *et al.*, 1995). Despite the uncertainty of the precise mechanism for elastin preservation, vitamin A maintains elastin levels in the lung.

H. MECHANISMS FOR PROTECTION AGAINST EMPHYSEMA BY VITAMIN A

Several mechanisms have been proposed to describe how vitamin A may protect against emphysema. Progressive breakdown of elastin within the alveoli, which is a key feature in development of emphysema (Cardoso *et al.*, 1993), is thought to be caused by excessive activity of elastase (Cichy *et al.*, 1997) and/or metalloproteinases (Seagrave, 2000). Cigarette smoke significantly increases lung and lavage macrophage elastolytic activity of smoke-exposed guinea pigs (Sansores *et al.*, 1997). RA inhibits the activity of human leukocyte elastase (Sklan *et al.*, 1990) and metalloproteinase activity (Varani *et al.*, 1994). During systemic infection or inflammation, increased neutrophil elastase activity is regulated by α1-proteinase inhibitor (α1-PI) that contributes most of the functional anti-elastase protection of the alveolar wall (Gadek *et al.*, 1981). Retinol and retinaldehyde, but not RA, increase α1-protease inhibitor in corneal epithelial cells (Boskovic and Twining, 1997). Thus, vitamin A could effectively inhibit breakdown of elastin.

Vitamin A not only protects against elastin degradation but also promotes elastin synthesis. In chick embryonic fibroblasts, RA stimulates elastin synthesis, increasing both elastin mRNA and protein (Tajima *et al.*, 1997). RA, but not retinol, increases transcription of the elastin precursor, tropoelastin (Hayashi *et al.*, 1995; Liu *et al.*, 1993; Tajima *et al.*, 1997) and promotes elastin protein synthesis (Hayashi *et al.*, 1995). This elastin induction appears to be mediated through the nuclear RA receptor gamma (RARγ) as evidenced by an increased content of RA paralleled with an increased RARγmRNA (McGowan *et al.*, 1995). Furthermore, in RARγ knockout mice, tropoelastin mRNA decreases, as does the number of alveoli, and their lungs morphologically resemble emphysema (McGowan *et al.*, 2000). On the basis of a number of studies, it appears that RA preserves the levels of elastin in a number of ways.

I. VITAMIN A, LUNG INFLAMMATION, AND EMPHYSEMA

Lung inflammation typically precedes the development of emphysema. The inflammatory cells release proteolytic enzymes that eventually degrade the matrix proteins, elastin and collagen, and eventually create areas of emphysema. In the lungs of rats that have developed emphysema there are also areas of inflammation in the same lung as observed in vitamin A deficiency and smoke exposure (Figs. 1 and 2). One way that vitamin A can protect against emphysema is by inhibiting inflammation.

Vitamin A has been found to exert anti-inflammatory properties within the lung in a number of different animal models. In one such model, monocrotaline, a pyrrolizadine alkaloid, is injected subcutaneously in rats to induce pulmonary inflammation. When monocrotaline-treated rats are fed a diet with eight times the amount of retinol of the control diet, the inflammatory response is significantly inhibited in the lung parenchyma and pulmonary arterioles (Swamidas *et al.*, 1999). Dietary retinol used in this study decreases the inflammatory responses within the alveolar septa, the vasculature, and the cardiac tissue, indicating a common anti-inflammatory effect of retinol on the response to monocrotaline toxicity. Another anti-inflammatory role for vitamin A was reported (Redlich et al., 1998), in which vitamin A reduced lung inflammation after thoracic radiation. Other studies have shown that high doses of retinol are anti-inflammatory in the lung of rats treated with 1-nitronaphthalene (Sauer et al., 1995) and bleomycin (Habib et al., 1993). In addition, a similar response is observed when rats are fed an enriched β-carotene diet (Baybutt and Molteni, 1999), as illustrated in Fig. 3. The observed effects of β-carotene likely are due to retinol because much of the ingested β-carotene is converted to retinol in the rat intestine (Wang, 1994). The rat is an efficient converter of β-carotene into retinol (a ratio of about 2:1).

Anti-inflammatory properties are also evident in a vitamin A-deficient model (Fig. 1). When weanling rats are deprived of dietary vitamin A for 6 weeks, lung inflammation occurs in certain areas of the lung, while emphysema occurs in other areas of the same lung (Baybutt et al., 2000). Thus, without vitamin A lung inflammation increases. The increased presence of pulmonary inflammation and emphysema in cigarette smokers may be related to vitamin A status.

Exposing rats to cigarette smoke creates another useful model for inducing lung inflammation. Cigarette smoke exposure depletes the anti-inflammatory vitamin A in the lung and increases pulmonary inflammation and emphysema in rats (Li et al., 2003). Providing additional dietary vitamin A inhibits the inflammatory response and thus the emphysema. As mentioned previously, provitamin A carotenoids do not always substitute for vitamin A and its functions.

At present, the precise mechanism underlying the anti-inflammatory effect of retinol is not known. However, one possible mechanism may involve the neutrophil, an important mediator and promoter of inflammation. Retinol has been shown to moderate the activity of the neutrophil in a number of different ways. Retinol inhibits the release of the superoxide anion (the oxygen free radical) that initiates the inflammatory response (Camisa et al., 1982; Sharma et al., 1990). In addition, retinol inhibits the conversion of arachidonic acid to leukotriene B_4 (Randall et al., 1987), which acts as a chemoattractant, amplifying the inflammatory response through recruitment of other neutrophils. Consistent with such a role of retinol, vitamin A deficiency in rats resulted in a 43% increase in leukocytes, over half of which are neutrophils (Wiedermann et al., 1996). Also vitamin A-deficient mice exhibited an enhanced inflammatory response to ozone (Paquette et al., 1996). Some investigators have found that inflammation accelerates depletion of lung retinol, resulting in a localized deficiency (Kanda et al., 1990). Dietary supplemental vitamin A may help prevent the localized deficiency, thereby blunting the response of the neutrophils and other inflammatory cells. The biochemical and cellular observations cited above clearly suggest that retinol plays a critical role in suppressing inflammatory responses.

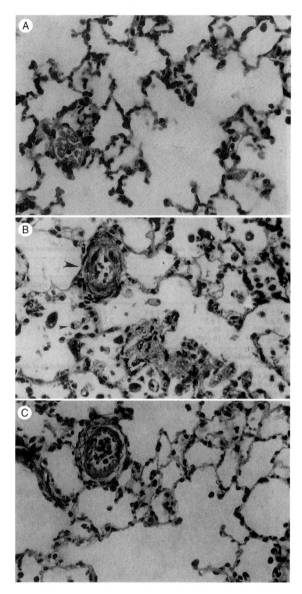

FIGURE 3. Histological sections of lungs of rats in response to dietary and MCT treatment. Staining: Masson Trichrome and Magnification: 600×. (A) Representative section of the lung of a control rat. No significant abnormalities. The β-carotene-supplemented rats did not differ from the rats fed the control diet (not shown). (B) Representative section of the lung of a rat receiving MCT. Severe and widespread inflammatory reaction with large number of inflammatory cells and macrophages (small black arrowhead). Red cells are spread in the septa and the alveolar lumen (large white arrowhead). Septal walls are thicker with increased collagen deposition (small white arrow). A small artery shows narrowing of the lumen and increased thickness of the media with significant collagen deposition and presence of inflammatory cells

J. EVIDENCE FOR LUNG RESTORATION BY VITAMIN A

At least three research groups, Massaro and Massaro (1997), Belloni *et al.* (2000), and Tepper *et al.* (2000), have found that intraperitoneally administered RA (all-*trans* and 9-*cis*) provide some restoration of lung epithelium in rats that have been treated intratracheally with elastase. These consistent protective effects of RA against elastase-induced emphysema observed in rats have not been observed in some other animal models. For younger mice less than 2 months of age, RA protects against elastase-induced emphysema (Massaro and Massaro, 2000), but does not protect in one study using older mice (Fujita *et al.*, 2006). In another study in cigarette smoke-exposed guinea pigs, RA did not prevent the development of emphysema (Meshi *et al.*, 2002). There are many factors to consider in evaluating the effectiveness of vitamin A in protecting against or reversing emphysematous lungs. Clearly more studies are needed to delineate the role of vitamin A in protecting against emphysema in animal studies.

There have been two published clinical studies that tested the effectiveness of RA in restoring airway tissue of former cigarette smokers (Kurie *et al.*, 2003; Mao *et al.*, 2002). In one of the studies, a high dose of RA (100 mg/person/day) was effective in reducing the degree of tracheal metaplasia (Kurie *et al.*, 2003), but a lower dose of RA (20 mg/person/day) showed no benefit for those with emphysema (Mao *et al.*, 2002). Although the different outcomes of the studies may have been due to dose of RA, the precise reason is not known.

II. CONCLUSIONS

The possibility of using RA as a potential therapy against emphysema is quite exciting, though perplexing. Although there are a number of potential mechanisms to explain the benefits of RA against emphysema, the animal data have not convincingly demonstrated its usefulness. In addition, the clinical studies in humans have not demonstrated clearly that RA may benefit the patient with emphysema. There are many unanswered questions that remain as to the appropriate dose of RA, form of RA (*cis* or *trans*), and route of administration. Whether RA will protect against emphysema or restore emphysematous tissue in humans is still uncertain. These are some of the challenging issues that remain to be resolved before a definitive statement

---

(large black arrowhead). Many inflammatory cells are also evident in the adventitia. (C) Representative section of the lung in which β-carotene was added to the diet of MCT-treated rats. The inflammation of the parenchyma is less severe, and collagen deposition in the septa is reduced. One small artery still shows an increased thickness of the media but few inflammatory cells. Adapted from Baybutt and Molteni (1999).

can be made concerning the use of vitamin A or, in particular, RA as a therapy against emphysema.

REFERENCES

al Senaidy, A. M., al Zahrany, Y. A., and al Faqeeh, M. B. (1997). Effects of smoke on serum levels of lipid peroxides and essential fat-soluble antioxidants. *Nutr. Health* **12,** 55–65.

Ammigan, N., Nair, U. J., Amonkar, A. J., and Bhide, S. V. (1990). Effect of tobacco extract and N'-nitrosonornicotine on the carcinogen metabolizing enzymes under different dietary vitamin B status. *Cancer Lett.* **52,** 153–159.

Antipatis, C., Ashworth, C. J., Grant, G., Lea, R. G., Hay, S. M., and Rees, W. D. (1998). Effects of maternal vitamin A status on fetal heart and lung: Changes in expression of key developmental genes. *Am. J. Physiol.* **275,** L1184–L1191.

Baybutt, R. C., and Molteni, A. (1999). Dietary β-carotene protects lung and liver parenchyma of rats treated with monocrotaline. *Toxicology* **137,** 69–80.

Baybutt, R. C., Hu, L., and Molteni, A. (2000). Vitamin A deficiency injures lung and liver parenchyma and impairs function of rat type II pneumocytes. *J. Nutr.* **130,** 1159–1165.

Belloni, P. N., Garvin, L., Mao, C. P., Bailey-Healy, I., and Leaffer, D. (2000). Effect of all-trans retinoic acid in promoting alveolar repair. *Chest* **117,** 235S–241S.

Biesalski, H. K., and Engel, L. (1996). An *ex vivo* model of the rat trachea to study the effect of inhalable toxic compounds. *Res. Exp. Med.* **196,** 195–210.

Boskovic, G., and Twining, S. S. (1997). Retinol and retinaldehyde specifically increase alpha1-proteinase inhibitor in the human cornea. *Biochem. J.* **322,** 751–756.

Camisa, C., Eisenstat, B., Ragaz, A., and Weissmann, G. (1982). The effects of retinoids on neutrophil functions *in vitro*. *J. Am. Acad. Dermatol.* **6,** 620–629.

Cardoso, W. V., Sekhon, H., Hude, D. M., and Thurlbeck, W. M. (1993). Collagen and elastin in human pulmonary emphysema. *Am. Rev. Respir. Dis.* **147,** 975–981.

Chelchowska, M., Laskowska-Klita, T., and Szymborski, J. (2001). Level of retinol and beta-carotene in plasma of smoking and non-smoking women. *Wiad. Lek.* **54,** 248–254.

Cichy, J., Potempa, J., and Travis, J. (1997). Biosynthesis of alpha1-proteinase inhibitor by human lung-derived epithelial cells. *J. Biol. Chem.* **272,** 8250–8255.

Edes, T. E., and Gysbers, D. S. (1993). Carcinogen-induced tissue vitamin A depletion: Potential protective advantages of β-carotene. *Ann. NY Acad. Sci.* **686,** 203–212.

Fujita, M., Ye, Q., Ouchi, H., Nakashima, N., Hamada, N., Hagimoto, N., Kuwano, K., Mason, R. J., and Nakanishi, Y. (2006). Retinoic acid fails to reverse emphysema in adult mouse models. *Thorax* **59,** 224–230.

Furr, H. C., Amedee-Manesme, O., Clifford, A. J., Bergen, H. R., III, Jones, A. D., Anderson, D. P., and Olson, J. A. (1989). Vitamin A concentrations in liver determined by isotope dilution assay with tetradeuterated vitamin A and by biopsy in generally healthy adult humans. *Am. J. Clin. Nutr.* **49,** 713–716.

Gadek, J. E., Fells, G. A., Zimmerman, R. L., Rennard, S. I., and Crystal, R. G. (1981). Antielastases of the human alveolar structures. Implications for the protease-antiprotease theory of emphysema. *J. Clin. Invest.* **68,** 889–898.

Habib, M. P., Lackey, D. L., Lantz, R. C., Sobonya, R. E., Grad, R., Earnest, D. L., and Bloom, J. W. (1993). Vitamin A pretreatment and bleomycin induced rat lung injury. *Res. Commun. Chem. Pathol. Pharmacol.* **81,** 199–208.

Haskell, M. J., Handelman, G. J., Peerson, J. M., Jones, A. D., Rabbi, M. A., Awal, M. A., Wahed, M. A., Mahalanabis, D., and Brown, K. H. (1997). Assessment of vitamin A status by the deuterated-retinol-dilution technique and comparison with hepatic vitamin A concentration in Bangladeshi surgical patients. *Am. J. Clin. Nutr.* **66,** 67–74.

Hayashi, A., Suzuki, T., and Tajima, S. (1995). Modulation of elastin expression and cell proliferation by retinoids in cultured vascular smooth muscle cells. *J. Biochem. (Tokyo)* **117**, 132–136.

Heger, R. J., and Baybutt, R. C. (1999). Regulation of polyamine synthesis and transport by retinoic acid and epidermal growth factor in cultured type II pneumocytes. *J. Nutr. Biochem.* **10**, 518–524.

Janoff, A., Pryor, W. A., and Bengali, Z. H. (1987). NHLBI workshop summary. Effects of tobacco smoke components on cellular and biochemical process in the lung. *Am. Rev. Respir. Dis.* **136**, 1058–1064.

Kanda, Y., Yamamoto, N., and Yoshino, Y. (1990). Utilization of vitamin A in rats with inflammation. *Biochim. Biophys. Acta* **1034**, 337–341.

Klann, R. C., and Marchok, A. C. (1982). Effects of retinoic acid on cell proliferation and cell differentiation in a rat tracheal epithelial cell line. *Cell Tissue Kinet.* **15**, 473–482.

Kurie, J. M., Lotan, R., Lee, J. J., Lee, J. S., Morice, R. C., Liu, D. D., Xu, X. C., Khuri, F. R., Ro, J. Y., Hittelman, W. N., Walsh, G. L., Roth, J. A., *et al.* (2003). Treatment of former smokers with 9-cis-retinoic acid reverses loss of retinoic acid receptor-beta expression in the bronchial epithelium: Results from a randomized placebo-controlled trial. *J. Natl. Cancer Inst.* **95**, 206–214.

Lancillotti, F., Darwiche, N., Celli, G., and De Luca, L. M. (1992). Retinoid status and the control of keratin expression and adhesion during the histogenesis of squamous metaplasia of tracheal epithelium. *Cancer Res.* **52**, 6144–6152.

Laskowska-Klita, T., Szymborski, J., Chelchowska, M., Czerwinska, B., and Chazan, B. (1999). Compensatory antioxidant activity in blood of women whose pregnancy is complicated by cigarette smoking. *Med. Wieku Rozwoj.* **3**, 485–494.

Li, T., Molteni, A., Latkovich, P., Castellani, W., and Baybutt, R. C. (2003). Vitamin A depletion induced by cigarette smoke is associated with development of emphysema in rats. *J. Nutr.* **133**, 2629–2634.

Lim, P. S., Wang, N. P., Lu, T. C., Wang, T. H., Hsu, W. M., Chan, E. C., Hung, W. R., Yang, C. C., Kuo, I. F., and Wei, Y. H. (2001). Evidence for alterations in circulating low-molecular-weight antioxidants and increased lipid peroxidation in smokers on hemodialysis. *Nephron* **88**, 127–133.

Liu, C., Wang, X. D., Bronson, R. T., Smith, D. E., Krinsky, N. I., and Russell, R. M. (2000). Effects of physiological versus pharmacological beta-carotene supplementation on cell proliferation and histopathological changes in the lungs of cigarette smoke-exposed ferrets. *Carcinogenesis* **21**, 2245–2253.

Liu, C., Russell, R., and Wang, X. D. (2003). Exposing ferrets to cigarette smoke and a pharmacological dose of β-carotene supplementation enhance *in vitro* retinoic acid catabolism in lungs via induction of cytochrome P_{450} enzymes. *J. Nutr.* **133**, 173–179.

Liu, R., Harvey, C., and McGowan, S. (1993). Retinoic acid increases elastin in rat lung fibroblast cultures. *Am. J. Physiol.* **265**, L430–L437.

Mao, J. T., Goldin, J. G., Dermand, J., Ibrahim, G., Brown, M. S., Emerick, A., McNitt-Gray, M. F., Gjertson, D. W., Estrada, F., Tashkin, D. P., and Roth, M. D. (2002). A pilot study of all-trans-retinoic acid for the treatment of human emphysema. *Am. J. Respir. Crit. Care Med.* **165**, 718–723.

Massaro, D., and Massaro, G. D. (2001). Pulmonary alveolus formation: Critical period, retinoid regulation and plasticity. *Novartis Found. Symp.* **234**, 229–236.

Massaro, G. D., and Massaro, D. (1996). Postnatal treatment with retinoic acid increases the number of pulmonary alveoli in rats. *Am. J. Physiol.* **270**, L305–L310.

Massaro, G. D., and Massaro, D. (1997). Retinoic acid treatment abrogates elastase-induced pulmonary emphysema in rats. *Nat. Med.* **3**, 675–677.

Massaro, G. D., and Massaro, D. (2000). Retinoic acid treatment partially rescues failed septation in rats and in mice. *Am. J. Physiol. Lung Cell Mol. Physiol.* **278**, L955–L960.

Meshi, B., Vitalis, T. Z., Ionescu, D., Elliott, W. M., Liu, C., Wang, X.-D., Hayashi, S., and Hogg, J. C. (2002). Emphysematous lung destruction by cigarette smoke. The effects of latent adenoviral infection on the lung inflammatory response. *Am. J. Respir. Cell Mol. Biol.* **26**, 52–57.

McDowell, E. M., Keenan, K. P., and Huang, M. (1984). Effects of vitamin A-deprivation on hamster tracheal epithelium. *Virchows Arch. B Cell Pathol. Incl. Mol. Pathol.* **45**, 197–219.

McDowell, E. M., Ben, T., Coleman, B., Chang, S., Newkirk, C., and DeLuca, L. M. (1987). Effects of retinoic acid on the growth and morphology of hamster tracheal epithelial cells in primary culture. *Virchows Arch. B Cell Pathol. Incl. Mol. Pathol.* **54**, 38–51.

McGowan, S., Jackson, S. K., Jenkins-Moore, M., Dai, H. H., Chambon, P., and Snyder, J. M. (2000). Mice bearing deletions of retinoic acid receptors demonstrate reduced lung elastin and alveolar numbers. *Am. J. Respir. Cell Mol. Biol.* **23**, 162–167.

McGowan, S. E., Harvey, C. S., and Jackson, S. K. (1995). Retinoids, retinoic acid receptors, and cytoplasmicretinoid binding proteins in perinatal rat lung fibroblasts. *Am. J. Physiol.* **269**, L463–L472.

McMenamy, K. R., and Zachman, R. D. (1993). Effect of gestational age and retinol (vitamin A) deficiency on fetal rat lung nuclear retinoic acid receptors. *Pediatr. Res.* **33**, 251–255.

McSorley, L. C., and Daly, A. K. (2000). Identification of human cytochrome P450 isoforms that contribute to all-trans-retinoic acid 4-hydroxylation. *Biochem. Pharmacol.* **60**, 517–526.

Morabia, A., Menkes, M. J., Comstock, G. W., and Tockman, M. S. (1990). Serum retinol and airway obstruction. *Am. J. Epidemiol.* **132**, 77–82.

Mukherjee, S., Nayyar, T., Chytil, F., and Das, S. K. (1995). Mainstream and sidestream cigarette smoke exposure increases retinol in guinea pig lungs. *Free Radic. Biol. Med.* **18**, 507–514.

Nabeyrat, E., Besnard, V., Corroyer, S., Cazals, V., and Clement, A. (1998). Retinoic acid-induced proliferation of lung alveolar epithelial cells: Relation with the IGF system. *Am. J. Physiol.* **275**(1, Pt. 1), L71–L79.

Osman, M., Cantor, J. O., Roffman, S., Keller, S., Turino, G. M., and Mandl, I. (1985). Cigarette smoke impairs elastin resynthesis in lungs of hamsters with elastase-induced emphysema. *Am. Rev. Respir. Dis.* **132**, 640–643.

Paiva, S. A., Godoy, I., Vannucchi, H., Favaro, R. M., Geraldo, R. R., and Campana, A. O. (1996). Assessment of vitamin A status in chronic obstructive pulmonary disease patients and healthy smokers. *Am. J. Clin. Nutr.* **64**, 928–934.

Pamuk, E. R., Byers, T., Coates, R. J., Vann, J. W., Sowell, A. L., Gunter, E. W., and Glass, D. (1994). Effect of smoking on serum nutrient concentrations in African-American women. *Am. J. Clin. Nutr.* **59**, 891–895.

Paquette, N. C., Zhang, L. Y., Ellis, W. A., Scott, A. L., and Kleeberger, S. R. (1996). Vitamin A deficiency enhances ozone-induced lung injury. *Am. J. Physiol.* **270**, L475–L482.

Pegg, A. E. (1986). Recent advances in the biochemistry of polyamines in eukaryotes. *Biochem. J.* **234**, 249–262.

Randall, R. W., Tateson, J. E., Dawson, J., and Garland, L. G. (1987). A commentary on the inhibition by retinoids of leukotriene B_4 production in leukocytes. *FEBS Lett.* **214**, 167–170.

Redlich, C. A., Rockwell, S., Chung, J. S., Sikora, A. G., Kelley, M., and Mayne, S. T. (1998). Vitamin A inhibits radiation-induced pneumonitis in rats. *J. Nutr.* **128**, 1661–1664.

Sansores, R. H., Abboud, R. T., Becerril, C., Montano, M., Ramos, C., Vanda, B., and Selman, M. L. (1997). Effects of exposure of guinea pigs to cigarette smoke on elastoltic activity of pulmonary macrophages. *Chest* **112**, 214–219.

Sauer, J. M., Hooser, S. B., and Sipes, I. G. (1995). All-trans-retinol alteration of 1-nitronaphthalene-induced pulmonary and hepatic injury by modulation of associated inflammatory responses in the male Sprague-Dawley rat. *Toxicol. Appl. Pharmacol.* **133**, 139–149.

Seagrave, J. (2000). Oxidative mechanisms in tobacco smoke-induced emphysema. *J. Toxicol. Environ. Health A* **61,** 69–78.

Sharma, A., Lewandoski, J. R., and Zimmerman, J. J. (1990). Retinol inhibition of *in vitro* human neutrophil superoxide anion release. *Pediatr. Res.* **27,** 574–579.

Sklan, D., Rappaport, R., and Vered, M. (1990). Inhibition of the activity of human leukocyte elastase by lipids particularly oleic acid and retinoic acid. *Lung* **168,** 323–332.

Swamidas, G. P., Basaraba, R. J., and Baybutt, R. C. (1999). Dietary retinol inhibits inflammatory responses of rats treated with monocrotaline. *J. Nutr.* **129,** 1285–1290.

Tajima, S., Hayashi, A., and Suzuki, T. (1997). Elastin expression is up-regulated by retinoic acid but not by retinol in chick embryonic skin fibroblasts. *J. Dermatol. Sci.* **15,** 166–172.

Takahashi, Y., Miura, T., and Takahashi, K. (1993). Vitamin A is involved in maintenance of epithelial cells on the bronchioles and cells in the alveoli of rats. *J. Nutr.* **123,** 634–641.

Tepper, J., Pfeiffer, J., Aldrich, M., Tumas, D., Kern, J., Hoffman, E., McLennan, G., and Hyde, D. (2000). Can retinoic acid ameliorate the physiologic and morphologic effects of elastase instillation in the rat? *Chest* **117,** 242S–244S.

US Surgeon-General (1984). "The Health Consequences of Smoking: Chronic Obstructive Lung Disease." US Department of Health and Human Services, Washington DC.

Varani, J., Burmeister, B., Sitrin, R. G., Shollenberger, S. B., Inman, D. R., Fligiel, S. E., Gibbs, D. F., and Johnson, K. (1994). Expression of serine proteinases and metalloproteinases in organ-cultured human skin. Altered levels in the presence of retinoic acid and possible relationship to retinoid-induced loss of epidermal cohesion. *Am. J. Pathol.* **145,** 561–573.

Villard, P. H., Herber, R., Seree, E. M., Attolini, L., Magdalou, J., and Larcarelle, B. (1988). Effect of cigarette smoke on UDP-glucuronosyltransferase activity and cytochrome P_{450} content in liver, lung, and kidney microsomes in mice. *Pharmacol. Toxicol.* **82,** 74–79.

Wang, X. (1994). Review: Absorption and metabolism of β-carotene. *J. Am. Coll. Nutr.* **13,** 314–325.

Wang, X. D., Liu, C., Bronson, R. T., Smith, D. E., Krinsky, N. I., and Russel, R. M. (1999). Retinoid signaling and activator protein-1 expression in ferrets given β-carotene supplements and exposed to tobacco smoke. *J. Natl. Cancer Inst.* **91,** 60–66.

Wiedermann, U., Chen, X. J., Enerback, L., Hanson, L. A., Kahu, H., and Dahlgren, U. I. (1996). Vitamin A deficiency increases inflammatory responses. *Scand. J. Immunol.* **44,** 578–584.

INDEX

Page numbers followed by f and t indicate figures and tables, respectively.

A

ABCG5, 8
ABCG8, 8
ABC1 protein, 20
A375 cell lines, 313
Acid-binding protein (aP2), 12
Acute myeloid leukemia, 19
Acute promyelocytic leukemia (APL), 303
Acyl-CoA:diacylglycerol acyltransferase (DGAT), 233
Acyl-CoA:retinol acyl-transferase (ARAT), 233, 242
Acyl-CoA synthase (ACS), 12
ADH4 mRNA expression, 106
Age-related macular degeneration (AMD), 124
AGGTCA hexamer, 4, 11, 262
Aging, of skin, 245–246
AGN 190730, 45
AGN 192837, 45
AGN 193109, 44–45
AhR/Arnt pathway, 40–45
AhR nuclear translocator (Arnt), 40
AhR pathway, 45
  cross talk with RA pathway, 52–56
AIDS epidemic, 357
Albendazole, 372
Albumin, 99
Alcohol dehydrogenases (ADHs), 47, 99

Aldehyde dehydrogenase (ALDH) family, 47
All-*trans* retinoic acid (atRA), 1, 106, 303
  pathways, 45–46
  steric interconversion of, 49–50
Alopex lagopus, 137
AM580, 306
δ-aminolevulinic acid synthetase, 43
AML1/ETO protein, 333
Androgen receptor (AR), 4
Angiotensin converting enzyme (ACE), 272
Angiotensin II (Ang II), 269
Anti-CD3/phorbol myristyl acetate, 207
Anti-measles antibodies, 203
α-1 antitrypsin protein, 388
AP-1, 244
ApoCIII expression, 14
aP2 protein, 12
Arachidonic acid, 11
ARA9 protein, 40, 42f
Arctic animals, vitamin A in, 137–141
Aryl hydrocarbon receptor (AhR)-signaling pathway, 35
Astaxanthine, 122
Atheresthes evermanni, 151
Atherosclerosis, 293–294
Atrial natriuretic peptide (ANP), 269
Atrioventricular septum, 260
5-aza-2'-deoxycytidine (5-aza-CdR), 332–334

B

Basal cell carcinoma, 303
Basic helix-loop-helix per-Arnt-sim (bHLH-PAS), 40
Bax gene, 43
Bayley Scales of infant development, 374
BCDO isoform 2, 104
B-cell differentiation, 244
B16 cells, 330
BCNU, 320
Bexarotene, 19
Bile acid, 9
Bile salt export pump (BSEP), 10
Bishop, Katherine S., 287
BMP-4, 77, 82
Brahma/SWI-related gene protein (Brg-1), 43
Breast cancer, 311
Breast feeding and HIV transmission, 360
Bromodeoxyuridine (BrdU), 142
B12 vitamin, 364

C

CACA box, 123
CAC-1 cells, 330
CALU-1, 327
cAMP-dependent protein kinase (PKA), 318
cAMP pathway, 318–320
Cancer, 19
Carbaprostacyclin (cPGI), 12
Cardiac arrhythmias, 269
Cardiac chamber formation process, 260
Cardiac hypertrophy, 269
Cardiac jelly, 260
Cardiac septation, 260
Cardiogenic area, 260
Cardiomyocytes, 263
Cardiomyopathy, 13
Carnitine palmitoyl transferase-I, 272
β-carotene, 104, 370–371, 376, 386
  in retinal pigment epithelial cells, 124–126
Carotene-15,15' dioxygenase mRNA, 108
β,β-carotene-9',10'-monooxygenase (Bcmo2), 121
β,β-carotene 15,15'-monooxygenase (Bcmo1), 119–120, 122
Carotenoids, 386
CAR/PXR/PPAR-signaling pathways, 47
CBP/p300, 38
CCAAT box, 42
CCR5 receptors, 376
CD437, 306
CD80/CD86, 208
CD4 cells, 360–362, 365–367
CD8 cells, 364–365, 370–372
Cellular retinoic acid-binding protein II (CRABPII), 236
Cellular retinoic acid-binding proteins (CRABPs), 41f, 51, 261f
Cellular retinol-binding protein I (CRBPI), 50, 178, 231
Cellular retinol-binding protein II (CRBPII), 36f, 123
CEP-1347 compound, 324
c-Fos, 43
Chemopreventatives, 43
Chenodeoxycholic acid, 9
C3H/HE mice, 75
Cholesterol 7α-hydroxylase (CYP7A), 7
Cholesterol metabolism, 5, 7
  effluxed, 8
Cholesteryl ester transfer protein (CETP), 8
CHOP gene, 323
Chronic myeloid leukemia (CML), 315
Chylomicrons (CM), 136f, 229
Cigarette smoking and vitamin A, 389–395
9-cis-4-oxo-13,14-dihydroRA, 39, 40f
c-Jun N-terminal kinase (JNK), 270
[^{14}C]-labeled retinyl acetate, 164
Colchicine, 147–148
Collagenase-1 (MMP-1), 228
Collagen gels, 147
Collagen production, 143–144
Comet assay, 248
Constitutive androstane receptor (CAR), 5
CRAF protein, 246–247
CRBP II expression, 108
[^{14}C]Retinol, 165
c-Src phosphorylation, 286
CYP26A1 protein, 99
CYP26a1, 84
CYP1A1 gene expression, 47, 54–55, 393
CypA gene, 86
CYP7A gene, 8
Cyp26A1 gene, 107
CYP26a1 gene, 84
CYP4A1 gene, 13
CYP4A6 gene, 13
Cyp7A protein, 20
Cyp26B1 enzyme, 99
Cyp26B1 mRNA gene, 108
CYP2C39 enzyme, 48
Cyp26C1 enzyme, 99
CYP450 protein, 41f
CYP2S1 gene, 48
Cystophora cristata, 137

Cytochrome P450 (CYP450), 43, 48, 99, 243, 261, 393
Cytodifferentiation, 304, 310
Cytokeratin, 311
Cytoplasmic-binding proteins (CRABPs), 100
Cytosol, 291
Cytosolic retinoic acid-binding protein 1 (CRBP1), 327

D

Dab2 gene marker, 82
Danio rerio, 123
Darier's disease, 240
D407 cell lines, 125
Deoxycholic acid, 9
Dermal collagen, 246
Dermal fibroblasts, 227
Desmethyl (D-P21-T3), 287
Desmin, 142
Diacylglycerol acyltransferase (DGAT1), 104
Dibutyryl-cAMP, 318
3,4-didehydroretinol, 234
Didesmethyl (D-P25-T3), 287
Dietary-derived all-*trans* RA (atRA), 35
Dietary phytol metabolites, 16
Differentiation therapy, 334–336
1,25-dihydroxy vitamin D3, 307
Diphtheria–pertussis–tetanus (DPT) vaccines, 199, 202
Disabled-2 gene, 83
DNA-binding domain (DBD), 3
DNA methylation, 331–333
Doxorubicin, 320
Drosophila spp., 83
  D. melanogaster, 120
  D. wingless gene, 85

E

ECM proteolysis, 43
Eicosanoids, 5, 11
Elaeis guineensis fruit, 292
Elastin metabolism, 393
Embryonal carcinoma cells (EC cells), 73–75
Embryonic stem cells (ES cells), 75–77, 312
Emphysema, 388
  and cigarette smoking, 389–393
  and elastin metabolism, 393
  and lung inflammation, 394–395, 396*f*
  and vitamin A, 389, 393–394
Endobiotic compounds, 5
Endothelial nitric oxide synthase (eNOS), 294
Engelbreth-Holm-Swarm (EHS) tumor, 146

EPAC factor, 318
Erignathus barbatus, 137
ERK pathway, 321–322
Escherichia coli, 120
Estrogen receptor (ER), 3, 311
Evans, Herbert M., 287
Evx-2 protein, 82
Expanded Program on Immunization (EPI), 199
Expressed sequence tag (EST) databases, 79, 121
Extracellular-regulated kinases (ERKs), 262, 320
Extracellular signal-regulated kinase 1/2 (ERK1/2), 270
Extrahepatic stellate cells, 151–152

F

Factor X, 287
Farnesoid X receptor (FXR), 5
Farnesyl pyrophosphate, 293
Fatty acid synthase (FAS) expression, 9
Fatty acid transport protein (FAT/CD36), 12, 20
F9 cells, 80–81, 312
  EC cells, 73
FDA's Voluntary Cosmetics Registration Program, 225
Fletcher, William, 286
FOG-2, 270
Folic acid, 364
Forkhead receptor in rhabdomyosarcoma (FKHRL1), 327–328
6-formylindolo[3,2-*b*]carbazole (FICZ), 44*f*
Fratercula arctica, 137
French American British (FAB) classification of AML, 308
Fulmarus glacialis, 137
Functional homodimers, 1
Funk, Casimir, 286

G

GADD153 gene, 323
GAP 43, 312
GATA-4, 208, 270
Gata2 gene marker, 82
Gata6 gene marker, 82
Gbx-2 protein, 82
Gelatinase B (MMP-9), 228
Geranylgeranyl pyrophosphate, 293
GF109203X, 330
Glucocorticoid receptor (GR), 3

Glucuronidation, 99
Glutathione S-transferase (GST), 41f, 49
Glut4 expression, 14
G-protein–coupled receptor
   (GPCR)-mediated hypertrophy, 268
Granulocyte colony-stimulating factor
   (G-CSF), 314–315
Granulocyte macrophage colony-stimulating
   factor (GM-CSF), 314–315
GST-Ya enzyme, 43
GW501516 compound, 13

H
HAIR-62, 82
Head and neck cancer, 310–311
Head and neck squamous cell carcinomas
   (HNSCC), 310–311
Heart development and retinoic acid.
   See Retinoic acid (RA)
Heart diseases, 13, 269, 293, 295
Heart morphogenesis, 259
Heart progenitor cells, 259–260
Hepa-1c1c7 cells, 44
Hepatic stellate cells (HSCs)
   in arctic animals, 137–141
   and 3D structures of ECM, 146–149
   morphology of, 133–134
   in production and degeneration of
      ECM, 143–146
   in regulation of vitamin A
      homeostasis, 134–137
   role in liver regeneration, 141–143
   stimulation of proliferation of, 149–151
Her2/Neu, 327
High-density lipoprotein (HDL), 8, 294
High performance liquid chromatography
   (HPLC), 103, 137
Hippocrates, 286
Histone acetylation, 331–333
Histone acetyltransferase (HAT), 43
Histone deacetylases (HDACs), 55, 305
Histone methyltransferases (HMTs), 55
HIV-infection
   in children, 372–374, 375t
   disease progression in adults, 366–372
   pregnant women, 362
   transmission and pregnancy
      outcomes, 361–365
HL-60 cells, 319, 321, 323, 326, 329–330
HMG-1, 82
HMG-CoA reductase enzyme, 293–294
HNF4, 4

HoloRBP4, 229–230
Hopkins, Sir Frederick Gowland, 286
Hoxa-1 gene, 81–83
Hoxb-1 gene, 81–83
[^{3}H]retinol, 132–133, 146
[^{3}H]retinol/RBP/transthyretin, 169
[^{3}H]retinyl acetate, 165
HT4 neuronal cells, 268
3-hydroxy-3-methylglutaryl-CoA synthase
   gene, 13
3-hydroxy-3-methylglutaryl CoenzymeA
   (HMG-CoA) reductase, 292
3-hydroxy-3-methylglutaryl coenzyme A
   (Hmgcs2) gene, 16
14-hydroxy-4,14-retro-retinol, 236
Hypercholesterolemia, 19
Hypertrophic markers, 268
Hypolipidemic drugs, 11
Hypothyroidism, 19

I
IgG1/IgG2a plasma antibody levels, 210
Ileal bile acid-binding protein (IBABP), 10
Immune response system, of skin, 244–245
Insulin growth factor I (IGF-I), 327
Insulin-like growth factor-1, 245
Insulin sensitizers, 11
Interferons (IFNs), 316–317
   IFN γ, 204
   responsive factor 1 (IRF1), 317
Interleukin-1β, 43
Interleukin-6 (IL-6), 187
Interleukin-12 (IL-12), 204
Intracellular retinol-binding proteins
   (IRBP), 51
Ischemia heart disease, 295
Isoproterenol, 269

J
JB6-Cl 41-5a cell line, 54
JEG-3 cell line, 103
Jun-B, 43
Jun-D, 43
Jun N-terminal kinase (JNK), 320–323
Jurkat T-cells, 247

K
Keratinization of epithelia, 224
Keratinocytes, 226, 230, 244
Keratosis follicularis, 240

KLF5 factor, 272
Kupffer cells, 132, 143–144

L

lacZ gene, 105
Lagopus mutus hyperboreus, 137
Lama1 gene marker, 82
Laminin, 134
Lampetra japonica, 151
Langerhans cells, 227, 244
Larus argentatus, 137
Larus hyperboreus, 137
l-ascorbic acid 2-phosphate (Asc 2-P), 149–151
Lecithin:retinol acyltransferase (LRAT), 41*f*, 50, 104, 233
Left ventricular (LV) remodeling, 271
Leukemia inhibitory factor (LIF), 75
Ligand/AhR/HSP90 complex, 40, 42*f*
Ligand-binding domain (LBD), 3
Linoleic acid, 11
Lipopolysaccharides (LPS), 187, 200
Lipoprotein lipase (LPL), 12
Lipoteichoic acid, 201
12-lipoxygenase (12-Lox), 291
Lithocholic acid, 9
Liver X receptors, 5
Long-Evans cinnamon (LEC)-like colored rats, 144
Lovastatin, 293–294
Low-density lipoprotein (LDL), 289, 294
Lung inflammation and vitamin A, 394–395, 396*f*
Lutein, 122
Lycopene, 122
LY294002 gene, 326
Lysophosphatidylcholine, 269
L5178Y/Tk$^{+/-}$ mouse lymphoma cells, 247

M

Matrix metalloproteinases (MMPs), 54, 145, 228
MCF-7, 332
  breast carcinoma cells, 316
MC3T3-E1 fibroblasts, 323
MDA-MB-231, 332
Medium-chain acylcoenzyme A dehydrogenase (MCAD), 272
Melanocytes, 226
Melanoma, 312
Melphalan, 320
Mental Development Index, 374

3-methylcholanthrene (3-MC), 44*f*
Mineralocorticoid receptor (MR), 4
Mitogen-activated protein kinase kinases (MKKs), 270
Mitogen-activated protein kinase (MAPK), 270
  p38, 270, 317–318, 323–326
Mitogen activated protein kinase phosphatase 1 (MKP-1), 270
Mitogen-activated protein kinases (MAPKs), 320
Mono-oxidized derivatives, of cholesterol, 7
Mother-to-child transmission (MTCT) of HIV, 360
Multicenter AIDS Cohort study (MACS), 366
Multiple myeloma (MM), 325
MYCN gene, 310
Myeloid leukemia, 308–309
Myocardial infarction (MI), 293
α-myosin heavy chain (α-MHC), 269
Myosin light chain 2 (MLC-2) genes, 264

N

NADPH-quinone-oxido-reductase, 43
α-naphthoflavone, 45
Natural killer (NK) cells, 201, 212*f*, 213
  T cells, 213–215
NB4 cells, 319
NB4 cellular model, 309
N-Cadherin gene, 86
Neoplastic disease management. *See* Retinoic acid (RA)
Neuroblastoma, 309–310
NeuroD gene, 85–86
Neurogenin-1 gene, 85–86
Nevirapine, 360
N-(4-hydroxyphenyl)retinamide, 181
Niacin, 364, 367
Niemann-Pick C1-like 1 (NPC1L1) protein, 15
NIH3T3 cells, 262
Nitric oxide (NO), 294
N′-nitrosonornicotine, 391
Nonpermissive heterodimers, 1
Nonsmall cell lung carcinoma (NSCLC) cells, 322
Npm–RARα fusion protein, 308
Nuclear receptors corepressor (N-CoR), 38
Nuclear receptor superfamily
  classification, 4–7
  overview, 34
Nurr1 LBD, 5

Nurr/TR3/NGF1-B:RXR, 7
Nutrition for healthy living (NFHL), 358

O

Oct3/4 gene, 83
Oral leukoplakia, 303
Oral polio vaccine (OPV), 202–203
4-oxo-RA, 39
4-oxo-retinaldehyde, 39
4-oxo-ROH, 39
Oxysterols, 7

P

p23/ARA9 molecules, 40
Parenchymal cells (PCs), 132
Partial hepatectomy (PHx), 141
Pattern recognition receptors (PRRs), 201
Pbx family of proteins, 83–84
P19 cells, 77, 80–81, 312
  dimethylsulfoxide treatment of, 75
  EC cells of, 75
  RA treatment of, 75
People living with HIV/AIDS (PLWHA), 357
Permissive heterodimers, 2
  FXR:RXR physiological activities, 9–10
  LXR:RXR physiological activities, 7–9
  PPAR:RXR physiological activities, 10–15
Peroxisome proliferator-activated receptors, 4
Phenylbutyrate (PB), 332
Phoca hispida, 137
Phosphatidyl-inositide-dependente kinase 1 (PDK1), 325
Phosphatidylinositol-diphosphate (PIP2), 325
Phosphodiesterase IV inhibitors, 318
Phosphoinositide 3-kinase (PI3K), 262
Phospholipid transfer protein (PLTP), 10
Phytanic acid, 16
PI3K/AKT pathway, 325–328
PKC-zeta, 329
p27kip1 gene, 43
Plasminogen activator inhibitor-2 (PAI-2), 43
PLZF–RARα fusion protein, 308–309, 333
Polyamines, 388
Polycyclic and halogenated aromatic hydrocarbons (PAH/HAH), 34
Polymerase chain reaction (PCR), 374
PPAR$\gamma^{+/-}$ heterozygous mice, 12
PPAR response elements (PPRE), 11
PPAR:RXR physiological activities
  in fatty acid oxidation, 12–14
  general, 10–11
  in lipoprotein metabolism, 14–15
  as a regulator of fatty acid storage and adipogenesis, 11–12
Pregnane X receptor (PXR), 5
Prevention of mother-tochild transmission (PMTCT) services, 358
Progesterone receptor (PR), 4
Promyelocytic leukemia cells, 262
Prostacyclin PGI$_2$, 12
Protease/antiprotease theory, 388
α1-proteinase inhibitor (α1-PI), 394
Protein kinase C (PKC) family, 328–331
Provitamin A, 50
Psychomotor Development Index, 374
PubMed database, 162

R

RAC65, 77
RALDH family, 107
RALDH2 gene, 47
Rangifer tarandus platyrhynchus, 137
RARα2 gene, 80–81
RARβ2 gene, 80–81
RAR/RXR heterodimers, 38, 264, 270
RAR–RXR heterodimers, 262
RBP$^{(-/-)}$/Bcmo1$^{(-/-)}$ double knockout mice, 127
Reactive oxygen species (ROS), 313
Recommended dietary allowance (RDA), 367
Red palm oil, 286
Renin–angiotensin system (RAS), 272–273
Retinaldehyde dehydrogenase 2 (Raldh2) gene, 266
Retinaldehyde dehydrogenases (RALDHs), 105, 261
Retinal dehydrogenase type 2 (RALDH2), 47
Retinal pigment epithelium (RPE), 119
Retinoblastoma-binding protein-2, 82
Retinoic acid (RA)
  based differentiation therapy, 334–336
  catabolism, 48–49
  in chemoprevention and treatment of neoplastic disease, 303–304
  in combination therapy
    along cAMP pathway, 318–320
    along MAP kinase pathway, 320–325
    along PI3K/AKT pathway, 325–328
    with growth factors and cytokines, 314–318
  EC cells model systems, 73–75
  embryonic metabolism during mammalian development, 105–107

ES cells model systems, 75–77
  in histone acetylation and DNA
    methylation, 331–333
  inactive products of, 261
  molecular mechanism of, 71–73
  placental transport during mammalian
    development, 101–104
  postnatal development effects in heart
    in cardiac remodeling, 271–272
    in neonatal cardiomyocytes, 269–270
    regulation of renin–angiotensin
      system, 272–273
  regulated genes of, 78–80
    CypA, 86
    CYP26a1, 84
    Disabled-2, 83
    Hoxa-1 and Hoxb-1, 81–83
    N-Cadherin, 86
    Neurogenin-1 and NeuroD, 85–86
    Pbx family of proteins, 83–84
    RARβ2 and RARα2, 80–81
    Rex-1 (Zfp-42), 83
    Sox6, 84
    Wnt-1, 85
  role in heart development and congenital
    heart defects
    for embryonic and fetal heart
      development, 263–268
    normal heart development, 259–260
    signaling process, 260–263
  role in terminal differentiation of neoplastic
    cells
    breast cancer, 311
    head and neck cancer, 310–311
    melanoma, 312–313
    myeloid leukemia, 308–309
    neuroblastoma, 309–310
    teratocarcinoma, 311–313
  role of RARs, 77–78, 305–307
  storage and transport, 50–52
    in mammalian species, 98–100
  synthesis, 46–48
  treatment in vitamin A models, 205–208
Retinoic acid receptors (RARs), 1, 38, 119,
  243, 258, 304
Retinoic acid response elements (RAREs), 38,
  79, 262
Retinoic X receptor (RXRs), 258
Retinoid-related molecules (RRMs), 307
Retinoid-signaling pathways, 38–40
  cross talk with AhR pathway, 52–56
  nature of endogenous ligands, 16–17
  nature of functional complexes, 17–18

receptors, 15
  RXR functional activities, 18–21
Retinoid X receptors (RXRs), 38
Retinol-binding protein (RBP), 41f, 50, 132,
  163, 176
Retinol dehydrogenase (RDH), 47, 99
Retinol dehydrogenases (RolDHs), 41f
Retinol ester hydrolases (REH), 41f
Retinol (ROH). See Vitamin A
Retinosomes, 241
Retinyl ester hydrolase (REH), 104
Retinyl ester hydrolases (REHs), 34, 36f, 50
Retinyl esters (REs), 41f, 119
  cutaneous absorption and deposition
    of, 235–240
  physiological and environmental factors
    affecting, 240–241
  mobilization and metabolism of, 241–243
Retinyl ester-storage particles, 241
Retinyl linoleate, 232
Retinyl myristate, 232
Retinyl oleate, 232
Retinyl palmitate, 232. See also Retinyl esters
  (REs); Vitamin A (VA)
Retinyl stearate, 232
Rexinoid receptors (RXRs), 1
Rex-1 (Zfp-42) gene, 83
Riboflavin, 364
Ro 41-5253, 332
Roaccutane®, 103
Rox-1 gene, 83
RPE65, 121
RPE65 gene, 230–231
RXR/RAR heterodimers, 320
RXR/RAR transcriptional complexes, 329
RXR/RXR homodimers, 305
RXR$\alpha^{-/-}$/RXR$\gamma^{-/-}$ mutants, 265

S

San Francisco Men's Health Study
  (SFMHS), 366
SAP18, 82
Scavenger receptor BI (SR-BI)/CLA-1, 14
SCC-4 keratinocytes, 51
SCC12Y cells, 55
S91 cell lines, 313
SCID animals, 320
SDS-polyacrylamide slab gel
  electrophoresis, 144
Secreted retinol:RBP complexes, 163
Ser61, 322
Ser75, 322

Ser96, 327
Ser265, 322
Ser/473, 326
Ser-473 kinase, 262
Sertoli cells, 15
Sexually transmitted disease (STD), 362
Short-chain alcohol dehydrogenases (SCADs), 47
Short heterodimerization partner (SHP-1), 9–10
SH-SY5Y cells, 262, 315
Signal transducer and activator of transcription (STAT)-1, 206
Signal transduction and activator of transcription (STAT), 315
Silencing mediator of retinoid and thyroid receptors (SMRT), 38, 55–56
Simulation, Analysis and Modeling computer program (SAAM), 168–170, 172, 179, 189
Sinusoidal endothelial cells (SECs), 132
SKH-1 mice, 237
Skin
  biological responses of
    aging, 245–246
    immune responses, 244–245
    response to UV light, 246–247, 248f
    wound healing, 245
  penetration and percutaneous absorption of retinol and retinyl palmitate effects on, 235–240
  physiological functions of, 228
  structure and physiological functions of, 226–228
Skin-associated lymphoid tissues (SALT), 228
SK MEL 28, 313
SMAD2, 316
Sodium taurocholate cotransporting polypeptide (NTCP), 9
Solanaceae species, 292
Sox6 gene, 84
Sox17 gene marker, 82
SP600125 animal model, 324
Spinal cord development and vitamin A, 106
Spontaneously hypertensive rats (SHR), 268, 271, 273
Squamous cell carcinoma of skin, 303
SRC3, 324
Src kinase, 295
SRC-1/TIF2/RAC3, 39
SREBP-1c expression, 8
ST1346, 322, 325
ST1926, 306
STAT5b–RARα fusion protein, 308

Stratum basale, 226
Stratum corneum, 226, 228
Stratum granulosum, 226
Stratum lucidum, 226
Stratum spinosum, 226
Stromelysin-1 (MMP-3), 228
Subordination mechanism, 6
Sun protection factor (SPF), 247
Superficial bladder carcinoma, 303
Superficial corneoscleral slices (SCSS), 76
Suprabasal keratinocytes, 16

T

Targretin® capsules, 19
TATAAA box, 55
TATA box, 42, 123
T-bet, 208
TCDD exposure lesions, 37t
TCDD exposures, 45–46, 53
T-cell-dependent (TD) antigens, 200
Teratocarcinoma, 311–312
Tetanus toxoid, 206
2,3,7,8-tetrachlorodibenzofuran (TCDF), 44f
2,3,7,8-tetrachlorodibenzo-p-dioxin (TCDD), 34, 181–183
TGF-α, 43
TGF-β₂ gene, 43
Thiamine, 364
Thiazolidinediones, 11
Thyroid hormone receptors, 4
Tissue inhibitor of metalloproteinases (TIMP), 145
TLR4, 204
T-lymphocytes, 244
Tocotrienol-rich fraction (TRF), 286
  in rice bran oil, 292
Tocotrienols, 285–286
  and cardioprotection, 292–293
  in free radical scavenging and antioxidant activity, 294–295
  role in ischemic heart disease, 295–296
  sources of, 292
  vs tocopherols, 289–292
Toll-like receptors (TLRs), 201
Transactivation domain sequences (TADs), 40
Transforming growth factors (TGFs), 314
Transforming growth factor-β1 (TGF-β1), 36
Transforming growth factor-β (TGF-β), 147
Transthyretin, 163

Trial of Vitamins (TOV) study, 364
Tricostatin A, 333
Triglyceridemia, 19
Triglyceride-rich absorptive lipoproteins, 181
Triiodothyronine, 5
Trophoblastic cells, 101
Truncus arteriosus, 264
Tumor necrosis factor (TNF)-α, 205, 289
Type II pneumocyte, 387–388
Type IV collagen, 134

U

UDP-glucuronosyltransferase 1A1
    (UGT1A1), 41f
UDP-glucuronosyltransferases (UGTs), 49
UGT1A1 enzyme, 43
UGT1A1 expression, 49
Ultraviolet (UV) light effects on skin, 228,
    231, 241, 244, 246–247, 248f
Uria lomvia, 137
Ursus arctos, 137
Ursus maritimus, 137

V

Vascular smooth muscle cells (VSMCs), 273
Visceral endoderm, 74
Vitamin A_2, 234
Vitamin A-deficient (VAD) rats, 258
Vitamin A (VA), 1, 35, 364, 372
    compartmental analysis, 166–169
        whole-body models, 169–173
    conversion of β-carotene to
        background, 120
        cleavage mechanisms, 121–122
        cloning of enzymes in formation
            of, 120–121
        eccentric cleavage of, 121
        regulatory mechanisms, 122–123
        in tissue development, 123–124
    cutaneous absorption and deposition of
        dietary form of, 229–234
    deficiency in animal models, 37t
    deficiency syndrome, 263
        cardiovascular defects, 265
    embryonic metabolism during mammalian
        development, 105–107
    Expanded Program on Immunization
        (EPI), 199
    experimental models, 204–205
    functions of, 386–387
    and immunity in young children, 202–204
    kinetic studies of metabolism, 164–166
    as a function of vitamin A
        status, 173–176
        in humans, 189–190
        in liver, 176–178
        other organs, 179–180
    and lung, 387
    metabolism, 118–119
    physiological and environmental factors
        affecting cutaneous levels
        of, 240–241
    placental transport during mammalian
        development, 101–104
    RA treatment effects, 205–208
    regulation by HSCs, 134–137
    and regulation of innate immune cells and
        factors, 211–215
    role in emphysema, 389, 393–394
    role in lung restoration, 397
    supplementation and antibody
        production, 199–201
        in neonatal model, 209–210
    supply routes, 126–128
    Th1/Th2 cytokines effects, 210–211
    and type II pneumocyte, 387–388
    whole body metabolism of, 163–164
        dietary RA effects, 184–186
        4-HPR effects, 181
        inflammation effects, 186–188
        interpretation of three and four
            compartment models, 188–189
        iron deficiency effects, 184
        TCDD effects, 181–183
Vitamin B, 367
Vitamin C, 364, 367, 372
Vitamin D receptor, 4
Vitamin E, 364, 372
Vitamin E stereoisomers. *See* Tocotrienols
Vitamins. *See also* Vitamin A (VA)
    history of, 286
        tocotrienols vs tocopherols, 289–292
        sources of tocotrienols, 292
        vitamin E, 287–289
    in HIV disease progression in adults
        evidence from observational
            studies, 366–370
        evidence from trials, 370–372
    in HIV-infected children and HIV-negative
        children, 372–374, 375t
    in HIV transmission and pregnancy
        outcomes, 376–377
        evidence from observational
            studies, 361–362
        evidence from trials, 362–365

Vitamins (*continued*)
   and immune function, 358, 359*t*
Vulpes vulpes, 137

W

Wnt-1 gene, 85
World Health Organization (WHO), 199, 293, 367, 372, 376
Wound healing process, of skin, 245

X

Xenobiotic compounds, 5
Xenobiotics, 140
Xenopus, 10, 16, 39, 267
Xeroderma pigmentosum, 303

Z

Zidovudine, 360, 367

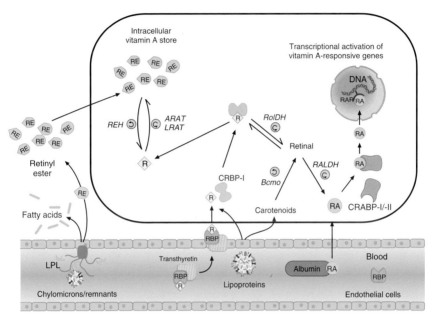

BIESALSKI ET AL., FIGURE 2. Vitamin A supply of extrahepatic tissue from the blood. Compensation of RBP deficiency by REs, RA, and carotenoids. Normally, vitamin A is delivered to the cells by the RBP–TTR–retinol complex. After binding to a receptor, retinol can be released from the complex originally formed in the liver. After absorption, there is either intracellular binding to CRBP-I, oxidation to RA, or reesterification through either ARAT or LRAT. The resulting REs constitute an intracellular pool that can be assessed through hydrolysis by REH. REs can also be derived directly from lipid metabolism. During degradation of chylomicrons to remnants, LPL releases not only fatty acids but also REs, all of which are absorbed by cells, enabling vitamin A supply to target cells independently of the controlled hepatic release of the RBP–TTR–retinol complex. In addition, retinol and carotenoids from plasma lipoproteins may enter the cell. Retinol is directly bound to CRBP-I, provitamin A carotenoids may partially cleaved to retinal by intracellular BCO and subsequently oxidized to RA by an RALDH. Small amounts of RA (bound to albumin) in the blood can pass the cell membrane and attach to specific cellular retinoic acid-binding proteins (CRABP-I and II) within the cell. RA can act within the nucleus of a cell to regulate transcription of vitamin A-responsive genes by binding to specific transcription factors. (From Biesalski and Grimm, 2005.)

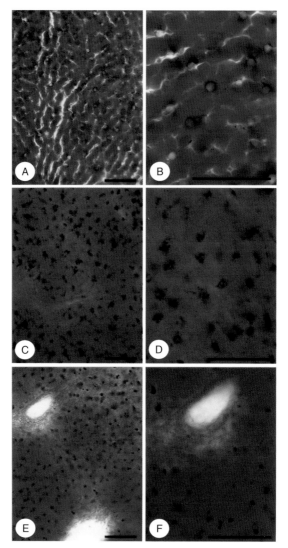

SENOO ET AL., FIGURE 5. Gold chloride staining specifically demonstrating black-stained HSCs of polar bears (A and B), Arctic foxes (C and D), and rats (E and F). Scale bars indicate 100 μm.

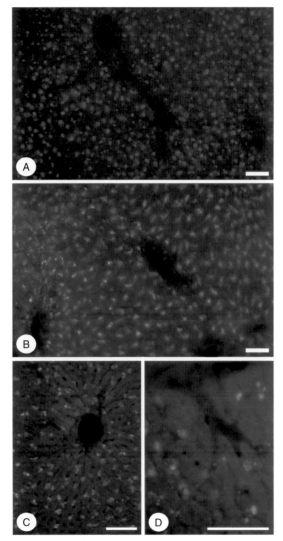

SENOO ET AL., FIGURE 6. Fluorescence micrographs demonstrating vitamin A autofluorescence in HSCs of polar bears (A), Arctic foxes (B), and rats (C and D). Scale bars indicate 100 μm.

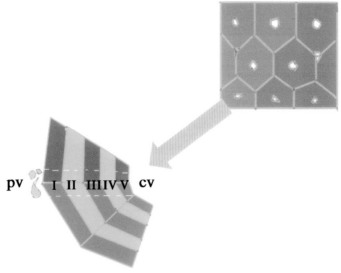

SENOO ET AL., FIGURE 7. Zonal division of the liver lobule. To make a zonal morphometric analysis, the liver lobule was divided histologically into five zonal areas (zones I–V) of equal widths from the portal vein (pv) to the central vein (cv).

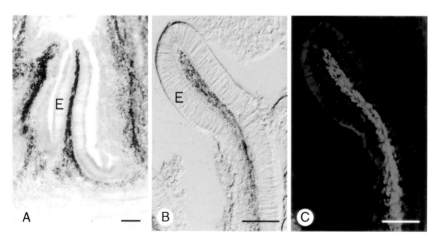

SENOO ET AL., FIGURE 15. Stellate cells in the pyloric cecum. The pyloric cecum of the arrowtooth halibut (*Atheresthes evermanni* Jordan et Starks) was observed by Sudan III staining (A), differential interference microscopy (B), and fluorescence microscopy (C) for detecting autofluorescence of vitamin A. Scale bars indicate 100 mm.

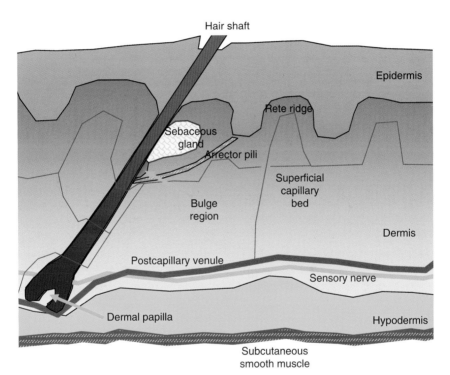

FU ET AL., FIGURE 2. Major structural components of the skin.

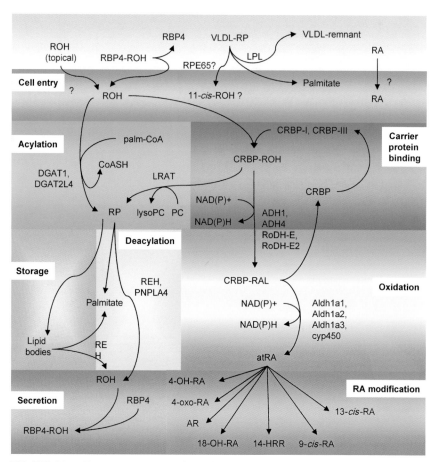

FU ET AL., FIGURE 3. Uptake and fate of retinoids in the skin. *Uptake*: Retinol (ROH), retinyl palmitate (RP), or retinoic acid (RA) can be taken up by the skin in a receptor-dependent or independent process. *Storage*: ROH, bound to CRBP, may be acylated and stored in lipid bodies, or *Metabolism*: ROH undergoes a two-step oxidation to retinaldehyde to all-*trans* retinoic acid (atRA). *Retinoic acid modification*: Levels of intracellular atRA are modulated by oxidation to less bioactive metabolites. *Mobilization*: Retinyl esters can be hydrolyzed to form ROH for use in the cell or return to serum.

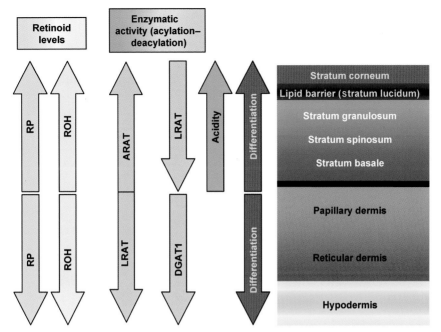

FU ET AL., FIGURE 4. Acylation of retinol (ROH). ROH may be acylated by LRAT, ARAT which may be indistinguishable from DGAT. The relative importance of each acyltransferase may vary with pH and location in the skin.

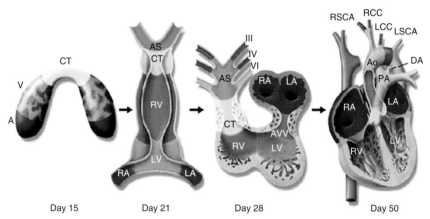

PAN AND BAKER, FIGURE 1. Steps of heart morphogenesis. The illustrations depict cardiac development with color coding of morphologically related regions, seen from a ventral view. Cardiogenic precursors form a crescent (left-most panel) that forms specific segments of the linear heart tube, which is patterned along the anterior–posterior axis to form the various regions and chambers of the looped and mature heart. Each cardiac chamber balloons out from the outer curvature of the looped heart tube in a segmental fashion. Neural crest cells populate the bilaterally symmetrical aortic arch arteries (III, IV, and VI) and aortic sac (AS) that together contribute to specific segments of the mature aortic arch. Mesenchymal cells form the cardiac valves from the conotruncal (CT) and atrioventricular valve (AVV) segments. Corresponding days of human embryonic development are indicated. A, atrium; Ao, aorta; DA, ductus arteriosus; LA, left atrium; LCC, left common carotid; LSCA, left subclavian artery; LV, left ventricle; PA, pulmonary artery; RA, right atrium; RCC, right common carotid; RSCA, right subclavian artery; RV, right ventricle; and V, ventricle.

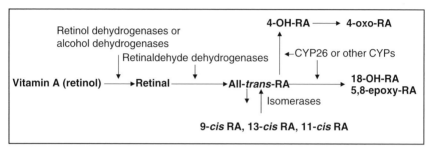

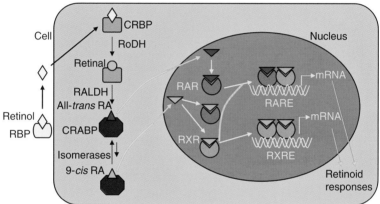

PAN AND BAKER, FIGURE 2. The metabolic pathway and the cellular mechanism of RA action. The metabolic pathway of RA. Retinol is taken up from the blood, where it binds to retinol-binding proteins (RBPs) and rebinds intracellularly to cellular retinol-binding proteins (CRBPs). Intracellularly, retinol is converted to retinal by the retinol or alcohol dehydrogenases (RoDH), and retinal is further metabolized to RA by RALDHs. RA is bound in the cytoplasm by cellular RA-binding protein (CRABP). Alternatively, RA and its 9-*cis* isomer enter the nucleus and bind to RARs or RXRs, respectively. On dimerization of these receptors, RAR/RXR heterodimer or RXR/RXR homodimer, the activated receptors bind to a sequence of DNA known as the RA-response element (RARE). This latter action either activates or represses transcription of the target gene.